Integrated Approaches for Sustainable E-Waste Management

Aşkıner Güngör
Pamukkale University, Turkey

Olcay Polat
Pamukkale University, Turkey

IGI Global
Scientific Publishing
Publishing Tomorrow's Research Today

Vice President of Editorial — Melissa Wagner
Managing Editor of Acquisitions — Mikaela Felty
Managing Editor of Book Development — Jocelynn Hessler
Production Manager — Mike Brehm
Cover Design — Phillip Shickler

Published in the United States of America by
IGI Global Scientific Publishing
701 East Chocolate Avenue
Hershey, PA, 17033, USA
Tel: 717-533-8845
Fax: 717-533-7115
Website: https://www.igi-global.com E-mail: cust@igi-global.com

Library of Congress Cataloging-in-Publication Data

Names: Gungor, Askiner, 1971- editor | Polat, Olcay, 1983- editor
Title: Integrated approaches for sustainable e-waste management / edited by Askiner Gungor, Olcay Polat.
Description: Hershey, PA : IGI Global Scientific Publishing, 2025. | Includes bibliographical references and index. | Summary: "This book aims to present a comprehensive examination of the challenges and opportunities associated with the sustainable management of electronic waste (e-waste)"-- Provided by publisher.
Identifiers: LCCN 2024052845 (print) | LCCN 2024052846 (ebook) | ISBN 9798369373835 hardcover | ISBN 9798369373842 paperback | ISBN 9798369373859 ebook
Subjects: LCSH: Electronic waste--Management | Electronic apparatus and appliances--Environmental aspects | Electronic waste--Recycling
Classification: LCC TD799.85 .I58 2025 (print) | LCC TD799.85 (ebook) | DDC 363.72/88--dc23/eng/20250609
LC record available at https://lccn.loc.gov/2024052845
LC ebook record available at https://lccn.loc.gov/2024052846

British Cataloguing in Publication Data
A Cataloguing in Publication record for this book is available from the British Library.

Table of Contents

Preface xv

Chapter 1
A Review on the Management of End-of-Life Solar Photovoltaic Panels: Challenges, Opportunities, and Future Directions 1
Ozan Capraz, Pamukkale University, Turkey
Zeynep Ozsut Bogar, Pamukkale University, Turkey
Askiner Gungor, Pamukkale University, Turkey

Chapter 2
Advancing E-Waste Management: Integrative Methods and Practices 53
Dankan Gowda V., Department of Electronics and Communication Engineering, BMS Institute of Technology and Management, India
Vijay C. R., Department of Management Studies, Dr. N.G.P. Institute of Technology, Coimbatore, India
Renu, TMIMT, Teerthanker Mahaveer University, Moradabad, India
N. M. G. Kumar, Department of Electrical and Electronics Engineering, Mohan Babu University, Tirupathi, India
Ratidev Samal, School of Management, Centurion University of Technology and Management, Bhubaneswar, India

Chapter 3
Advancing E-Waste Management Through Circular Economy Principles: A Pathway to Sustainable Innovation 75
P. Deepika, R.V. College of Engineering, India
Neeta B. Malvi, R.V. College of Engineering, India
B. Sahana, R.V. College of Engineering, India
Likith Gowda, R.V. College of Engineering, India
B. J. Harshitha, R.V. College of Engineering, India

Chapter 4
Current and Forthcoming Viewpoints to Generate and Manage Sustainable E-Waste 105
Sumanta Kuila, Haldia Institute of Technology, India
Namrata Dhanda, Amity University, India
Subhankar Joardar, Haldia Institute of Technology, India
Amrita Sarkar, Haldia Institute of Technology, India
Angana Chakraborty, Haldia Institute of Technology, India

Chapter 5
Towards a Greener Future: Integrated Strategies for E-Waste Management..... 133
B. Yamini, SRM Institute of Science and Technology, India
Anish T. P., R.M.K. College of Engineering and Technology, India
B. Maheswari, R.M.K. Engineering College, India
J. Praveenkumar, Rajalakshmi Institute of Technology, India
Siva Subramanian R., R.M.K. College of Engineering and Technology, India

Chapter 6
Chemical and Biological Processes for E-Waste Recycling 151
P. Selvakumar, Department of Science and Humanities, Nehru Institute of Technology, India
Ritu Dahiya, Chhotu Ram Arya College, India
T. C. Manjunath, Rajarajeswari College of Engineering, India
Santosh Kumar Nathsharma, Stewart Science College, India

Chapter 7
The Role of Blockchain Technology With IoT for Revolutionizing Waste Management 177
S. Vidhya, Department of Computing Technologies, SRM Institute of Science and Technology, India
Balaji Mahadevan, Department of Electrical and Electronics Engineering, Sri Sivasubramaniya Nadar College of Engineering, India
V. Kamaraj, Department of Electrical and Electronics Engineering, Sri Sivasubramaniya Nadar College of Engineering, India

Chapter 8
Carbon Footprint Analysis of Electronic Products 215
P. Selvakumar, Science and Humanities, Nehru Institute of Technology, India
Ritu Dahiya, Chhotu Ram Arya College, India
T. C. Manjunath, Rajarajeswari College of Engineering, India
Mohit Sharma, Maharshi Dayanand University, Rohtak, India

Chapter 9
Ecotoxicity and Human Health Risks of Improper E-Waste Handling 243
Odangowei Inetiminebi Ogidi, Bayelsa Medical University, Nigeria
Nwanne Dike Ijere, College of Arts, Science, and Professional Studies, Aba, Nigeria
Iniebiyo Felagha, University of Africa, Nigeria

Chapter 10
Impact of E-Waste Processing on Health and Environment 283
Benu Kesar, Symbiosis International University, India
Shaji Joseph, Symbiosis International University, India

Chapter 11
The Growing Environmental Risks of E-Waste: A Review 309
Subhadra Rajpoot, Amity Institute of Environmental Toxicology Safety & Management, Amity University, Noida, India

Chapter 12
E-Waste Recycling Intention Among Generation Z: An Integration of the Theory of Planned Behaviour 337
Tengku Nurshazwina Tengku Sahrum, Universiti Teknologi MARA, Malaysia
Safwah Malyanah Sahariddun, Universiti Teknologi MARA, Malaysia
Nurul Hidayana Mohd Noor, Universiti Teknologi MARA, Malaysia

Chapter 13
The Link Between E-Waste and Linear Economy Through a Bootstrap Rolling Window Analysis 365
Mikail Pehlivan, Trakya University, Turkey
Şahin Akkaya, Istanbul University, Turkey

Chapter 14
Understanding E-Waste Disposal Behavior Across Diverse Groups: A Social Cognitive Perspective 413
Varanasi Rahul, Jain University, India
N. Sathyanarayana, Jain University, India
Usha Prabhu, Jain University, India
Y. Fathima, Jain University, India
V. Lava Kumar, GITAM University, India

Compilation of References 441

About the Contributors 515

Index 527

Detailed Table of Contents

Preface xv

Chapter 1
A Review on the Management of End-of-Life Solar Photovoltaic Panels: Challenges, Opportunities, and Future Directions 1
Ozan Capraz, Pamukkale University, Turkey
Zeynep Ozsut Bogar, Pamukkale University, Turkey
Askiner Gungor, Pamukkale University, Turkey

The rapid increase in the use of solar photovoltaic (PV) panels emphasizes the crucial role of renewable energy sources in meeting energy needs while reducing the environmental impact of fossil fuels. Consequently, managing the end-of-life (EOL) of these panels has gained importance, as the growing number of EOL PV panels brings significant waste management concerns. Major solar energy markets are projected to generate considerable PV waste by 2050; however, recycling infrastructure remains inadequate in many regions. Given the high value of materials in PV panels, including base metals like aluminum, critical metals such as indium and gallium, and hazardous metals like lead and cadmium, adopting recycling approaches that align with circular economy is increasingly essential. This chapter provides a comprehensive review of EOL solar PV panel management, examining studies on PV waste estimation, exploring recycling processes, material composition and relevant policies, analyzing key challenges and opportunities, and offering recommendations for sustainable practices.

Chapter 2
Advancing E-Waste Management: Integrative Methods and Practices 53
Dankan Gowda V., Department of Electronics and Communication Engineering, BMS Institute of Technology and Management, India
Vijay C. R., Department of Management Studies, Dr. N.G.P. Institute of Technology, Coimbatore, India
Renu, TMIMT, Teerthanker Mahaveer University, Moradabad, India
N. M. G. Kumar, Department of Electrical and Electronics Engineering, Mohan Babu University, Tirupathi, India
Ratidev Samal, School of Management, Centurion University of Technology and Management, Bhubaneswar, India

This chapter focuses on the strategies regarding e-waste management involving the use of technology, policy, and community intervention for reduction of environmental and health effects, as well. Taking insights in state-of-art recycling approaches, policies including EPR and engaging the public, the authors discuss ways to enhance collection, resource recovery, and pollution. Examples suggest that respective policy and advanced technologies result in better rates of recyclability and decreased emissions. Some major highlights reveal efficiency improvement in the context of using AI in sorting programs and cuts in emissions with circular economy strategies. The third section discusses future trends such as online tracking of e-waste using blockchain and sustainable design. They provide practical information about how to design an e-waste framework that is both sustainable in the long run and capable of being reproduced across the globe.

Chapter 3
Advancing E-Waste Management Through Circular Economy Principles: A Pathway to Sustainable Innovation 75
P. Deepika, R.V. College of Engineering, India
Neeta B. Malvi, R.V. College of Engineering, India
B. Sahana, R.V. College of Engineering, India
Likith Gowda, R.V. College of Engineering, India
B. J. Harshitha, R.V. College of Engineering, India

Our world has been seriously disturbed by the problem of electronic waste, or e-waste, due to the fast pace of our digital world. The inclination of consumers has increased due to faster and newer innovations, which have accelerated more quickly than expected. This has led to ever-decreasing product lifespans and higher electronic consumption. As a consequence, e-waste is among the wastes that is rapidly accumulating in the world. By 2019, the global generation of e-waste already exceeded 53.6 million metric tons, and by 2030, it is predicted to increase up to 74 Mt. Such an increase of e-waste raises critical environmental and health hazards as well as economic losses because when hazardous materials are not disposed of, resource content gets wasted. Therefore, there is a rapidly increasing desire to introduce the circular economy model to transform the design, production, use, and end-of-life processing of electronic products. The circular economy emphasizes product longevity, longer life cycles, and encourages the reuse and recycling of product materials.

Chapter 4
Current and Forthcoming Viewpoints to Generate and Manage Sustainable E-Waste 105
Sumanta Kuila, Haldia Institute of Technology, India
Namrata Dhanda, Amity University, India
Subhankar Joardar, Haldia Institute of Technology, India
Amrita Sarkar, Haldia Institute of Technology, India
Angana Chakraborty, Haldia Institute of Technology, India

This study examines the growing global challenge of electronic waste (e-waste), driven by shorter product life cycles and technological advancements. E-waste, a major source of contaminants and valuable materials, requires proper recycling and disposal. Despite efforts by developed nations, such as the Group of Seven (G7) countries' Reuse, Recycling, and Recovery (3Rs) program, the high cost of recycling has led to the transfer of e-waste to developing nations with inadequate recycling infrastructure and weak regulatory enforcement. This research focuses on identifying and analyzing the interactive barriers faced by local governments and e-waste recycling businesses in both industrialized and developing countries. It emphasizes the direct and indirect impacts of these barriers on sustainable e-waste management practices. While prior studies have addressed individual barriers, this study explores how they interact and the aggregate effect of these interactions on long-term sustainability.

Chapter 5
Towards a Greener Future: Integrated Strategies for E-Waste Management..... 133
B. Yamini, SRM Institute of Science and Technology, India
Anish T. P., R.M.K. College of Engineering and Technology, India
B. Maheswari, R.M.K. Engineering College, India
J. Praveenkumar, Rajalakshmi Institute of Technology, India
Siva Subramanian R., R.M.K. College of Engineering and Technology, India

Technology has evolved and advanced over the years, and as a result, more electronic items are subscribed to and disposed of, turning into e-waste, a big threat to the environment and health. This chapter explores an integrated look into sustainably managing e-waste drawing from the related sources of the product, classification of e-waste, and the more traditional steps in the e-waste management life cycle, which include collection, sorting, and recycling. It articulates the present practices and discusses ideas such as utilizing models of circular economy and the extended producer responsibility. Recycling and the technology of automated sorting processes are included together with the corresponding legislation on the national and international levels. Best practices are seen through case studies of successful e-waste management programs while the issues raised include technological limitations, financial constraints, or limitations in the policy framework. Some of the recommendations include international cooperation and innovation for change.

Chapter 6
Chemical and Biological Processes for E-Waste Recycling 151
P. Selvakumar, Department of Science and Humanities, Nehru Institute of Technology, India
Ritu Dahiya, Chhotu Ram Arya College, India
T. C. Manjunath, Rajarajeswari College of Engineering, India
Santosh Kumar Nathsharma, Stewart Science College, India

The proliferation of technology-driven consumption coupled with shortened product lifecycles contributes significantly to the escalating volumes of e-waste generated worldwide. As societies become increasingly dependent on electronic gadgets for communication, entertainment, and daily tasks, the disposal and recycling of these devices pose critical environmental and health challenges. However, within these challenges lie opportunities for innovation, sustainable practices, and economic growth through the establishment of effective e-waste recycling systems. Furthermore, the complexity of electronic devices, characterized by a wide variety of materials and components, complicates the recycling process. Moreover, the lack of standardized e-waste collection systems and regulations in many regions contributes to informal recycling practices, often carried out under unsafe conditions by informal workers in developing countries. Despite these challenges, e-waste recycling presents significant opportunities for sustainable development and economic benefits.

Chapter 7
The Role of Blockchain Technology With IoT for Revolutionizing Waste Management 177
S. Vidhya, Department of Computing Technologies, SRM Institute of Science and Technology, India
Balaji Mahadevan, Department of Electrical and Electronics Engineering, Sri Sivasubramaniya Nadar College of Engineering, India
V. Kamaraj, Department of Electrical and Electronics Engineering, Sri Sivasubramaniya Nadar College of Engineering, India

Modern approaches for waste management assure effectiveness, openness, and sustainability and hence can be adopted with the aim of handling increasing waste problem globally. The use of blockchain, which is well known for its decentralization and immutability, may help to revolutionalize this area. The chapter focuses on the uses of blockchain in waste management. Better traceability is one of the primary benefits of blockchain's application in waste management. Thus, it is possible to monitor the movement of waste garbage from the time it leaves the source to the time it gets to the waste garages. The integration of the internet of things (IoT) with blockchain technology is useful in the following ways in enhancing the functionality of the system. In this chapter, the blockchain-enabled smart waste management system is proposed to change the traditional waste management system in the urban areas by using the IoT devices, smart contracts, and data analytics.

Chapter 8
Carbon Footprint Analysis of Electronic Products 215
P. Selvakumar, Science and Humanities, Nehru Institute of Technology, India
Ritu Dahiya, Chhotu Ram Arya College, India
T. C. Manjunath, Rajarajeswari College of Engineering, India
Mohit Sharma, Maharshi Dayanand University, Rohtak, India

The significance of carbon footprint analysis lies in its ability to inform and drive sustainability efforts across various sectors. It serves as a metric for assessing environmental performance, guiding organizations and policymakers towards informed decision-making aimed at reducing emissions. Understanding the carbon footprint of products helps consumers make environmentally conscious choices, thereby influencing market demand towards more sustainable practices. Moreover, for businesses, it can be a strategic tool for improving operational efficiencies, reducing costs associated with energy use, and enhancing corporate social responsibility (CSR) efforts. Carbon footprints encompass both direct emissions (e.g., from burning fossil fuels for heating or transportation) and indirect emissions (e.g., from the production and transportation of goods consumed). This comprehensive approach ensures that all significant sources of greenhouse gases are accounted for, offering a holistic view of environmental impact.

Chapter 9
Ecotoxicity and Human Health Risks of Improper E-Waste Handling 243
Odangowei Inetiminebi Ogidi, Bayelsa Medical University, Nigeria
Nwanne Dike Ijere, College of Arts, Science, and Professional Studies, Aba, Nigeria
Iniebiyo Felagha, University of Africa, Nigeria

Globally, the management of solid waste encompasses electronic waste. E-waste, which makes up a substantial part of solid waste, contains several dangerous substances, including halogenated compounds like polychlorinated biphenyls (PCBs), tetrabromobisphenol A (TBBPA), and polybrominated biphenyl (PBB). These substances exert detrimental impacts on both human beings and the environment. Heavy metals constitute a significant portion of electronic waste, so it is crucial to exercise caution while dismantling them. The involvement of the informal sector in underdeveloped countries exacerbates the issue of handling e-waste. The existing methods for disposing and treating electronic waste are insufficient, and the materials have both immediate and indirect impacts on human health and the environment. This chapter examines the diverse types of electronic waste and their effects on different aspects of the environment and human health. It also discussed information on the procedures for managing electronic waste and effective handling methods.

Chapter 10
Impact of E-Waste Processing on Health and Environment 283
Benu Kesar, Symbiosis International University, India
Shaji Joseph, Symbiosis International University, India

This chapter explores innovative e-waste management solutions focusing on technology and sustainability. It introduces the universal risk assessment framework (URAF) for e-waste sites, a model to evaluate contamination. The chapter highlights urban mining as a sustainable alternative to traditional resource extraction, along with supercritical water oxidation (SCWO) and thermal plasma technology, which transform hazardous e-waste into harmless byproducts and clean energy sources. Technological advancements, including blockchain-enabled waste tracking, AI-powered automated sorting, and IoT-integrated smart waste bins are revolutionizing e-waste management efficiency. Additionally, policy-driven strategies such as right-to-repair legislation and modular electronics design aim to extend product lifespans and reduce overall waste. However, challenges remain in implementation, particularly in developing nations where regulatory enforcement and economic viability pose obstacles to formal recycling.

Chapter 11
The Growing Environmental Risks of E-Waste: A Review 309
Subhadra Rajpoot, Amity Institute of Environmental Toxicology Safety & Management, Amity University, Noida, India

One of the most current problems to emerge in the rich as well as the poor globally is electronic garbage, or e-waste. Trash is one of the most challenging and quickly growing problems in the industrial world. Electronic trash, sometimes referred to as e-waste, is a sort of garbage created by electronics. Old or end-of-life electronic equipment, including computers, laptops, TVs, generators, DVD players, cell phones, refrigerators, and other devices that are usually thrown away by their original owners because of their limited lifespan are referred to as e-waste. They have several potentially harmful ingredients that, if not adequately handled, might harm human health as well as the environment. It is difficult to deal with e-waste since it contains hazardous chemicals. According to previous studies, India produced 0.4 million tons of e-waste in 2010, and by 2013–2014, that amount might rise to 0.5–0.6 million tons.

Chapter 12
E-Waste Recycling Intention Among Generation Z: An Integration of the Theory of Planned Behaviour 337
Tengku Nurshazwina Tengku Sahrum, Universiti Teknologi MARA, Malaysia
Safwah Malyanah Sahariddun, Universiti Teknologi MARA, Malaysia
Nurul Hidayana Mohd Noor, Universiti Teknologi MARA, Malaysia

The practice of recycling reduces the amount of waste produced, reduces the use of natural resources, and provides opportunities to produce new products using recycled materials. This study identified the determinants of e-waste recycling intention among Gen Z using the theory of planned behavior (TPB). This preliminary study used a quantitative method involving a total of 251 respondents. The analysis consists of descriptive and correlation analysis and multiple regression analysis. The result shows that Gen Z's intention to recycle e-waste is high. Next, significant relationships exist between attitude, social norms, moral norms, information, incentives, and e-waste recycling intention. Finally, the regression results indicate that information strongly influences e-waste recycling intention. This study is essential to expand the understanding of e-waste recycling among the younger generation.

Chapter 13
The Link Between E-Waste and Linear Economy Through a Bootstrap Rolling Window Analysis 365
Mikail Pehlivan, Trakya University, Turkey
Şahin Akkaya, Istanbul University, Turkey

It is known that the dominant economic doctrine in the current economic system is neoclassical economics. The fact that waste electrical and electronic equipment (e-waste) is the fastest growing solid waste in the world may be an important indicator of this linear production and consumption economy model. Therefore, this study empirically tests the relationship between e-waste and (linear) economic growth. According to this analysis, the relationship between total e-waste and economic growth in the countries included in the study (USA, Germany, Australia, Brazil, China, Ghana, South Korea, India, Switzerland, Japan, Türkiye) was tested with panel bootstrap causality analysis using annual data for the period 1960-2020. According to the findings of the analysis, a bidirectional causality is found between e-waste and economic growth in all sub-samples for all countries. In this respect, it can be stated that within the framework of the sustainable production and consumption paradigm, a transition from a linear economy to a circular economy seems essential.

Chapter 14
Understanding E-Waste Disposal Behavior Across Diverse Groups: A Social Cognitive Perspective 413
Varanasi Rahul, Jain University, India
N. Sathyanarayana, Jain University, India
Usha Prabhu, Jain University, India
Y. Fathima, Jain University, India
V. Lava Kumar, GITAM University, India

This chapter seeks to examine the factors and effects of e-waste disposal behaviour among various demographic groups. Social cognitive theory (SCT) is utilised to formulate and evaluate a conceptual model that examines the impact of psychological factors, observational learning, and self-regulation on individuals' e-waste disposal behaviour. This study employs a structured survey methodology via a questionnaire to investigate the e-waste disposal behaviours of various demographic groups across different regions of India. Multiple regression analysis is utilised as the statistical method for data examination. The results of the multiple regression analysis reveal that psychological factors, observational learning, and self-regulation are positively correlated with the e-waste disposal behaviours of various demographic groups. This study has constraints, including common method bias and social desirability bias, owing to its dependence on self-report measures.

Compilation of References 441

About the Contributors 515

Index 527

Preface

The rapid evolution of technology has led to an unprecedented increase in electronic waste (e-waste), presenting significant environmental and public health challenges. As editors of this volume, Integrated Approaches for Sustainable E-Waste Management, we recognize the urgent need for innovative and interdisciplinary strategies to address the complexities of e-waste management. This book aims to provide a comprehensive examination of the challenges and opportunities associated with the sustainable handling of e-waste, offering holistic and integrated solutions.

E-waste is one of the fastest-growing waste streams globally, driven by the increasing consumption and disposal of electronic devices. Managing this waste stream effectively requires a multidisciplinary approach that spans the entire lifecycle of electronic products—from design and manufacturing to usage and eventual disposal. This volume brings together experts from various fields, including industrial engineering, environmental engineering, computer engineering, electronic engineering, and management science, to present strategies that minimize environmental impact while maximizing resource recovery and reuse.

In the chapters that follow, readers will find a structured exploration divided into four thematic categories, each addressing crucial facets of sustainable e-waste management.

Beginning with policy frameworks and integrated management approaches, Chapter 1 by Dr. Ozan Capraz, Dr. Zeynep Ozsut Bogar, and Prof. Askiner Gungor, titled "A Review on the Management of End-of-Life Solar Photovoltaic Panels: Challenges, Opportunities, and Future Directions," emphasizes the significance of adopting circular economy principles in managing photovoltaic waste. Building upon this perspective, Chapter 2 by Dr. Dankan Gowda V, "Advancing E-Waste Management Integrative Methods and Practices," examines integrative frameworks employing innovative technologies, effective policies, and active community involvement to enhance e-waste management globally.

Continuing the exploration of integrated management strategies, Chapter 3, authored by Mrs. Deepika P, Dr. Sahana B, Mrs. Neeta B Malvi, Mr. Likith Gowda B S, and Ms. Harshitha B J, "Advancing E-Waste Management through Circular Economy Principles: A Pathway to Sustainable Innovation," delves into sustainable innovations focused on product life-cycle extension and resource recovery. In Chapter 4, "Current and Forthcoming Viewpoints to Generate and Manage Sustainable E-Waste," Dr. Sumanta Kuila, Dr. Namrata Dhanda, Dr. Subhankar Joardar, Mrs. Amrita Sarkar, and Ms. Angana Chakraborty present insights on global trends and future opportunities to sustainably manage the rising volumes of e-waste. Chapter 5, "Towards a Greener Future: Integrated Strategies for E-Waste Management," by Siva Subramanian R, Dr. Nalini M, Mr. Anish T P, Mrs. Maheswari B, Mr. Ezhilvendan M, Mrs. Girija P, and Dr. Beslin Pajila P J, provides an insightful synthesis of integrated approaches to e-waste management, emphasizing practical policies, recycling models, and best practices.

Transitioning to technological interventions, Chapter 6 by Dr. Selvakumar P, "Chemical and Biological Processes for E-Waste Recycling," explores innovative methods to recover valuable materials from complex electronic waste streams. In Chapter 7, "The Role of Blockchain Technology with IoT for Revolutionizing Waste Management," Vidhya S, Dr. Balaji Mahadevan, and Dr. Kamaraj V discuss cutting-edge digital technologies designed to enhance transparency, efficiency, and accountability in waste management.

Shifting to the environmental and health impacts of e-waste, Chapter 8 by Dr. Selvakumar P, "Carbon Footprint Analysis of Electronic Products," examines methodologies for evaluating greenhouse gas emissions associated with electronic devices, highlighting their significance for sustainable decision-making. Chapter 9, authored by Dr. Odangowei Inetiminebi Ogidi, Dr. Nwanne Dike Ijere, and Dr. Iniebiyo Felagha, "Ecotoxicity and Human Health Risks of Improper E-Waste Handling," investigates the hazardous effects of improper e-waste disposal on public health and ecosystems. Building further, Chapter 10, "Impact of E-Waste Processing on Health and Environment," by Dr. Benu Kesar, provides an analysis of novel technological solutions and policies aimed at minimizing health risks and environmental contamination. Chapter 11 by Dr. Subhadra Rajpoot, "The Growing Environmental Risks of E-Waste: A Review," synthesizes current research highlighting the increasing global environmental hazards posed by unmanaged e-waste.

In the concluding thematic section focusing on behavioral, social, and economic dimensions, Chapter 12 by Miss Tengku Nurshazwina Tengku Sahrum, Miss Safwah Malyanah Sahariddun, and Dr. Nurul Hidayana Mohd Noor, "E-Waste Recycling Intention among Generation Z: An Integration of the Theory of Planned Behaviour," examines motivational factors shaping e-waste recycling behaviors among youth. In Chapter 13, "The Link Between E-Waste and Linear Economy through a Bootstrap

Rolling Window Analysis," Dr. Mikail Pehlivan and Dr. Şahin Akkaya empirically analyze how linear economic paradigms exacerbate e-waste generation. Lastly, Chapter 14 by Mr. Varanasi Rahul, N. Sathyanarayana, Dr. Usha Prabhu, Dr. Y Fathima, and Dr. V Lava Kumar, "Understanding E-Waste Disposal Behavior Across Diverse Groups: A Social Cognitive Perspective," leverages social cognitive theory to understand demographic variations in e-waste disposal practices.

Collectively, these chapters underscore the interconnectedness of policy development, technological innovation, environmental stewardship, and behavioral dynamics in shaping sustainable e-waste management. As the complexities surrounding e-waste grow, integrated and interdisciplinary strategies become increasingly critical.

As we conclude this volume, we reaffirm the importance of addressing e-waste through integrated, innovative, and sustainable approaches. The chapters presented collectively emphasize the urgent need for collaborative efforts across disciplines, industries, and policymakers to devise effective e-waste solutions. By harnessing technological advancements, promoting responsible consumption, and instituting progressive policies, we can significantly mitigate the environmental and societal impacts of electronic waste.

We extend our sincere gratitude to all contributors for their valuable insights and dedication to advancing this crucial field. We also thank our readers for engaging with this work and their ongoing commitment to fostering sustainable and responsible e-waste management practices worldwide.

Aşkıner Güngör
Pamukkale University, Turkey

Olcay Polat
Pamukkale University, Turkey

Chapter 1
A Review on the Management of End-of-Life Solar Photovoltaic Panels:
Challenges, Opportunities, and Future Directions

Ozan Capraz
https://orcid.org/0009-0005-4080-9538
Pamukkale University, Turkey

Zeynep Ozsut Bogar
https://orcid.org/0000-0002-9089-2764
Pamukkale University, Turkey

Askiner Gungor
https://orcid.org/0000-0002-1223-6796
Pamukkale University, Turkey

ABSTRACT

The rapid increase in the use of solar photovoltaic (PV) panels emphasizes the crucial role of renewable energy sources in meeting energy needs while reducing the environmental impact of fossil fuels. Consequently, managing the end-of-life (EOL) of these panels has gained importance, as the growing number of EOL PV panels brings significant waste management concerns. Major solar energy markets are projected to generate considerable PV waste by 2050; however, recycling infrastructure remains inadequate in many regions. Given the high value of materials

DOI: 10.4018/979-8-3693-7383-5.ch001

in PV panels, including base metals like aluminum, critical metals such as indium and gallium, and hazardous metals like lead and cadmium, adopting recycling approaches that align with circular economy is increasingly essential. This chapter provides a comprehensive review of EOL solar PV panel management, examining studies on PV waste estimation, exploring recycling processes, material composition and relevant policies, analyzing key challenges and opportunities, and offering recommendations for sustainable practices.

INTRODUCTION

Throughout history, humanity has continuously worked to meet its energy requirements, which are essential for social development, technological advancement, and the maintenance of daily life (Demir & Yakışık, 2024). Since the beginning of the Industrial Revolution, the demand for energy has significantly increased globally over the years (IEA, 2023). Consequently, due to this increasing global energy demand, fossil fuels such as coal, oil, and natural gas are considered the primary energy source due to their high energy density and ease of access (IEA, 2023). However, this rising demand is rapidly depleting oil and natural gas reserves while also generating environmental problems, particularly greenhouse gas emissions that contribute to global climate change (IPCC, 2023). Studies show that fossil fuels have negative impacts, including air, water, and soil pollution, as well as dependence on energy imports in many regions worldwide (Bertrand, 2021). According to the Global Renewables Outlook: Energy Transformation 2050 report published by IRENA (2020), currently, it is estimated that air pollution results in 7 million premature deaths annually. Furthermore, around 840 million people lack access to electricity, and 2.6 billion lack access to clean cooking fuels (IRENA, 2020). Over the years, global CO_2 emissions have risen significantly, with most of this increase coming from fossil fuels (Ritchie & Rosado, 2022; Ritchie et al., 2023). As illustrated in Figure 1, the continuous increase in fossil fuel consumption use has directly contributed to the increase in CO_2 emissions (Ritchie & Rosado, 2022; Ritchie et al., 2023). Consequently, there has been a global interest in the use of clean and sustainable renewable energy (RE) sources in recent years because of the concerns related to environmental issues and climate change caused by fossil fuels (IPCC, 2023).

The data used in this figure was adapted from the Our World in Data: Fossil Fuels (Ritchie & Rosado, 2022) and the Our World in Data: CO_2 and Greenhouse Gas Emissions (Ritchie et al., 2023).

Figure 1. Global fossil fuel consumption and CO_2 emissions from fossil fuels (1965-2022)

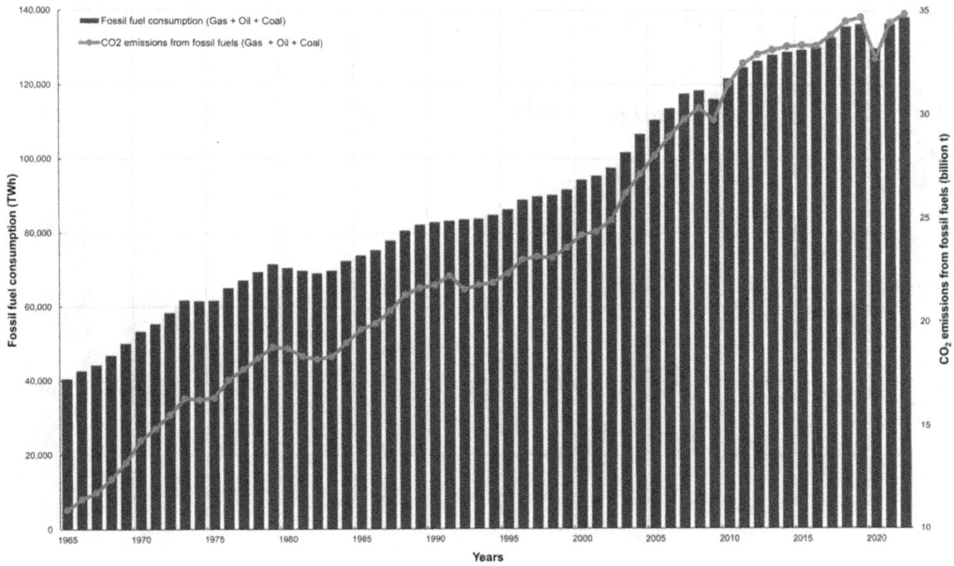

RE can be defined as energy obtained from sources that are naturally replenished and are expected to remain available for the long term (Al Mubarak et al., 2024). The RE sources include solar, wind, hydro, geothermal, biomass, tidal and wave energy, which are cleaner alternatives with lower greenhouse gas emissions compared to conventional fossil fuels (Panwar et al., 2011). As the negative impacts of fossil fuel consumption become more evident, there has been a global trend in the use of RE sources (Capraz et al., 2024; Karlilar Pata & Balcilar, 2024). Moreover, with the limited availability of fossil fuels, rising energy prices, population growth, and increasing energy demand, RE is expected to play a significant role in the meeting energy needs and is also considered essential for addressing environmental challenges such as climate change (IPCC, 2023). According to the Global Landscape of Renewable Energy Finance 2023 report (IRENA & CPI, 2023), global investments in RE technologies have been steadily increasing. In 2022, total global investment in RE reached \$499 billion, a rise of nearly 16% compared to 2021. As shown in Figure 2, solar energy technologies have become a major focus of RE investments (IRENA & CPI, 2023). In 2020, solar energy investments reached \$162 billion, representing an 11% increase from 2019. Investments in solar technologies continued to grow, reaching \$226 billion in 2021, which represents an increase of nearly 40% from 2020. The trend also continued in 2022, with solar energy investments reaching \$308 billion, showing a 36% increase compared to the previous year (IRENA

& CPI, 2023). As a result, solar energy has become one of the most prominent RE sources globally (IRENA, 2024).

The data used in this figure was adapted from the Global Landscape of Renewable Energy Finance 2023 (IRENA & CPI, 2023).

Figure 2. The global annual investments for renewable energy technologies (2013-2022)

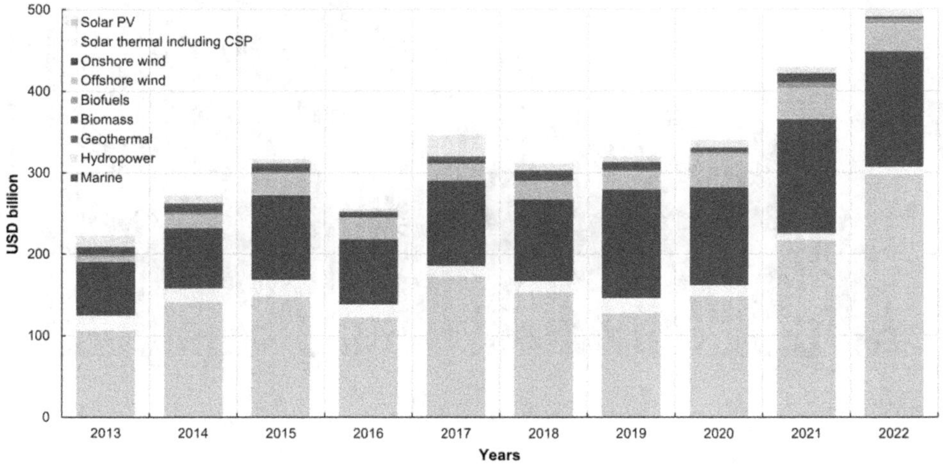

Solar energy has gained a significant share in energy production with technological developments, decreasing costs, and consistent investment growth (IRENA, 2024). As reported in the Renewable Energy Statistics 2024 report by IRENA (2024), the total RE capacity worldwide reached 3,864,522 megawatts (MW) in 2023, while solar energy accounting for 36.7% of this capacity, reaching 1,418,016 MW. Specifically, of this solar capacity, 1,411,139 MW came from solar photovoltaic (PV) and 6,876 MW from concentrated solar power (CSP). Solar PV is one of the most widely used RE sources today (IRENA, 2024). As highlighted in the Renewable Power Generation Costs in 2022 report (IRENA, 2023), solar PV has also experienced one of the most rapid cost reductions among the RE technologies since 2010. As shown in Figure 3, the global weighted-average Levelized Cost of Energy (LCOE) for utility-scale solar PV projects fell by 89%, from \$0.445/kWh in 2010 to \$0.049/kWh in 2022. This dramatic reduction in LCOE has been mainly driven by significant declines in module prices. The reduction in module prices has been driven by improved module efficiency, optimized manufacturing processes, increased economies of scale, and reduced material usage (IRENA, 2023). Additionally, the report from IRENA (2023) further notes that the global capacity weighted-average total installed cost of utility-scale solar PV projects decreased by approximately 83%, from \$5,124/

kW in 2010 to \$876/kW in 2022 (please see Figure 4). These cost reductions have made solar PV one of the most affordable and accessible RE technologies around the world (NREL, 2021).

The data used in this figure was adapted from the Renewable Power Generation Costs in 2022 (IRENA, 2023)

Figure 3. The global weighted-average LCOE for solar PV (2010-2022)

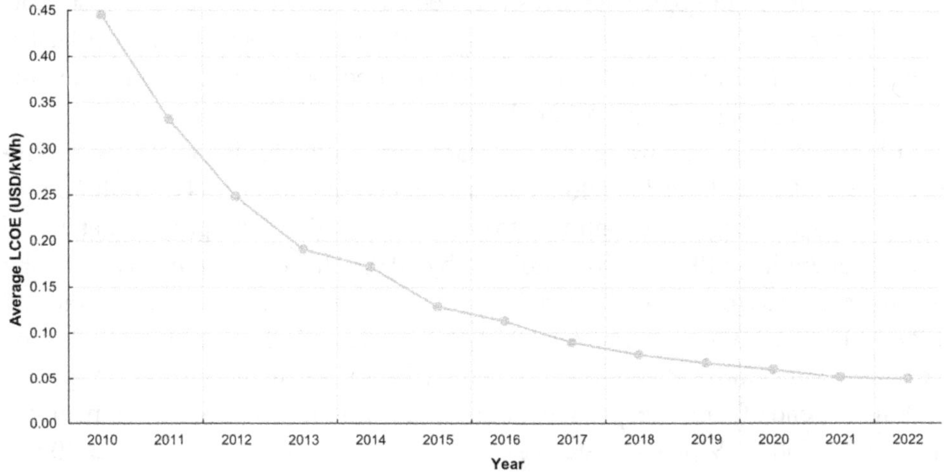

The data used in this figure was adapted from the Renewable Power Generation Costs in 2022 (IRENA, 2023)

Figure 4. The global weighted-average total installed cost for solar PV (2010-2022)

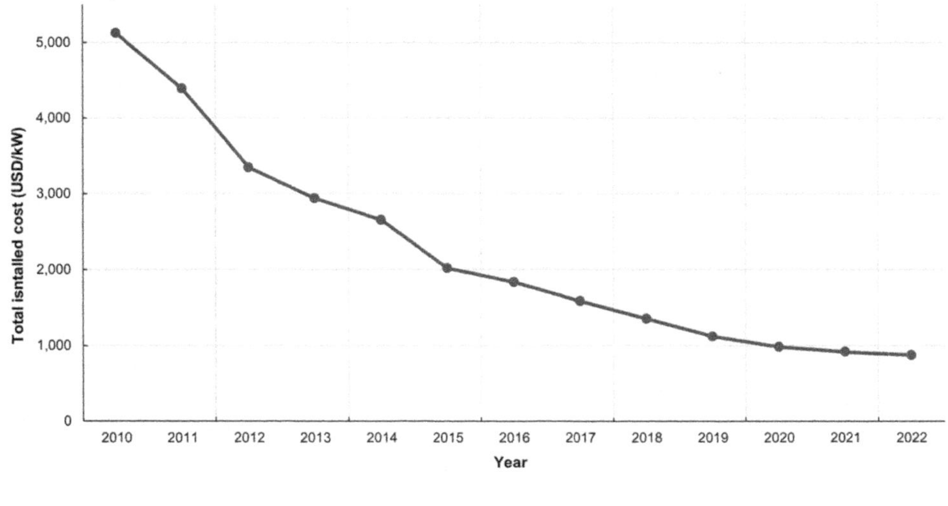

As discussed in the World Energy Outlook 2021 report (IEA, 2021), the solar PV panels are now attractive not only among individual households and businesses, but also for the large-scale energy projects. To meet increasing energy demands and global carbon reduction goals, the use of solar PV panels is steadily expanding and major economies such as China, the United States, India, and the European Union are boosting their investments in solar energy (IEA, 2021). Indeed, the volume of waste generated from solar PV panels reaching the end-of-life (EOL) or becoming obsolete due to technological changes is expected to rise in parallel to investments in solar energy (Weckend et al., 2016). Thus, the management of EOL solar PV panels is crucial for reducing the negative environmental impacts and ensuring the sustainability of solar energy (Ali et al., 2024).

Millions of electronic waste (e-waste) are generated worldwide each year. According to Baldé et al. (2017), global e-waste generation reached 44.7 million tons (t) in 2016, and this amount continues to grow. Forti et al. (2020) project that global e-waste generation will reach 74.7 million t by 2030. The technical lifespan of solar PV panels ranges from 25 to 30 years (EPA, 2024; Weckend et al., 2016). At the end of their lifespan, PV panels can be classified as e-waste, significantly contributing to the growing e-waste stream (Gautam et al., 2021). As the number of PV panels reaching the end of their lifespan increases along with the solar PV deployment, the cumulative global PV panel waste is expected to reach 1.7-8 million t by 2030 and 60-78 million t by 2050 (Weckend et al., 2016). Based on data from the End-of-Life Management: Solar Photovoltaic Panels report, the estimated cumulative global waste volumes of solar PV panels for both regular-loss and early-loss scenarios are shown in Figures 5 and 6, respectively (Weckend et al., 2016).

The data used in this figure was adapted from the End-of-Life Management: Solar Photovoltaic Panels (Weckend et al., 2016). The regional totals were calculated by summing the values provided under their respective subcategories.

Figure 5. Estimation of the cumulative global waste volumes of PV panels for regular-loss scenario

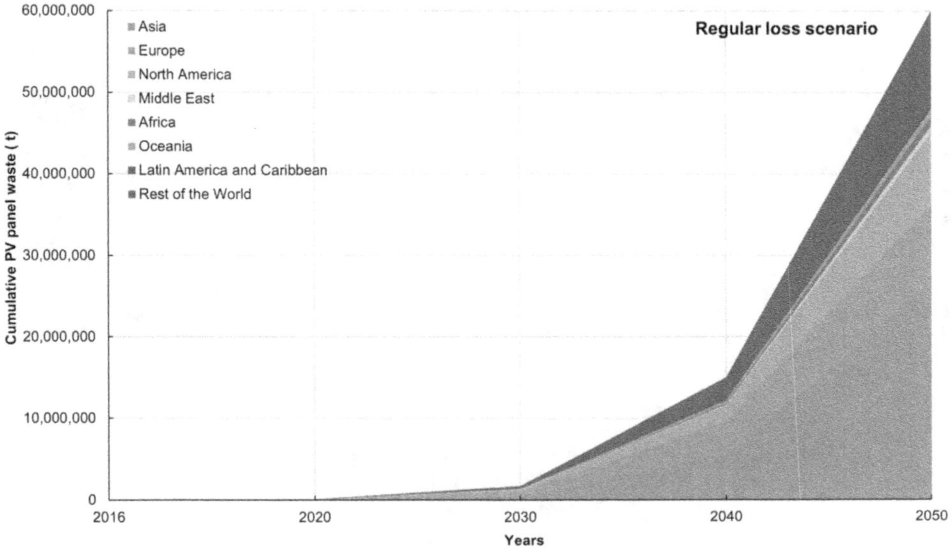

The data used in this figure was adapted from the End-of-Life Management: Solar Photovoltaic Panels (Weckend et al., 2016). The regional totals were calculated by summing the values provided under their respective subcategories.

Figure 6. Estimation of the cumulative global waste volumes of PV panels for early-loss scenario

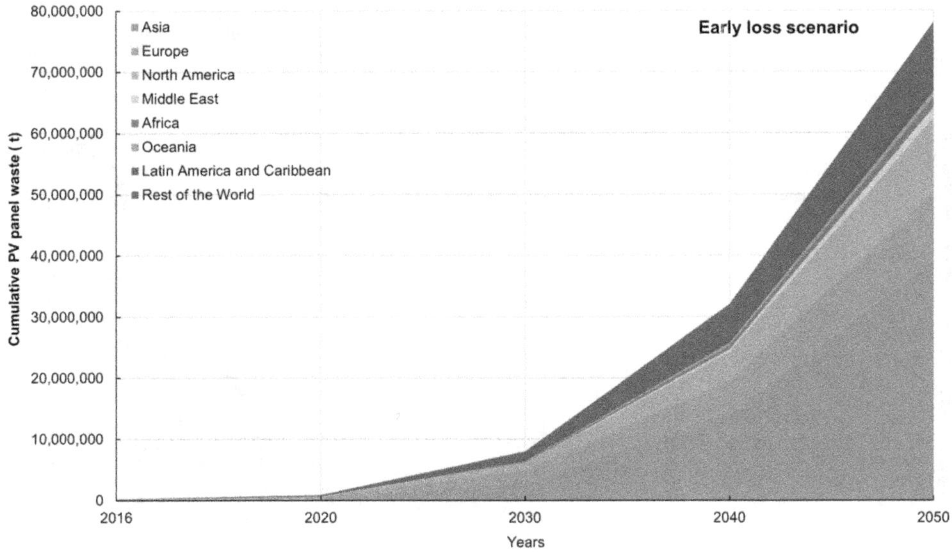

The report from Weckend et al. (2016) states that in the regular-loss scenario, it is assumed that solar PV panels have a typical 30-year lifetime, with no significant early failures. This scenario reflects a standard lifetime, and waste accumulates gradually when PV panels reach the end of their life. Conversely, the early-loss scenario accounts for the possibility of PV panels failing before completing their expected 30-year lifetime due to a variety of factors such as manufacturing defects, environmental damage, operational issues etc. This leads to a more rapid accumulation of waste compared to the regular-loss scenario, as solar PV panels reach the end of their lifespan earlier than expected (Weckend et al., 2016).

In the same report, Weckend et al. (2016) discuss that Asia, Europe, and North America are expected to be the largest contributors to solar PV panel waste, driven by their significant solar PV capacity. By 2030, Asia is projected to lead the global PV panel waste stream, with total waste potentially reaching up to 3.5 million t. Generating between 200,000 and 1.5 million t of waste, China is expected to be the largest contributor, due to its significant installed PV capacity. With projected waste volumes ranging from 200,000 to 1 million t and 50,000 to 325,000 t, respectively, Japan and India are also expected to contribute significantly. Europe is projected to become the second-largest region for PV waste, with accumulation reaching up to 3 million t by 2030. Germany is expected to contribute between 400,000 and 1 million t of this total. Italy and France are also expected to be key contributors to Europe's increasing PV waste. In North America, the United States is expected to generate between 170,000 and 1 million t of PV waste by 2030. Canada and Mexico are projected to have increasing waste volumes, with estimates reaching up to 80,000 t and 30,000 t, respectively. By 2050, the accumulation of PV waste is projected to continue increasing. China is expected to remain the leader in waste generation, accumulating between 13.5 and 20 million t of PV waste. The United States is projected to surpass Germany, generating between 7.5 and 10 million t, while Japan and India are estimated to contribute 6.5-7.5 million t and 4.4-7.5 million t, respectively (Weckend et al., 2016).

Furthermore, e-waste can consist of a variety of components in different sizes and materials, containing hazardous substances such as mercury, lead, arsenic, cadmium, and selenium, as well as valuable metals like gold, silver, and palladium (Cobbing, 2008; Widmer et al., 2005). The management and recovery of e-waste is essential for three key reasons: to extract valuable materials, to prevent environmental and health hazards from harmful substances, and to promote more efficient use of the world's limited natural resources. Therefore, recovering and reusing e-waste is vital for economic, environmental, and health benefits (Özgür Polat & Gungor, 2021; Özgür Polat & Güngör, 2019). According to a World Health Organization (WHO) report on e-waste, published in 2024, the total amount of e-waste generated globally reached over 62 million t in 2022, with only about 22% being properly recycled, mainly due to the lack of regulations and recycling infrastructure (WHO, 2024). Therefore, the management of e-waste has become a critical worldwide concern. A significant portion of this global e-waste collects in open dumpsites and landfills in African countries, where they are illegally recycled and disposed. This leads to contaminating water sources and causing environmental pollution and public health problems (Andeobu et al., 2023; Moyen Massa & Archodoulaki, 2023).

Given these challenges, recovering and reusing e-waste not only reduces environmental risks but also harnesses the economic value of raw materials. Extracting precious metals and other valuable components, can contribute to a circular economy, which supports both economic growth and environmental sustainability (Golsteijn & Valencia Martinez, 2017). Moreover, Duran et al. (2022) reported that less than 10% of solar panels are recycled in the United States. They observed that, given current capacity and waste levels, the recycling cost of a PV panel ranges from $20-30, whereas landfilling costs are only $1-2. As a result, many panels are being disposed of in landfills, despite containing hazardous substances and valuable metals. They also stated that the decrease in solar PV panel installation costs and improvements in their efficiency have led to rapid growth in the market. Due to the waste generated from early reached EOL PV panels, the LCOE for solar generation could potentially double and jeopardize its cost competitiveness (Duran et al., 2022). On the other hand, solar panels typically may contain valuable and scarce metals such as silver, copper, gallium, indium, and germanium, as well as common materials like aluminum and glass, and hazardous substances such as lead and cadmium (Fiandra et al., 2019). A proper EOL management of solar PV panels promotes circular economy by reusing and recovering resources (Duran et al., 2022; Weckend et al., 2016).

To sum up, developing a sustainable management strategy for EOL solar PV panels is of great importance both environmentally and economically. As the global adoption of solar energy continues to expand, there is an urgent need to address the growing waste from PV panels. In addition to the environmental impact of discarded panels due to their hazardous materials, they also contain valuable materials with considerable potential for reuse within a circular economy. Furthermore, in the light of international climate goals (UNFCCC, 2015) and sustainable development goals (UN, 2023), effective management of the lifecycle of solar PV panels is essential for ensuring sustainability. Motivated by the current status and anticipated needs and challenges in PV waste generation, this chapter aims to provide an extensive review of current practices in managing waste PV panels, examine the existing literature, evaluate the challenges, and explore the opportunities related to circular economy strategies and policy developments. It focuses on the technical, environmental, and economic considerations to systematically address key research gaps in the field. To organize the flow of this chapter, the given research questions are addressed:

- Which technological and environmental factors affect the lifespan of solar PV panels?
- How can the quantity of waste PV panels be accurately assessed, and what methods can provide more precise estimations and forecasting?
- What is the potential contribution of recovering and recycling waste PV panels?
- Which policies and regulations exist for managing waste PV panels?
- What are the challenges and future opportunities in managing waste PV panels?

This chapter aims to provide a comprehensive review on the management of EOL solar PV panels by evaluating the current situation and exploring future opportunities and challenges. The rest of this chapter is organized as follows: The second section addresses EOL PV waste, starting with environmental and technological factors that affect panel lifespan. It reviews studies on waste generation and examines methods for estimating PV waste quantities. The third section focuses on recovery and recycling processes that address material composition, recovery potential, and current policies and legislatives. The fourth section discusses challenges and opportunities by analyzing technical and cost issues, and environmental impacts. It also provides recommendations for enhancing waste management practices and suggestions for future directions. Finally, the fifth section presents the main findings of the study and offers conclusions on solar PV panel waste management.

E-WASTE FROM SOLAR PV PANELS

With the rapid increase in the use of PV panels, the quantity of generated PV panel waste is reaching considerable levels. While developed countries have management systems in place, developing countries are still installing solar panels without plans for managing the resulting waste (Yadav et al., 2023). Consequently, the effective management of PV waste has become critical due to its potential environmental and humanitarian impacts, which are similar to other types of e-waste. Although existing studies generally focus on increasing efficiency and analyzing solar PV panels to meet economic expectations, the urgency of addressing PV waste management is rising as these panels approach the end of their lifespan. Therefore, the economic potential and environmental impacts of PV panel waste have become increasingly important and urgent issues today. The amount of potentially generated PV waste is a key input for all stakeholders in designing an effective PV waste management system. As a result, the literature on PV waste estimation has advanced significantly in recent years. Quantifying the potential waste volume helps address gaps in potential revenues, environmental risks, facility and capacity planning, and the need for qualified labor in recycling and recovery processes (Ali et al., 2024; EPA, 2024; Gautam et al., 2021; Weckend et al., 2016).

Environmental Factors and Technological Advancements Affecting the Lifespan of Solar PV Panels

Countries take varying approaches to solar PV waste management. Generally, panel waste can be classified as industrial or general waste (Weckend et al., 2016). Although substantial research exists on the recycling and reuse of electronic items, such as computers, mobile phones, and televisions; the disposal, recycling, management, and reuse of solar PV waste have not yet been adequately quantified or analyzed (Arora et al., 2018). Moreover, the improper disposal of solar panels can lead to significant environmental issues, including the leaching of harmful substances like cadmium and lead, as well as the loss of valuable metals and conventional resources (Gautam et al., 2021).

The lifespan of PV panels directly affects the amount of waste generated over a given period. Thus, different lifespan profiles are used to analyze PV panel-related issues, such as power generation, waste generation, and maintenance needs. Generally, the average lifespan of a PV panel ranges between 20 and 35 years (Achillas et al., 2013; El-Khawad et al., 2022; Panda et al., 2021; Souza et al., 2024; Vinayagamoorthi et al., 2024; Yadav et al., 2023). PV waste arises for various reasons. The primary source is panels reaching EOL phase, which accounts for 90% of the waste generated (Yadav et al., 2023). Additionally, PV waste may also be generated during transportation, installation, maintenance, or due to adverse weather conditions (Bošnjaković et al., 2023).

In this regard, various optimization problems focus on increasing the effectiveness of PV panels (Bogar, 2023; Khoshnazm et al., 2023; Kolamroudi et al., 2023; Sangeetha et al., 2024).These improvements may address performance, quality, maximum power point tracking issues, or effective design and maintenance practices to extend the panels' lifespan. Such changes affect both the lifespan of PV panels, and the quantity of waste generated over time. The lifespan of PV panels is influenced by several factors (Kaaya et al., 2021). The most important is the type of PV panel technology. While mono-crystalline panels have a longer lifespan, thin-film panels generally have a shorter one. However, mono-crystalline panels are more expensive than thin-film panels, so consumers might consider a trade-off between lifespan and cost. Moreover, environmental factors also impact the lifespan of PV panels, which are temperature, humidity, UV radiation, and the uncertainties about these variables (Kaaya et al., 2021). For example, the efficient lifetime of PV panels decreases under high temperatures and excessive UV exposure. To address these effects, various cooling technologies exist to enhance PV panel performance and extend their lifespan (Nižetić et al., 2021). In addition, lifespan is influenced by maintenance and monitoring activities (Orosz et al., 2024), as the surfaces of PV panels can wear and deteriorate due to dust and other external factors. Preventive maintenance and regular servicing can have a positive impact on the lifespan of PV panels (Baklouti et al., 2020). Consequently, it is challenging to define an exact lifespan expectancy.

The Amount of Generated Solar PV Panel Waste

The literature on PV waste generation differs from that of general e-waste estimation. Most studies in the e-waste estimation literature focus on large household appliances (such as refrigerators, washing machines, and ovens), IT and telecommunications equipment (including computers, laptops, printers, and telephones), as well as consumer equipment like TVs (Ozsut Bogar & Gungor, 2023). Gautam et al. (2021) examine the assessment of e-waste generated from decommissioned solar PV panels in India, analyzing the recycling processes, existing regulatory frameworks, and sustainability goals associated with solar panel waste management. Additionally, the study highlights the necessity of effective strategies to mitigate the environmental impact of the solar energy sector while promoting resource efficiency within the context of a circular economy. Mahmoudi et al. (2019) discuss the assessment of waste generated by PV systems in Australia throughout their life cycle. In another study, Mahmoudi et al. (2021) estimate PV waste for OECD countries. Furthermore, Chowdhury et al. (2020) analyze the environmental impacts of PV panels and their components, as well as potential recycling and management methods. They also discuss how solar panels can be recycled at the end of their life cycle and emphasize the importance of recycling processes to reduce the environmental impact of solar panels. Finally, in a recent local study in the PV waste estimation literature, Desai and Pathak (2024) assess the waste management and recycling potential of EOL solar panels in Gujarat, India and estimate the number of solar panels that will reach the end of their useful lifetimes by 2030.

Table 1 summarizes the studies discussed above, and the other related studies in the PV waste generation literature. This literature can be categorized in several ways. Specifically, the following parameters are analyzed in Table 1 in order to illustrate the general status of PV waste estimation literature: geographical region, type of PV technology, lifespan of PV panels, estimation methodologies, data sources, forecasting periods, and materials analysis. PV waste estimation studies are geographically dispersed around the world.

However, it is clear that these studies are often conducted in countries with large populations or in regions where the WEEE Directive is in force. The most commonly estimated PV types are crystalline silicon and thin-film technologies, as these technologies dominate the PV panel market. Therefore, existing studies generally focus on estimating the waste generated from these technologies. Product lifespans are typically considered in two groups: fixed or distribution based. In studies that assume a fixed lifespan, product lifespans are generally set between 20 and 30 years. Conversely, distribution-based lifespan profiles consider a product's life dynamically, with the Weibull distribution commonly used to model PV panel lifespan profiles (V. Tan et al., 2022; Weckend et al., 2016).

Moreover, estimation methods are directly related to the lifespan profile since the type of data is critical to the method chosen. Simple delay and distribution delay methods are frequently employed to quantify PV waste generation. Data is generally obtained or adapted from sector reports and installed solar capacity. Estimation periods vary from case to case, as some studies estimate current PV waste quantities based on historical data, while others forecast future PV waste generation using sectoral PV installation capacity projections. In addition to estimating PV waste generation, most studies also analyze the potential amount of recoverable material for recycling and recovery processes (Franzoni et al., 2024; Paiano, 2015).

Table 1. Overview of PV waste estimation and forecasting studies in the literature

Author(s) (Year)	Region	PV Type	Lifespan	Method	Data	Forecasting period	Material analysis
Paiano (2015)	Italy	c-Si a-Si	Fixed: 25 years	Simple delay	1987-2013 installed solar power capacity 2014-2025 projected	2012-2038 2038-2050	Yes
Domínguez and Geyer (2017)	Mexico	c-Si a-Si CdTe CIGS	Fixed: 30 years	Simple delay	2014	2044	Yes
Mahmoudi et al. (2019)	Australia	c-Si a-Si CdTe CIGS OPV CPV Advanced c-Si Others	Distribution	Distribution delay	2001-2018	2031-2047 2047-2060	Yes
Domínguez and Geyer (2019)	USA	c-Si a-Si CdTe CIGS	Fixed: 30 years	Simple delay	2003-2030	2028-2060	Yes
Sharma et al. (2019)	EU USA Japan China India Australia	c-Si Thin film	Fixed: 20-25 years	Collecting and reviewing the reports			

continued on following page

Table 1. Continued

Author(s) (Year)	Region	PV Type	Lifespan	Method	Data	Forecasting period	Material analysis
Chowdhury et al. (2020)	Global	c-Si Thin film	Fixed: 25 years	Simple delay	2014-2030	2050	Yes
Gautam et al. (2021)	India	Si-based CdTe CIGS	Fixed and distribution / Fixed: The average life of the system is considered 25 years by 2014 and 30 years after this / Distribution: Weibull reliability function	Simple delay and distribution delay	2010-2019 installed solar power capacity 2020-2030 projected	2020-2047	Yes
Mahmoudi et al. (2021)	OECD countries	c-Si / a-Si CdTe / CIGS OPV CPV Advanced c-Si / Others	Distribution: Weibull	Annual installed PV capacity The combination of PV deployment in MW / PV market share / Panel specification / Materials inventory of the panels / PV waste generation / Distribution delay	1980-2030 collected from various sources	2001-2058	Yes
De Sousa et al. (2023)	Brazil	c-Si Thin films DSSC Hybrid		Collecting and reviewing the reports			

continued on following page

Table 1. Continued

Author(s) (Year)	Region	PV Type	Lifespan	Method	Data	Forecasting period	Material analysis
Yadav et al. (2023)	India	Si-based Thin film	Fixed: 25 years and 30 years scenario	Simple delay	Data published online / Reports of government bodies and nodal agencies Expert's opinion	2020-2050	Yes
Desai and Pathak (2024)	Gujarat, India	General	Distribution	Distribution delay	2009-2020	2009-2022	
Romel et al. (2024)	Canada	General	Distribution	Linear regression /Polynomial regression / Pover regression	2005-2021	2022-2045	

The Methods for Estimating the Amount of Generated PV Panel Waste

The origins of the e-waste estimation literature are relatively recent, beginning in the 1990s with studies by Matthews et al. (1997), as well as Lohse et al. (1998). Interest in this field has increased over time due to growing attention to the e-waste problem. Various estimation methods are used to determine the amount of generated e-waste (Bogar et al., 2019). These methods can be grouped into three main categories: input-output analysis, time series analysis, and hybrid analysis methods. Most estimation studies are based on input-output analysis. In early studies, data on sales or products were considered to estimate e-waste generation. Commonly used input-output analysis methods include sales (Ikhlayel, 2016; Li et al., 2022), simple delay (Jain & Sareen, 2006; Streicher-Porte et al., 2005), distribution delay (Kalmykova et al., 2015; Oguchi et al., 2008; Polák & Drápalová, 2012; Sakthivel et al., 2022; Wang et al., 2013; Zhang et al., 2021), consumption and use (Chung et al., 2011; Gonda et al., 2019; Robinson, 2009), Carnegie Mellon (Dwivedy & Mittal, 2010; Peralta & Fontanos, 2006), and sales-stock-lifespan (Golev et al., 2016). These methods use sales or stock data and product lifespan to estimate e-waste generation.

While sales, simple delays, and Carnegie Mellon methods consider a fixed lifespan value, the distribution delay method uses a dynamic lifespan profile for products. These methods are particularly useful when data is limited. However, it is worth noting that these methods are unable to make future projections. With the increase in e-waste generation and public awareness regarding legislation, future projections have become necessary for effective decision-making, targeting, and planning. To meet this need, time series analysis has also been applied to quantify e-waste generation along with improved data recording quality in recent years. Commonly used time series methods include logistic (Guo & Yan, 2017), ARIMA (Petridis et al., 2016), Bass (Kosai et al., 2020) and Holt's (Islam & Huda, 2019) models, which are applied to estimate for different e-waste types in the literature. Additionally, grey models are increasingly used in time series forecasts due to their successful forecasting ability under limited data conditions (Duman et al., 2019, 2020; Wang et al., 2021).

The third category, also called as hybrid methods, combines both input-output analysis and time series methods (Kastanaki & Giannis, 2022; Koshta et al., 2021; Owusu-Sekyere et al., 2022; Wang et al., 2018). In these methods, sales, stock or other factors such as population and GDP, which may affect sales or lifespan, are first forecast using time series methods. Then, classical input-output analysis methods are applied to forecast e-waste generation. Furthermore, if e-waste generation data are initially obtained via input-output analysis, future projections can subsequently be made using time series methods.

In conclusion, all the different types of estimation methods mentioned above can be applied to PV waste estimation and forecasting. The main inputs, such as sales or products in use, are similar or easily convertible PV systems to panel sales or installed PV capacity. Additionally, PV panels have a lifespan profile, which may vary and thus require analysis to assess the lifespan profile for each type of PV panel. Therefore, the general concept of e-waste estimation methods can be easily adapted to PV waste estimation. As shown in Table 1, the PV waste estimation literature follows a similar progress to that of commonly used electronic products. However, the PV waste estimation literature has developed more slowly than that of other e-waste because the history of PV panels is young, and specific regulations for PV waste management were implemented only after those for large household appliances and IT and telecommunications equipment.

CURRENT PV WASTE MANAGEMENT PRACTICES

PV cells are the basic units that convert sunlight into electricity. These cells are interconnected to form a PV panel that is the larger unit used in solar power systems (Singh, 2013). The overall efficiency of the system depends on the combined performance of the individual PV cells (Al-Ezzi & Ansari, 2022). PV cell technologies are typically categorized based on their main light-absorbing material into three categories: silicon based, thin-film based, and others (Domínguez & Geyer, 2019). They are generally classified into four generations based on their technological advancements. Crystalline silicon (c-Si) cells are classified as first generation, thin-film modules as second generation, emerging technologies as third generation and hybrid technologies as fourth generation (Chawla et al., 2020; Dambhare et al., 2021; Pastuszak & Węgierek, 2022):

- First generation: Crystalline silicon PV cells include mono-crystalline silicon (m-Si) and poly-crystalline silicon (p-Si), which are known for their high efficiency and durability.
- Second generation: Thin film PV cells include amorphous silicon (a-Si), copper indium gallium selenide (CIGS), cadmium telluride (CdTe), and gallium arsenide (GaAs), which offer a thinner and flexible alternative.
- Third generation: Emerging technologies include innovative approaches like dye-sensitized solar cells (DSSC), organic solar cells (OSC), perovskite solar cells (PSC), quantum dots (QDs), and concentrated solar cells (CPV), which are still in development and have potential for higher efficiency and lower costs.

- Fourth generation: Hybrid technologies involve advanced materials such as metal nanoparticles and oxides, carbon nanotubes, and graphene, which offer affordability and flexibility.

Additionally, PV cell technologies can be classified as wafer-based (semiconductor wafers) or thin-film cells (semiconducting films on substrates) (Domínguez & Geyer, 2017). Figure 7 illustrates the manufacturing technologies of solar PV cells, drawing from the most common categories in the literature (Chawla et al., 2020; Dambhare et al., 2021; Pastuszak & Węgierek, 2022).

Initially, the first generation of PV technology is mainly characterized by crystalline silicon (c-Si) PV cells (Arul Prishya et al., 2023), which include m-Si and p-Si. These technologies continue to be the most widely used in the solar energy market today due to their proven performance and reliability. m-Si PV cells, made from pure silicon, are generally considered durable and efficient, typically with confirmed solar cell efficiencies between 20-26%, though they can be expensive to produce. On the other hand, p-Si PV cells, which have lower purity, are less expensive but offer lower efficiency, usually confirmed solar cell efficiencies ranging between 14-20% (Chawla et al., 2020; Muthusamy et al., 2022). c-Si PV cells are the one of the oldest PV technologies and still dominate the market, accounting for approximately 92% of the market share in 2014. However, projections suggest that this market share could decrease to around 44.8% by 2030, as Figure 8 demonstrates (Chowdhury et al., 2020; Weckend et al., 2016). First-generation technologies can often be found in large-scale solar installations due to their mature status and relatively high efficiency (Chawla et al., 2020).

This figure was adapted and synthesized from multiple sources in the literature, including Chawla et al. (2020), Dambhare et al. (2021) and Pastuszak and Węgierek (2022).

Figure 7. Overview of solar PV cell types covering four technological generations

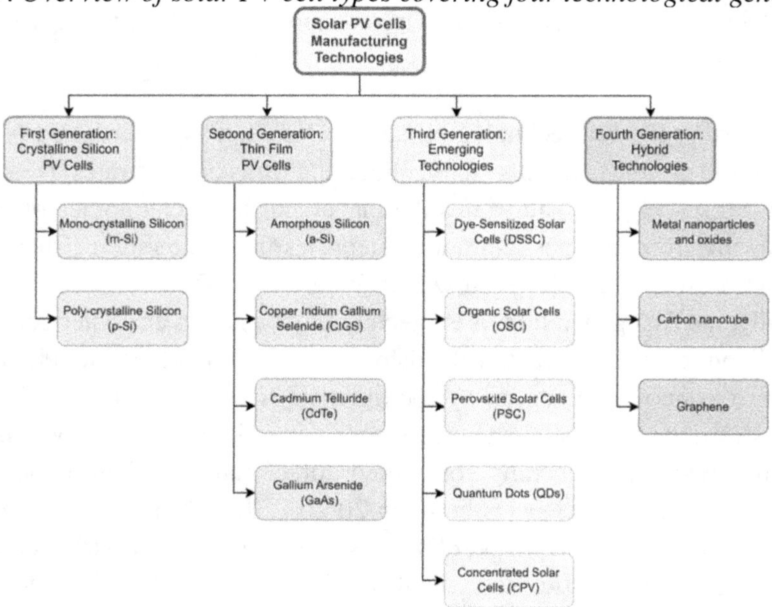

The data used in this graph was adapted from the End-of-Life Management: Solar Photovoltaic Panels (Weckend et al., 2016). The totals were calculated by summing the values provided under the respective PV technology categories.

Figure 8. Market share of PV cell technologies (2014-2030)

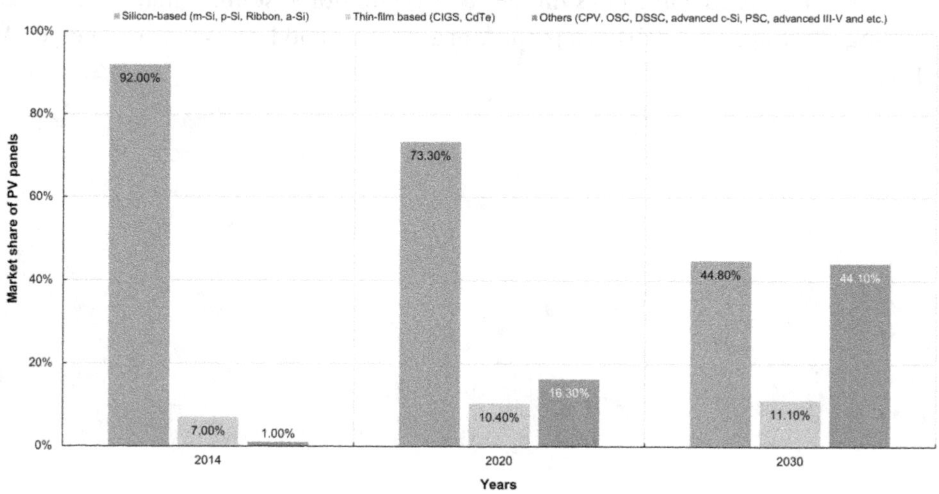

Subsequently, the second generation, which includes thin film PV technologies such as CdTe and CIGS, generally achieve efficiencies in the range of 10-15%, offering an economical alternative to first generation because they require less material and can still achieve appropriate efficiency levels (Chawla et al., 2020; Farji, 2021). CdTe is likely the dominant thin-film technology in the market with confirmed cell efficiencies ranging between 8-11% because of its low cost, but it is generally considered less reliable. Meanwhile, CIGS panels, while less common, can achieve higher efficiencies ranging from 10-16% and offer the advantage of lower costs (Chawla et al., 2020). According to the Photovoltaics Report of Fraunhofer ISE (2024), the highest laboratory efficiency for CIGS has reached 23.4%, while CdTe PV cells have achieved 21.0% in laboratory conditions. Furthermore, the efficiency of CdTe modules has improved significantly, increasing from 9% to nearly 20% in recent years (Fraunhofer ISE, 2024). Second generation technologies can be typically applied in medium-scale installations (Chawla et al., 2020). According to the report of IRENA and IEA-PVPS, CIGS has significant potential for higher efficiencies, and is projected to increase its market share from 2% in 2014 to 6.4% by 2030, while CdTe is expected to see a slight decline, from 5% to 4.7% over the same period (Weckend et al., 2016). According to the Fraunhofer ISE (2024)'s Photovoltaics Report, despite these projections, the thin film technologies contributed only 2.5% to the total PV market in 2023 (Fraunhofer ISE, 2024).

Furthermore, third-generation PV panels, such as Perovskite and Organic PV cells, are emerging technologies with the potential to achieve higher efficiencies due to their advanced materials (Kim et al., 2023; Krebs-Moberg et al., 2021). While still largely in the R&D phase or applied in small-scale projects, these technologies could hold significant promises for the future, particularly in terms of improving efficiency and cost-effectiveness. However, their high production costs currently limit commercial applications compared to first- and second-generation technologies. Despite this, they are believed to have great potential as development continues. This generation likely uses organic materials and may include highly efficient, expensive multi-junction solar PV cells (Chawla et al., 2020). As reported in the Photovoltaics Report by Fraunhofer ISE (2024), perovskite cells, for instance, have reached efficiencies of up to 25.2% in laboratory settings, while organic PV cells are rapidly advancing. Multi-junction concentrator solar cells, another promising technology, have achieved efficiencies of up to 47.6% in laboratory conditions (Fraunhofer ISE, 2024). The new generations of PV panels offer higher efficiencies and potential cost reductions due to technological advancements, and they may find broader applications in the future. As a result, these new generation PV panels are becoming competitive in both economic and technological terms (Kim et al., 2023).

Material Contents of Solar PV Panel Waste: Types, Quantities, and Recovery Potential

The rapid deployment of PV technologies over the last few decades has led to growing concerns regarding the EOL management of PV panels. The management of PV panel waste is essential not only for reducing environmental impacts but also for recovering valuable materials, such as silver, copper, aluminum, and silicon (Dias et al., 2016; Komoto et al., 2018). The metals used in various PV technologies and system components are shown in Table 2, adapted from Domínguez and Geyer (2017), highlighting the critical, valuable, base and hazardous metals.

PV panels typically contain around 10 metallic elements, including silver, gold, copper, nickel, zinc, and aluminum (Domínguez & Geyer, 2017). However, PV systems, which consist not only the panels but also additional components such as inverters and wiring, battery, magnet, and specialty categories can include up to 41 elements (Simandl et al., 2023). Some of which include valuable metals, such as silver and copper. Some are classified as critical raw materials, while others, like lead and cadmium, fall under hazardous materials (Domínguez & Geyer, 2017).

Table 2. Metal content used in various PV technologies and system components

Material	PV technologies				PV system components				
	c-Si	a-Si	CdTe	CIGS	Inverter	Transformer	Cabling	Mounting	Tracking
Precious metals									
Ag	×				×				
Au					×				
Base and special metals									
Al	×	×	×	×	×	×		×	×
Cu	×	×	×	×	×	×	×		×
Ni	×				×				
Fe	×	×	×		×	×		×	×
Ti	×		×						
Sn	×		×	×	×				
Zn	×	×	×	×	×			×	×
Cr		×	×						
Mn		×			×				
Mo				×					
Ta					×				
Hazardous metals									

continued on following page

Table 2. Continued

Material	PV technologies				PV system components				
	c-Si	a-Si	CdTe	CIGS	Inverter	Transformer	Cabling	Mounting	Tracking
Pb	×		×		×				
Cd		×	×	×					
Se				×					
Other metals									
Si	×	×	×						
Critical metals									
Mg	×	×		×	×				
Te		×	×						
In		×		×					
Ga				×					
Other materials									
Glass	×		×	×					
EVA	×		×	×					

This table has been adapted and modified from data presented by Domínguez and Geyer (2017).

According to the study by Muthusamy et al. (2022), most c-Si PV panels are composed of approximately 76% glass, 10% plastic, 8% aluminum, 5% silicon, and 1% metal composites. The recycling process for these panels generally focuses on the glass component, which can potentially be recycled with 100% efficiency. CdTe and CIGS thin-film technologies can use glass as a substrate but may also incorporate flexible materials depending on the application. Similarly, most thin-film PV panels may contain 89% glass, 4% plastic, 6% aluminum, and 1% metal composites. As with c-Si panels, recycling efforts for thin-film PV panels are often concentrated on the glass component due to its predominant share in the material composition. Figures 9 and 10 provide representations of the material composition in c-Si and thin-film PV panels, respectively, based on the data from Muthusamy et al. (2022).

This figure have been derived from data presented by Muthusamy et al. (2022).

Figure 9. Material composition of c-Si PV panels

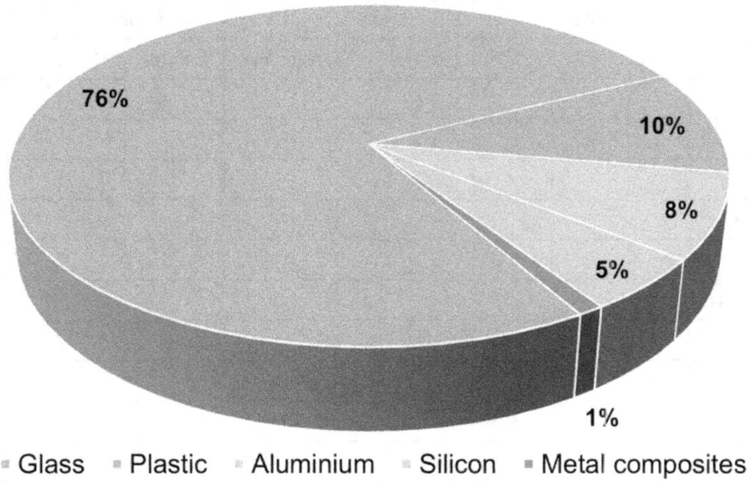

This figure have been derived from data presented by Muthusamy et al. (2022).

Figure 10. Material composition of thin-film PV panels

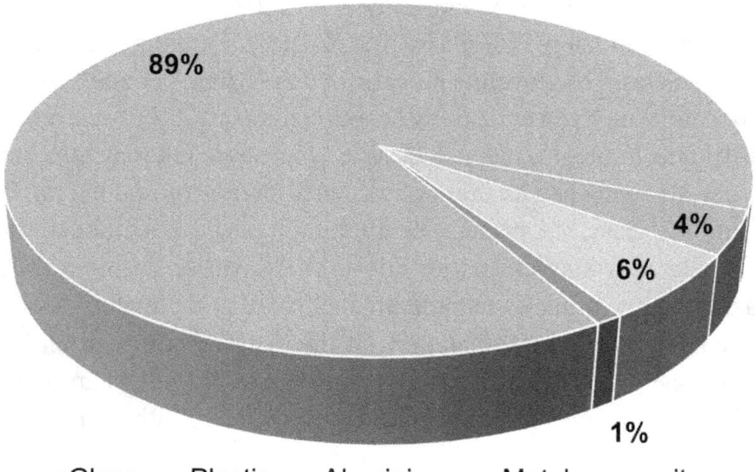

The composition of solar PV panels can vary depending on the manufacturer, the type of PV technology, and the application. For example, c-Si PV modules are typically composed of glass, an aluminum frame, solar cell, ethylene vinyl acetate (EVA) as an encapsulant, and a backsheet (Lee et al., 2018). In thin film PV modules, absorbing layers such as CdTe, CIGS, or a-Si are sandwiched between a transparent conductive oxide (TCO) coating and a back contact, which are deposited onto substrates like glass, polymers, or metal. (Gerold & Antrekowitsch, 2024). Table 3 provides an brief overview of the material and component composition in different PV technologies, adapting from Gerold and Antrekowitsch (2024).

Table 3. Material and component composition of different PV technologies

PV technology	Materials and components used
c-Si	Aluminum frame, Glass, Encapsulant (EVA), Solar cells, Encapsulant (EVA), Backsheet
Thin film (a)	Glass, TCO coating, Absorbing layers (CdTe, a-Si), Back contact, Encapsulant (EVA), Backsheet
Thin film (b)	Glass, Encapsulant (EVA), TCO coating, Absorbing layers (CIGS), Back contact, Blocking layer, Back substrate

This table has been derived from data presented by Gerold and Antrekowitsch (2024)

Various studies in literature have investigated the materials and components found in solar PV panels, identifying both valuable and hazardous components. These materials, such as silicon, aluminum, and silver in c-Si PV panels, or rare metals like tellurium in CdTe panels, present both recycling opportunities and environmental challenges. Table 4 presents the findings from the literature on the material and component composition of different PV technologies (Fiandra et al., 2019; Herceg et al., 2020; Maani et al., 2020; Paiano, 2015; J. Tan et al., 2022; Xu et al., 2018).

Based on the data from Paiano (2015), Italy is projected to generate a cumulative total of 8,238,967 t of PV waste by 2050, including 4,843,891 t from c-Si technology, 988,362 t from a-Si technology, 950,058 t from CIGS technology, 1,252,617 t from CdTe technology, and 204,040 t from emerging technologies and CPV. The authors projected that a significant amount of materials could be recovered from cumulative PV waste by 2050. This includes 6,108,742 t of glass, 617,658 t of aluminum, 131,490 t of silicon, 767 t of tellurium, 50,403 t of copper, 97 t of silver, 1,540 t of indium, 29 t of gallium, and 1,483 t of germanium. Furthermore, the authors have indicated that improper disposal of hazardous metals like cadmium (877 t) and lead (3,381 t) by 2050 could result in costs exceeding 2 billion euros for lead and nearly 14 million euros for cadmium. They also highlighted that Italy could recover 50,403 t of copper and 617,658 t of aluminum, reducing dependence on imports and providing significant economic benefits. Additionally, valuable metals such as silver, indium, gallium, and germanium could be recovered from PV waste, with the economic value of these metals reaching billions of euros by 2050.

Table 4. Material and component composition (%) of different PV technologies

Author(s)(year)	Fiandra et al. (2019)		Herceg et al. (2020)		Maani et al. (2020)	
PV technology	**c-Si**		**c-Si**		**c-Si**	
Material composition	Glass	74.16%	Glass	70.00%	Glass	67.8%
	Al frame	10.30%	Al frame	18.00%	EVA	6.73%
	Silicon wafer	3.48%	Junction box		Al	16.9%
	Encapsulation (EVA)	6.53%	Cu	0.33%	Ag	0.08%
	Backsheet	3.60%	Plastic	0.67%	Cu	0.77%
	Adhesives	1.16%	Encapsulant (EVA)	5.10%	Mg	0.58%
	Cu	0.57%	Backsheet (PET)	1.50%	Ni	<0.01%
	Ag	0.01%	Silicon	3.65%	Pb	<0.01%
	Sn	0.12%	Al	0.53%	Si	6.27%
	Pb	0.07%	Cu	0.11%	Sn	0.07%
			Ag	0.05%	Ti	<0.01%
			Other metals (Sn, Pb)	0.05%	Zn	<0.01%

Author(s)(year)	J. Tan et al. (2022)		Xu et al. (2018)		Paiano (2015)	
PV technology	**c-Si**		**c-Si**		**a-Si**	
Material composition	Glass	70%	Glass	54.721%	Glass	86%
	Al frame	18%	Al frame	12.771%	Al	0.035%
	Cu connector	1%	Si	3.101%	Adhesives	0.02%
	Encapsulation (EVA)	5.1%	Ag	0.03%	Polyol/MDI	12%
	Backsheet (PVF)	1.5%	Cu	0.451%	Cu	0.9%
	Silicon	3.56%	Plastic	17.091%	Pb	0.043%
	Ag	0.053%	Adhesives	10%	Si	0.0064%
	Al	0.53%			In	0.5%
	Cu	1.14%			Ge	0.5%
	Various metal (Sn, Pb)	0.053%				

Author(s)(year)	Maani et al. (2020)		Paiano (2015)		Paiano (2015)	
PV technology	**CdTe**		**CdTe**		**CIGS**	
Material composition	Glass	95.3%	Glass	95%	Glass	84%
	EVA	3.74%	Al	0.35%	Al	12%
	Al	0.05%	Polymers (EVA)	3.5%	Polymers (EVA)	3%
	Cd	0.12%	Cu	1%	Cu	0.8%
	Cr	0.02%	Zn	0.01%	Zn	0.12%
	Cu	0.68%	Cd	0.07%	Pb	0.05%
	Pb	<0.01%	Te	0.07%	Cd	0.0005%
	Si	0.30%			In	0.02%
	Sn	<0.01%			Se	0.03%
	Te	0.12%			Ga	0.01%
	Ti	<0.01%				
	Zn	<0.01%				

Current Methods, Processes, and Technologies for Recovering and Recycling Solar PV Panel Waste

Effective management of PV panel waste is essential to reduce environmental impacts and recover valuable materials, such as silver, copper, aluminum, and silicon, which contribute to reducing resource depletion (Gautam et al., 2021; Komoto et al., 2018). As PV waste continues to accumulate, recycling and recovery methods are becoming increasingly crucial for environmental sustainability and resource efficiency. The processes commonly employed in PV panel recycling consist of physical separation, thermal treatment, and chemical treatment, as shown in Figure 11, which was adapted from Chowdhury et al. (2020)Muthusamy et al. (2022) and Ndalloka et al. (2024). As a whole, these processes enable the recycling of key materials, thereby supporting the circular economy and promoting sustainable growth in the PV industry (Dias et al., 2016). With the expected increase in PV installations, efficient recycling strategies will become even more important. Further advancements in these methods could enhance resource recovery and provide additional environmental benefits (Weckend et al., 2016).

The stages depicted in this diagram were adapted from Chowdhury et al. (2020) Muthusamy et al. (2022) and Ndalloka et al. (2024).

Figure 11. The processes in the recycling and recovery of EOL solar PV panels

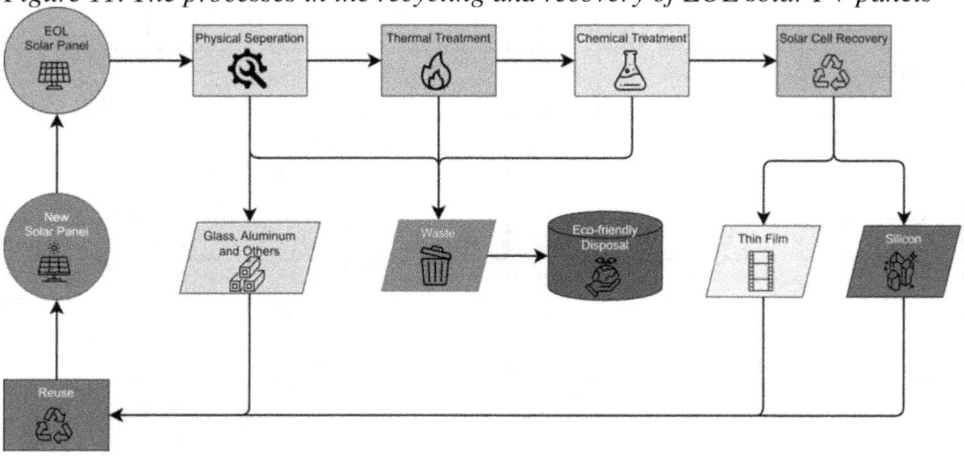

Trivedi et al. (2023) emphasize the need for efficient recycling of EOL c-Si PV modules to recover valuable materials and meet the rising demand for raw resources. They investigate combinations of chemical, physical, and thermal processes to optimize recycling, aiming for minimal cost and environmental impact. The study notes that methods like chemical treatment, incineration, and landfilling carry higher environmental impacts compared to pyrolysis and mechanical treatment. Consequently, the authors offer recommendations for future advancements in recycling technology. Cheema et al. (2024) address the environmental risks of decommissioned PV panels which include heavy metals like copper, cadmium, and lead, as well as hazardous polymer substances such as EVA, PVF, and PET, all of which can pose environmental threats if not disposed of properly. They assess existing recycling processes and their environmental and commercial viability and emphasize the need for advancements to improve these methods. Additionally, the authors note that despite challenges like the long lifespan of PV panels and their relatively lower waste volume compared to e-waste, first generation PV panels nearing EOL has increased interest in recycling. Furthermore, the review by Sivagami et al. (2024) provides an analysis of several recycling techniques, including physical, thermal, and chemical methods, which could be employed to manage solar PV waste effectively. Their study emphasizes the increasing importance of physical separation as an initial step to enhance the overall recycling process.

The recycling process generally begins with the physical separation of PV panel components. This stage involves the careful dismantling of panels to separate materials such as glass, aluminum, and other metallic elements (Chiang et al., 2024; Ndalloka et al., 2024). Physical separation is essential for preparing the materials for subsequent chemical or thermal treatments. Dias et al. (2016) report that silver content in c-Si PV modules can reach 600 g/t, thus making efficient physical separation critical for effective recovery. The physical separation process typically unfolds in the following stages. First, the PV module is carefully disassembled. Frames, junction boxes, and other easily removable parts are separated to prepare the module for further processing. Then, the module is broken down manually or mechanically into smaller pieces to make it easier to handle. These smaller pieces undergo size reduction through crushing or milling. The aim is to break them down into finer particles that are suitable for subsequent separation steps. Next, sieving is applied to classify the particles by their size by enabling the efficient separation of materials like glass and metal. The process then continues with the dense media separation, where a medium, typically a liquid or heavy suspension, is used to separate the materials by their density. This method is particularly effective for isolating precious metals from lighter materials, such as plastics and glass. Once the separation process is complete, precious metals such as silver and copper, are collected and prepared for recovery, as they are valuable for reuse in manufacturing. Finally, these recovered materials are cleaned, processed, and put back into the production cycle, contributing to the circular economy by reducing waste and the need for raw materials (Ali et al., 2024; Chowdhury et al., 2020; Sivagami et al., 2024; Strachala et al., 2017).

Solar panel recycling facilities employ various effective physical separation techniques, including mechanical processing, shredding, sieving, and magnetic separation, to achieve optimal material recovery (Ali et al., 2024; Chen et al., 2024; Ndalloka et al., 2024). Proper separation maximizes the value of recyclable resources and minimizes the environmental impact of discarded components (Akhter et al., 2024; Sivagami et al., 2024). Various physical separation and mechanical methods are employed in the recycling of PV modules. Crushing breaks down the modules into smaller pieces (Chen et al., 2024; Pagnanelli et al., 2017). High voltage pulse crushing effectively separates different material layers within waste PV modules (Akimoto et al., 2018). Electrostatic separation sorts materials with different electrical conductivities by separating conductive and non-conductive particles (Dias et al., 2018). Hot knife cutting method uses a heated, conductive blade to soften adhesives within PV module layers. This method allows clean separation of the backsheet from the silicon cells without damaging the sensitive PV layers to enable efficient material recovery (Wahman et al., 2024). Meanwhile, water jet cutting technology employs high pressure water to separate solar cells and the backsheet from the glass without causing damage for making recycling more efficient and eco-friendly (Deng et al., 2022). By integrating these methods, recycling facilities aim to enhance recovery rates and reduce adverse environmental impact.

Thermal and chemical methods play an essential role in PV module recycling, especially for separating and recovering materials that are not easily accessible through physical processes (Ndalloka et al., 2024). Thermal treatment, which involves processes like combustion or incineration at 500-600°C, is crucial for removing adhesive EVA layers. While thermal methods are generally better at preserving the integrity of silicon cells and glass compared to chemical methods, which may require hazardous and costly substances, they still present challenges due to high energy use and emissions (Chen et al., 2024). Additionally, the application of high heat enables the melting or vaporizing the materials, allowing for the recovery of metals like aluminum, copper, and silver, as well as purifying silicon wafers for reuse (Sivagami et al., 2024).

Fiandra et al. (2019) implemented a two-step thermal treatment for c-Si PV panel recycling. First, they removed the polymer layer without fluorinated emissions. Then, the remaining polymers were removed at 500°C with low volatile organic compounds emissions. This approach aims to support effective PV waste management by balancing material recovery with reduced environmental impact. Moreover, pyrolysis, seen as a more environmentally friendly alternative to incineration, has been shown to eliminate more than 99% of polymers from PV modules (Chen et al., 2024; Trivedi et al., 2023). Dias et al. (2017) achieved similar results by using pyrolysis at 500°C for 30 minutes to remove EVA layers, thus promoting sustainable recycling practices. Furthermore, Wang et al. (2019) investigated a two-stage heating method to separate c-Si PV modules, where they removed the TPT backsheet at 150°C and then applied pyrolysis at 500°C with acetic acid and hydrocarbon compounds to remove the EVA layer. These processes yielded undamaged TPT, glass, and silicon wafers that could be further processed for recycling. This method highlights the potential for reduced emissions and optimized material recovery. Finally, Birtürk et al. (2023) emphasize that thermal treatments play a crucial role in PV recycling by breaking down polymer structures effectively. As a result, thermal methods are widely applied in both traditional and innovative recycling approaches.

Chemical treatment in solar panel recycling is an essential process for efficiently recovering high-purity materials like silicon, silver, copper, and other metals (Chen et al., 2024). This approach employs chemical agents, solvents, and reactions to selectively dissolve or separate specific components of PV panels (Sivagami et al., 2024). Techniques like leaching and precipitation are used to effectively recover valuable metals, including silver, copper, gallium, and indium. Additionally, chemical treatment enables the dissolution or removal of layers such as encapsulation materials, semiconductor layers, and coatings (Chowdhury et al., 2020; Sivagami et al., 2024). Palitzsch and Loser (2011) highlighted the role of chemical treatment in recycling c-Si PV modules by focusing on de-metalizing broken cells and scrap. They developed an aluminum chloride solution to remove aluminum from the backs of silicon cells. Then, they dissolved the remaining silver from the front contacts using nitric acid and achieved significant concentrations of silver alongside copper and tin. The authors state that this method reduces waste and produces valuable by-products. Similarly, Marwede et al. (2013) highlighted the importance of chemical treatment in recycling thin-film PV modules such as CdTe and CIGS. They stated that various acids, including nitric acid, sulphuric acid, hydrochloric acid, and sodium hydroxide, have been utilized to leach and reclaim thin-film materials from the glass in PV module waste, and that these chemical methods enable the recovery of critical materials, including cadmium, tellurium, copper, indium, and gallium. Therefore, this approach is crucial for enhancing both economic and environmental sustainability. In the study by Dias et al. (2016), two methods for silver extraction from c-Si PV modules were tested, and the results demonstrated the efficiency of chemical methods in metal recovery from PV modules. The first method, which includes milling, sieving, and leaching in a 64% nitric acid solution with 99% sodium chloride, achieved a silver concentration yield of 94%, while the second method, which also added pyrolysis step at 500°C, resulted in a silver concentration yield of 92%. Savvilotidou et al. (2017) examined chemical treatment techniques for a-Si and CIS thin-film PV panels using various acids and acid mixtures. They found that sulfuric acid and lactic acid performed strong performance in achieving effective panel treatment at gentle temperatures with low energy requirements. The authors also indicated that greater emphasis is needed on the recycling of CIS panels, particularly regarding critical metals like indium and gallium.

Weckend et al. (2016) highlighted the potential of thin-film recycling to conserve scarce materials like indium and gallium that are essential for electronics and solar technologies, and noted high-value recycling as the preferred approach for all PV technologies due to its social benefits. Mahmoudi et al. (2019) compiled data from the literature on the recycling yields of materials in PV modules. Similarly, Sivagami et al. (2024) reviewed various literature source and reported the recycling rates of different materials in typical PV panels as follows: aluminum (100%), cadmium (98%), copper (78%), gallium (99%), indium (75%), molybdenum (99%), selenium (80%), silicon (85%), tin (99%), tellurium (80%), zinc (90%), silver (95%), and glass (97%). These data demonstrate the role of recycling in reclaiming critical materials, ensuring resource sustainability, and supporting the rising demand in electronic and solar industries.

The management of PV panels is a critical issue and waste generated from their process requires careful disposal. The recent studies of Babaei and Nasr Esfahani (2024), Biyouki et al. (2024) and Gerold and Antrekowitsch (2024) emphasize the need for effective recycling and disposal methods to reduce environmental impacts, especially due to toxic substances like cadmium and lead.

Current Policies and Legislative Frameworks for Solar PV Panel Waste Management

Effective frameworks and policies are essential for managing EOL solar PV panels, thereby making PV recycling more feasible. Consequently, several countries have begun developing strategies and conducting research to support sustainable recycling practices. For instance, the study by Ndalloka et al. (2024) examined global PV waste management policies and legislation and highlighted the importance of PV recycling for environmental benefits, while also emphasizing the need for regulations at both national and international levels.

In European Union, PV modules are classified as e-waste under the Waste Electrical and Electronic Equipment (WEEE) Directive, which requires specific collection and recycling targets, separate treatment of silicon and non-silicon PV panels, and decontamination of hazardous metals like cadmium, selenium, and lead during disassembly (Wang et al., 2024). The United Kingdom was the first country to adopt the directive, followed by Germany (Cheema et al., 2024; Ndalloka et al., 2024).

In the United States, PV modules are generally classified as solid waste. However, modules containing high levels of heavy metals, such as lead and cadmium, are considered hazardous and thus require compliance with the Resource Conservation and Recovery Act (RCRA) for safe recovery or disposal (Wang et al., 2024). Moreover, many states have introduced hazardous waste programs, universal waste rules, and e-waste recycling programs for EOL PV panels Nonetheless, due to varying regulations and guidelines across states, ensuring a consistent EOL PV waste management is challenging (Ali et al., 2023). Ndalloka et al. (2024) noted recent growth in PV recycling companies and research institutions in the United States, where interest in and focus on PV waste management are increasing.

Ali et al. (2023) and Wang et al. (2024) indicated that China's National Solid Waste Law establishes a framework for managing solid waste, including EOL PV panels; however, it does not explicitly address the specific requirements for handling EOL PV waste. Therefore, the authors emphasized the need for more targeted regulations to improve EOL PV waste management.

On the other hand, despite advancements in developed countries, some developing countries have yet to initiate discussions on PV waste management (Chowdhury et al., 2020; Ndalloka et al., 2024). In addition, the study by Duran et al. (2022) discusses the disposal challenges of EOL solar panels and the limited recycling infrastructure needed to address the increasing PV waste volume. They suggest establishing appropriate incentives for recycling infrastructures and implementing suitable legislation to ensure sustainable funding for recycling of PV technologies.

CHALLENGES, OPPORTUNITIES AND FUTURE DIRECTIONS IN SOLAR PV WASTE MANAGEMENT

The existing situation of the PV waste generation has challenges and opportunities in different ways. Moreover, these challenges and opportunities have the potential to drive future directions for effective PV waste management.

The major challenges of PV panel waste can be listed as increased waste quantity, material content complexity, challenges in recycling processes, lack of regulations, insufficient recycling infrastructure, and uncontrolled, informal recycling activities. The use of PV panels has been increasing in recent years, and with their expected lifespan of 20 to 30 years, a significant rise in PV panel waste generation is inevitable in the near future. On a global scale, not only PV module waste but also other types of e-waste are being generated at an increasing rate. When considering all categories of e-waste, the quantities already generated and those expected to be generated are substantial.

However, the existing e-waste management policies and recycling infrastructures for PV waste are insufficient, presenting a major challenge for sustainable PV panel waste management. Like other types of e-waste, PV panel waste needs to be controlled by effective laws and regulations. However, such a regulatory framework has not yet been fully established. Since PV panel waste is a relatively new category compared to other e-waste, the measures and regulations to control and monitor its generation remain inadequate.

Additionally, the recycling infrastructure for PV panels poses a significant challenge. High recycling costs discourage investors, manufacturers, and users. Furthermore, the responsibilities of consumers and producers are not yet clearly defined or enforceable. To design a closed-loop supply chain, collected PV panels must be recycled, and the materials recovered in efficient and cost-effective ways. If PV waste is not properly recycled, substantial economic benefits are lost, as PV panels contain valuable materials.

Beyond economic losses, improper recycling of PV panels can cause environmental harm and pose risks to human health. Most PV panels, like other types of e-waste, are recycled in uncontrolled conditions. Two major challenges arise from this inappropriate recycling technique: the leakage of hazardous substances into the environment and their potential entry into the human body. The release of hazardous materials into the air, soil, and water is a significant threat. The contamination of source water, air, and agricultural soil with toxic substances will negatively affect the well-being of all living organisms in the medium and long term. In the short term, exposure to these toxic materials during recycling, whether through inhalation or direct contact, presents serious health risks to workers.

PV panel waste presents many opportunities in addition to the challenges. The most promising opportunities lie in material recovery. Recovering valuable materials such as silver, silicon, and glass could reduce the demand for these materials in future solar panel production. Since resources are scarce and the cost of mining these materials is very high, their recovery is essential for developing a circular economy model.

Additionally, promoting strategies such as reuse and refurbishment can extend the lifespan of panel waste, while effective recovery and recycling strategies minimize waste and reduce the demand for new materials. To increase the quantities of recovered and recycled materials, there is a growing need for innovative and technological recycling solutions. Developing efficient and cost-effective methods for both mechanical, thermal and chemical recycling processes is crucial to supplying materials for new products.

Moreover, automation technologies, image processing, and artificial intelligence hold promises for improving recycling options. With the adoption of new methods and technologies in PV waste management, new investment opportunities emerge. These opportunities contribute to economic growth, green technologies, and the circular economy. As a natural result of these developments, new business lines and workforce needs are arising. The new jobs require specialized competencies in e-waste management, which in turn contributes to employment.

Lastly, considering material content, sustainability, the circular economy, and recycling, the design of PV panels also presents opportunities from a zero-waste approach. The selection of recyclable materials and ease of disassembly directly affect the quantities of panel waste generated, environmental impacts, recycling costs, recovered material quantities, and the associated economic benefits. These factors also influence the need for labor, energy, time, and financial resources.

The demand for RE sources is increasing due to the scarcity of natural resources. As the world's solar energy capacity continues to grow, the management of PV modules at the end of their life will become increasingly important. Several key future directions are likely to focus on PV waste management. Firstly, recycling technologies need to be improved. This is urgently required both to recover valuable materials from PV waste and to reduce the harmful effects of the chemicals used in panel production and recycling. Another important area is the development of modular dismantling designs for PV panels. If the panels can be disassembled more easily thanks to well-designed structures, the recycling process can be carried out more quickly. Additionally, future PV panels can be produced with more sustainable and recyclable materials. The use of these materials may help reduce the hazardous environmental impacts of PV panels in the future. In addition to technical future directions, global systematic directions should be put into action. Governments will need to implement strict policies and regulations to ensure the recycling of PV panel waste. Also, the responsibilities of producers and consumers could be addressed in both purchasing decisions and EOL management. An integrated global coordination of waste management systems will play a key role in future. If the principles of the circular economy can be applied and sustainability can be promoted, environmentally friendly named energy production systems can be fully realized in the future.

CONCLUSIONS

In this chapter, the current situation of PV panel waste and related issues are reviewed in different ways, examining the challenges, opportunities and future directions. The number of PV panel systems installed has been increasing all over the world due to increased energy supply needs. While the main source of energy, the sun, is eco-friendly in this system, PV panels reach their end of life between 20-30 years after installation. As a result, it is inevitable that the volume of PV panel waste will increase (Bajagain et al., 2020; Weckend et al., 2016). This study examines the demand for PV panels and the process by which these panels become waste, by reviewing relevant literature and current practices. It also addresses issues related to the management of future PV panel waste.

The findings show that the amount of PV panel waste generated will increase rapidly between 2030 and 2050. So, the recycling processes of PV panel waste are crucial in terms of reducing the environmental effects of hazardous materials and recovering valuable materials. Different attempts are driven with physical, thermal, and chemical methods to recover materials in recycling phases (Chowdhury et al., 2020; Muthusamy et al., 2022; Ndalloka et al., 2024). However, high recycling costs and the need for technological advancements in recycling processes are major challenges in PV waste management (Duran et al., 2022). In addition to technical challenges, the inadequacy of e-waste management legislation and enforcement, along with the lack of global standards, also presents critical challenges (Ali et al., 2023).

This chapter also reveals the potential open ways to improve PV waste management. Considering the circular economy, different approaches and business models need to be established, from the design to the recycling of PV panels. In the near future, with the adoption of new technologies with artificial intelligence and automation systems to recycling processes, the recovered material quantities could be increased, and the environmental damage could be reduced. In addition, appropriate materials can be selected, and easy disassembly strategies can be developed during the design phase of new types of PV panels to increase the amount of recycled materials and to reduce recycling costs.

The findings and results in this chapter may provide valuable input and recommendations and gain a different perspective to all stakeholders. The PV waste management issue is not only a theoretical concern for academia but also affects all stakeholders in the world, including producers, consumers, governments, and entrepreneurs. For a sustainable future, it is essential that all parties contribute to establish an effective PV waste management system.

REFERENCES

Achillas, C., Aidonis, D., Vlachokostas, C., Karagiannidis, A., Moussiopoulos, N., & Loulos, V. (2013). Depth of manual dismantling analysis: A cost–benefit approach. *Waste Management, 33*(4), 948-956. https://doi.org/https://doi.org/10.1016/j.wasman.2012.12.024

Akhter, M., Al Mansur, A., Islam, M. I., Lipu, M. S. H., Karim, T. F., Abdolrasol, M. G. M., & Alghamdi, T. A. H. (2024). Sustainable Strategies for Crystalline Solar Cell Recycling: A Review on Recycling Techniques, Companies, and Environmental Impact Analysis. *Sustainability*, *16*(13), 5785. https://www.mdpi.com/2071-1050/16/13/5785

Akimoto, Y., Iizuka, A., & Shibata, E. (2018). High-voltage pulse crushing and physical separation of polycrystalline silicon photovoltaic panels. *Minerals Engineering*, *125*, 1–9. https://doi.org/https://doi.org/10.1016/j.mineng.2018.05.015

Al-Ezzi, A. S., & Ansari, M. N. M. (2022). Photovoltaic Solar Cells: A Review. *Applied System Innovation*, *5*(4), 67. https://doi.org/https://doi.org/10.3390/asi5040067

Al Mubarak, F., Rezaee, R., & Wood, D. A. (2024). Economic, Societal, and Environmental Impacts of Available Energy Sources: A Review. *Eng, 5*(3), 1232-1265. https://doi.org/https://doi.org/10.3390/eng5030067

Ali, A., Malik, S. A., Shafiullah, M., Malik, M. Z., & Zahir, M. H. (2023). Policies and regulations for solar photovoltaic end-of-life waste management: Insights from China and the USA. *Chemosphere*, *340*, 139840. https://doi.org/https://doi.org/10.1016/j.chemosphere.2023.139840 PMID: 37597624

Ali, A., Shahid, M., Qadir, S. A., Islam, M. T., Khan, M. W., & Ahmed, S. (2024). Solar PV End-of-Life Waste Recycling: An Assessment of Mechanical Recycling Methods and Proposed Hybrid Laser and High Voltage Pulse Crushing Method. *Preprints*. https://doi.org/http://dx.doi.org/10.20944/preprints202411.0560.v1

Andeobu, L., Wibowo, S., & Grandhi, S. (2023). Informal E-waste recycling practices and environmental pollution in Africa: What is the way forward? *International Journal of Hygiene and Environmental Health*, *252*, 114192. https://doi.org/https://doi.org/10.1016/j.ijheh.2023.114192 PMID: 37348165

Arora, N., Bhattacharjya, S., Bakshi, S. K., Anand, M., Dasgupta, D., Gupta, A., Prasad, N. S., Nanda, N., Pandey, S., Garud, S., & Bineesan, M. K. (2018). *Greening the Solar Power PV value chain*. European Union's Resource Efficiency Initiative (EU-REI) Project, Deutsche Gesellschaft für Internationale Zusammenarbeit (GIZ) GmbH www.resourceefficiencyindia.com

Arul Prishya, A. S., Chopra, L., & Manikanika. (2023). Comprehensive review on uses of silicon dioxide in solar cell. *Materials Today: Proceedings, 72*, 1471-1478. https://doi.org/https://doi.org/10.1016/j.matpr.2022.09.348

Babaei, A., & Nasr Esfahani, A. (2024). A Review of Photovoltaic Waste Management from a Sustainable Perspective. *Electricity, 5*(4), 734-750. https://www.mdpi.com/2673-4826/5/4/36

Bajagain, R., Panthi, G., An, Y.-J., & Jeong, S.-W. (2020). Current Practices on Solar Photovoltaic Waste Management: An Overview of the Potential Risk and Regulatory Approaches of the Photovoltaic Waste. *Journal of Korean Society of Environmental Engineers, 42*(12), 690–708. https://doi.org/https://doi.org/10.4491/KSEE.2020.42.12.690

Baklouti, A., Mifdal, L., Dellagi, S., & Chelbi, A. (2020). An Optimal Preventive Maintenance Policy for a Solar Photovoltaic System. *Sustainability, 12*(10), 4266. https://doi.org/https://doi.org/10.3390/su12104266

Baldé, C. P., Forti, V., Gray, V., Kuehr, R., & Stegmann, P. (2017). *The global e-waste monitor 2017: Quantities, flows and resources*. United Nations University (UNU), International Telecommunication Union (ITU), International Solid Waste Association (ISWA). collections.unu.edu/

Bertrand, S. (2021). *Fact Sheet | Climate, Environmental, and Health Impacts of Fossil Fuels*. Environmental and Energy Study Institute (EESI). Retrieved 04/03/2025 from https://www.eesi.org/papers/view/fact-sheet-climate-environmental-and-health-impacts-of-fossil-fuels-2021

Birtürk, A., Keskin, T., & Çeliktaş, M. S. (2023). The effect of thermal treatments on photovoltaic module recycling. In Aloui, F., Varuvel, E. G., & Sonthalia, A. (Eds.), *Handbook of Thermal Management Systems* (pp. 659–671). Elsevier. https://doi.org/https://doi.org/10.1016/B978-0-443-19017-9.00015-5

Biyouki, Z. A., Zaman, A., Marinova, D., Minunno, R., & Askari Shayegan, M. (2024). Solar Photovoltaics Value Chain and End-of-Life Management Practices: A Systematic Literature Review. *Sustainability, 16*(16), 7038. https://www.mdpi.com/2071-1050/16/16/7038

Bogar, E. (2023). Chaos Game Optimization-Least Squares Algorithm for Photovoltaic Parameter Estimation. *Arabian Journal for Science and Engineering, 48*(5), 6321–6340. https://doi.org/https://doi.org/10.1007/s13369-022-07364-6

Bogar, Z. O., Capraz, O., & Güngör, A. (2019). An Overview of Methods Used for Estimating E-waste Amount. In Prasad, M. N. V., & Vithanage, M. (Eds.), *Electronic Waste Management and Treatment Technology* (pp. 53–75). Butterworth-Heinemann. https://doi.org/https://doi.org/10.1016/B978-0-12-816190-6.00003-0

Bošnjaković, M., Stojkov, M., Katinić, M., & Lacković, I. (2023). Effects of Extreme Weather Conditions on PV Systems. *Sustainability, 15*(22), 16044. https://doi.org/https://doi.org/10.3390/su152216044

Capraz, O., Gungor, A., Mutlu, O., & Sagbas, A. (2024). Optimal sizing of grid-connected hybrid renewable energy systems without storage: A generalized optimization model. *Energy Sources. Part A, Recovery, Utilization, and Environmental Effects, 46*(1), 13317–13350. https://doi.org/https://doi.org/10.1080/15567036.2020.1803454

Chawla, R., Singhal, P., & Garg, A. K. (2020). Photovoltaic Review of all Generations: Environmental Impact and Its Market Potential. *Transactions on Electrical and Electronic Materials, 21*(5), 456-476. https://doi.org/https://doi.org/10.1007/s42341-020-00217-9

Cheema, H. A., Ilyas, S., Kang, H., & Kim, H. (2024). Comprehensive review of the global trends and future perspectives for recycling of decommissioned photovoltaic panels. *Waste Management, 174*, 187-202. https://doi.org/https://doi.org/10.1016/j.wasman.2023.11.025

Chen, P.-H., Chen, W.-S., Lee, C.-H., & Wu, J.-Y. (2024). Comprehensive Review of Crystalline Silicon Solar Panel Recycling: From Historical Context to Advanced Techniques. *Sustainability, 16*(1), 60. https://www.mdpi.com/2071-1050/16/1/60

Chiang, P. F., Han, S., Claire, M. J., Maurice, N. J., Vakili, M., & Giwa, A. S. (2024). Sustainable Treatment of Spent Photovoltaic Solar Panels Using Plasma Pyrolysis Technology and Its Economic Significance. *Clean Technologies, 6*(2), 432-452. https://doi.org/https://doi.org/10.3390/cleantechnol6020022

Chowdhury, M. S., Rahman, K. S., Chowdhury, T., Nuthammachot, N., Techato, K., Akhtaruzzaman, M., Tiong, S. K., Sopian, K., & Amin, N. (2020). An overview of solar photovoltaic panels' end-of-life material recycling. *Energy Strategy Reviews, 27*, 100431. https://doi.org/https://doi.org/10.1016/j.esr.2019.100431

Chung, S.-s., Lau, K.-y., & Zhang, C. (2011). Generation of and control measures for, e-waste in Hong Kong. *Waste Management, 31*(3), 544-554. https://doi.org/https://doi.org/10.1016/j.wasman.2010.10.003

Cobbing, M. (2008). *Toxic Tech: Not in Our Backyard Uncovering the Hidden Flows of e-Waste*. Greenpeace International. https://www.greenpeace.org/global/

Dambhare, M. V., Butey, B., & Moharil, S. V. (2021). Solar photovoltaic technology: A review of different types of solar cells and its future trends. *Journal of Physics: Conference Series*, *1913*(1), 012053. https://doi.org/https://doi.org/10.1088/1742-6596/1913/1/012053

De Sousa, N. M., Oliveira, C. B., & Cunha, D. (2023). Photovoltaic electronic waste in Brazil: Circular economy challenges, potential and obstacles. *Social Sciences & Humanities Open*, *7*(1), 100456. https://doi.org/https://doi.org/10.1016/j.ssaho.2023.100456

Demir, M. A., & Yakışık, H. (2024). Historical Development of Energy. *Journal of the Human and Social Science Researches*, *13*(3), 1333–1353. https://doi.org/https://doi.org/10.15869/itobiad.1500506

Deng, R., Zhuo, Y., & Shen, Y. (2022). Recent progress in silicon photovoltaic module recycling processes. *Resources, Conservation and Recycling*, *187*, 106612. https://doi.org/https://doi.org/10.1016/j.resconrec.2022.106612

Desai, N., & Pathak, M. (2024). End of Life Solar Panels: Assessment for Gujarat, India. https://doi.org/https://doi.org/10.2139/ssrn.4693717

Dias, P., Javimczik, S., Benevit, M., & Veit, H. (2017). Recycling WEEE: Polymer characterization and pyrolysis study for waste of crystalline silicon photovoltaic modules. *Waste Management, 60*, 716-722. https://doi.org/https://doi.org/10.1016/j.wasman.2016.08.036

Dias, P., Javimczik, S., Benevit, M., Veit, H., & Bernardes, A. M. (2016). Recycling WEEE: Extraction and concentration of silver from waste crystalline silicon photovoltaic modules. *Waste Management, 57*, 220-225. https://doi.org/https://doi.org/10.1016/j.wasman.2016.03.016

Dias, P., Schmidt, L., Gomes, L. B., Bettanin, A., Veit, H., & Bernardes, A. M. (2018). Recycling Waste Crystalline Silicon Photovoltaic Modules by Electrostatic Separation. *Journal of Sustainable Metallurgy*, *4*(2), 176–186. https://doi.org/https://doi.org/10.1007/s40831-018-0173-5

Domínguez, A., & Geyer, R. (2017). Photovoltaic waste assessment in Mexico. *Resources, Conservation and Recycling*, *127*, 29–41. https://doi.org/https://doi.org/10.1016/j.resconrec.2017.08.013

Domínguez, A., & Geyer, R. (2019). Photovoltaic waste assessment of major photovoltaic installations in the United States of America. *Renewable Energy, 133*, 1188–1200. https://doi.org/https://doi.org/10.1016/j.renene.2018.08.063

Duman, G. M., Kongar, E., & Gupta, S. M. (2019). Estimation of electronic waste using optimized multivariate grey models. *Waste Management, 95*, 241-249. https://doi.org/https://doi.org/10.1016/j.wasman.2019.06.023

Duman, G. M., Kongar, E., & Gupta, S. M. (2020). Predictive analysis of electronic waste for reverse logistics operations: A comparison of improved univariate grey models. *Soft Computing, 24*(20), 15747–15762. https://doi.org/https://doi.org/10.1007/s00500-020-04904-w

Duran, A. S., Atasu, A., & Van Wassenhove, L. N. (2022). Cleaning after solar panels: Applying a circular outlook to clean energy research. *International Journal of Production Research, 60*(1), 211–230. https://doi.org/https://doi.org/10.1080/00207543.2021.1990434

Dwivedy, M., & Mittal, R. K. (2010). Future trends in computer waste generation in India. *Waste Management, 30*(11), 2265-2277. https://doi.org/https://doi.org/10.1016/j.wasman.2010.06.025

El-Khawad, L., Bartkowiak, D., & Kümmerer, K. (2022). Improving the end-of-life management of solar panels in Germany. *Renewable & Sustainable Energy Reviews, 168*, 112678. https://doi.org/https://doi.org/10.1016/j.rser.2022.112678

EPA. (2024). *End-of-Life Solar Panels: Regulations and Management*. United States Environmental Protection Agency. Retrieved 04/03/2024 from https://www.epa.gov/hw/end-life-solar-panels-regulations-and-management

Farji, M. (2021). Development of Photovoltaic Cells: A Materials Prospect and Next-Generation Futuristic Overview. *Brazilian Journal of Physics, 51*(6), 1916–1928. https://doi.org/https://doi.org/10.1007/s13538-021-00981-w

Fiandra, V., Sannino, L., Andreozzi, C., & Graditi, G. (2019). End-of-life of silicon PV panels: A sustainable materials recovery process. *Waste Management, 84*, 91-101. https://doi.org/https://doi.org/10.1016/j.wasman.2018.11.035

Forti, V., Balde, C. P., Kuehr, R., & Bel, G. (2020). *The Global E-waste Monitor 2020: Quantities, flows and the circular economy potential*. United Nations University (UNU)/United Nations Institute for Training and Research (UNITAR), International Telecommunication Union (ITU), International Solid Waste Association (ISWA). https://collections.unu.edu/

Franzoni, A., Leggerini, C., & Bannò, M. (2024). A Predictive Framework for Photovoltaic Waste Quantities and Recovery Values: Insights and Application to the Italian Context. *Environmental and Climate Technologies, 28*(1), 243-257. https://doi.org/https://doi.org/10.2478/rtuect-2024-0020

Fraunhofer, I. S. E. (2024). *Photovoltaics Report*. Fraunhofer Institute for Solar Energy Systems, ISE with the support of PSE Projects GmbH. https://www.ise.fraunhofer.de/

Gautam, A., Shankar, R., & Vrat, P. (2021). End-of-life solar photovoltaic e-waste assessment in India: A step towards a circular economy. *Sustainable Production and Consumption*, *26*, 65–77. https://doi.org/https://doi.org/10.1016/j.spc.2020.09.011

Gerold, E., & Antrekowitsch, H. (2024). Advancements and Challenges in Photovoltaic Cell Recycling: A Comprehensive Review. *Sustainability*, *16*(6), 2542. https://doi.org/https://doi.org/10.3390/su16062542

Golev, A., Schmeda-Lopez, D. R., Smart, S. K., Corder, G. D., & McFarland, E. W. (2016). Where next on e-waste in Australia? *Waste Management, 58*, 348-358. https://doi.org/https://doi.org/10.1016/j.wasman.2016.09.025

Golsteijn, L., & Valencia Martinez, E. (2017). The Circular Economy of E-Waste in the Netherlands: Optimizing Material Recycling and Energy Recovery. *Journal of Engineering*, *2017*(1), 8984013. https://doi.org/https://doi.org/10.1155/2017/8984013

Gonda, L., D'Ans, P., & Degrez, M. (2019). A comparative assessment of WEEE collection in an urban and rural context: Case study on desktop computers in Belgium. *Resources, Conservation and Recycling*, *142*, 131–142. https://doi.org/https://doi.org/10.1016/j.resconrec.2018.11.008

Guo, X., & Yan, K. (2017). Estimation of obsolete cellular phones generation: A case study of China. *The Science of the Total Environment*, *575*, 321–329. https://doi.org/https://doi.org/10.1016/j.scitotenv.2016.10.054 PMID: 27744159

Herceg, S., Pinto Bautista, S., & Weiß, K.-A. (2020). Influence of Waste Management on the Environmental Footprint of Electricity Produced by Photovoltaic Systems. *Energies*, *13*(9), 2146. https://doi.org/https://doi.org/10.3390/en13092146

IEA. (2021). *World Energy Outlook 2021*. International Energy Agency. https://www.iea.org/

IEA. (2023). *World Energy Outlook 2023*. International Energy Agency. https://www.iea.org/

Ikhlayel, M. (2016). Differences of methods to estimate generation of waste electrical and electronic equipment for developing countries: Jordan as a case study. *Resources, Conservation and Recycling*, *108*, 134–139. https://doi.org/https://doi.org/10.1016/j.resconrec.2016.01.015

IPCC. (2023). *Climate Change 2023: Synthesis Report* (Team, C. W., Lee, H., & Romero, J., Eds.)., DOI: 10.59327/IPCC/AR6-9789291691647

IRENA. (2020). *Global Renewables Outlook: Energy Transformation 2050 (Edition: 2020)*. International Renewable Energy Agency. https://www.irena.org/

IRENA. (2023). *Renewable power generation costs in 2022*. International Renewable Energy Agency. https://www.irena.org/

IRENA. (2024). *Renewable energy statistics 2024*. International Renewable Energy Agency. https://www.irena.org/

IRENA & CPI. (2023). *Global landscape of renewable energy finance 2023*. International Renewable Energy Agency and Climate Policy Initiative. https://www.climatepolicyinitiative.org/

Islam, M. T., & Huda, N. (2019). E-waste in Australia: Generation estimation and untapped material recovery and revenue potential. *Journal of Cleaner Production*, *237*, 117787. https://doi.org/https://doi.org/10.1016/j.jclepro.2019.117787

Jain, A., & Sareen, R. (2006). E-waste assessment methodology and validation in India. *Journal of Material Cycles and Waste Management*, *8*, 40–45. https://doi.org/https://doi.org/10.1007/s10163-005-0145-2

Kaaya, I., Ascencio-Vásquez, J., Weiss, K.-A., & Topič, M. (2021). Assessment of uncertainties and variations in PV modules degradation rates and lifetime predictions using physical models. *Solar Energy*, *218*, 354–367. https://doi.org/https://doi.org/10.1016/j.solener.2021.01.071

Kalmykova, Y., Patrício, J., Rosado, L., & Berg, P. E. (2015). Out with the old, out with the new – The effect of transitions in TVs and monitors technology on consumption and WEEE generation in Sweden 1996–2014. *Waste Management, 46*, 511-522. https://doi.org/https://doi.org/10.1016/j.wasman.2015.08.034

Karlilar Pata, S., & Balcilar, M. (2024). Decarbonizing energy: Evaluating fossil fuel displacement by renewables in OECD countries. *Environmental Science and Pollution Research International*, *31*(21), 31304–31313. https://doi.org/https://doi.org/10.1007/s11356-024-33324-8 PMID: 38630397

Kastanaki, E., & Giannis, A. (2022). Forecasting quantities of critical raw materials in obsolete feature and smart phones in Greece: A path to circular economy. *Journal of Environmental Management*, *307*, 114566. https://doi.org/https://doi.org/10.1016/j.jenvman.2022.114566 PMID: 35091243

Khoshnazm, M. J., Marzban, A., & Azimi, N. (2023). Performance enhancement of photovoltaic panels integrated with thermoelectric generators and phase change materials: Optimization and analysis of thermoelectric arrangement. *Energy*, *267*, 126556. https://doi.org/https://doi.org/10.1016/j.energy.2022.126556

Kim, S., An, J., Choi, H., & Hong, T. (2023). Assessment the technical and economic performance of a window-integrated PV system using third-generation PV panels. *Energy and Building*, *286*, 112978. https://doi.org/https://doi.org/10.1016/j.enbuild.2023.112978

Kolamroudi, M. K., Ilkan, M., Egelioglu, F., & Safaei, B. (2023). Feature selection by ant colony optimization and experimental assessment analysis of PV panel by reflection of mirrors perpendicularly. *Renewable Energy*, *218*, 119238. https://doi.org/https://doi.org/10.1016/j.renene.2023.119238

Komoto, K., Lee, J.-S., Zhang, J., Ravikumar, D., Sinha, P., Wade, A., & Heath, G. (2018). *End-of-Life Management of Photovoltaic Panels: Trends in PV Module Recycling Technologies*. International Energy Agency Photovoltaic Power Systems Program (IEA-PVPS). https://iea-pvps.org/

Kosai, S., Kishita, Y., & Yamasue, E. (2020). Estimation of the metal flow of WEEE in Vietnam considering lifespan transition. *Resources, Conservation and Recycling*, *154*, 104621. https://doi.org/https://doi.org/10.1016/j.resconrec.2019.104621

Koshta, N., Patra, S., & Singh, S. P. (2021). Estimation of E-waste at micro level for reverse logistics: A case of Delhi. *Journal of Cleaner Production*, *314*, 128063. https://doi.org/https://doi.org/10.1016/j.jclepro.2021.128063

Krebs-Moberg, M., Pitz, M., Dorsette, T. L., & Gheewala, S. H. (2021). Third generation of photovoltaic panels: A life cycle assessment. *Renewable Energy*, *164*, 556–565. https://doi.org/https://doi.org/10.1016/j.renene.2020.09.054

Lee, J.-K., Lee, J.-S., Ahn, Y.-S., Kang, G.-H., Song, H.-E., Kang, M.-G., Kim, Y.-H., & Cho, C.-H. (2018). Simple pretreatment processes for successful reclamation and remanufacturing of crystalline silicon solar cells. *Progress in Photovoltaics: Research and Applications*, *26*(3), 179–187. https://doi.org/https://doi.org/10.1002/pip.2963

Li, A., Li, B., Lu, B., Yang, D., Hou, S., & Song, X. (2022). Generation estimation and material flow analysis of retired mobile phones in China. *Environmental Science and Pollution Research International, 29*(50), 75626–75635. https://doi.org/https://doi.org/10.1007/s11356-022-21153-6 PMID: 35657548

Lohse, J., Winteler, S., & Wulf-Schnabel, J. (1998). *Collection targets for waste from electrical and electronic equipment (WEEE).* Report compiled for the Directorate General (DG XI Evironment, Nuclear Sfety and Civil Potection of the Cmmission of the European Communities. https://oekopol.de/

Maani, T., Celik, I., Heben, M. J., Ellingson, R. J., & Apul, D. (2020). Environmental impacts of recycling crystalline silicon (c-SI) and cadmium telluride (CDTE) solar panels. *The Science of the Total Environment, 735*, 138827. https://doi.org/https://doi.org/10.1016/j.scitotenv.2020.138827 PMID: 32464407

Mahmoudi, S., Huda, N., & Behnia, M. (2019). Photovoltaic waste assessment: Forecasting and screening of emerging waste in Australia. *Resources, Conservation and Recycling, 146*, 192–205. https://doi.org/https://doi.org/10.1016/j.resconrec.2019.03.039

Mahmoudi, S., Huda, N., & Behnia, M. (2021). Critical assessment of renewable energy waste generation in OECD countries: Decommissioned PV panels. *Resources, Conservation and Recycling, 164*, 105145. https://doi.org/https://doi.org/10.1016/j.resconrec.2020.105145

Marwede, M., Berger, W., Schlummer, M., Mäurer, A., & Reller, A. (2013). Recycling paths for thin-film chalcogenide photovoltaic waste – Current feasible processes. *Renewable Energy, 55*, 220–229. https://doi.org/https://doi.org/10.1016/j.renene.2012.12.038

Matthews, H. S., McMichael, F. C., Hendrickson, C. T., & Hart, D. J. (1997). *Disposition and End-of-Life Options for Personal Computers*. Carnegie Mellon University.

Moyen Massa, G., & Archodoulaki, V.-M. (2023). Electrical and Electronic Waste Management Problems in Africa: Deficits and Solution Approach. *Environments, 10*(3), 44. https://doi.org/https://doi.org/10.3390/environments10030044

Muthusamy, P. D., Velusamy, G., Thandavan, S., Govindasamy, B. R., & Savarimuthu, N. (2022). Industrial Internet of things-based solar photo voltaic cell waste management in next generation industries. *Environmental Science and Pollution Research International, 29*(24), 35542–35556. https://doi.org/https://doi.org/10.1007/s11356-022-19411-8 PMID: 35237911

Ndalloka, Z. N., Nair, H. V., Alpert, S., & Schmid, C. (2024). Solar photovoltaic recycling strategies. *Solar Energy, 270*, 112379. https://doi.org/https://doi.org/10.1016/j.solener.2024.112379

Nižetić, S., Jurčević, M., Čoko, D., & Arıcı, M. (2021). A novel and effective passive cooling strategy for photovoltaic panel. *Renewable & Sustainable Energy Reviews, 145*, 111164. https://doi.org/https://doi.org/10.1016/j.rser.2021.111164

NREL. (2021). *Documenting a Decade of Cost Declines for PV Systems.* The National Renewable Energy Laboratory. Retrieved 04/03/2025 from https://www.nrel.gov/news/program/2021/documenting-a-decade-of-cost-declines-for-pv-systems.html

Oguchi, M., Kameya, T., Yagi, S., & Urano, K. (2008). Product flow analysis of various consumer durables in Japan. *Resources, Conservation and Recycling, 52*(3), 463–480. https://doi.org/https://doi.org/10.1016/j.resconrec.2007.06.001

Orosz, T., Rassõlkin, A., Arsénio, P., Poór, P., Valme, D., & Sleisz, Á. (2024). Current Challenges in Operation, Performance, and Maintenance of Photovoltaic Panels. *Energies, 17*(6), 1306. https://doi.org/https://doi.org/10.3390/en17061306

Owusu-Sekyere, K., Batteiger, A., Afoblikame, R., Hafner, G., & Kranert, M. (2022). Assessing data in the informal e-waste sector: The Agbogbloshie Scrapyard. *Waste Management, 139*, 158-167. https://doi.org/https://doi.org/10.1016/j.wasman.2021.12.026

Özgür Polat, L., & Güngör, A. (2019). Network Design Problems in E-waste Management. In Prasad, M. N. V., & Vithanage, M. (Eds.), *Electronic Waste Management and Treatment Technology* (pp. 77–102). Butterworth-Heinemann., https://doi.org/https://doi.org/10.1016/B978-0-12-816190-6.00004-2

Özgür Polat, L., & Gungor, A. (2021). WEEE closed-loop supply chain network management considering the damage levels of returned products. *Environmental Science and Pollution Research International, 28*(7), 7786–7804. https://doi.org/https://doi.org/10.1007/s11356-020-10249-6 PMID: 33040288

Ozsut Bogar, Z., & Gungor, A. (2023). Forecasting Waste Mobile Phone (WMP) Quantity and Evaluating the Potential Contribution to the Circular Economy: A Case Study of Turkey. *Sustainability, 15*(4), 3104. https://doi.org/https://doi.org/10.3390/su15043104

Pagnanelli, F., Moscardini, E., Granata, G., Abo Atia, T., Altimari, P., Havlik, T., & Toro, L. (2017). Physical and chemical treatment of end of life panels: An integrated automatic approach viable for different photovoltaic technologies. *Waste Management, 59*, 422-431. https://doi.org/https://doi.org/10.1016/j.wasman.2016.11.011

Paiano, A. (2015). Photovoltaic waste assessment in Italy. *Renewable & Sustainable Energy Reviews*, *41*, 99–112. https://doi.org/https://doi.org/10.1016/j.rser.2014.07.208

Palitzsch, W., & Loser, U. (2011, 19-24 June 2011). A new and intelligent de-metalization step of broken silicon cells and silicon cell production waste in the recycling procedure of crystalline si modules. *37th IEEE Photovoltaic Specialists Conference*, Seattle, WA, USA.

Panda, S. K., Behera, S., & Patnaik, S. S. (2021, September 20-22). 2021). Lifetime Estimation of DC-Link Capacitor in a Standalone PV System with an Integrated Backup Battery. 2021 International Symposium of Asian Control Association on Intelligent Robotics and Industrial Automation (IRIA). https://doi.org/https://doi.org/10.1016/j.rser.2010.11.037

Panwar, N. L., Kaushik, S. C., & Kothari, S. (2011). Role of renewable energy sources in environmental protection: A review. *Renewable & Sustainable Energy Reviews*, *15*(3), 1513–1524. https://doi.org/https://doi.org/10.1016/j.rser.2010.11.037

Pastuszak, J., & Węgierek, P. (2022). Photovoltaic Cell Generations and Current Research Directions for Their Development. *Materials (Basel)*, *15*(16), 5542. https://doi.org/https://doi.org/10.3390/ma15165542 PMID: 36013679

Peralta, G. L., & Fontanos, P. M. (2006). E-waste issues and measures in the Philippines. *Journal of Material Cycles and Waste Management*, *8*(1), 34–39. https://doi.org/https://doi.org/10.1007/s10163-005-0142-5

Petridis, N. E., Stiakakis, E., Petridis, K., & Dey, P. (2016). Estimation of computer waste quantities using forecasting techniques. *Journal of Cleaner Production*, *112*, 3072–3085. https://doi.org/https://doi.org/10.1016/j.jclepro.2015.09.119

Polák, M., & Drápalová, L. (2012). Estimation of end of life mobile phones generation: The case study of the Czech Republic. *Waste Management, 32*(8), 1583-1591. https://doi.org/https://doi.org/10.1016/j.wasman.2012.03.028

Ritchie, H., & Rosado, P. (2022). *Fossil fuels*. Our World in Data. Retrieved 26/09/2024 from https://ourworldindata.org/fossil-fuels

Ritchie, H., Rosado, P., & Roser, M. (2023). *CO_2 and Greenhouse Gas Emissions*. Our World in Data. Retrieved 26/09/2024 from https://ourworldindata.org/co2-and-greenhouse-gas-emissions

Robinson, B. H. (2009). E-waste: An assessment of global production and environmental impacts. *The Science of the Total Environment*, *408*(2), 183–191. https://doi.org/https://doi.org/10.1016/j.scitotenv.2009.09.044 PMID: 19846207

Romel, M., Kabir, G., & Ng, K. T. W. (2024). Prediction of photovoltaic waste generation in Canada using regression-based model. *Environmental Science and Pollution Research International, 31*(6), 8650–8665. https://doi.org/https://doi.org/10.1007/s11356-023-31628-9 PMID: 38182949

Sakthivel, U., Swaminathan, G., & Anis, J. (2022). Strategies for Quantifying Metal Recovery from Waste Electrical and Electronic Equipment (WEEE/E-waste) Using Mathematical Approach. *Process Integration and Optimization for Sustainability, 6*, 1-10. https://doi.org/https://doi.org/10.1007/s41660-022-00250-6

Sangeetha, B., Manjunatha, K., Thirusenthil Kumaran, P., Sheela, A., Yamuna, K. S., & Sivakumar, S. (2024). Performance Optimization in Photovoltaic Systems: A Review. *Archives of Computational Methods in Engineering, 31*(3), 1507–1518. https://doi.org/https://doi.org/10.1007/s11831-023-10023-0

Savvilotidou, V., Antoniou, A., & Gidarakos, E. (2017). Toxicity assessment and feasible recycling process for amorphous silicon and CIS waste photovoltaic panels. *Waste Management, 59*, 394-402. https://doi.org/https://doi.org/10.1016/j.wasman.2016.10.003

Sharma, A., Pandey, S., & Kolhe, M. (2019). Global review of policies & guidelines for recycling of solar PV modules. *International Journal of Smart Grid and Clean Energy, 8*(5), 597–610. https://doi.org/https://doi.org/10.12720/sgce.8.5.597-610

Simandl, G. J., Paradis, S., & Simandl, L. (2023). Future of photovoltaic materials with emphasis on resource availability, economic geology, criticality, and market size/growth. *CIM Journal, 14*(3), 133–157. https://doi.org/https://doi.org/10.1080/19236026.2023.2168419

Singh, G. K. (2013). Solar power generation by PV (photovoltaic) technology: A review. *Energy, 53*, 1–13. https://doi.org/https://doi.org/10.1016/j.energy.2013.02.057

Sivagami, K., Bose, S., Vinayak, A. K., Sreenivas, M., Ghosh, A., Narasimhan, M., & Gurumoorthy, A. V. P. (2024). Solar Panel Recycling from Circular Economy Viewpoint: A Review. *Applied Solar Energy, 60*(2), 328–345. https://doi.org/https://doi.org/10.3103/S0003701X23601862

Souza, V., Rodrigues Figueiredo, A. M., & Santos Bortolocci Espejo, M. M. d. (2024). Challenges and strategies for managing end-of-life photovoltaic equipment in Brazil: Learning from international experience. *Energy Policy, 188*, 114091. https://doi.org/https://doi.org/10.1016/j.enpol.2024.114091

Strachala, D., Hylský, J., Jandova, K., Vaněk, J., & Cingel, Š. (2017). Mechanical Recycling of Photovoltaic Modules. *ECS Transactions*, *81*(1), 199. https://doi.org/https://dx.doi.org/10.1149/08101.0199ecst

Streicher-Porte, M., Widmer, R., Jain, A., Bader, H.-P., Scheidegger, R., & Kytzia, S. (2005). Key drivers of the e-waste recycling system: Assessing and modelling e-waste processing in the informal sector in Delhi. *Environmental Impact Assessment Review*, *25*(5), 472–491. https://doi.org/https://doi.org/10.1016/j.eiar.2005.04.004

Tan, J., Jia, S., & Ramakrishna, S. (2022). End-of-Life Photovoltaic Modules. *Energies*, *15*(14), 5113. https://doi.org/https://doi.org/10.3390/en15145113

Tan, V., Dias, P. R., Chang, N., & Deng, R. (2022). Estimating the Lifetime of Solar Photovoltaic Modules in Australia. *Sustainability*, *14*(9), 5336. https://doi.org/https://doi.org/10.3390/su14095336

Trivedi, H. K., Meshram, A., & Gupta, R. (2023). Recycling of photovoltaic modules for recovery and repurposing of materials. *Journal of Environmental Chemical Engineering*, *11*(2), 109501. https://doi.org/https://doi.org/10.1016/j.jece.2023.109501

UN. (2023). *Global Sustainable Development Report 2023: Times of crisis, times of change: Science for accelerating transformations to sustainable development.* United Nations. https://sdgs.un.org/gsdr

UNFCCC. (2015). *The Paris Agreement.* United Nations Framework Convention on Climate Change. https://unfccc.int/process-and-meetings/the-paris-agreement

Vinayagamoorthi, R., Bhargav, P. B., Ahmed, N., Balaji, C., Aravinth, K., Krishnan, A., Govindaraj, R., & Ramasamy, P. (2024). Recycling of end of life photovoltaic solar panels and recovery of valuable components: A comprehensive review and experimental validation. *Journal of Environmental Chemical Engineering*, *12*(1), 111715. https://doi.org/https://doi.org/10.1016/j.jece.2023.111715

Wahman, M., Surowiak, A., Ebin, B., & Berent, K. (2024). PV back sheet recovery from c-Si modules using hot knife technique. *Solar Energy Materials and Solar Cells*, *276*, 113067. https://doi.org/https://doi.org/10.1016/j.solmat.2024.113067

Wang, F., Huisman, J., Stevels, A., & Baldé, C. P. (2013). Enhancing e-waste estimates: Improving data quality by multivariate Input–Output Analysis. *Waste Management, 33*(11), 2397-2407. https://doi.org/https://doi.org/10.1016/j.wasman.2013.07.005

Wang, F., Yu, L., & Wu, A. (2021). Forecasting the electronic waste quantity with a decomposition-ensemble approach. *Waste Management, 120*, 828-838. https://doi.org/https://doi.org/10.1016/j.wasman.2020.11.006

Wang, J., Feng, Y., & He, Y. (2024). The research progress on recycling and resource utilization of waste crystalline silicon photovoltaic modules. *Solar Energy Materials and Solar Cells*, *270*, 112804. https://doi.org/https://doi.org/10.1016/j.solmat.2024.112804

Wang, M., You, X., Li, X., & Liu, G. (2018). Watch more, waste more? A stock-driven dynamic material flow analysis of metals and plastics in TV sets in China. *Journal of Cleaner Production*, *187*, 730–739. https://doi.org/https://doi.org/10.1016/j.jclepro.2018.03.243

Wang, R., Song, E., Zhang, C., Zhuang, X., Ma, E., Bai, J., Yuan, W., & Wang, J. (2019). Pyrolysis-based separation mechanism for waste crystalline silicon photovoltaic modules by a two-stage heating treatment. *RSC Advances*, *9*(32), 18115–18123. https://doi.org/https://doi.org/10.1039/C9RA03582F PMID: 35515232

Weckend, S., Wade, A., & Heath, G. (2016). *End of life management: Solar Photovoltaic Panels*. International Renewable Energy Agency (IRENA), International Energy Agency Photovoltaic Power Systems (IEA-PVPS). https://www.irena.org/

WHO. (2024). *Electronic waste (e-waste)*. World Health Organization. Retrieved 10/10/2024 from https://www.who.int/news-room/fact-sheets/detail/electronic-waste-(e-waste)

Widmer, R., Oswald-Krapf, H., Sinha-Khetriwal, D., Schnellmann, M., & Böni, H. (2005). Global perspectives on e-waste. *Environmental Impact Assessment Review*, *25*(5), 436–458. https://doi.org/https://doi.org/10.1016/j.eiar.2005.04.001

Xu, Y., Li, J., Tan, Q., Peters, A. L., & Yang, C. (2018). Global status of recycling waste solar panels: A review. *Waste Management, 75*, 450-458. https://doi.org/https://doi.org/10.1016/j.wasman.2018.01.036

Yadav, D., Kumar Saraf, A., & Singh Rathee, N. (2023). Assessment of PV waste generation in India. *Materials Today: Proceedings*. https://doi.org/https://doi.org/10.1016/j.matpr.2023.03.734

Zhang, S., Gu, Y., Tang, A., Li, B., Li, B., Pan, D., & Wu, Y. (2021). Forecast of future yield for printed circuit board resin waste generated from major household electrical and electronic equipment in China. *Journal of Cleaner Production*, *283*, 124575. https://doi.org/https://doi.org/10.1016/j.jclepro.2020.124575

KEY TERMS AND DEFINITIONS

Circular Economy: An economic system focused on reducing waste and promoting the reuse and recycling of materials to enhance sustainability.

E-Waste: Discarded electronic devices and components that have reached the end of their useful life and require proper disposal or recycling to reduce environmental impact.

Hazardous Substances: Materials in PV panels that can cause environmental or health risks if not managed properly.

Material Recovery: The process of extracting valuable and reusable materials from waste PV panels.

PV Panels: Physical devices in solar energy systems that convert sunlight into electricity, including silicon-based, thin-film, and emerging types.

PV Waste Management: The process of collecting, recycling, reusing, and disposing of end-of-life PV panels in an environmentally responsible manner.

Recycling Technologies: The use of physical, thermal, and chemical processes to recover materials and components from PV waste.

Solar PV Waste: PV panels that have reached the end of their operational lifespan and are no longer effective at generating electricity.

Chapter 2
Advancing E-Waste Management:
Integrative Methods and Practices

Dankan Gowda V.
https://orcid.org/0000-0003-0724-0333
Department of Electronics and Communication Engineering, BMS Institute of Technology and Management, India

Vijay C. R.
https://orcid.org/0000-0001-6739-935X
Department of Management Studies, Dr. N.G.P. Institute of Technology, Coimbatore, India

Renu
https://orcid.org/0009-0001-8592-6771
TMIMT, Teerthanker Mahaveer University, Moradabad, India

N. M. G. Kumar
https://orcid.org/0000-0003-1494-5737
Department of Electrical and Electronics Engineering, Mohan Babu University, Tirupathi, India

Ratidev Samal
https://orcid.org/0000-0002-4022-6085
School of Management, Centurion University of Technology and Management, Bhubaneswar, India

ABSTRACT

This chapter focuses on the strategies regarding e-waste management involving the use of technology, policy, and community intervention for reduction of environmen-

DOI: 10.4018/979-8-3693-7383-5.ch002

tal and health effects, as well. Taking insights in state-of-art recycling approaches, policies including EPR and engaging the public, the authors discuss ways to enhance collection, resource recovery, and pollution. Examples suggest that respective policy and advanced technologies result in better rates of recyclability and decreased emissions. Some major highlights reveal efficiency improvement in the context of using AI in sorting programs and cuts in emissions with circular economy strategies. The third section discusses future trends such as online tracking of e-waste using blockchain and sustainable design. They provide practical information about how to design an e-waste framework that is both sustainable in the long run and capable of being reproduced across the globe.

INTRODUCTION

Electronic scrap, or e-scraps, is one of the most rapidly accruing waste fractions all over the world due to increasing technological development, shrinking life spans of commodities and consumer demand for electronic products. This waste consists of numerous end-of-life products – small appliances and tools, abandoned industrial machinery and much more. Lots of electronic equipment contain one or more precious metals such as gold, silver and copper alongside hazardous substances like lead, mercury and cadmium if not disposed off correctly (Johnson, L., & Roberts, P. 2019). Thus, toxic components can pollute soil and water, and, in regions with increased informal recycling, negatively affect people's health and the environment.

It is noteworthy, that all consequences of e-waste fulfil their missions not only on a local level but also exert influence on all environmental spheres throughout the world. For example, recycling informal sector in developing countries discharges toxic fumes of gases into the air, which affects air quality and health of communities living close to initiation sites (Chen, Y., & Wang, X. 2020). Moreover, lead in efficient segregated processing of e-waste which is a cause a loss of the value of raw materials which need to be reclaimed. In this regard, the management of e-waste is a topical issue that is in integration to the sustainable development aims and objectives which include the protection of resources, control of pollution and the health of the public. This chapter examines the multi-approach strategies in e-waste management, including technology, policy and society strategies in this way (Ahmed, S., & Kumar, R. 2018). With these dimensions in mind, it seeks to offer a broader and more environmentally friendly outlook to the problem of e-waste management, together with its negative impact on society.

Overview of E-Waste Management

E-waste management covers the recovery, treatment and disposal of end-of-life electronic devices or hardware sub-assemblies. Since almost every person or company uses digital devices in their processes and operations, e-waste has emerged as one of the most pressing problems of the world because the materials of electronic devices are very diverse and heterogeneous. E-waste is different from other waste types, because it must be processed carefully, as it comprises both precious metals and toxic chemicals (Brown, M., & Smith, T. 2017). Some of such constituents include gold, copper, silver, and other special metals that may be recycled and reused; other difficult constituents are lead, mercury, and flame retardants which should not get into the wrong hands or the environment.

The social cost attributable to e-waste environmental problem and its effects on human health are disastrous and complex. Chemical compounds are known to find its way into the soil and water which are some of the key sources of foods and water in a region. Public health hazards include breathing in toxic fume from informal dismantling and open burning of hazardous components within close proximity electronic waste recycling sites among residents within such community. More so, there is not a sustainable method of disposing the e-waste hence it calls for efficient way of recycling and recovering these non-renewable resources (Zhang, H., & Li, D. 2021).

Figure 1. E-waste management workflow

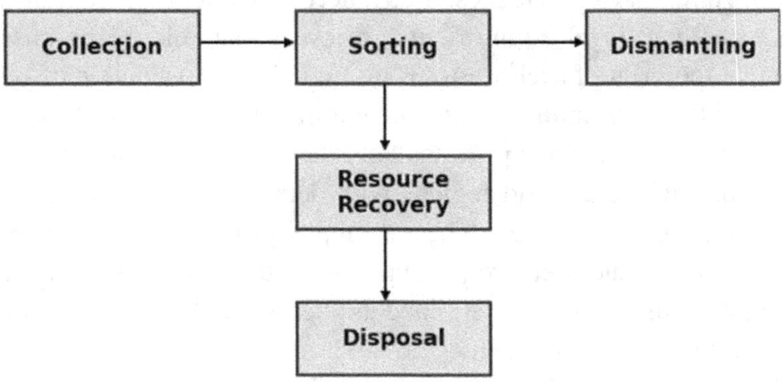

Figure.1. presents the E-waste management workflow provides an ideal workflow for e-waste management as it depicts a workflow of the e-waste management process from collection of e-waste to disposal of the same. This workflow is usually kicked start at Collection where obsolete electronic gadgets are collected using bins,

recycling facilities or drives. The subsequent phase is Sorting in which gathered items are sorted in categories that are based on their material type and recoverability (Kim, S., & Park, J. 2019). The next stage after sorting, which is the Dismantling stage entails taking apart devices with a view of seizing parts that are useful or can be recycled. The Resource Recovery phase takes proper efforts to recover metals, plastics and other useful items by adopting recycling technologies so that end product manufacturers do not rely much on raw products and thus the effect on environment is minimized (Sharma, A., & Gupta, N. 2020). The final of the process is Disposal with intents relating to non-recyclable and hazardous materials and components. This workflow offers an appropriate success model for handling e-waste so as to help maximize resource collection alongside minimizing the correct disposal of hazardous material.

In the context of sustainable consumption both reduction of e-waste and its proper disposal or recycling can be viewed as an important element of sustainability. It focuses on discussing the rationale for enhancing the notion of integrated e-waste management schemes considering technological, legislative, and community-based strategies.

Objectives and Scope of Chapter

Specifically, this chapter seeks to equip the reader with a clear perspective into the current state of e-waste management through an understanding of effective integrative management approaches that involve technology, policy, and the people. The objectives are threefold. First, to present advanced technologies of recycling e-waste as well as techniques involved in resource recovery that hold the promise to slash the ecological footprint of such wastes. This contains an evaluation of sustainable design aspects that can minimize the creation of waste at the product's life cycle first stage. Second, this chapter provides brief descriptions of e-waste management practices across different world regions with detailed case studies of successful sectors (Liu, F., & Chen, Z. 2020). These examples provide an overview of real-life measures that have reduced environmental effects successfully and it is possible to single out those aspects which determined its success, and which directions should be focused on further development.

Last, the chapter aims at discussing about challenges and prospects of Integrative e-waste management system. Some of the challenges that are normally associated with pharmacogenomics include; Regulatory disparities, Economic restraints and low health literacy. This discussion highlights shortcomings and challenges in meeting the goals of governments, manufacturers, and consumers to embrace best practices proposed in this discussion. With the analysis of the current state and perspectives in e-waste management, this chapter shall help the stakeholders to find out the preferable and efficient strategies for sustainable and sustainable e-waste management which will decrease the negative impact on resources and the environment.

LITERATURE SURVEY

Current State of E-Waste Management

E-waste management over the world depends on methodology as well as practices that have emerged due to the increased rates of the e-waste generation and its heterogeneous composition. Managing e-waste is conventionally followed a number of techniques, which include collection, disassembly, recycling, and disposal (Evans, K., & Walker, G. 2019). However, these steps continue facing other challenges including low collection rates, a weak recycling system, and informal flow recycling that dominates developing regions. These fragmented recycling services entail scrapping, lobbying and burning, with inadequate protection of toxic products, thus emitting serious health and environmental hazards (Patel, R., & Singh, L. 2018). On the other hand, more developed waste management nations use technologies of sorting material and equipment for sorting material which reduce time and health implications of people who deal with wastes by sorting them for recycling.

In the policy area, there are many frameworks set by governments to contain e-waste, encourage recycling and disposal. These policies include regulation of emissions by factories, mandated use of catalytic converters, and rules intended to prevent fuel slicing and other methods of cheating on emissions tests (Kim, Y., & Lee, J. 2021). Some of the countries that exhibited improved handling of e-waste have closed regulatory gaps and inadequate facilities though some countries lag behind. With the future advancement of technology, the tends in e-waste management showed that an integrated approach is still needed through technological inventions in the management of e-waste, policy – making framework as well as the support of the community.

Traditional vs. Modern Approaches

E-waste management has changed from arbitrary disposal to complex innovative methods. Earlier, e-waste was disposed as common waste, and many of it went to the dumpsites where its hazardous components seep into the ground water and later affect environment and health significantly. Previous approaches to recycling included disassembly by hand – this was time-consuming, and physically hazardous to the operators (Ahmed, M., & Jones, C. 2020). Nonetheless, in areas most especially where formal structures of recycling are not yet dominant, these measures persist mainly due to inadequate regulation and the revenue that informal segregation of metals delivers.

Figure 2. Comparative analysis of traditional and modern recycling approaches

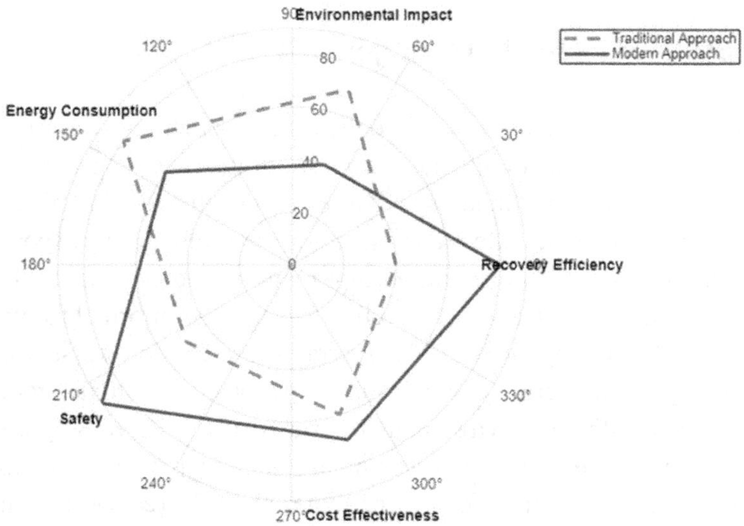

Figure 2 presents the radar chart of an Appropriate Assessment of Traditional Recycling against Modern Recycling ideas assessing them on five important parameters: Recovery Efficiency, Environmental Impact, Energy Consumption, Safety and Cost Effectiveness. Unmasking the progression of evaluation, it becomes clear that the existing recycling methods are superior to the old ones in such aspects as recovery efficiency, safety, among which it is possible to identify higher MRR rates and more secure recycling processes (Martinez, D., & Anderson, B. 2018). Also, methods used today are less destructive to the environment because of new technology that reduces emission of pollutants (Wang, J., & Zhou, H. 2019). While initial cost and technological demands can afford some conventional techniques to be used at a lower cost than innovations in certain areas, these conforming techniques will usually be more energy-intensive and detrimental to the human health and environment. Comparing the two, one is in a position to see the importance of implementing the new recycling technologies in order to encourage a proper management of e-waste (Chen, L., & Zhang, Y. 2021).

In contrast sustainable modern practices focus on high tech recycling and designs. Automated processes such as automated shredding and sorting have supplanted manual dismantling which advances both effectiveness and lessening of dangers.

Also, mechanical and biochemical techniques for materials recycling have been developed, which may be used to recover precious metals orange elements such as gold and rare earths without straining the environment (Gummadi, A. K., & Damacharla, P. 2024). As another significant transition, sustainable design is a practice in which manufacturers build items to be easily disassembled and recycled. Unlike traditional approaches, the current methods aim at developing a closed loop for the management of electronics to minimize e-waste consequences.

Global Standards and Regulatory Frameworks

The legal system of managing e-waste has not received equal implementation and enforcement throughout the world with some countries adopting stringent legal policies and the others lacking behind. Many developed countries today have developed legislation to control e-waste, but the European Union's Waste Electrical and Electronic Equipment (WEEE) directive stands out as the most elaborate as it includes objectives of collection, reuse and recycling of e-waste (Ramesha, M., & Pai, G. N. 2020). This has been made possible by the WEEE regulations that require manufacturers to –organize the management and disposal of their products which promotes the concept of a producer take back. Based on the above model, similar policies around the globe have been inspired to adopt the EPR principle as the central part of e-waste management (Shruthi, M., & Naveen K., B. 2020).

Several countries in Asia, including Japan and South Korea have set down e-waste policies that have policy requirement on recycling and recovery of resources. Indian government has initiated e-waste management rules that promotes EPR but there are hitch in enforcing those rules and regulation because informal sector dominates recycling of e-wastes. Currently there is no federal e- waste regulation in the United States, with the different states have put in place different policies which give a rather patchy picture (Puneeth Kumar, B. S., & Srinivasulu Raju, S. 2021). Taken together, these frameworks underscore the central role of policy in setting up systematic e-waste management frameworks; it however demonstrates variation in enforcement, which in turn calls for harmonisation in the management of e-waste flows across borders.

Technological Innovations in E-Waste Processing and Resource Recovery

Modern technologies have greatly enhanced efficacy of e-waste processing and resource recovery among the various techniques that were conventional. A breakthrough in the technologies of automatic sorting and shredding in the recycling process has led to better mechanical separation of materials and extraction of metals permitting minimizing human contact with hazardous substances (Kudari, J. M., & Samal, A. 2021). Some of the kinds of sorting and recovery systems include, Optical and sensor-based sorting systems, which are used to sort particular objects, thus allowing better sorting of recyclable material. Moreover, there are two other extraction processes which are relatively new in the extraction of metals from e-waste, namely hydrometallurgical and bioleaching processes (S, K., & Arun, M. R. 2022).

There is also some development of new technologies for the management of e-waste including AI and Robotics. Use of AI in the sorting increases efficiency and minimizes contamination errors and loss of the valuable material (Singla, A., & Sharma, N. 2022). Robotics and especially their use in dismantling allow faster disassembling of complexity electronics to get more metals and rare earth elements back. In this as these technologies develop, they enhance efficiency in the recovery of resources while trying to come up with ways of handling e-waste recycling across the world.

Challenges Identified in Literature

In the extant literature regarding e-waste management, there are several clusters of challenges that the authors raise in relation to the efficiency of e-waste processing and disposal. On this account, a primary challenge is the global heterogeneity in regulatory requirements and enforcement, causing disjointed and patchy management systems. Such a contradiction is especially inconvenient when it comes to e-waste since countries with weak policies inevitably become receiving points for waste from countries with stricter requirements (Pavankumar, P., & Darwante, N. K. 2022). Furthermore, small scale recycling continues to pose a significant concern in developing countries because of inadequate structures and financial barriers to proper handling.

Some other challenges include paucity of consumer sensitivity, general lack of adequate facilities for effective collection and recycling of e-waste. The general public still lacks the necessary knowledge concerning the effect their electronics have on the environment, or even where and how to properly get rid of the waste (Reddy, S. P., & Patwal, P. S. 2022). Moreover, availability of resources in a developing country poses a restrictive factor towards the setting up of sophisticated recycling plants thus the continued use of backwoods techniques. The mentioned challenges are best solved by serious policy amendments in the sphere of e-waste management, further investment in the ecosystem infrastructure and the increase of concern of the general public.

INTEGRATIVE METHODS FOR E-WASTE MANAGEMENT

Technological Innovation

There is need to foster e-waste technological advancement in order to optimise the recovery of resources and prevent pollutive impact. Currently there are so many facilities that technological approaches such as mechanical, chemical, and biotechnological means for recycling of valuable components from discarded electronics exist. Magnetic and manual sorting, shredding, and separation techniques, organize e-waste required for the recovery of useful copper or aluminum-metal objects. Hydrometallurgical processes involve using water soluble agents to selectively leach specific metals and, hence the extraction of metals such as gold, silver and the rare earth metals requires little power usage (Agrawal, R. 2023). Bioleaching, an overall biotechnological method in recycling, uses microorganisms to dissolve and recover metals from their compounds compared to conventional processes that rely on chemical solvents, heat or pressure.

The two further important aspects of technological breakthrough include the use of artificial intelligence and robotics in the handling of e-waste. Morphological differences of AI sorted materials and robot dismantled processes also enhance the aesthetics and effectiveness of recycling to ensure that more than 70% is recovered and the residual contamination rate is lowered (Hegge, K. K., & Agrawal, R. 2023). Given that recycling technology is a progressively more environmentally sound approach to these problems, recycling recovery appears to represent a steadily improving solution to e-waste management.

Design for Sustainability (DfS)

Design for Sustainability (DfS) means following sustainable principles during the design of a product, thus trying to reduce as much as possible e-waste generation at source. DfS is to design products that have long life cycle, little or no repairability and if no longer useful can be readily disassembled and reclaimed. It is often a requirement of DfS that items are designed in a modular fashion, certain materials are used, and it can help lead to longer product lifetimes. Sustained innovation to a degree entails such concepts as modular design, electronics made from parts that can be replaced or altered and do not require the buyer to get a new gadget. This way helps to minimize e-waste production due to emphasis on the use of the same as well as upgrades.

In sustainable design, selection of material is another point that has to be decided carefully. This way manufacturers reduce the negative impact of their products on the environment and help recycling plants to process them more easily. Another key DfS strategy is extended product lifespan where companies are urged to come up with products that are longer lasting so as to reverse the trend prompted by technology and its PN (Cell 5). Accompanying these principles of DfS, not only minimize the production of e-waste but also advance a circular economy for products and material flows.

Policy Development and Regulatory Support

Traditionally, policy is a more important factor in determining how e-waste is managed, recycled or disposed since it provides general guidelines for governing potential waste. Laws and regulations sometimes form the framework for progressive management of e-waste by setting the benchmarks for producers, users and collectors. Eco-responsibility policies such as the Extended Producer Responsibility (EPR) require manufacturers to take the full scope of their products' life cycle and ensure the product is designed in a way that will encourage recycling. By and large, EPR has been successful in raising the collection rate for e-waste and fostering recycling in many regions.

Nevertheless, that is where support of regulation can be useful, and, on the same hand, be disadvantageous. In areas where the enforcement of legislation might be wanting, laws can hardly translate to positive change in e-waste practices and thus lots of informal recycling and illicit disposal practice may persist. Furthermore, the differences in the laws of different countries make it easier to dispose of waste in a country where the rules are less strict compared to where the waste generating country's rules are much stricter. Overcoming these issues implies the integrated approach where the national and international legislation matches to ban the cross-border waste export/dumping and to facilitate the international e-waste management targets.

Global and National Policies

International and domestic laws and regulations present guidelines for e-waste management, as well, they serve to shape countries' approach regarding their electronic waste. For instance, The European Union's Waste Electrical and Electronic Equipment (WEEE) Directive set high recycling and reuse standards for electronic products and implements the EPR model that compels manufacturers to subsidize the collection and recycling of their products. It has increased recycling figures among EU member countries and prompted other areas to approve similar regulation. The Home Appliance Recycling Law in Japan requires consumers to return their electrical products to specific retailers for recycle hence a well-constructed infrastructure for the recycle of e-waste is established in Japan.

Unlike the ITAR, the countries such as the United States do not have an overall national policy regarding e-waste, instead, e-waste management is implemented on state level. It leads to the variation of the availability of recycling programs from one state to the other, while some states have proper recycling programs, others have no e-waste recycling programs at all. While existing in developing nations, there are policies mostly landfilled, but few policies are enforced, and as such, informal recycling occurs with health and environmental consequences. juggling the policies at the international level alongside agreeing at the policy level can minimize the issue of illegal dumping of e-waste and enforce sustainable process worldwide.

Community Engagement and Public Awareness

This paper therefore proposes engagement of the members of the society and awareness that is required in the management of e-waste since this would ensure that people practice proper disposal and take an active involvement in recycling. Organized approached an e-waste donation campaign for collection at local centres, they organize e-waste disposing events and offer recycling awareness to the society. Local groups, including non-profit organizations, are very important in their role in scheduling the collection campaigns, and championing for proper e-waste disposal, and co-operations with recyclers to properly dispose off the collected electronics.

Other strategies are also used with high efficiency for raising public awareness to the proper e-waste recycling issue. Officials and organization can use different channels to pass information to the people on the dangers of e-waste, the benefits of recycling and the right methods to recycle them. In schools and workplaces e-waste education programs can make people aware of the right approach in dealing with the e-waste from the age of ideas development. Collectively, all these broad efforts enhance community participation in e-waste management, decrease electronic waste's effects to the environment, and create a better loop for using and disposing of electronic products.

CASE STUDIES OF E-WASTE MANAGEMENT PROGRAMS

Successful Global Programs

There are successful e-waste management systems in several countries which provide that it is possible to gather, sort, transport and dispose e-waste for its optimum benefit with structured regulated models. A good example is Switzerland, that has a very efficient take-back scheme whereby many electronic producers and retailers provide for the collection and recycling of e-waste. It supplements consumer charges and EPR principles, which make it mandatory for the e-waste to be dropped off, collected from retailers and recycling centres. This system has seen Switzerland among the highest e-waste recycling nations in the globe; this proves that EPR can work in e-waste management.

In Japan, the Home Appliance Recycling Law mandates that manufacturers and retailers facilitate the recycling of four main types of household appliances: appliances such as air conditioners, refrigerator, television, and washing machines. According to this law consumers give back their old appliances to retailers who in turn take them to certified recycling centers. This structured process also means that useful effluence is retrieved and the unwanted waste removed averted the environment. Despite the issues have been observed due to the limitations in the far-reaching logistics in the rural regions of Japan, it has taken this screensaver program towards the greater heights in recycling by eliminating the environmental issues associated with the rapid growth of the e-waste industry.

South Korea has also created a good system on EPR specially in the field of e-waste. Korean program requires financial targets for recycling from the manufacturers and there are body like government to supervise how companies operate. For this purpose, there are recycling stations in South Korea and people are paid to return used electronics. Due to the government's engaging multiple strategies, which this paper has highlighted, South Korea has been able to realised high collection rates, low landfill disposal, and minimal health harm from wrong e-waste disposal.

Comparative Analysis

The management programs of e-waste are dissimilar from each other due to policies and legislations, industrial engagement, and public participation. For instance, EU WEEE directive is a typical government led program, which has specific legal obligations to the collection and recycling of the e-waste. The WEEE directive requires manufacturers of electronics to cover the cost of recycling as well as encourage the recycling of electronics in every EU country, as the highlighted concept of EPR. It has been used in Europe mainly because the collection rates are high, and the recycling industry is highly developed.

On the other hand, manufacturer-based initiatives like the Apple and Dell's programmes are more structured around enterprise-based recognition initiatives and collection returns. Apple's "GiveBack" program persuades consumers to return their used devices back to.bp Apple can recycle most of these materials internally and reclaim valuable metals for manufacturing new ones. This closed-loop plays part and partial to Apple's sustainable development goals, and shows how the private sector is taking up the mantle in e-waste recycling where it might not be funded by the government.

Other stakeholders involved in the management of e-waste are the non-governmental organisations; especially in developing countries. pdf In nations where there is no formal recycling mechanism, advocacy groups such as Toxics Link in India educate the public, collection services are arranged, and recyclers are identified to reclaim e-waste properly. Despite these limitations, the cases show that NGOs should continue to play the leading role in tackling e-waste problems where government actions are scarce, and community engagement is crucial.

Results and Impact Analysis

Studies on some of the effective e-waste managements have shown that they have triggered positive environmental and economic impacts, a testimony of the fact that structured e-waste recycling has the potential to lower the global e-waste index. In Switzerland, the take-back system based on EPR has produced a recycling rate higher than 75% of such WEEE, thus lowering the quantity that is dumped in the disposal areas while preserving useful metals such as copper and aluminum and metals regarded as rare. Furthermore, the consequences of recycling as compared to landfilling regarding e-waste are significant because the recycled materials diminish the emission of hazardous substances and preserve sources that need a lot of energy for their provision. In Japan, the Home Appliance Recycling Law has been effective to promote recycling millions of appliances each year it has been implemented and there are evidence that support resource recovery. This situation extends to metals, and analysing the recycling attribute, it becomes clear that about 95% of metals from appliances fall under this category, hence supporting Japan's recycle oriented economy to cut on importation of the raw materials. This has not only helped environmentally to minimize impacts such as pollution but has also brought economically benefits as it has started creating employment in the facilities that deal with recycling and has thus developed an independent recycling market within the country.

Collection rates of e-waste in South Korea have also shown environment returns with the country's EPR based system that has continued to record increases, since the initiation of the system. Reducing the e-waste disposed of in South Korea landfills through achieving high recycling standards helps avoid pollution and lower greenhouse gas emissions otherwise emitted from the normal dumped waste. The economic returns are also remarkable because the call for the recycling exercise relieves the pressure on the raw material demand while creating employment opportunities and saving on resources. The two cases demonstrate that highly structured e-waste management systems will hold great returns, environmentally and financially.

Employing action-oriented economic incentives, these programs present other countries to aspire to by engaging in or refining their pursuit of e-waste collection and recycling programs to mitigate the persisting dependency on landfills for disposal of e-waste and to develop economically efficient recycling systems.

RESULTS AND DISCUSSION

Analysis of Integrative Methods

This paper has also shown that when technological support, legal frameworks, and people's involvement augment e-waste management efforts, the impacts improve considerably. Paper under analysis and case studies based on such countries as Japan, Switzerland, and South Korea show that there is no single solution that can help to collect, recycle, and dispose of e-waste effectively in the shortest amount of time. The end-use sector recovery separated using automated sorting and the development of a recycling method enhances the process and reduces environment outcomes. Some of the endorsements include legal instruments such as the EPR policy and legislation that mandates recycling targets from producer-end drives interest in e-waste management. As illustrated in the Figure 3 depicting the view of the collection rate by country The rate of collection and recycling of e-waste is considerably high in the countries that have rigid framework for regulating e-waste and countries having advance technologies.

Figure 3. Collection rate comparison across countries

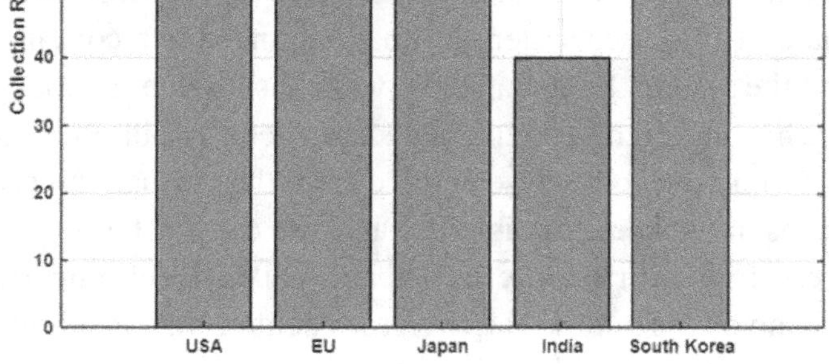

Public awareness also helps a lot, as depicted in Figure 4 which displays awareness level per each region. Therefore, high public awareness ensures appropriate uses of the products and disposal that escalates the rates of recycling. There are also other approaches that support the main policies involving regulations and technologies: collections and awareness programs at lower levels of the communities. Complementing these approaches copes with diverse aspects of e-waste management, including minimizing waste production through sustainable design to make discards recyclable and restricting dangerous disposal. Not only does this strategy enhance as much resource recovery as is evident in Figure 5 but also promotes economic returns in terms of employment in the recycling industry.

Figure 4. Public awareness level on e-waste in different regions

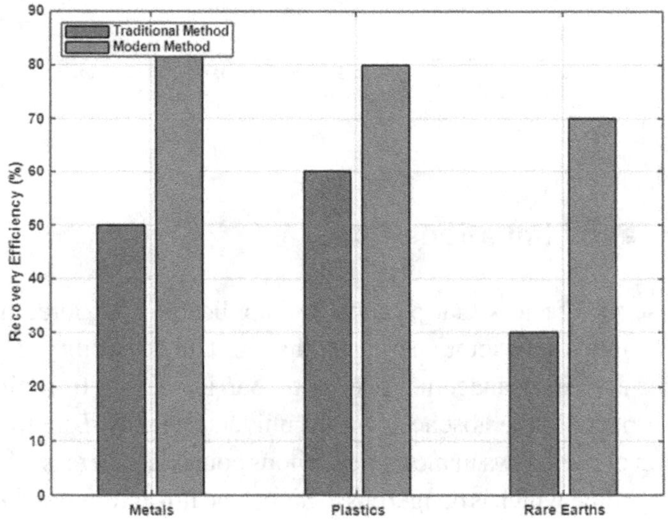

Figure 5. Resource recovery efficiency for different materials

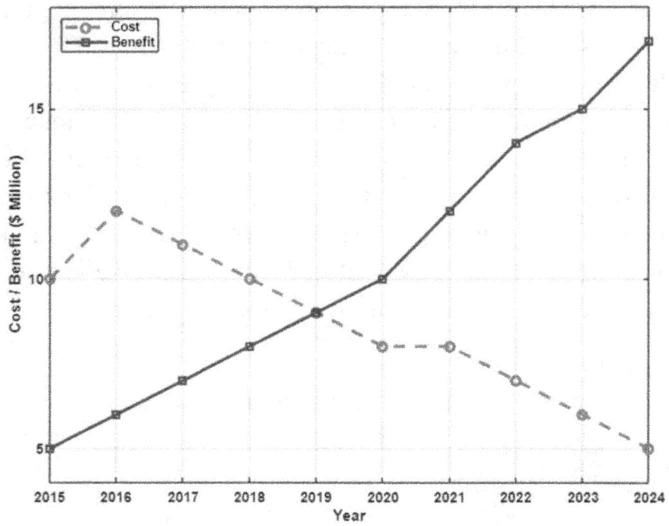

Challenges and Limitations

However, several factors that promote the applicability of integrative e-waste management methods have some challenges that limit the scalability of the solutions. In cross-border e-waste management common barriers ensue from the regulatory requirement, especially the absence of a set universal policy. Due to laxity in the implementation of these laws in most endogenous countries, there is widespread use of informal recycling, which is dangerous to human health and causes environmental pollution. There are also economic challenges that hinder the e-waste programmes, because; creating elaborate collection infrastructures and adopting smart sorting systems such as the ones depicted in Figure 8 demands a massive amount of capital investment. For this reason, these economic factors can limit the implementation of today's progressive techniques of recycling in the regions of shortages in financial and material support.

Free social challenges make the development process even more difficult in the management of e-waste. This awareness helps but remains low in different parts of the globe. Accordingly, the public awareness level is low, meaning that consumers are ill informed about how to dispose e-waste correctly and, therefore, recycling rates are low too, as Figure 5 illustrates. Further, these ISRs underpin formal recycling industries; formally, they decrease the collection shown in Figure 3 and resort to environmentally unsafe practices that escape regulation. All these constitute the global imperatives of policy harmonization, demands for economic assistance and large-scale sensitization campaigns to overcome the qualities of the barriers to the use of integrative e-waste management methods thereof.

Future Trends

This chapter established that current trends in e-waste management are characterized by improved, technological, and utica processes. AI is making sorting and processing different as; in Figure 6, advanced technologies are handy in the categorization and deconstruction of the electronic waste. With sorting done using AI, recovery efficiency is boosted as is evidenced by Figure 5 above; in addition, the process eliminates contamination that takes place in the recovered material therefore boosting the quality of recycled products. Also, the use of a blockchain system to track e-waste through its life cycle is on the increase. In this respect, the idea of accountability that has originated from the use of the blockchain contributes to such regulatory issues as combating against the process of illicit dumping that has been described in Challenges and Limitations.

Figure 6. E-waste processed per year by technology type

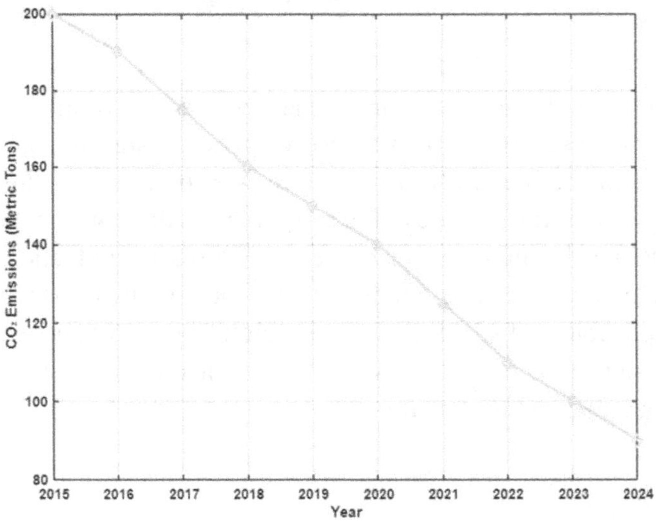

The principles of circular economy as in the conception of sustainable design for reuse and efficiency in resource utilisation are also becoming central to the present-day e-waste management framework. Because circular practices are implemented around the manufacturing loop, manufacturers design their products with a longer lifespan and think about their recyclability, limiting the amount of e-waste produced. From the life-cycle perspective it is clear that circular economy practices have the potential of reducing emission levels which is illustrated in the graph in Figure 7 below. Such outcomes show the emerging better practices and management of e-waste by focusing on resource productivity and environmental sustainability.

Figure 7. Reduction in environmental footprint (emissions) over time

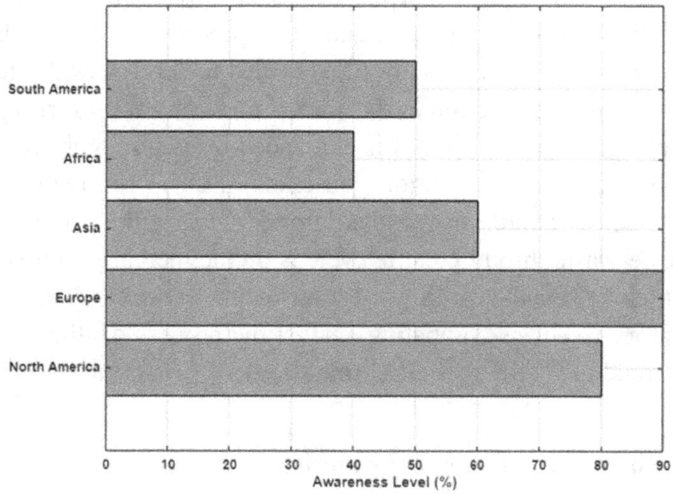

In addition, as indicated in Figure 8, the revenue opportunities of integrating modern technologies and CE principles in e-waste sector are coming into focus. The formal analysis of such trade-offs brings out the idea that although investments into the technology and infrastructure at the early stages will be substantial, the sum of the value generated by minimisation of resource losses and reduction of dependency on raw materials in the long run exceeds the total implied costs. In general, the integrated applicability of AI, blockchain, and circular economy novelties proclaims an optimistic future direction in e-waste management to serve sustainable ideologies to the existing issues in this domain.

Figure 8. Cost-benefit analysis of recycling processes

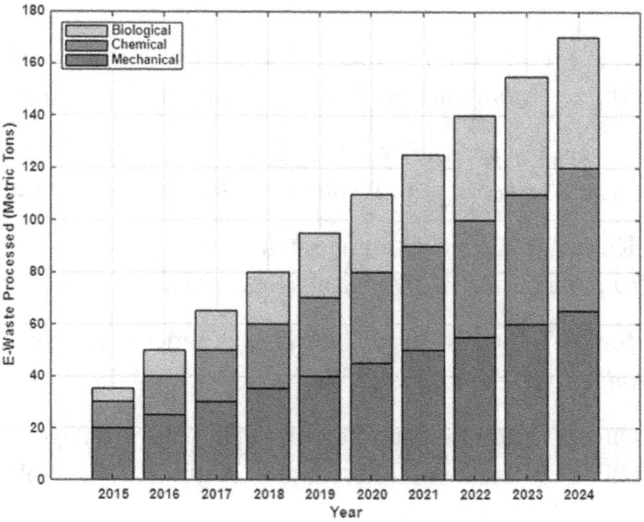

CONCLUSION

In Conclusion, this chapter shows that applying advanced technology, legal measures and the involvement of the community to manage e-waste in an integrated manner leads to improvement in both the performance and the impacts on the environment. This result shows that countries with sound and integrated policies and technologies concerning recycling collect more resources and with higher returns, thus emphasizing the significance of formatted strategies. Some of the outcome includes; efficiency gains in recovering more material using today's technology, dramatic reduction in emission in the long-run, high levels of economic returns accruing from material reclamation. There is much potential for new developments in e-waste management for instance the application of AI and blockchain to aid sorting and tracking and the use of circular economy principles. The aimed developments also not only enhance the performance of recycling but also prioritize sustainable design, aiming to minimize wastes. Future innovations in these fields can therefore mitigate present challenges by providing better sustainable, scalable, and globally synchronized e-waste management avert environmental and human health degradation related to resource misuse.

REFERENCES

Agrawal, R. (2023). Industrial Automated Multipurpose Robot Using WIFI. *Proceedings of the 4th International Conference for Emerging Technology (INCET)*, 1-8.

Ahmed, M., & Jones, C. (2020). E-Waste in Developing Countries: A Challenge for Sustainability. *The Journal of Development Studies*, *36*(4), 104–121.

Ahmed, S., & Kumar, R. (2018). Community Involvement in E-Waste Management: Case Studies from Asia. *Journal of Environmental Sciences (China)*, *32*(5), 573–589.

Brown, M., & Smith, T. (2017). Sustainable Design and E-Waste Reduction. *Journal of Cleaner Production*, *142*(1), 327–339.

Chen, L., & Zhang, Y. (2021). AI and Robotics in E-Waste Management: Enhancing Recovery Efficiency. *Robotics and Automation Journal*, *29*(2), 189–204.

Chen, Y., & Wang, X. (2020). Recycling Technologies for E-Waste: Challenges and Opportunities. *Waste Management & Research*, *38*(4), 445–460.

Evans, K., & Walker, G. (2019). Blockchain for E-Waste Tracking and Transparency. *Journal of Technology Management*, *56*(4), 288–302.

Gummadi, A. K., & Damacharla, P. (2024). An Intelligent Framework for Energy Management in IoT Networks Using Machine Learning. *Proceedings of the 5th International Conference on Smart Electronics and Communication (ICOSEC)*, 765-771.

Hegge, K. K., & Agrawal, R. (2023). Wireless and Solar-Powered Multipurpose Robot for Industrial Automation: A Sustainable Solution. *Proceedings of the 8th International Conference on Communication and Electronics Systems (ICCES)*, 146-151.

Johnson, L., & Roberts, P. (2019). The Role of Extended Producer Responsibility in E-Waste Management. *Journal of Environmental Management*, *25*(3), 256–267.

Kim, S., & Park, J. (2019). Extended Producer Responsibility and Its Impact on E-Waste Collection Rates. *Environmental Policy Review*, *35*(2), 98–112.

Kim, Y., & Lee, J. (2021). Public Awareness and Participation in E-Waste Recycling. *The Journal of Environmental Education*, *42*(1), 77–90.

Kudari, J. M., & Samal, A. (2021). Smart Agriculture and Smart Farming Using IoT Technology. *Journal of Physics: Conference Series*, *2089*(1), 012038. DOI: 10.1088/1742-6596/2089/1/012038

Liu, F., & Chen, Z. (2020). The Influence of AI on E-Waste Sorting and Processing. *AI and Environmental Science*, *44*(1), 23–37.

Martinez, D., & Anderson, B. (2018). The Environmental Impact of E-Waste Disposal. *Journal of Environmental Protection*, *22*(3), 355–371.

Patel, R., & Singh, L. (2018). Regulatory Frameworks in E-Waste Management: Comparative Analysis. *International Journal of Environmental Law*, *19*(3), 150–168.

Pavankumar, P., & Darwante, N. K. (2022). Performance Monitoring and Dynamic Scaling Algorithm for Queue Based Internet of Things. *Proceedings of the International Conference on Innovative Computing, Intelligent Communication and Smart Electrical Systems (ICSES)*, 1-7.

Puneeth Kumar, B. S., & Srinivasulu Raju, S. (2021). Smart Villages: IoT Technology Based Transformation. *Journal of Physics: Conference Series*, *2070*(1), 012128. DOI: 10.1088/1742-6596/2070/1/012128

Ramesha, M., & Pai, G. N. (2020). Internet of Things: Internet Revolution, Impact, Technology Road Map, and Features. *Advances in Mathematical Sciences Journal*, *9*(7), 4405–4414.

Reddy, S. P., & Patwal, P. S. (2022). Data Analytics and Cloud-Based Platform for Internet of Things Applications in Smart Cities. Proceedings of the International Conference on Industry 4.0 Technology (I4Tech), 1-6.

S, K., & Arun, M. R. (2022). Priority Queueing Model-Based IoT Middleware for Load Balancing. Proceedings of the 6th International Conference on Intelligent Computing and Control Systems (ICICCS), 425-430.

Sharma, A., & Gupta, N. (2020). The Circular Economy in Electronic Waste Management: A Global Perspective. *Sustainability*, *12*(14), 5678.

Shruthi, M., & Naveen, K. B. (2020). Internet of Things: A Survey on Devices, Ecosystem, Components, and Communication Protocols. *Proceedings of the 4th International Conference on Electronics, Communication and Aerospace Technology (ICECA)*, 611-616.

Singla, A., & Sharma, N. (2022). IoT Group Key Management Using Incremental Gaussian Mixture Model. *Proceedings of the 3rd International Conference on Electronics and Sustainable Communication Systems (ICESC)*, 469-474.

Wang, J., & Zhou, H. (2019). Economic Benefits of Advanced E-Waste Recycling Technologies. *Journal of Economic Sustainability*, *15*(5), 502–515.

Zhang, H., & Li, D. (2021). Advances in Resource Recovery Technologies for E-Waste. *Resources, Conservation and Recycling*, *168*(1), 105254.

Chapter 3
Advancing E-Waste Management Through Circular Economy Principles: A Pathway to Sustainable Innovation

P. Deepika
https://orcid.org/0000-0003-2009-5935
R.V. College of Engineering, India

Neeta B. Malvi
https://orcid.org/0000-0002-5477-5878
R.V. College of Engineering, India

B. Sahana
https://orcid.org/0000-0002-7232-7286
R.V. College of Engineering, India

Likith Gowda
https://orcid.org/0009-0004-3090-5174
R.V. College of Engineering, India

B. J. Harshitha
https://orcid.org/0009-0004-0885-9669
R.V. College of Engineering, India

DOI: 10.4018/979-8-3693-7383-5.ch003

ABSTRACT

Our world has been seriously disturbed by the problem of electronic waste, or e-waste, due to the fast pace of our digital world. The inclination of consumers has increased due to faster and newer innovations, which have accelerated more quickly than expected. This has led to ever-decreasing product lifespans and higher electronic consumption. As a consequence, e-waste is among the wastes that is rapidly accumulating in the world. By 2019, the global generation of e-waste already exceeded 53.6 million metric tons, and by 2030, it is predicted to increase up to 74 Mt. Such an increase of e-waste raises critical environmental and health hazards as well as economic losses because when hazardous materials are not disposed of, resource content gets wasted. Therefore, there is a rapidly increasing desire to introduce the circular economy model to transform the design, production, use, and end-of-life processing of electronic products. The circular economy emphasizes product longevity, longer life cycles, and encourages the reuse and recycling of product materials.

1. INTRODUCTION

1.1 The Global Rise of E-Waste: Scope and Consequences

In recent decades, it has been demonstrated that advances in electronics have dramatically changed the face of industries, lifestyles, and economies around the world. No longer just simple mobile phones and laptops, these electronic devices in households and medical tools have mushroomed in numbers. Such reliance on electronic devices, however, comes with a stupendous cost-the rapidly piling up electronic waste. E-waste is defined as discarded electronic devices no longer usable, wanted, or functional. This category of waste ranges from smaller personal gadgets to large electrical appliances and industrial equipment.

In 2019, 53.6 million metric tons of e-waste were generated worldwide, which will increase by 30% within the next decade (Forti et al., 2020). The fast growth of e-waste is attributed to various factors, such as having short lifecycles of modern electronics, a boom in consumer demand for new technology, and the last but not least constant innovation cycle that encourages frequent upgrading. For example, the life cycle of smartphones is just about two or three years; therefore, people require new models to outperform the previous ones. The progression in technology requires millions of more aged gadgets and literally chucks them to the e-waste annually.

Exponential growth in e-waste has serious environmental and health implications. E-waste contributes to hazardous decays in landfills and incineration, releasing lead, mercury, cadmium, and brominated flame retardants into the ecosystems where it increases toxicities in soils, water, and air, hence causing serious health impacts on human and wildlife health. Moreover, most of the discarded electronics contain valuable resources such as gold, silver, palladium, and important raw materials, like cobalt and rare earth elements. Economic losses amount to enormous sums because unutilized values end up in landfills or through basic and harmful recyclable processes.

Issues generated from e-waste are not only environmental but also economic. An estimation report reveals that the raw materials contained in global e-waste in 2019 had an economic value of around $57 billion (Forti et al., 2020). Yet only 17.4% of global e-waste is formally collected and recycled. The remaining parts of the economic potential of these resources go unrecovered. Such depletion of natural resources, increase in the costs of production, and increased vulnerabilities in the supply chain especially to manufacturers of rare and critical elements indeed result from failure to recover valuable materials from discarded electronics.

1.2 The Limitations of the Linear Economic Model

A linear economic model defines the traditional approach of managing resources and production. In this model, raw materials are extracted, processed into products, sold to consumers, and eventually discarded after being used. Such a "take-make-dispose" system has been fueling industrial growth for centuries; however, its intrinsic inefficiencies have created a serious environment as well as economy chockablock. So, as industries grow and global consumption increases, the reliance of the linear model on finite resources and its inability to manage waste effectively is quickly becoming unsustainable.

With e-waste, the constraints of the linear model are particularly obvious. Technological innovation moves at a pace that is fast; the price of electronics has come tumbling down, thereby initiating a cycle of frequent product obsolescence. Instead of repairing or upgrading used products, new products replace much of the old stock. Manufacturers continue to look for ways to capitalize on such consumer trends. Devices with shorter lifespans and less repairability are designed, aggravating the problem of e-waste. For instance, proprietary components and sealed designs make it difficult or impossible for consumers to get a repair for their smartphone or a laptop, which makes it hard for consumers to replace particular parts. In the end, minor problems often lead to the disposal of the whole device.

Another aspect adding to the problem is the fact that proper e-waste recycling infrastructure is often lacking in many parts of the world. The developed economy-land countries like the USA or Japan have good collection and recycling systems for e-waste, while the rest of the world, especially the low- and middle-income countries, depends on the informal recycling sectors. Manual processing of e-waste in such informal settings exposes workers to hazardous chemicals and also causes environmental pollution. This is because the linear model tends to contribute to the growing accumulation of e-waste through short product lifecycles, whereas inefficient recycling systems coupled with poor options for repair can lead to loss in valuable resources.

1.3 Circular Economy: A New Framework for Sustainable E-Waste Management

The CE is a novel paradigm in the use and management of resources. While the traditional linear model "take-make-dispose," the circular economy aims at "closing the loop through extension of product life cycles, recovery of valuable materials, and avoidance of waste." In short, products should be used up and repaired but then recycled: that lowers the environmental impact of production and disposal based on the principles of CE.

CE promises to solve the problem of e-waste, changing the design, use, and handling of electronics at their life end. The main focus areas of the CE approach are:

1.Design for durability and recyclability: Electronics must be designed as modular, repairable, and recyclable. This increases the life cycle of products and eases the recovery of valuable materials at the end of a product's life.
2.Promote refurbishment and re-packaging: CE enables the emergence of a 'used or second-hand' market for refurbished and updated products. It helps to decrease demand for new products and thus reduces the overall quantity of e-waste.
3. Recycling systems: CE-based models highlight "the development of closed-loop technologies" that can be utilized in an effort to recover value-added material from waste electronics. Recycling efficiency increase minimizes the extraction and processing of raw materials, while also reducing environmental footprints for the production and manufacturing process.

By applying all these principles, CE is offering an all-encompassing holistic approach to sustainable e-waste management without causing any further damage to the environment and appropriate economic capture of the value in used electronics. The use of CE in e-waste management also requires dramatic changes in the design of products, consumer behavior, and recycling infrastructure. Importantly, it is also looking for entry support from regulatory frameworks and policy initiatives that create incentives for manufacturers and consumers to undertake circular practice.

2. OVERVIEW OF THE E-WASTE CRISIS

2.1 Global E-Waste Trends

The sheer scale of the global e-waste crisis itself calls for urgent need management solutions. In 2019, the world produced 53.6 million metric tons of e-waste. With the expected growth to 74 Mt by 2030, as calculated by Forti et al. (2020) several related factors are driving this increase:

1. Technological Innovation and Short Product Lifecycles Since technology innovates rapidly, so does the upgrading of new devices featuring more capabilities in the market. Added to this is the desire of consumers for the newest gadget, which makes product replacement very common. Devices like smartphones, tablets, and laptops often get discarded after mere years of usage, adding significantly to the rising e-waste stream.
2. Higher Consumption of Electronics Globally: The middle class is growing each day worldwide, primarily in developing nations. Greater development and higher incomes lead to greater consumption of electronics. The more electronic items that people buy, the higher the e-waste that is going to be produced. In 2019, the average weight of e-waste produced per capita in high-income countries was 16.2 kg whilst the low-income countries generated much lesser (Vanessa Forti, 2020). In the future, however, as the low-income countries continue to develop economically, e-waste generation will also increase.
3. Offer Less Opportunities for Repair and Rehabilitation Modern electronics designs pay a much higher attention towards visually appealing aesthetics and size reduction than towards possible repair options. Some units are even sealed using proprietary units of components that cannot really be fixed or can be infeasible. It means that the devices will mostly be replaced rather than repaired, adding onto the e-waste pile.

2.2Motivation

The increased use of electronic devices has resulted in an enormous volume of electronic waste (e-waste) that is causing serious environmental and economic issues. Conventional methods of waste management are insufficient to address the intricacy of e-waste, which is comprised of toxic materials and worth recoverable resources.

The driving force for this study is to search for sustainable e-waste management practices that are in accordance with circular economy principles.

The main reason behind this principle is to reduce environmental pollution, maximize resource recovery, and ensure responsible disposal. The adoption of circular economy approaches—reuse, recycling, and refurbishment—has the potential to decrease landfill waste, reduce carbon footprints through raw material demand, and curb the negative impacts of uncontrolled e-waste disposal. The role of new technologies like AI, IoT, and blockchain in providing effective tracking and management solutions is also critical in streamlining e-waste recycling.

2.3 Environmental and Health Impacts of E-Waste

Improper disposal of e-waste could badly damage the environment and human health. A lot of e-devices contain harmful substances including lead, mercury, cadmium, and brominated flame retardants. In such cases leakage of these elements into the environment because of dumping in landfills or unsafe processing would contaminate the soil and water with atmospheric air. For example, lead, which is a compound contained in most old electronics, can be harmful neurotoxin to children once the compound enters the human body through infected water or air.

In most of these low- and middle-income countries, e-waste is usually processed informally with basic methods such as open burning or acid baths to extract valuable materials. Such practices expose the workers to toxic fumes and chemicals and other serious health consequences, such as respiratory problems, cancers, and reproductive problems. It has also been connected to environmental degradation because hazardous chemicals are released into the atmosphere and waterways, hurting ecosystems and biodiversity.

3. LITERATURE REVIEW

3.1 Sustainable E-waste Management and Circular Economy Practices

The article "A Systematic Review on Sustainable E-waste Management: Challenges, Circular Economy Practices, and a Conceptual Framework" by Gaur, Yadav, Mittal, and Sharma (2024) performed a systematic review of literature to determine challenges and circular economy (CE) practices in sustainable e-waste management. The research identified seven main challenges namely regulatory policies, infrastructure shortcomings, consumerism, informal recycling sector, cultural influences, technology constraints, and economic limitations (Guar, T.S., 2024). In efforts to resolve these challenges, eight CE practices were considered: reuse, recycling, remanufacturing, refurbishment, repair, reduction, recovery, and repurposing. The authors developed a conceptual framework combining the practices to provide increased sustainability in the management of e-waste.

3.2 E-waste Management in India: A Circular Economy Perspective

Veena, Kumar, and Singh (2023) conducted the research "E-Waste Management: A Circular Economy Perspective," which focused on e-waste management in India from a circular economy perspective. The research highlighted the economic significance of materials contained in electrical and electronic equipment and how these materials needed to be reused to stimulate economic growth. Besides, the paper outlined methods of reducing land waste through recycling electromechanical system waste and sustainable solar panel recycling processes as part of environmental sustainability (N. Veena, 2023).

3.3 Enabling Technologies for Effective E-waste Management

The study "Enabling Technologies for Effective E-waste Management" by Chatterjee and Kumar (2023) examined how emerging technologies contribute to increasing e-waste management efficiency. The paper explained the environmental risks of the improper disposal of e-waste and recognized the enabling technologies including the Internet of Things (IoT), Wireless Sensor Networks (WSN), blockchain, and artificial intelligence (AI) as possible solutions (S. Chatterjee, 2023). The authors have suggested a framework combining these technologies to solve the problems in managing e-waste and proposed best practices and research directions for the future.

3.4 Global E-Waste Management: Urban Mining and Circular Economy

Murthy and Ramakrishna (2022) presented the paper "Global E-Waste Management: Urban Mining and Circular Economy." They carried out an extensive review on global e-waste management with a focus on urban mining as a steppingstone towards sustainability and the circular economy. The research focused on the environmental risks associated with illegal e-waste disposal and proposed incorporating urban mining strategies to reuse valuable materials from abandoned electronic waste (S. Murthy,2022). The review also emphasized the need for policy enforcement, technological innovation, and social consciousness in attaining a circular economy in e-waste management.

3.5 Digital Twin Technology for E-waste Tracking and Reduction

The research "Digital Twin Technology for E-waste Tracking and Reduction" by Singh, Gupta, and Mehta (2022) proposed the usage of Digital Twin Technology (DTT) in tracking and optimizing the process of e-waste management. The research showed how DTT can be used to achieve real-time monitoring, predictive analytics, and lifecycle assessments of electronic products, thus enabling proactive e-waste reduction strategies. The authors suggested further research on the scalability of DTT for global e-waste management systems (P. Singh, 2022).

3.6 The Role of AI and Machine Learning in E-waste Sorting

Fernandez, Kim, and Park (2023) conducted the paper "The Role of AI and Machine Learning in E-waste Sorting" that examined the application of artificial intelligence (AI) and machine learning (ML) to enhance e-waste sorting and recycling effectiveness (L. Fernandez, 2023). The research delved into AI-based automated sorting systems that sort e-waste materials by composition and toxicity content. The authors emphasized the promise of AI-based sorting to increase recycling rates and decrease dependency on manual labour but also noted issues of high costs of implementation and additional training of AI models.

3.7 Research Gaps and Future Directions

- Urban Mining Implementation: Although urban mining has been identified for its ability to recover valuable resources, its effective implementation and scalability are still in its infancy. Additional research must be conducted to establish effective urban mining processes and incorporate them into existing waste management infrastructures.
- Technological Integration: The enabling role of technologies like IoT, blockchain, and AI in e-waste management is recognized; nonetheless, their mass adoption and incorporation into proper recycling streams remains in early development stages.
- Policy Enforcement: Although e-waste legislation exists in many regions, enforcement and compliance remain challenges. Developing standardized global policies and ensuring their implementation is crucial for effective e-waste management.
- Consumer Involvement: Increasing consumer involvement and awareness about e-waste recycling programs is critical. More work needs to be done on strategies to encourage responsible disposal and recycling habits.

- Scalability of AI-Based Sorting: Although AI and ML technologies have indicated potential for sorting e-waste, their prohibitively high installation costs and lack of adequate training data are deterrents to broad implementation (L. Saha, 2021).

4. CIRCULAR ECONOMY PRINCIPLES APPLIED TO E-WASTE MANAGEMENT

Transition from a linear economy to a CE in the management of e-waste is important to mitigate the environmental and social impacts associated with electronic waste. A circular economy is about thinking this linear economic model to reduce waste and increase efficiency in the use of resources, not creating a cyclic system where material from the same product can be reused or remanufactured or recycled. Industrial application of CE principles will assist industries in reducing their ecological footprint and discovering new business opportunities. The following section presents an overview of how the circular economy principles are applied while designing and innovating electronic products and of creating new business models and technological innovations that will revolutionize the recycling of e-waste.

4.1 Product Design and Innovation: Enabling a Circular Approach

One of the most important features to construct circular economy principles in furthering e-waste management right into the design of electronic products. The linear model of "take, make, dispose" needs an end and replacement by sustainability all through the product lifecycle through durable, modular, recyclable, and sustainably sourced electronic devices.

Design for Durability

The most apparent but also the most efficient method of having less e-waste is by designing for durability. It expands the product's lifetime because it will last for a longer time. This can actually be achieved through either using more durable materials, longer warranty periods, or even repair services. Social enterprise Fairphone already builds smartphones with durability in mind. Its devices are built from high-quality materials that are less likely to fail (C. Cole, 2019). The company also offers a five-year warranty; thus, this will make the customers keep their phones for longer periods. Therefore, in the general demand for new devices, it is declined and thus reduces the generation of e-waste.

Another aspect of durability is the software aspect. Most electronic products become obsolete not because the hardware fails, but because of incompatibility or because of a software update. Ensuring that the software can be updated without the use of new hardware is one other means through which the product life of electronic products could be increased, which would again translate to a reduction in e-waste.

Modular Design

The other critical aspect of circular design is modularity because it allows for the replacement of many parts of the product instead of wasting the whole device. The approach toward the modular design of products enables the replacement of parts like the battery, screen, or storage for memory. It thus avoids the replacement of the whole product and the waste of e-waste.

Another excellent example of modular design is the series of laptops developed by Dell Latitude 5000, which ensured exchange and upgradation of the hard drive, memory, and battery without tooling; hence, the product would remain relevant for a long time without wasting its value when one of the components fails. From an environmental perspective, the modular product undoubtedly does well since it reduces material extraction, manufacturing energy, and e-waste disposal.

End-of-Life Recyclability

Design for recyclability secures that the products are recycled safely and efficiently at the end of their lives. Materials combined in electronic devices include plastics, metals, and hazardous materials which could raise complications in the recycling process. The designers need to opt for separable materials and reduce hazardous contents. HP's DfE program is an excellent illustration of how companies can make their products recyclable (B. A. Hakami,2018). HP will design its printers and computers with recyclable plastics and very minimal toxic parts. For example, HP makes its ink cartridges from recovered plastics, while its printers are easily taken apart at the end of their useful life for end-of-lifecycle recycling. This will minimize the impact of e-waste towards the environment and will encourage material circulation.

Closing the Loop in Electronic Product Manufacturing

The concept of closing the loop is used to describe the generation of new products using recycled material. Demand for virgin resources is reduced, energy consumption in the extraction process of raw materials decreases, and consequently the connected environmental impacts of the production process are minimized. Setting an example within the industry, Apple has committed itself to using 100 percent recycled aluminium in every MacBook model. Recycling aluminium from old devices reduces Apple's dependence on mining and processing virgin material. In 2020, Apple produced 30% fewer carbon emissions in producing devices from recycled material. Such interventions underscore the integration of recycled material in product manufacturing as enablers for a circular economy.

Design for durability, modularity, recyclability, and use of recycled materials must be the foundation of all changes in how we do business with e-waste. These principles will ensure that the products become stable, more readily repaired and upgraded, and even recycled-helping to reduce the generation of e-waste globally.

4.2 Business Models for a Circular Economy

While product design for sustainability is important, an equally important and in many ways more important strategy is that of innovative business model that facilitates circularity throughout all stages of a product's life. Ownership-based models conceived by traditional thinking favor the generation of more waste: consumers buying and owning products until obsolescence and then discarding them, while circular business models serve to extend product lifecycles, reward

Collection Programmes and EPR

Extended Producer Responsibility is a form of producer responsibility tool through regulation wherein producers are made liable for the whole lifecycle that includes post-consumer disposal. Under such an EPR arrangement, producers are mandated to collect back the product at the end-of-life or pass it on for collection and proper treatment either by recycling or refurbished products.

The best example of EPR would be the Waste Electrical and Electronic Equipment (WEEE) Directive of the European Union. The WEEE Directive sets a legal requirement that demands these companies to collect as well as recycle electronic waste in an environmentally sound manner. Application of EPR creates favourable conditions for companies to design products that are easier to disassemble and recycling, thus business conduct lying with circular economy principle (M. Ikhlayel,2017). Take-back schemes reduce e-waste and provide useful materials to the companies in such a way that the materials acquired can be reused in new products.

Refurbishment and Resale

The other business model which supports circular economy principles is refurbishing old electronics with an intent to resell them. It ensures that used devices are repaired and refurbished to be resold, usually at a lower price point.

Some of them have managed to gain only recently. It is said that Best Buy and Dell have created successful refurbishment programs. Dell's Refurbished Products Division collects returned laptops and desktops, reconditions them, and then sells them at extremely low prices. This model reaps all benefits for the consumer-the source for lower-cost electronics-while keeping devices in use much longer, thereby reducing the demand for new products and generating e-waste.

Material Recovery as Business Model

Besides the above advantages, circular economy helps companies who carry out material recovery and recycling processes since they extract valuable materials from e-waste, such as gold, silver, and platinum to reintroduce it to the cycle of production.

Umicore is one of the biggest companies working in that area. It is a Belgium-based company for materials technology, there using very advanced technologies extract precious metals from electronic wastes. This way, by recycling materials that otherwise would be disposed of, Umicore shows that e-waste recycling may both have an ecological security and be profitable. Once companies specialized in material recovery take advantage of constantly growing demand for precious metals and the source of profit these offer in opportunities opened up by e-waste.

The core business models of the circular economy are Product-as-a-Service, take-back schemes, refurbishment, and material recovery. These represent a transformation from consumption and disposal to sustainability and resource efficiency, ushering in a more resilient economic system and ultimately helping businesses, consumers, and the environment.

4.3 Technological Advancements in E-Waste Recycling

This paper explores a list of technological innovations that can handle e-waste as given in Table1. Advanced technologies can optimize recycling processes, recover valuable materials more efficiently and provide greater transparency and accountability in e-waste supply chains. Some emerging technologies transforming the recycling of e-waste are discussed in this section.

Table 1. Comparative study of E-waste management technologies

Technology	Key Features	Advantages	Challenges	Applications
Artificial Intelligence (AI) & Machine Learning (ML)	Automated sorting of e-waste, predictive analytics, material classification	Increases sorting efficiency, reduces manual labour, improves recycling rates	High implementation costs, requires extensive training data	Automated recycling plants, AI-based sorting facilities
Internet of Things (IoT)	Smart bins, real-time tracking of e-waste movement	Enables efficient collection and tracking, improves logistics	High initial setup costs, requires stable network infrastructure	Smart e-waste collection systems, inventory management
Blockchain Technology	Secure and transparent e-waste tracking	Ensures data integrity, prevents illegal disposal, improves traceability	Scalability issues, high energy consumption	Digital e-waste tracking, supply chain management
Digital Twin Technology (DTT)	Virtual representation of devices, lifecycle analysis	Real-time monitoring, predictive maintenance, better resource management	High computational requirements, integration challenges	Product lifecycle management, smart waste tracking
Urban Mining	Extraction of valuable metals from e-waste	Reduces dependence on raw materials, promotes sustainability	Requires advanced processing techniques, high energy consumption	E-waste recycling plants, material recovery facilities

Robotics and Automation

This is where dismantling of electronic devices in e-waste recycling presents a significant challenge: a collection of components and materials mostly not separable by hand. But robotics and automation technologies have proven to answer such a challenge.

Companies like Apple, for example, have led the march into the futuristic automation of e-waste recycling, as exemplified in systems like Daisy, which physically break down iPhones to help recover higher-valued material.". Using Daisy 200 iPhones can be taken apart within an hour. Among the materials that are potentially recoverable from this list, cobalt, lithium, and aluminum can probably be reused as new products. The automation system of Daisy will minimize human labor time and even save much time for the recycling of electronic products. Hence, it will increase efficiency and reduce the cost associated with this process.

5. BLOCKCHAIN FOR SUPPLY CHAIN TRANSPARENCY

Blockchain-based technology has come out as an epitome solution to the critical problem of transparency in the supply chain for e-waste recycling. The collection, processing, and recycling of e-waste in responsible and ethical manners are among the leading concerns in e-waste management. In general, the antiquated management systems of supply chains are not transparent or traceable enough in order to prevent mishandling or clandestine exportation to areas not sufficiently protected with regard to environmental safety (R. Rautela, 2021). It is here that blockchain technology shines through with its decentralized, irreversibly immutable ledger.

5.1 Benefits of Blockchain for E-Waste Management

The most beneficial aspect of blockchain, in this regard, is the traceability of materials-from the source all the way to the destination-from collection points or manufacturers, for example, all the way to recycling centers or refurbishment plants. Companies can ensure that every step in the e-waste recycling process through blockchain, making it less likely that materials lose their way, get mishandled, or otherwise illegally exported.

Improved Traceability

Blockchain technology enables the traceability of e-waste from collection to final recycling stages. Every transaction starting with e-waste collection by a consumer or transport waste to recycling centre/processing of material is recorded and verified in real-time on the blockchain. This brings transparency at every step on the supply chain. Example: A firm can look at whether a discarded cellular phone that was gathered in Europe was indeed recycled or diverted to a developing country with liberal environmental laws.

Reducing Illegal Dumping

E-waste illegal dumping is one of the main issues globally, and massive amounts of e-waste are shipped in large quantities to developing nations for unsafe recycling. The benefit blockchain technology will offer in this respect is an open system of tracing the waste materials such that illegal shipments can be easily caught and stopped. In the European Union, known for high regulation on e-waste exportation, blockchain will also serve to ensure compliance by leaving a record of exactly where the e-waste materials are going and how they are being processed (M. Sharma, S. Luthra,2022).

Encourage Ethical Practices

Blockchain can encourage better practices of handling e-waste through the responsibility instilled in businesses on how they handle their waste. For instance, recycling companies can use blockchain to show that they are doing what is expected of them and abiding by the environmental standards. This gives consumers and investor's confidence in how e-waste is handled because it is now becoming transparent.

Promoting Circular Economy End

Blockchain supports traceability, which guarantees the recycling of valuable material from e-waste into the production cycle for the circular economy. Companies producing products with recycled materials can prove on blockchain that those recycled materials were sourced responsibly. This may assist companies in sustainability goals and improves their brand image and reputation.

5.2 Blockchain Pilots in the European Union

Blockchain technology is also under experimentation in the European Union (EU) to allow for better traceability of recycled materials while ensuring that electronic waste is processed under environmental compliance. The pilots are directed at establishing blockchain-based systems tracing e-waste from collection points to recycling facilities, thereby averting legally-exported or processed under unsafe conditions.

The second EU pilot project involves a blockchain-based certification of e-waste recyclers. This will enable the recyclers to show effectively that they adhere to the best practices while also being compliant with the EU regulations. Such is of significant value to the consumers and businesses in ensuring increased levels of confidence in the recycling processes. Successful tests show that applying blockchain can considerably enhance the level of transparency within the whole supply chain of the wastes, as electronic materials are recycled in an honest manner and opportunities for illegal dumping are minimized.

5.3 Promoting Trust in the E-Waste Recycling Process

The beauty of this is that transparency built by blockchain with stakeholders such as consumers, recyclers, manufacturers, and regulators in the e-waste management ecosystem builds this sort of trust. Blockchain allows for a secure record of how e-waste is processed, providing verifiability that can alleviate some concerns around the ethical and environmental impact of recycling e-waste. Such trust can actually facilitate greater participation in recycling programs as well as develop more environmentally friendly practices.

It also enables the consumers to participate in take-back or recycling programs if they can be sure about the responsible processing of their used devices. Similarly, companies can demonstrate commitment to sustainability with blockchain, through which they enhance their brand image and thus consumer loyalty. The technology of blockchain will have a very paramount role in achieving more sustainable or circular management of e-wastes through much higher trust and accountability.

6. CASE STUDIES OF CIRCULAR ECONOMY IN E-WASTE MANAGEMENT

Some firms in the world have already led successful practices through the implementation of the circular economy principle in e-waste management practices. Case studies will provide a clear indication of how businesses can close the loop for electronic products to avoid waste and minimize adverse effects on the environment as well as create new economic opportunities.

6.1 Case Study: Dell's Closed-Loop Recycling Program

Dell Technologies has been always a leader in sustainability and circular economy initiatives, mainly in the sphere of e-waste management. One of the most successful examples of CE in the electronics sector is Dell's Closed-Loop Recycling Program, which focuses on reclaiming valuable materials from used electronics and reintroducing them into the production cycle. This program proves to be quite promising inside a circular economy and is promising, thereby preventing waste and stimulating innovation while promoting resource efficiency.

Key Features of Dell's Program:

Consumer Engagement

Some of the essential characteristics of Dell's program include active involvement in working with its consumers. For example, Dell encourages its consumers to return old electronic products so that the laptops, desktops, among other peripherals, can be recycled through free recycling services. To improve consumer participation, Dell gives customers discounts on future purchases if they recycle their old products. This reduces e-waste and also promotes participation in the circular economy.

Closed-Loop Plastics

In this closed-loop recycling program, Dell utilizes recycled plastics to produce new products. The company collects used electronics, retorts out the plastics from them, processes these into raw material and then uses them within their products. By 2019, Dell used over 100 million pounds of recycled plastics within its products and also saved virgin plastic usage by over 70%. This has many impacts on the environment through lowering the demand for fossil fuels, reducing carbon emissions, and minimizing the amount of plastic waste in landfills.

Global Reach

Dell's e-waste recycling footprint stretches to more than 75 countries, which thus constitutes the largest collection network globally. Because Dell has a global presence, it can collect e-waste from any scope of markets and old electronics irrespective of their point of origin shall be recycled or refurbished properly. Through operating at a global scale, Dell demonstrates that an approach of large-scale implementation of circular economy principles is feasible and worthwhile when applied in the management of e-waste.

Sustainability Goals

Dell sets challenging sustainability goals to strengthen circularity commitments. The company has committed to recycling an equal product for every new one sold by 2030, thus producing a fully closed lifecycle for its products. This commitment will mark the leadership position of the company in sustainability as well as show the example followed by other manufacturers. The closed-loop programme for Dell has not only taken the environmental burden from its products but rather also be able to show business benefits to adopting a circular economy model (D. Zhang, 2022).

6.2 Case Study: Fairphone – A Modular Approach to Electronics

A Dutch social enterprise, Fairphone, has been revolutionary in its approach to conventional electronics design with a focus on modularity, repairability, and ethics sourcing. In fact, the Fairphone design falls into a category of electronics that are increasingly hard to fix, with most electronic products designed to have a short lifetime. This philosophy in the world of product design doesn't only extend the lifetime of a product but also reduce e-waste generation, making Fairphone a leading company in circular electronics.

Key Features of Fairphone:

Modular Design Modular design

It is one of the most significant features of Fairphone-it allows individual components such as the battery, camera, and screen to be easily swapped out. This cuts down on the requirement of totally replacing the device in case one component fails. This is one of the significant reasons for electronic waste generated by traditional electronics. The modular design also makes it easier to upgrade the phone to extend its useful life. For example, users can attach their camera modules as a way of customizing the phone if they require more sophisticated module parts. This modularity completely aligns with the principles of circular economy as the entire aspect encourages the longevity of products and minimum generation of waste.

Sustainable Sourcing

Fairphone sources its materials responsibly, including its conflict-free minerals and recycled plastics. It will source its products, with the smallest harm for people and the environment, directly from the suppliers. For example, Fairphone sources its tin and tantalum from the conflict-free mines in the Democratic Republic of Congo and also sources recycled plastics for its phone casings. This focus on ethical sourcing differentiates Fairphone from nearly every electronics producer and shows the brand's commitment to sustainability.

Repairability

Fairphone advises consumers to fix their devices rather than throw them away by providing consumers with considerable instructions as to how it can be done and providing spares that are relatively easy to get hold of and can be fitted at home. Fairphone also partnered with iFixit, an online repair guide platform with tools and more, to sustain consumers further in extending the life of devices for a longer time. This repairing commitment not only reduces e-waste but also inspires consumers to remain active participants in lengthening the life of products.

Transparency

Transparency is another value Fairphone emphasizes. It often releases reports on its environmental and social impacts, and with detailed information on where its materials are sourced and the manufacture of its phones. Fairphone is encouraging all other manufacturers to follow suit, nudging the industry toward more sustainable and ethical production methods. It is very rare for the electronics industry to witness such a feature. Thus, Fairphone brings it out very distinctly and highly stands out among the front lines of examples reflecting the principles of the circular economy.

It serves as a model for other hardware manufacturers wishing to make their products more environmentally friendly and interactive with the circular economy. Fairphone is showing it is possible to produce high-quality electronics that minimize harm to people and the environment, thereby extending product life and promoting responsible sourcing.

7. EXTENDED PRODUCER RESPONSIBILITY (EPR)

A very strong policy in the circular economy for managing e-waste is Extended Producer Responsibility (EPR), which takes the responsibility of a manufacturer from the design and production stage to the end of life, including requirements for manufacturers to have designed products with longevity, recyclability, and sustainability (P. Bradu, 2023).

The EPR policies of the European Union are most vividly expressed in the WEEE, or Waste Electrical and Electronic Equipment. It was adopted in 2003 and has since been amended several times. The WEEE makes it obligatory for the producers of electrical and electronic equipment to ensure the collection, recycling, and proper disposal of e-waste through proper collection systems in the EU. This policy forces manufacturers to finance the collection and recycling of e-waste through setting up separate take-back systems or as part of collective recycling schemes.

An ambitious target of recycling 65% of e-waste by weight in every member state of the European Union has been set under the WEEE Directive. This, in turn, has resulted in the massive growth of thousands of collection points throughout the EU where consumers can deposit their unwanted electronic devices for correct recycling. Manufacturers are then required to fund these collection systems to help consumers easily recycle with environmental responsibility.

Benefits of EPR:

1. Design for Circular

Economy EPR inspires electronics that would be easy to repair, refurbish, and recycle by enforcing end-of-life management responsibility from the producers' side. Manufacturers are likely to implement modular designs, separable materials, and non-toxic components to reduce the impact of these devices on the environment.

2. Financial Responsibility

EPR shifts financial responsibility of waste management from the wallet of a taxpayer to producers. It also provides an excellent incentive for manufacturers to reduce wastage and make recycling processes more efficient among themselves (F. D. Castro, 2022).

3. Higher Recycling

It would have led to higher rates of recycling of waste electronics. Germany, one of the first movers of the EPR, has one of the highest recycling rates in the world. In the year 2020, their e-waste recycling rate reached as high as nearly 80%. Producers, therefore, add up all costs and make sure that they invest better in recycling infrastructure as well as recycling technologies.

4. Consumer Awareness

EPR policies always include public education programs that help increase consumer awareness on the need for electronics recycling. Increased consumer involvement in take-back programs will result and there will be less e-waste in the landfills.

While such success is remarkable, EPR faces significant challenges in low-and middle-income countries, where regulatory capacity for enforcement is weak. In places where there are nearly or completely no structures in place for collection and recycling of e-waste, illegal dumping and informal recycling continue to pose environmental and public health risk. Where work is in progress, governments in developing countries are taking EPR frameworks with substantial assistance from international organizations and NGOs.

7.1 Right to Repair Legislation

A move to Right to Repair is now gaining a foothold around the world, as consumers and advocacy groups push through legislative action that allows people to simply repair their own devices. Most electronic products, nowadays, are built in such a way as to have planned obsolescence-that is, they are difficult or impossible to repair except using specific tools or knowledge. Such leads to a shorter lifecycle of the product, with increased e-waste generation. This problem may be addressed by the Right to Repair laws compelling manufacturers to make the consumer and independent repair shops access spare parts, repair manuals, and diagnostic tools.

The European Union implemented the Right to Repair Directive in 2021. All producers are supposed to make available items with enhanced reparability and spare parts availability for a period of more than 10 years from the date of sale of devices, such as washing machines and refrigerators, electronic devices including smartphones and laptops. This should elongate the life expectancy of electronic products while reducing the e-waste quantity generated every year.

Key Aspects of Right to Repair Legislation:

1. Readily available spare parts: The company should provide the necessary spare parts for both consumer and independent repair technicians. This will ensure that the product is repaired and extended in lifespan.
2. Repair Manuals and Equipment: Give access to the equipment, repair manuals, diagnostics, and any software updates that may have to be applied to fix devices so that consumers and third-party repair firms are not deprived of warranty when making repairs.
3. Standardized Parts Some of the Right to Repair legislation requires standardized parts, in which case there are easier substitutes for those. For example, standardized chargers for smartphone and laptop will save on electronic waste in not having multiple incompatible power adapters (O. J. Oyebode, 2013).

Global Right to Repair Initiatives:

The Right to Repair legislation is winning ground beyond the borders of the EU. Across the Atlantic, with particular regards to the United States, several bills for states have been presented so consumers can now be allowed to repair their electronics. The Federal Trade Commission also commented in favor of the movement, saying restrictions on repairs merely result in unnecessary waste and obstruct competition in the repair market.

A similar movement is germinating in India, where the Indian government is exploring the prospects of enactments of Right to Repair laws to address India's growing problem of e-waste. India's informal sector of repair, currently servicing a whole variety of electronic products at affordable prices, will have access to more parts and equipment through such legislations.

Right to Repair laws would greatly reduce e-waste as the lifespan of products would be increased, a culture of repair and reuse would be promoted, and the demand for new electronics would be decreased. However, the majority of manufacturers oppose these policies, stating that if freely repaired, devices might compromise safety and security aspects. Balancing the need for repairability with intellectual property concerns and those regarding security would go a long way in making sure the success of such policies.

7.2 International Cooperation and Standards

E-waste is a global problem that requires international collaboration to handle it properly. One of the main challenges on the governments' and environmentalists' part is transboundary movement, mainly from developed to developing countries. Most developing countries lack proper infrastructure for recycling e-waste safely, and as such, informal recycling activities prevail in these places, thus exposing the exposed workers and communities to chemicals.

The Basel Convention

Toxic waste, including e-waste, and the processes how it is as well regulated by one of the best international covenants known as the Basel Convention. The Basel Convention was adopted in 1989 primarily to mitigate the hazardous waste trade between countries, especially from the developed to the developing world. It also provides for the safe and environmentally friendly disposal of hazardous wastes and recognizes the need for safe management of e-wastes in developing countries.

It was in 2019 that the Basel Convention was amended to add specific e-waste provisions in the treaty itself; and, hence, transboundary movement of this waste finds strength through this amendment. The so-called “Basel Ban” which prevents the hazardous waste being exported from developed OECD nations to other developing non-OECD states (R. Akram,2019).The concern brought by this amendment is to ensure that reckless regions where e-waste exportation is processed do not encourage illegal e-waste trade because of lack of proper regulations.

International Standards for E-Waste Recycling

Besides, a system for e-waste recycling within its set guidelines can also be worldwide as in the case of international conventions promoting the e-waste trashing. The ISO (International Organization for Standardization) and IEC (International Electrotechnical Commission) have standards that discuss the various issues involved in the operationalisation of e-waste management strategies, inclusive of those devoted to the environmentally friendly recycling processes. For example, areas such as the use of toxic agents, salvage operations, and occupational health practices are well covered.

National and Local Policies for Circular E-Waste Management

As significant as international collaboration can be, the same cannot be said of the bottom-up approach to the problem since national and local governments are equally important in the implementation of circular economy model for the disposal of e-waste. In countries such as Japan, South Korea and China, these efforts have been backed by sound national legislation to advance e-waste collection and recycling.

- In Japan, the Home Appliance Recycling Law (HARL) which came into force in 2001 prohibits consumers from discarding specified household electronics devices (air-conditioners, refrigerating cabinets, washing machines and the like) but rather brings them back to appointed intake centers. The obligation to recycle these appliances rests on the manufacturers causing them to design for recycle.
- In South Korea, the Act on Resource Circulation of Electrical and Electronic Equipment and Vehicles mandates the disassembly and recycling of electrical products and the retrieval of ferrous and non-ferrous metals contained in them. The legislation places an obligation on producers for the collection and recycling of e-waste, defining also the percentage of certain materials, especially noble metals such as gold and silver, that should be reclaimed.
- China too has adopted a number of measures to solve the problem of e-waste within its territories. This Regulations for the Administration of the Recovery and Disposal of Waste Electrical (O. J. Oyebode, 2022).

Incentives for Circular Business Models

In addition, governments could advance the transition towards circular business models in e-waste management through various levers, such as tax concessions, subsides, grants and other financial aid. Such policies can help induce businesses and clientele to adopt environmentally friendly practices, upgrade their systems with recycling technology and to repair or refurbish goods.

Grants for Research and Development

Research and development activities focusing on the recycling of electronic waste can be enhanced by the use of government grants and funding programs. In the United States, for instance, the National Science Foundation (NSF) has sponsored several initiatives seeking to enhance technologies for e-waste recycling under its Small Business Innovation Research (SBIR) program. These grants assist industries in finding inventive ways to recover materials, efficient recycling, and designing for the environment.

Most importantly, funds should also be allocated towards R&D as this would help in the progression of recycling technologies and enhance the effects of e-waste management systems. The Governments can facilitate this by funding and supporting such research endeavours.

7.3 Public-Private Partnerships (PPPs)

PPPs will promote Governments, enterprises, and NGOs together to come up with circular e-waste management (R. Nithya, 2021). Through this collaboration, the hybrid model can access all the attributes from the resource sector and the skill sector, such as creating desirable recycling programs, launching education campaigns, and advancing innovation in e-waste management.

The government can partner with electronics manufacturers, electronics recycling firms, and other e-waste-related companies to spearhead an integrated e-waste collection and recycling program. Partnerships of this kind will help enhance the effectiveness of recycling programs and widen public education and enlightenment on issues of e-waste.

7.4 Case Studies of Successful Incentive Programs

In recent times, several countries have set up positive incentive programs to promote circular e-waste management.

1. Sweden: It has already been mentioned that Sweden refunds taxes paid to consumers in case they repair their electronic devices. This incentive for the repairing process of electronics leads to lesser carbon footprint in the consumption of electronics. As stated by the Swedish Tax Agency, this incentive policy has led to a tremendous surge in the amount of repair activities. Such a scenario highlights the fact that financial incentives are effective weapons for inducing sustainable behaviour.
2. France: France has instituted the "Eco-participation" fee paid at the point of sale of new electronic equipment to consumers; it will be applied for waste-management and recycling programs. The country has even developed a "Repairability Index," where items are ranked according to ease of repair, encouraging manufacturers to make products more repair-friendly (Y. A. Fatimah, 2020).
3. United States: Many states in the United States have placed several initiatives promoting recycling and proper disposal of electronic waste. An example is California, where e-waste recycling fee was instituted to fund programs for

state e-waste recycling. Grants are also given to local governments and organizations to install a collection program and build improvements on recycling infrastructure.

Efficient policies, effective regulatory frameworks, and incentive programs are provided by governments to make circular e-waste management. The development of an environment for a more sustainable practice in the electronics industry can be created through measures like EPR, Right to Repair legislation, and international cooperation.

There are incentives like tax rebates for repairs and grants for research and development that encourage the businesses and consumers towards sustainability of circular business models. In the case studies presented, examples of good practice have been identified from various initiatives implemented in promoting responsible e-waste management and reducing environmental impact.

The increased demand for electronics nowadays creates a greater challenge to the effective and meaningful management of e-waste globally. Governments can, therefore, help advance the transition towards a circular economy and thereby reduce the generation of e-waste, through this very use of policy and regulatory frameworks and financial incentives.

8. FUTURE SCOPE

1. Re-Designing the Product

One of the foundation stones from which the circular economy springs is in rethinking design for products. The more recent approach to design, focusing instead on durability, modularity, and recyclability reduces e-waste at its cores. Products designed for longevity also leave an easy path to maintenance while manufactured.

2.New Business Models

Changing to new business models is one of the key steps toward improving e-waste management. Models like PaaS push consumers into leasing or renting electronic devices, rather than buying them outright. Ownership dynamics shift in such a model, and product lifecycle management will go back to the manufacturer's side. For instance, Philips has ensured exactly this by making its lighting business one of offering lighting as a service rather than selling products. This creates incentives for manufacturers to design their products durable, upgradable, and recyclable, since they are responsible for the same product through its lifecycle.

3.Technological Innovations

The role of technological innovation itself can be quite very pivotal in revolutionizing the way e-wastes are managed. This incorporation of automation, artificial intelligence, and advanced material recovery methods makes a difference in the recycling process. For example, via Apple's Daisy robotic disassembly, the alleged convenient recycling of iPhones to their core would extract valuable material without much loss. Similarly, AI-sorted systems by startups, such as AMP Robotics, will make recycling processes so fast and efficient that people leave disbelief for what happens inside recycling. These not only enhance the effectiveness of recycling operations but also increase overall recovery rates of valuable materials-a more circular economy.

4.Enabling Supportive Policy Frameworks

A supportive policy framework will be required to nudge the management of e-waste in the direction of a circular economy. This may also be solicited regarding governments and their use of regulations, especially for governments with a role that tends to facilitate extended producer responsibility plus the right-to-repair legislation. International cooperation in e-waste management should also be encouraged.

9. CONCLUSION

The path toward enhancing e-waste management based on the principles of a circular economy is not just desirable but an imperative for the sustainability of our planet. With an unprecedented rise in electronic waste, or e-waste, due to the rapid proliferation of electronic devices, the challenge it poses has never been witnessed before, and this growing problem harbors unprecedented environmental, economic, and social ramifications. But the circular economy presents the opportunity to flip our design, manufacturing, usage, and waste patterns for electronics.

The linear economy has been unsustainable since it sets the production that follows the "take-make-dispose" model. As a result, it generates tremendous amounts of e-waste due to worldwide demand for electronics. That means 53.6 million metric tons of e-waste is generated globally, and, shockingly, only 17.4% of that waste got recycled in 2019. That number portrays an enormous gap between what we do and the current mechanism of e-waste management. It urgently requires a paradigm shift. Toward this end, the circular economy offers a much-needed holistic framework for prioritizing sustainability, efficiency, and recovery of resources.

REFERENCES

Akram, R., Natasha, F. S., Hashmi, M. Z., Wahid, A., Adnan, M., & Nasim, W. (2019). Trends of electronic waste pollution and its impact on the global environment and ecosystem. *Environmental Science and Pollution Research International*, *26*(17), 16923–16938. DOI: 10.1007/s11356-019-04998-2 PMID: 31025281

Bradu, P., Biswas, A., Nair, C., Sreevalsakumar, S., Patil, M., Kannampuzha, S., Mukherjee, A. G., Wanjari, U. R., Renu, K., Vellingiri, B., & Gopalakrishnan, A. V. (2023). Recent advances in green technology and Industrial Revolution 4.0 for a sustainable future. *Environmental Science and Pollution Research International*, *30*(60), 124488–124519. DOI: 10.1007/s11356-022-20024-4 PMID: 35397034

Castro, F. D., & Bassin, J. P. (2022). *"Electronic waste: Environmental risks and opportunities," in Hazard. Waste Manag*. Elsevier.

Chatterjee and Kumar. (2023). S. Murthy and S. Ramakrishna, "Global E-Waste Management: Urban Mining and Circular Economy,". *Springer Environmental Science and Pollution Research*, *29*(8), 11367–11385.

Cole, C., Gnanapragasam, A., Cooper, T., & Singh, J. (2019). Assessing barriers to reuse of electrical and electronic equipment: A UK perspective. *Resources, Conservation and Recycling, X. 1*, 100004. DOI: 10.1016/j.rcrx.2019.100004

Fatimah, Y. A., Govindan, K., Murniningsih, R., & Setiawan, A. (2020). Industry 4.0 based sustainable circular economy approach for smart waste management system to achieve sustainable development goals: A case study of Indonesia. *Journal of Cleaner Production*, *269*, 122263. DOI: 10.1016/j.jclepro.2020.122263

T.S. Gaur, V. Yadav, S. Mittal, and Sharma, M.K. (2020) "A Systematic Review on Sustainable E-waste Management: Challenges, Circular Economy Practices, and a Conceptual Framework,". *IEEE Transactions on Sustainable Computing*, *9*(1), 45–58.

N. Veena, R. Kumar, and A. Singh, "E-Waste Management: A Circular Economy Perspective,". *Springer Journal of Waste Management & Research*, *41*(3), 321–340.

Gupta, S., & Singh, R. (2020). E-waste management: A review on technologies, challenges and future prospects. *Journal of Cleaner Production*, *265*, 121955.

Hakami, B. A. (2018). Electronic waste management approaches in Saudi Arabia. *American Journal of Environmental Engineering*, *8*, 4–10.

Ikhlayel, M. (2017). Environmental impacts and benefits of state-of-the-art technologies for E-waste management. *Waste Management (New York, N.Y.)*, *68*, 458–474. DOI: 10.1016/j.wasman.2017.06.038 PMID: 28662843

P. Singh, R. Gupta, and Mehta, S. (2020). "Digital Twin Technology for E-waste Tracking and Reduction,". *IEEE Transactions on Industrial Informatics*, *19*(5), 7896–7908.

Nithya, R., Sivasankari, C., & Thirunavukkarasu, A. (2021). Electronic waste generation, regulation, and metal recovery: A review. *Environmental Chemistry Letters*, *19*(2), 1347–1368. DOI: 10.1007/s10311-020-01111-9

Oyebode, O.J. (2013). "Solid waste management for sustainable development and public health: A case study of Lagos State in Nigeria," Univ. J. Public Health.

Oyebode, O. J. (2022). *"Sustainable waste management towards circular economy in Nigerian context: Challenges, prospects and way forward," in Effective Waste Management and Circular Economy*. CRC Press.

Rautela, R., Arya, S., Vishwakarma, S., Lee, J., Kim, K. H., & Kumar, S. (2021). E-waste management and its effects on the environment and human health. *The Science of the Total Environment*, *773*, 145623. DOI: 10.1016/j.scitotenv.2021.145623 PMID: 33592459

Saha, L., Kumar, V., Tiwari, J., Rawat, S., Singh, J., & Bauddh, K. (2021). Electronic waste and their leachates impact on human health and environment: Global ecological threat and management. *Environmental Technology & Innovation*, *24*, 102049. DOI: 10.1016/j.eti.2021.102049

Sharma, M., Joshi, S., & Kumar, A. (2020). Assessing enablers of e-waste management in circular economy using DEMATEL method: An Indian perspective. *Environmental Science and Pollution Research International*, *27*(12), 13325–13338. DOI: 10.1007/s11356-020-07765-w PMID: 32020449

Sharma, M., Luthra, S., Joshi, S., & Kumar, A. (2022). Developing a framework for enhancing survivability of sustainable supply chains during and post-COVID-19 pandemic. *International Journal of Logistics*, *25*(4–5), 433–453. DOI: 10.1080/13675567.2020.1810213

L. Fernandez, H. Kim, and Park J. (2022). "The Role of AI and Machine Learning in E-waste Sorting,". *Springer Journal of Cleaner Production*, *402*(2), 129874.

Forti, V., Baldé, C., Kuehr, R. and Bel, G. "The Global E-waste Monitor 2020: Quantities, Flows, and the Circular Economy Potential", United Nations University/ United Nations Institute for Training and Research, International Telecommunication Union, and International Solid Waste Association.

S. Chatterjee and A. Kumar, (2023). "Enabling Technologies for Effective E-waste Management,". *IEEE Internet of Things Journal*, *10*(2), 2104–2118.

Zhang, D., Ding, L., & Zhang, L. (2022). The challenges and strategies for e-waste management in the era of the Industrial Internet of Things. *Journal of Cleaner Production*, *275*, 123548.

Chapter 4
Current and Forthcoming Viewpoints to Generate and Manage Sustainable E-Waste

Sumanta Kuila
https://orcid.org/0000-0002-6088-6806
Haldia Institute of Technology, India

Namrata Dhanda
Amity University, India

Subhankar Joardar
Haldia Institute of Technology, India

Amrita Sarkar
https://orcid.org/0009-0002-9206-6535
Haldia Institute of Technology, India

Angana Chakraborty
Haldia Institute of Technology, India

ABSTRACT

This study examines the growing global challenge of electronic waste (e-waste), driven by shorter product life cycles and technological advancements. E-waste, a major source of contaminants and valuable materials, requires proper recycling and disposal. Despite efforts by developed nations, such as the Group of Seven (G7) countries' Reuse, Recycling, and Recovery (3Rs) program, the high cost of recycling has led to the transfer of e-waste to developing nations with inadequate

DOI: 10.4018/979-8-3693-7383-5.ch004

recycling infrastructure and weak regulatory enforcement. This research focuses on identifying and analyzing the interactive barriers faced by local governments and e-waste recycling businesses in both industrialized and developing countries. It emphasizes the direct and indirect impacts of these barriers on sustainable e-waste management practices. While prior studies have addressed individual barriers, this study explores how they interact and the aggregate effect of these interactions on long-term sustainability.

INTRODUCTION

The waste streams with the fastest global growth is e-waste which is a result of shorter product life cycles and improving technology. Due to its status as one of the primary sources of contaminants in municipal waste and a secondary source of valuable metals and materials the e-waste is intended for proper recycling and disposal. Numerous nations and areas of the world have expressed interest in the e-waste issue(An et al. 2015). The developed world has worked very hard to manage e-waste. The Reuse, Recycling, and Recovery (3Rs) program aims to encourage sustainable production and consumption among the Group of Seven (G7) countries. The high expense of recycling e-waste is the reason that, despite laws and regulations, 80% of the world's e-waste is reportedly transferred to poor nations like China and India. The infrastructure for recycling is poor that's why the managerial resources are insufficient for good recycling and regulations are not properly enforced in many developing nations. There are many obstacles in the way of e-waste management practices meeting sustainability standards which include being socially, ecologically, and economically sustainable. To determine the difficulties and obstacles local governments and e-waste recycling businesses groups confronts numerous studies have been carried out to find the solutions. Previous works on the subject of obstacles to e-waste management practices has mostly looked at different areas to identify and analyse the causes of barriers. We specifically examined and contrasted the current state of the e-waste sector and the regulatory frameworks in industrialized and developing nations to pinpoint obstacles unique to each nation. These studies don't, however, specifically look at how the interconnections between the barriers would affect how long-term the e-waste management technique is sustainable. It is imperative that we not only ascertain the ways in which the barriers interact with one another to impact sustainable e-waste management practices but also evaluate the aggregate effect of these interactions (Awasthi et al. 2017; Bakhiyi et al. 2018). There are two components to a barrier's overall effect: direct and indirect. Because direct and indirect effects are measured or acquired differently, it is preferable to distinguish between them. This emphasizes the dual nature of the interaction between a barrier and sustainable e-waste treatment. As a result of the barrier's direct impact implements the barrier's indirect influence from other interacting barriers which are related in an indirect manner. The amount that a barrier's effective mitigation alone will contribute to the adoption of sustainable e-waste management practices is known as the barrier's direct impact. The degree to which a barrier's successful mitigation will affect the mitigation of other interacting barriers that are advancing sustainable e-waste management practices which is known as the barrier's indirect impact. The degree to which a barrier's successful mitigation will affect the mitigation of other interacting barriers works with advanced sustainable e-waste management practices is known as the barrier's indirect impact. This is so that the mitigation of one barrier will help to mitigate the obstacles that interact with it. Barriers such as the competitiveness of informal recyclers and restrictions that are not successfully enforced are well acknowledged. Since e-waste handling exposes people to dangerous materials, it poses serious health hazards especially in areas where informal recycling procedures are common. The following chemicals are common in electronic waste such as lead, mercury, cadmium, and brominated flame retardants. Among the many health problems that can result from exposure to these poisons are kidney and nerve damage, developmental delays in children and possible cancerous effects. The natural environment and public health are put at risk by the massive volume of e-waste that is gathered by unofficial recyclers and handled crudely. The unofficial recyclers spread their materials and resist being put under control. Informal recycling creates a shadow economy and is a profit-seeking enterprise with no constraints on worker safety or environmental protection which makes it difficult to police laws. To reform or outlaw informal recycling, a mechanism should be created which will work to enhance and implement e-waste management laws and policies more successfully and consistently. One well-known example is the transformation of “Guiyu” a town infamous for being a dump for e-waste into a brand-new formal recycling facility run by the Chinese government. This happened after China's authorities placed strict restrictions on the town's informal recycling operations. This study intends to analyze the total influence of interactive obstacles on e-waste management practice to re-cycled it in a productive manner. The degree to which a barrier can be effectively overcome in the mitigation level. The existing e-waste management practice settings and available resources must be taken into consideration while managing interaction barriers under the restricted resources. To manage interacting barriers with variable overall impacts and mitigation levels it is advised to use modern techniques (Awasthi et al. 2016; Bhatia et al. 2018).

RELATED STUDY

Research has indicated that the generation of e-waste is increasing as a result of the quick development of technology and the rising use of consumer devices. Only 17.4% of the 53.6 million metric tons of worldwide e-waste that were generated in 2019 were collected and recycled, according to the worldwide E-Waste Monitor (2020). This emphasizes how urgently we need efficient management techniques. The significance of circular economy concepts in e-waste management is emphasized by research on sustainable recycling methods (Chai et al. 2011). The function of laws in the sustainable management of e-waste is covered in a number of studies. The creation of e-waste can be decreased by encouraging manufacturers to design goods with end-of-life in mind through Extended Producer Responsibility (EPR) laws according to research published in Waste Management. Improving e-waste management procedures requires an understanding of consumer behaviour (Gabus et al. 1973). According to a report published in the Journal of Cleaner Production, higher recycling rates are correlated with greater public knowledge of the effects of e-waste. It is crucial to teach customers how to dispose of things properly. Innovations for recovering metals from e-waste that can be more ecologically friendly than conventional techniques include hydrometallurgy and microbiological techniques according to a study published in Waste Management & Research(He et al. 2006). Enhancing e-waste management can involve integrating informal recyclers and local communities. Formalizing the informal sector may improve e-waste collection and recycling rates since these workers frequently have local networks and knowledge, according to research published in Environmental Science & Policy (Kumar et al. 2019). Effective e-waste management involves both global frameworks and local methods. A comparative analysis in sustainable cities and society shows that, although international standards like the Basel Convention are important the implementation strategies must take local context into account for successful results (Liu et al. 2006) . The course of existence of lifecycle assessments (LCA) assist in assessing the environmental effects of e-waste. Better design decisions and recycling procedures can be informed by life cycle assessments (LCAs) which can lead to more sustainable e-waste management practices.

E-WASTE MANAGEMENT PROCEDURE

Figure 1 describes the workflow diagram of E-Waste management system. Here the potential effects that incorrect disposal of electronics can have on the environment and human health make managing electronic trash or e-waste imperative mechanism. List every electronic item in the catalog, including phones, laptops, monitors, printers, and accessories. Evaluate each item's state to ascertain if it is broken, functional, or outdated. Reusable or refurbished devices that are still in working order. The damaged or outdated electronics that need to be properly disposed of or recycled. Areas that are designated as the places of collection for electronic waste in your organization. E-waste is collected and transported to the storage location according to a predetermined collection schedule. To avoid leakage and damage dispose of electronic garbage in a dry secure location. Indicate the categories of e-waste and their status, which is used for recycling or reuse, on storage containers and locations with clear labels. Assess useful objects for possible repurposing or restoration (Cao et al. 2016; Chang et al. 2013). Collaborate with companies that specialize in second hand electronics refurbishment and resale. Assist accredited e-waste recycling businesses to guarantee safe handling and processing. Make sure that before recycling all data has been safely removed from devices. If required, use data wiping software or actual physical destruction techniques. When it comes to disposing of e-waste abide by local laws and regulations. Maintain records which includes certificates of destruction or recycling, of the amounts and kinds of e-waste disposed. Acquire the required permissions for the handling, transportation and disposal of electronic waste and keep abreast of local state, and federal rules pertaining to electronic waste. Keep thorough records of all the operations related to managing e-waste such as collection, storage, recycling and disposal. This provides internal and external stakeholders with frequent reporting on the procedures which creates the efficient e-waste management program. Educate staff members about e-waste management practices to manage the significance of appropriate disposal and clearly identify the roles involved in managing e-waste. Also educate staff members about the effects of e-waste on the environment and promote responsible behaviour by holding awareness programs (Chang et al. 2007; Chen et al. 2012). Review and assess the efficacy of the waste management process on a regular basis. Revise protocols and guidelines in light of emerging laws with new technology and industry best practices. Keep up with new developments in environmentally responsible and effective e-waste treatment techniques and technology. Establishing a strong e-waste management protocol not only aids in regulatory compliance but also encourages sustainability and lessens the negative effects of electronic waste on the environment (Chen et al. 2019).

Figure 1. Workflow diagram of e-waste management system

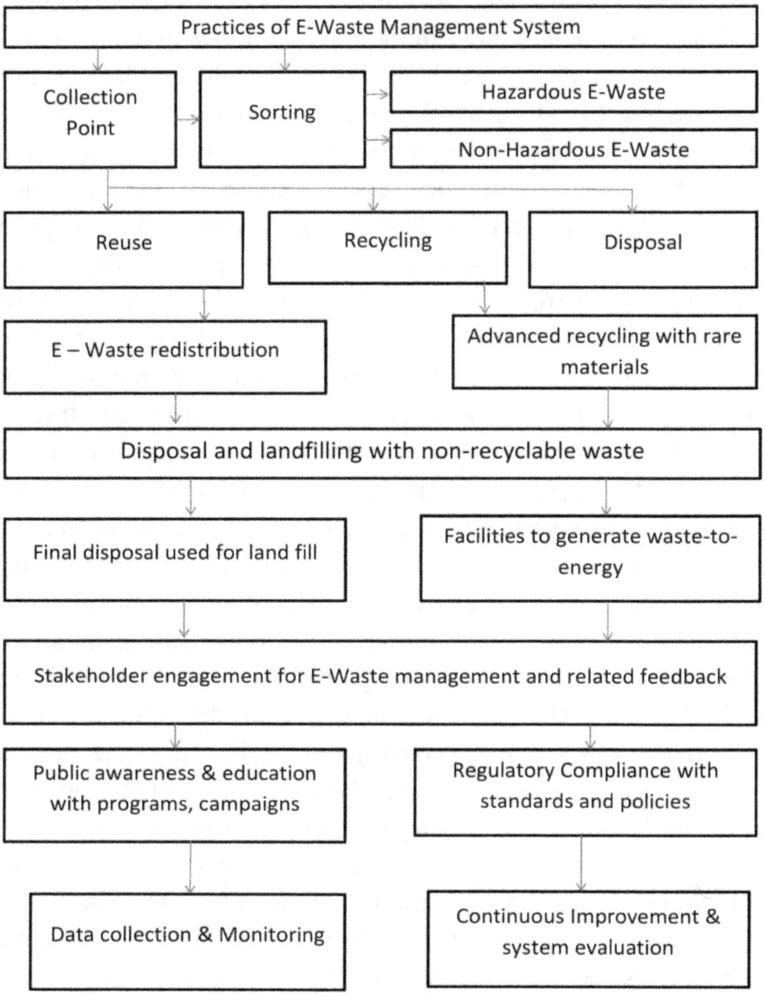

Collection Point

To make sure that electronic trash is collected, processed and managed appropriately e-waste collection entails a number of well-defined processes. Here's a thorough examination of the different e-waste collection techniques. Drop-off centres are sites set aside for people to bring their electronic garbage for recycling or appropriate disposal. Recycling facilities, garbage management firms or municipal governments may run these centres. They may operate alone or as a component of bigger recycling facilities. Ideal for people with limited e-waste requirements that enables appropriate disposal places. Items accepted and operating hours are subject to change. The kinds of e-waste that are accepted should be confirmed in advance. Convenient for locals who are unable to transfer their own electronic debris. It might be included in a more comprehensive waste management package. Programs for managing waste in the community will determine availability. It could be necessary to package or arrange the items in accordance with rules (Chi et al. 2011; Chi et al. 2014).

Sorting the Waste

Garbage's are categorically 2 types of Hazardous E-Waste and non –Hazardous E-Waste.

Hazardous E-Waste

Electronic garbage classified as hazardous e-waste includes harmful materials that can seriously endanger both human health and the environment. Electronics such as computers, cell phones, televisions, and appliances that are thrown away are referred to as e-waste. Essential details on dangerous electronic waste that are present in batteries solder and cathode ray tubes (CRTs) should be properly traced. Some poisonous elements are found in some fluorescent lights and several batteries and various kinds of circuit boards. When these harmful compounds are exposed to susceptible people in particular, it can cause major health concerns such as cancer, kidney damage, neurological damage and developmental disorders. Improper disposal can cause pollution of the soil and water which can impact plant and animal life and possibly make its way into the food chain (Chou et al. 2003).

Non-Hazardous E-Waste

Electronic garbage that does not include a large number of harmful compounds that are normally associated with hazardous e-waste is referred to as non-hazardous e-waste. Even if it might not be considered harmful it still needs to be recycled and disposed of properly to reduce its negative effects on the environment. Features of non-hazardous e-waste include having fewer or no dangerous materials like lead mercury or cadmium can be categorized on that which includes components of more recent electronics devices and accessories without a lot of dangerous elements. Instances of non-dangerous electronic waste basic electronics components like mercury-containing parts or CRT panels have less threat than the regular components of electronics devices. A few more recent or simpler circuit boards with less potentially dangerous parts. Hazardous substances including lead, mercury, cadmium and brominated flame retardants are found in a lot of electronic products. These hazardous materials can contaminate groundwater and soil when e-waste is dumped in a landfill. When decomposing in landfills e-waste can occasionally release harmful gasses, especially when it burns or is subjected to high temperatures. Valuable materials found in e-waste include rare earth elements, silver and gold. The resources that may be obtained through recycling are wasted when these things are disposed of in landfills (Cole et al. 2019; Dou et al. 2013).

Saving Valuable Components

Saving valuable components from the electrical devices is important. Electronic or electrical gadgets that are thrown away are referred to as e-waste or electronic garbage's. Because incorrect disposal of e-waste can have negative effects on the environment and human health, the swift growth of technology has made it a major worry. Reuse and recycling are key tactics for sustainable e-waste management. Let's examine these strategies in more detail. Refurbished gadgets can prolong the life of many electronic items, including printers, computers, and smartphones (Gabus et al. 1972). To restore the device's functionality, refurbishment entails updating, changing, or repairing its parts. Professional refurbishes can handle this, or for simpler gadgets, you can do it yourself. It is possible to sell or donate devices that are still functional. For individuals who require electronics but are unable to purchase new ones, this helps minimize e-waste and offers reasonably priced alternatives. Upgrading individual parts (such as giving a computer extra RAM) can sometimes greatly increase a device's usefulness rather than throwing it away (Gabus et al. 1973; Gardas et al. 2018).

Disposal

Electronic garbage, or "e-waste," must be disposed of properly to save the environment and public health. This is a guide about responsible e-waste disposal. Electronics such as computers, cell phones, TVs, printers, and other gadgets that are out of service are categorized as e-waste. Certain locations have rules about how to dispose of waste. To find out the regulations in your area, contact your local waste management body. Search for facilities or initiatives that recycle waste. Numerous merchants and manufacturers of electronics provide recycling services or take-back schemes. Verify the recycler's certification from groups such as e-Stewards or Responsible Recycling (R2) to make sure they adhere to eco-friendly standards (Geng et al. 2008; Govindan et al. 2016).

Disposal and Landfilling

Because these materials frequently contain hazardous elements, disposing of electronic waste in a landfill presents serious concerns to both human health and the environment. TVs, phones, computers, and other electronic equipment can all be considered pieces of electronic waste. Regarding disposing of e-waste in landfills, keep in mind the communities close to landfills may be exposed to dangerous substances from e-waste, which could cause brain damage, respiratory disorders, and other severe illnesses (Gumus et al. 2016). Hazardous material exposure is a possibility for employees who oversee landfills where e-waste is dumped. To reduce the negative environmental effects of e-waste disposal and recycling, support and abide by local rules. E-waste landfilling is not a safe or sustainable technique overall. Protecting the environment and public health can be achieved by giving recycling and appropriate disposal methods first priority (He et al. 2006).

Generate Waste-to-Energy

A new field that aims to meet the twin concerns of growing e-waste and rising energy needs is that of converting electronic garbage, or e-waste, to energy. An outline of how e-waste can be converted into electricity is provided. When e-waste is cooked without oxygen the components separate into gases, liquids and solid residues. The liquids and gases can be utilized as fuel to produce heat or power while the solid wastes can be disposed of or processed further. Heat-producing e-waste is burned. The incineration process produces heat that may be utilized to create steam which powers turbines to produce energy. It produces energy and drastically reduces the amount of garbage produced (Hicks et al. 2008; Y et al. 2016). Through a reaction with a regulated amount of oxygen or steam at high temperatures, e-waste is transformed into synthetic gas or syngas. Syngas has two possible uses first one is it can be burned to provide energy or processed further to create chemicals and fuels and second one less hazardous emission and more efficient than burning. A high-temperature plasma arc is applied to e-waste, dissolving things into their constituent parts. The procedure can yield vitrified slag or syngas, which can be used as a building material or as a source of energy. creates less leftover garbage and is capable of handling complex waste.

DIGITAL PLATFORMS FOR E-WASTE MANAGEMENT

Software for managing e-waste is made to make the handling, recycling and disposal of electronic trash easier and more efficient. Due to the environmental risks that electronic waste poses and the fact that it frequently contains toxic materials, effective waste management is essential.

E-Waste Management Software

Some important attributes and capabilities that e-waste management software may have. After an item is collected the mechanism to keep track it until it is recycled or disposed. Organize and maintain the location where the dumping places of e-waste materials can be maintained properly. Make sure that the rules for the recycling and disposal of e-waste are followed on a local, national and worldwide level. For audits and regulatory compliance create reports to maintain it in a systematic way. Organize e-waste's processing transportation and collection of the components. Plan pickups and oversee the logistics of moving e-waste to recycling facilities. Keep track of all the e-waste products in a database which will put together with comprehensive details on their nature, origin and processing history. Tracking, managing and properly disposing of electronic garbage (e-waste) is made easier for businesses with e-waste management software. Businesses and organizations usually use this kind of software to ensure correct management of electronic equipment which optimize the recycling process. This process complies with local legislation regarding the disposal of e-waste. Here are a few of the main attributes and capabilities that such software used for the e-waste management. Several leading software used for this work are AMCS Group, Optoro, Rheaply and Recycle Track Systems (Kong et al. 2012) .Data on electronic waste should be tracked and managed for reporting and analysis. Produce comprehensive reports on the amount, processing and recycling rates of e-waste. Analyse data to find patterns, streamline procedures to boost productivity. Give various stakeholders such as administrators, recycling centres and collection teams to access controls and user roles. All these activities monitor the user behaviour and privileges (Kumar et al. 2018).

E-Waste Tracking and Traceability Systems

Systems for tracking and tracing e-waste are necessary for efficient management of electronic waste, environmental protection and regulatory compliance. Here is a summary of the advantages and operation of various systems. To monitor e-waste from disposal to recycling, each piece is given a unique identification, such as an RFID tag or barcode. Information about the type, manufacturer and disposal date is written on the labels of the items. Throughout the e-waste lifecycle data is gathered at several points including collection sites for transportation, processing and recycling. Systems frequently interface with databases and Internet of Things devices to gather data in real time. These programs offer a way to keep an eye on and control electronic waste. They are able to produce reports to monitor the development of the adherence to rules. Proficient platforms provide insights to enhance workflows and boost productivity. Devices aid in guaranteeing compliance with national and worldwide e-waste laws like the EU's Waste Electrical and Electronic Equipment (WEEE) Directive and the Basel Convention. They help with the production of reports that stakeholders and regulatory agencies need to ensure that e-waste-related sensitive data is shielded from breaches and illegal access (Kumar et al. 2019; Lin et al. 2018). keeping records for verification and auditing needs to reduce the adverse effects to ensure that e-waste is handled and disposed of in an environmentally acceptable manner. The goal is to decrease the use of resources, encouraging recycling and recovering valuable materials from e-waste. The idea spreads for businesses and organizations in adhering to rules and staying out of trouble.

INTEGRATION OF E-WASTE IN THE CIRCULAR ECONOMY

Developing procedures and systems to handle electronic waste in a way that optimizes resource recovery and reduces environmental effects is necessary to incorporate e-waste into the circular economy. An outline on how to do this integration is provided here. Urge producers to create devices that are easy to maintain scalable and long-lasting. As a result, the product's life cycle is prolonged, and disposal frequency is decreased. Make items modular so that parts can be readily changed or replaced without having to throw away the entire gadget. Incorporating e-waste management procedures with Enterprise Resource Planning (ERP) systems helps guarantee regulatory compliance to boost operational effectiveness and encourage sustainable business practices. The management of electronic waste is one of the many company tasks that ERP systems optimize with their centralized platforms. Choose materials that pose less of a threat to the environment and are simpler to recycle. Provide e-waste collection locations that are easily accessible and convenient. These could be drop-off sites gatherings or initiatives for take-backs. Suitable processing works to distinguish valuable materials from electronic garbage which are used for sophisticated sorting of the system. Hazardous materials, polymers and metals must all be separated. Investment in and development of cutting-edge recycling technologies that extract precious elements from e-waste such as gold, rare earth metals and other high-value components is advised to preserve safely (Lin et al. 2018; Lue et al. 2006).

Economic and Social Impact

Discarded electrical or electronic equipment is referred to as "e-waste," or electronic garbage. Global e-waste has increased as a result of the quick development of technology and the rising demand for electronics. E-waste has substantial effects on society and the economy. Valuable materials found in e-waste include rare earth elements, silver, and gold. By recovering these minerals properly recycling can lessen the demand for new mining. This generates income and jobs in the recycling sector and opens up economic opportunities. It can be expensive to handle environmental contamination and health hazards caused by improper e-waste disposal. The costs of cleaning up and controlling e-waste are typically very significant for both businesses and governments. Innovation in recycling technology and sustainable design has been sparked by the need to manage e-waste. In order to spur economic growth and open up new markets, businesses are investing in producing items that are longer-lasting or easier to recycle. Poor working conditions are common in informal e-waste recycling operations found in many underdeveloped nations. E-waste management affects the technological, social and economic domains and includes the collecting, recycling, and disposal of electronic waste. Rare earth metals like gold and silver are among the valuable commodities found in e-waste. By recovering these minerals through effective recycling, fewer new mining activities are required, promoting a circular economy. These labourers frequently operate in dangerous conditions and without proper safety equipment. This circumstance has the potential to worsen exploitation and socioeconomic inequality. Distressed and impoverished neighbour hoods which are frequently situated close to recycling centres or landfills are disproportionately impacted by the dumping of waste. Environmental justice problems may result from this, as certain communities may be more affected by pollution and health hazards than others as well (Martilla et al. 1977). Several nations have put laws into place to handle e-waste in an ethical manner. One initiative to enhance e-waste recycling and recovery is the European Union's Waste Electrical and Electronic Equipment (WEEE) directive.

Recycling and Resource Recovery

Managing the effects of abandoned electronic devices on the environment and human health requires e-waste recycling and resource recovery. A summary of these ideas is provided below. Computers, cell phones, televisions, refrigerators, and other electrical and electronic items that are thrown away are referred to as "e-waste," or electronic garbage. E-waste must be recycled since electronic gadgets frequently include rare earth elements gold, silver and mercury, along with other valuable resources and potentially dangerous materials including lead, mercury and cadmium. Recycling properly aids in recovering these priceless resources and preventing environmental harm. There are several ways to gather e-waste including drop-off locations, carbide pickups and manufacturer return policies. E-waste is classified by material and type after it has been collected. Either manual labour or automated systems may be used in this. Devices are frequently shredded to reduce their size and facilitate the separation of various materials. Material separation involves the use of devices like air classifiers, magnets and eddy current separators to separate various materials, including glass, metal and plastic. Material separation results in additional processing. Glass and plastics can be recycled into new goods while metals are frequently melted down and refined. Toxic materials must be disposed of and handled carefully in order to avoid damaging the environment. Due to the complexity and variety of materials used in electronic gadgets, recycling procedures are difficult and expensive. Informal Recycling believes that because of risky methods and a lack of environmental safeguards some regions may find informal recycling activities to be harmful. The goal of improving the process effectiveness with economy and environmental friendliness acted as emerging technologies for recycling e-waste are developing quickly. Some noteworthy technologies that are influencing the future of recycling e-waste are Hydrometallurgical Processes, Bioleaching, Robotic Sorting and Artificial Intelligence, Advanced Shredding and Separation Techniques, Plasma Arc Technology, losed-Loop Recycling Systems, Supercritical Fluid Extraction, Magnetic and Electrostatic Separation etc. Using electric charges and magnetic fields these methods extract metals and other elements from electronic waste. Non-ferrous metals, polymers and other components can be separated by electrostatic means whereas ferrous materials (iron-based) are separated via magnetic separation. High effectiveness, little contamination and material separation according to physical characteristics. Utilizing fungi or bacteria, bioleaching breaks down e-waste materials to extract valuable metals. When bacteria or fungi interact with metals in e-waste they oxidize and break them down into a more soluble form, which makes it easier to remove the metals. Extracting valuable materials from e-waste and repurposing them for manufacturing or other uses is known as resource recovery. In addition to lowering the need for raw materials helps preserve natural resources (Moktadir et al. 2018]. There are a number of recovery techniques available, such as hydrometallurgical processes, which dissolve and remove valuable metals from electronic waste using chemicals. High-temperature procedures like smelting are employed in pyro metallurgical processes and recovery systems to extract metals and to extract metals from e-waste, the bioleaching approach uses microorganisms.

STAKEHOLDER ENGAGEMENT FOR WASTE MANAGEMENT

Involving stakeholders in trash management is essential for creating strategies that work, winning over the community and guaranteeing that waste management initiatives are carried out successfully. This is a methodical strategy for successfully involving stakeholders. Employees, management and board members are examples of internal stakeholders in an organization. Residents, nearby companies and institutions make up the community's membership of external stakeholders. Government agencies are national, regional and local bodies in charge of enforcing laws and regulations. Waste management organizations such as service providers and contractors. Environmental and advocacy groups are examples of non-governmental organizations (NGOs). Academic organizations that collaborate with local and national news outlets include research institutions universities and the media. Evaluate Interests which establishes the priorities for each stakeholder group with regard to cost convenience and environmental impact. Determine their capacity to influence or be impacted by waste management policies which includes public opinion and regulatory authority evaluate their influence. Numerous non-governmental organizations (NGOs) in are actively attempting to address the serious environmental and health problems posed by electronic garbage, or "e-waste," through recycling, collection, awareness-raising, and sustainable waste management techniques. These well-known NGOs are involved in the management of e-waste. A plan is required for communication among the stake holders. Customize communication tactics for various stakeholder groups through tailored messaging is important. Select channels that take advantage of relevant channels such as newsletters, social media, open forums and in-person consultations. The purpose of the communication is to inform, consult or involve stakeholders. The concern people should set objectives lay out specific goals for each attempt. That establish trust through openness, sincerity and consistency in interactions is the first of numerous strategies to build relationships. Addressing Concerns is another where feedback and concerns from stakeholders are acknowledged and addressed. Express gratitude for the accomplishments and contributions of stakeholders by recognizing and celebrating them. Track Engagement entails keeping an eye on how interested parties are interacting with your waste management initiatives. Measuring Impact evaluates the success of your engagement tactics in relation to program results and stakeholder satisfaction. Continuous engagement initiatives such as adjust tactics in response to evaluation results and feedback are entertained. The opinions of stakeholders have impacted program execution and decision-making. Highlight the successes and advancements brought about by stakeholder involvement. More sustainable practices, more community support, and improved project outcomes are all possible with effective stakeholder engagement in the e-waste management (Ongondo et al. 2011; Orlins et al. 2016).

Public Awareness

To promote sustainable behaviours and lessen environmental harm, it is imperative that the public is aware of e-waste management. Substances that are dangerous and can contaminate soil and water include lead, mercury, and cadmium found in e-waste. When dangerous substances are improperly disposed of, populations may be exposed to situations that pose a substantial risk to their health. Recyclable resources like glass, plastic, and metals can be found in abundance in e-waste. To raise awareness of e-waste and its effects, schools, non-profits and community groups can host workshops and seminars. By using these networks, we can swiftly and efficiently reach a larger audience with valuable material. Setting up recycling or e-waste collection drives can involve the community and offer easy ways to dispose of rubbish. Encourage individuals to recycle their waste by using authorized recycling facilities. the Raise awareness of manufacturers' take-back initiatives for ethical disposal. At the local and federal levels, it is required a push for stricter e-waste management policies and regulations. Programs that offer incentives for appropriate e-waste disposal and recycling should be highlighted. Community work is required to develop sustainable e-waste solutions with nearby companies and authorities (Ramzan et al. 2019). Volunteer programs involve locals in clean-up efforts and educational campaigns. Through raising public awareness and promoting responsible behaviour, communities may effectively mitigate the adverse effects of e-waste and foster a more sustainable future.

Standards and Policies

Manufacturers are required by Extended Producer Responsibility (EPR) to retrieve their products at the end of their useful lives, making sure that correct recycling and disposal are done. The European Union's Waste Electrical and Electronic Equipment (WEEE) Directive establishes goals for e-waste recovery, recycling, and collecting. National Environmental Policy Acts: Several nations have enacted laws that specifically address the disposal of e-waste, frequently mandating proper recycling practices. Guidelines for Recycling and Discarding Environmental management systems to enhance the performance of the environment, ISO 14001. ISO 45001: Management systems for occupational health and safety to safeguard employees in e-waste recycling establishments. The electronics recycling industry's Responsible Recycling Standard or R2, is a certification standard that emphasizes ethical recycling methods. Encourage actions to lessen the production of dangerous e-waste as part of pollution prevention policies. Programs for resource recovery encourages the removal of valuable elements from electronic waste, therefore lowering the demand for virgin resources. Public awareness campaigns teach the public about safe disposal practices and the value of recycling electronic waste. Take-back programs are campaigns that let customers return used devices to be recycled or repaired (Robinson et al., 2009; Salhofer et al. 2016). Working together and forming partnerships promote sustainable e-waste management practices, governments, NGOs, industry and communities collaborate in a process known as stakeholder engagement. International collaborations help global e-waste management is intended to be improved by programs such as the Partnership for Action on Computing Equipment (PACE). Technology development and innovation research and development assists to develop new technologies where the toxicity of e-waste and increase recycling efficiency. Green design guideline motivates producers to incorporate end-of-life concerns into their designs to facilitate easier recycling. Putting in place procedures to monitor and report on e-waste tracking from collection to recycling will help to guarantee appropriate handling. Reporting requirements evaluate the efficacy of e-waste management practices where regular reporting is required. Economic Incentives Tax breaks and subsidies helps financial incentives for businesses that use sustainable practices are used to promote recycling. Government loans or grants provide funding for recycling programs that help local communities take up recycling projects (Singh et al. 2016 ; Tansel et al. 2017).

Data Collection and Monitoring

Verify conformity to rules and guidelines to enhance streamline for recycling and pickup procedures. Increase public knowledge of e-waste issues through raising awareness. Methods of data collection utilizes surveys and questionnaires to learn more about the disposal habits of e-waste from customers. Mobile apps provide channels for users to submit recycling and disposal choices for e-waste. Smart bins install Internet of Things-enabled bins that monitor collected e-waste types and fill levels. The recycling facilities and retailers to get information on the amount of electronic waste that is consumed (Widmer et al. 2005). Measure reductions in resource recovery and landfill contributions using impact metrics for e-waste management. Technology Consolidation Databases are used by data management systems to store and process acquired data. Geographic Information Systems are able to display trends and collect locations for the disposal of e-waste. Tracking and documentation regular audits deliver review for the procedures and results of collecting on a regular basis. Dashboards provide stakeholders with real-time dashboards where they can view performance information. Public Reports encourages the openness and accountability to publish yearly reports on the status of e-waste management. Outreach and education programs for the community used to utilize data to customize courses on the effects of e-waste and the advantages of recycling. These provide programs with incentives to encourage people and companies to get involved in recycling. Data privacy confirms action to safeguard user information gathered from applications and surveys. On-going enhancement reaction cycles establishes channels for stakeholders and users to provide input on how to improve the system (Zhang et al. 2019).

METHODOLOGIES TO GENERATE AND MANAGE E-WASTE

Sustainable electronic waste management, or "e-waste," requires a multimodal strategy that incorporates data analysis, mathematical modeling, and sustainability concepts. This organized overview combines e-waste management concepts with mathematical derivation. One can use a function to simulate the development of e-waste that accounts for a number of variables, including population growth, technical advancements, and patterns of consumption.

Model to generate E-Waste

G(t): Total e-waste generated at time t.
P: Population affected by e-waste generation.
I: Rate of technological innovation.
D: Rate of device disposal.
L: Average lifespan of devices

The generation of e-waste can be described as

$$G(t)=P\times (R+I-D)\times t \qquad (1)$$

where R can be described as the replacement rate of devices.

Re Cycling process model

$R_{e:}$ Rate of the e-waste recycled.
E: Efficiency of the recycling process.
C: Capacity of the recycling facilities.

Now the amount of e-waste recycled over time can be described as

$$R_e(t) = E\times G(t)\times \min(1,C/G(t)) \qquad (2)$$

Resource recovery model can be described as

R_m is described as the rate of material recovery.

M is the total materials in e-waste.
R_f is described as rate of loss during recovery.

The entire amount of recyclable materials collected can be written as follows:

$$Rm(t) = (1-R_f)*R_e(t) \times M \quad (3)$$

Impact on Environment

E_{impac} is described as environmental impact from e-waste.
C_{toxic} is described as concentration of the toxic materials.
$R_{pollution}$ is described as the rate of pollution from improper disposal.

The environmental impact can be modeled as:

$$E_{impact}(t) = C_{toxic} \times (G(t)-R_e(t))+R_{pollution} \times G(t) \quad (4)$$

Index of Sustainability can be described as

S is Sustainability index.
$E_{positive}$ is the positive effects of recycling which is resource savings.
$E_{negative}$ is the negative effects which is nothing but pollution.

The simple sustainability index can be described as:

$$S(t)= E_{positive}(t)/ E_{negative}(t)+1 \quad (5)$$

final optimized process can be described as a function that minimizes environmental effect and maximizes resource recovery might be developed to optimize e-waste management.

$$\text{Minimize } E_{impact}(t) \text{ subject to } R_e(t) \leq C \quad (6)$$

Employ numerical techniques to simulate the model over a predetermined period of time in order to examine different situations, such as population shifts, technology development, and recycling effectiveness. Employ numerical techniques to simulate the model over a predetermined period of time in order to examine different situations, such as population shifts, technology development, and recycling effectiveness. A thorough foundation for producing and handling sustainable e-waste is provided by this mathematical model. Decisions that strike a balance between environmental sustainability and technology advancement can be made by corporations and policymakers by utilizing real-world data.

Algorithm to Implement the Methodology

```
// Step 1: Data Collection
function collect_data():
population = get_population_data()
device_usage = get_device_usage_data()
avg_lifespan = get_average_lifespan()
tech_innovation_rate = get_technology_innovation_rate()
disposal_rate = get_device_disposal_rate()
recycling_capacity = get_recycling_capacity()
recycling_efficiency = get_recycling_efficiency()
toxic_concentration = get_toxic_material_concentration()
pollution_rate = get_pollution_rate()
return {
"population": population,
"device_usage": device_usage,
"avg_lifespan": avg_lifespan,
"tech_innovation_rate": tech_innovation_rate,
"disposal_rate": disposal_rate,
"recycling_capacity": recycling_capacity,
"recycling_efficiency": recycling_efficiency,
"toxic_concentration": toxic_concentration,
"pollution_rate": pollution_rate
}
// Step 2: E-Waste Generation Prediction
function predict_e_waste_generation(data, time_periods):
e_waste_generation = []
for t in range(time_periods):
G = data["population"] * (data["device_usage"] + data["tech_
innovation_rate"] - data["disposal_rate"]) * t
e_waste_generation.append(G)
return e_waste_generation
// Step 3: Recycling Management
function manage_recycling(data, e_waste_generation):
recycling_estimates = []
for G in e_waste_generation:
```

```
R_e = data["recycling_efficiency"] * G * min(1, data["recycling_capacity"] / G)
recycling_estimates.append(R_e)
// Adjust capacity if needed
if R_e > data["recycling_capacity"]:
adjust_recycling_capacity()
return recycling_estimates
// Step 4: Resource Recovery Calculation
function calculate_resource_recovery(data, recycling_estimates):
material_recovery = []
for R_e in recycling_estimates:
R_m = (1 - data["recovery_loss_rate"]) * R_e * data["total_materials"]
material_recovery.append(R_m)
return material_recovery
// Step 5: Environmental Impact Assessment
function assess_environmental_impact(data, e_waste_generation, recycling_estimates):
environmental_impact = []
for G, R_e in zip(e_waste_generation, recycling_estimates):
E_impact = data["toxic_concentration"] * (G - R_e) + data["pollution_rate"] * G
environmental_impact.append(E_impact)
return environmental_impact
// Step 6: Sustainability Index Calculation
function calculate_sustainability_index(data, material_recovery, environmental_impact):
sustainability_index = []
for R_m, E_impact in zip(material_recovery, environmental_impact):
S = R_m / (E_impact + 1) // Avoid division by zero
sustainability_index.append(S)
return sustainability_index
// Step 7: Optimization and Decision Making
function optimize_management(data, e_waste_generation, recycling_estimates):
// Implement optimization logic here
// This could be linear programming or genetic algorithms to minimize environmental impact
```

```
// Step 8: Reporting
function generate_report(e_waste_generation, recycling_esti-
mates, material_recovery, environmental_impact, sustainabili-
ty_index):
print("E-Waste Generation:", e_waste_generation)
print("Recycling Estimates:", recycling_estimates)
print("Material Recovery:", material_recovery)
print("Environmental Impact:", environmental_impact)
print("Sustainability Index:", sustainability_index)
// Main Execution
function main():
data = collect_data()
time_periods = 10 // Example: for 10 years
e_waste_generation = predict_e_waste_generation(data, time_
periods)
recycling_estimates = manage_recycling(data, e_waste_genera-
tion)
material_recovery = calculate_resource_recovery(data, recy-
cling_estimates)
environmental_impact = assess_environmental_impact(data, e_
waste_generation, recycling_estimates)
sustainability_index = calculate_sustainability_index(data,
material_recovery, environmental_impact)
optimize_management(data, e_waste_generation, recycling_es-
timates)
generate_report(e_waste_generation, recycling_estimates, ma-
terial_recovery, environmental_impact, sustainability_index)
```

Reducing the hazards that electronic waste poses to the environment and human health requires efficient e-waste treatment. Numerous creative projects around the globe have shown effective methods for recycling and reusing electrical waste. There are a few noteworthy case studies which can be implemented by this algorithm. Australian engineer Professor Veena Sahajwalla was the first to propose the idea of micro-factories for recycling e-waste. In 2018, the first micro-factory was established at the University of New South Wales with the goal of recovering precious elements from discarded electronics, such as gold and rare earth metals. This technology promotes a circular economy by recovering valuable resources and repurposing glass and plastics into 3D printing filaments and industrial-grade ceramics. Using antiquated techniques that seriously contaminated the environment, Guiyu, a town in Guangdong province, China, was formerly the largest e-waste processing hub in the world. The local government responded by starting clean up activities, which included moving workshops to the National Circular Economy Pilot Industry Park. The goal of this change was to reduce health hazards and improve air and water quality by using safer recycling processes. (Zhang et al. 2015).

CONCLUSION

Sustainable e-waste management requires a comprehensive approach for the rapid evolution of technology and the rising amount of e-waste. For a circular economy E-waste can be turned from a liability into a useful resource by adopting a circular economy paradigm. Reducing waste and increasing resource recovery are achieved by designing products with lifetime, reparability and recycling in mind. Regulation and policy followed by the governments are essential in enacting strict e-waste regulations. The impact of e-waste on the environment delivers proper disposal, recycling and environmentally friendly activities . Customer awareness and education inform stakeholders about the negative effects that electronic trash. The environment and encouraging the appropriate disposal methods where the quantity of electronic waste produced can be greatly decreased. Equally crucial is empowering customers to make knowledgeable decisions regarding their technological purchases. Novel recycling technologies process the waste that can be made more efficient by investing in cutting-edge recycling technologies. Valuable resources can be recovered with minimal damage thanks to innovations like automated sorting systems and eco-friendly extraction techniques. Collaboration among industry stakeholders develop complete e-waste management systems, producers, merchants and recyclers must work together. Exchanging optimal methodologies and the implementation of Extended Producer Responsibility (EPR) legislation imposes accountability on producers for the complete lifecycle of their products. EPR promotes businesses to invest in recycling programs and design for sustainability by requiring them to manage their products after customer use. Community involvement Includes nearby communities in programs to recycle e-waste can promote involvement and a sense of duty. Community-led activities can help guarantee that e-waste is disposed of appropriately and also function as instructional platforms. International cooperation is necessary to address the worldwide burden of waste. Sharing knowledge, technology and best practices across borders can increase global capacity.

REFERENCES

An, D., Yang, Y., Chai, X., Xi, B., Dong, L., & Ren, J. (2015). Mitigating pollution of hazardous materials from WEEE of China: Portfolio selection for a sustainable future based on multi-criteria decision making. *Resources, Conservation and Recycling*, *105*, 198–210. DOI: 10.1016/j.resconrec.2015.10.025

Awasthi, K. A., & Li, J. (2017). Management of electrical and electronic waste: A comparative evaluation of China and India. *Renewable & Sustainable Energy Reviews*, *76*, 434–447. DOI: 10.1016/j.rser.2017.02.067

Awasthi, K. A., Zeng, X., & Li, J. (2016). Comparative examining and analysis of e-waste recycling in typical developing and developed countries. *Procedia Environmental Sciences*, *35*, 676–680. DOI: 10.1016/j.proenv.2016.07.065

Bakhiyi, B., Gravel, S., Ceballos, D., Flynn, M. A., & Zayed, J. (2018). Has the question of e-waste opened a Pandora's box? An overview of unpredictable issues and challenges. *Environment International*, *110*, 173–192. DOI: 10.1016/j.envint.2017.10.021 PMID: 29122313

Bhatia, M. S., & Srivastava, R. K. (2018). Analysis of external barriers to remanufacturing using grey-DEMATEL approach: An Indian perspective. *Resources, Conservation and Recycling*, *136*, 79–87. DOI: 10.1016/j.resconrec.2018.03.021

Cao, J., Lu, B., Chen, Y., Zhang, X., Zhai, G., Zhou, G., Jiang, B., & Schnoor, J. L. (2016). Extended producer responsibility system in China improves e-waste recycling: Government policies, enterprise, and public awareness. Renewable and Sustainable Energy Reviews, 62, 882–894. ttps://DOI: 10.1016/j.rser.2016.04.078

Chang, Y.-H., Yeh, C.-H., & Chang, Y.-W. (2013). A new method selection approach for fuzzy group multicriteria decision making. *Applied Soft Computing*, *13*(4), 2179–2187. DOI: 10.1016/j.asoc.2012.12.009

Chang, Y.-H., Yeh, C.-H., & Wang, S.-Y. (2007). A survey and optimization-based evaluation of development strategies for the air cargo industry. *International Journal of Production Economics*, *106*(2), 550–562. DOI: 10.1016/j.ijpe.2006.06.016

Chen, I.-S. (2012). A revised Inno-Qual performance system for higher education: The integrated applications of DEMATEL and ANP. *The Journal of the Operational Research Society*, *63*(4), 478–488. DOI: 10.1057/jors.2011.57

Chen, Z., Ming, X., Zhang, X., Yin, D., & Sun, Z. (2019). A rough-fuzzy DEMATEL-ANP method for evaluating sustainable value requirement of product service system. *Journal of Cleaner Production*, *228*, 485–508. DOI: 10.1016/j.jclepro.2019.04.145

Chi, X., Streicher-Porte, M., Wang, M. Y. L., & Reuter, M. A. (2011). Informal electronic waste recycling: A sector review with special focus on China. *Waste Management (New York, N.Y.), 31*(4), 731–742. DOI: 10.1016/j.wasman.2010.11.006 PMID: 21147524

Chi, X., Wang, W. Y. L., & Reuter, M. A. (2014). E-waste collection channels and household recycling behaviors in Taizhou of China. *Journal of Cleaner Production, 80*, 87–95. DOI: 10.1016/j.jclepro.2014.05.056

Chou, C. C. (2003). The canonical representation of multiplication operation on triangular fuzzy numbers. *Computers & Mathematics with Applications (Oxford, England), 45*(10–11), 1601–1610. DOI: 10.1016/S0898-1221(03)00139-1

Cole, C., Gnanapragasam, A., Cooper, T., & Singh, J. (2019). An assessment of achievements of the WEEE Directive in promoting movement up the waste hierarchy: Experiences in the UK. *Waste Management (New York, N.Y.), 87*, 417–427. DOI: 10.1016/j.wasman.2019.01.046 PMID: 31109542

Dou, Y., & Sarkis, J. (2013). A multiple stakeholder perspective on barriers to implementing China RoHS regulations. Resources Conservation and Recycling, 81, 92–104. ttps://DOI: 10.1016/j.resconrec.2013.10.004

Gabus, A., & Fontela, E. (1972). *World problems, an invitation to further thought within the framework of DEMATEL*. Battelle Geneva Research Centre.

Gabus, A., & Fontela, E. (1973). Perceptions of the world problematique: Communication procedure, communicating with those bearing collective responsibility (DEMATEL Report No. 1). Battelle Geneva Research Centre.

Gardas, B. B., Raut, R. D., & Narkhede, B. (2018). Modelling the challenges to sustainability in the textile and apparel (T&A) sector: A Delphi-DEMATEL approach. *Sustainable Production and Consumption, 15*, 96–108. DOI: 10.1016/j.spc.2018.05.001

Geng, Y., & Doberstein, B. (2008). Developing the circular economy in China: Challenges and opportunities for achieving ‘leapfrog development. *International Journal of Sustainable Development and World Ecology, 15*(3), 231–239. DOI: 10.3843/SusDev.15.3:6 PMID: 37309529

Govindan, K., Shankar, K. M., & Kannan, D. (2016). Sustainable material selection for construction industry – a hybrid multi criteria decision making approach. *Renewable & Sustainable Energy Reviews, 55*, 1274–1288. DOI: 10.1016/j.rser.2015.07.100

Gumus, S., Egilmez, G., Kucukvar, M., & Park, Y. S. (2016). Integrating expert weighting and multi-criteria decision making into eco-efficiency analysis: The case of US manufacturing. *The Journal of the Operational Research Society, 67*(4), 616–628. DOI: 10.1057/jors.2015.88

He, W., Li, G., Ma, X., Wang, H., Huang, J., Xu, M., & Huang, C. (2006). WEEE recovery strategies and the WEEE treatment status in China. *Journal of Hazardous Materials, 136*(3), 502–512. DOI: 10.1016/j.jhazmat.2006.04.060 PMID: 16820262

Hicks, C., Dietmar, R., & Eugster, M. (2005). The recycling and disposal of electrical and electronic waste in China-legislative and market responses. *Environmental Impact Assessment Review, 25*(5), 459–471. DOI: 10.1016/j.eiar.2005.04.007

Kojima, M., Yoshida, A., & Sasaki, S.Y. XU ET AL. (2009). Difficulties in applying extended producer responsibility policies in developing countries: Case studies in e-waste recycling in China and Thailand. *Journal of Material Cycles and Waste Management, 11*(3), 263–269. DOI: 10.1007/s10163-009-0240-x

Kong, S., Liu, H., Zeng, H., & Liu, Y. (2012). The status and progress of resource utilization technology of ewaste pollution in China. *Procedia Environmental Sciences, 16*, 515–521. DOI: 10.1016/j.proenv.2012.10.071

Kumar, A., & Dixit, G. (2018a). Evaluating critical barriers to implementation of WEEE management using DEMATEL approach. *Resources, Conservation and Recycling, 131*, 101–121. DOI: 10.1016/j.resconrec.2017.12.024

Kumar, A., & Dixit, G. (2018b). An analysis of barriers affecting the implementation of e-waste management practices in India: A novel ISM-DEMATEL approach. *Sustainable Production and Consumption, 14*, 36–52. DOI: 10.1016/j.spc.2018.01.002

Lin, K. P., Tseng, M.-L., & Pai, P.-F. (2018). Sustainable supply chain management using approximate fuzzy DEMATEL method. *Resources, Conservation and Recycling, 128*, 134–142. DOI: 10.1016/j.resconrec.2016.11.017

Liu, X., Tanaka, M., & Matsui, Y. (2006). Electrical and electronic waste management in China: Progress and the barriers to overcome. Waste Management & Research. *Waste Management & Research, 24*(1), 92–101. DOI: 10.1177/0734242X06062499 PMID: 16496875

Long, E., Kokke, S., Lundie, D., Shaw, N., Ijomah, W., & Kao, C.-C. (2016). Technical solutions to improve global sustainable management of waste electrical and electronic equipment (WEEE) in the EU and China. *Journal of Remanufacturing, 6*(1), 1–27. DOI: 10.1186/s13243-015-0023-6

Martilla, J. A., & James, J. C. (1977). Importance-performance analysis. *Journal of Marketing*, *41*(1), 77–79. DOI: 10.1177/002224297704100112

Moktadir, M. A., Ali, S. M., Rajesh, R., & Paul, S. K. (2018). Modeling the interrelationships among barriers to sustainable supply chain management in leather industry. *Journal of Cleaner Production*, *181*, 631–651. DOI: 10.1016/j.jclepro.2018.01.245

Ongondo, F. O., Williams, I. D., & Cherrett, T. J. (2011). How are WEEE doing? A global review of the management of electrical and electronic wastes. *Waste Management (New York, N.Y.)*, *31*(4), 714–730. DOI: 10.1016/j.wasman.2010.10.023 PMID: 21146974

Orlins, S., & Guan, D. (2016). China's toxic informal e-waste recycling: Local approaches to a global environmental problem. *Journal of Cleaner Production*, *114*, 71–80. DOI: 10.1016/j.jclepro.2015.05.090

Ramzan, S., Liu, C., Munir, H., & Xu, Y. (2019). Assessing young consumers' awareness and participation in sustainable e-waste management practices: A survey study in Northwest China. *Environmental Science and Pollution Research International*, *26*(19), 20003–20013. DOI: 10.1007/s11356-019-05310-y PMID: 31102225

Robinson, H. B. (2009). E-waste: An assessment of global production and environmental impacts. *The Science of the Total Environment*, *408*(2), 183–191. DOI: 10.1016/j.scitotenv.2009.09.044 PMID: 19846207

Salhofer, S., Steuer, B., Ramusch, R., & Beigl, P. (2016). WEEE management in Europe and China - A comparison. *Waste Management (New York, N.Y.)*, *57*, 27–35. DOI: 10.1016/j.wasman.2015.11.014 PMID: 26626812

Singh, N., Li, J., & Zeng, X. (2016). Global responses for recycling waste CRTs in e-waste. *Waste Management (New York, N.Y.)*, *57*, 187–197. DOI: 10.1016/j.wasman.2016.03.013 PMID: 27072617

Tansel, B. (2017). From electronic consumer products to e-wastes: Global outlook, waste quantities, recycling challenges. *Environment International*, *98*, 35–45. DOI: 10.1016/j.envint.2016.10.002 PMID: 27726897

Widmer, R., Oswald-Krapf, H., Sinha-Khetriwal, D., Schnellmann, M., & Böni, H. (2005). Global perspectives on e-waste. *Environmental Impact Assessment Review*, *25*(5), 436–458. DOI: 10.1016/j.eiar.2005.04.001

Zhang, B., Du, Z., Wang, B., & Wang, Z. (2019). Motivation and challenges for e-commerce in e-waste recycling under "Big data" context: A perspective from household willingness in China. *Technological Forecasting and Social Change*, *144*, 436–444. DOI: 10.1016/j.techfore.2018.03.001

Zhang, S., Ding, Y., Liu, B., Pan, D., Chang, C., & Volinsky, A. A. (2015). Challenges in legislation, recycling system and technical system of waste electrical and electronic equipment in China. [Y. XU ET AL.]. *Waste Management (New York, N.Y.)*, *45*, 361–373. DOI: 10.1016/j.wasman.2015.05.015 PMID: 26059074

Chapter 5
Towards a Greener Future:
Integrated Strategies for E-Waste Management

B. Yamini
https://orcid.org/0000-0003-3531-108X
SRM Institute of Science and Technology, India

Anish T. P.
https://orcid.org/0000-0002-4308-9645
R.M.K. College of Engineering and Technology, India

B. Maheswari
https://orcid.org/0000-0001-9711-7915
R.M.K. Engineering College, India

J. Praveenkumar
Rajalakshmi Institute of Technology, India

Siva Subramanian R.
https://orcid.org/0000-0002-7509-9223
R.M.K. College of Engineering and Technology, India

ABSTRACT

Technology has evolved and advanced over the years, and as a result, more electronic items are subscribed to and disposed of, turning into e-waste, a big threat to the environment and health. This chapter explores an integrated look into sustainably managing e-waste drawing from the related sources of the product, classification of e-waste, and the more traditional steps in the e-waste management life cycle,

DOI: 10.4018/979-8-3693-7383-5.ch005

which include collection, sorting, and recycling. It articulates the present practices and discusses ideas such as utilizing models of circular economy and the extended producer responsibility. Recycling and the technology of automated sorting processes are included together with the corresponding legislation on the national and international levels. Best practices are seen through case studies of successful e-waste management programs while the issues raised include technological limitations, financial constraints, or limitations in the policy framework. Some of the recommendations include international cooperation and innovation for change.

INTRODUCTION

Background

Electronics waste on the other hand is abbreviated as e-waste which is a term used to refer to all the damaged or unwanted electrical and electronic items. It covers a broad category of products which include the smartphones, laptops, washing machine, electric vehicles, motor, industrial and medical equipment among others (Khatun, A., & Dhara, N. 2022). With advancement in technology, the life cycle of electrical and electronic equipment remains shorter, thus result in increased e-waste. Current studies suggest that the global e-waste production totals at over 50 million metric tonnes generated per year and is projected to increase further with the growing trend of digitalization, increased consumption, and the speed of technological change (Jaiswal et al 2023). The problem with electronic waste is the fact that the majority of it is either recycled or dumped in an uncontrolled manner thus leading to terrible effects on the environment and health of the affected individuals (Yeruva et al 2024).

Importance of Sustainable Management

Appropriate management of e-waste has emerged as a critical concern for every country today due to effects on the environment and people's health. The electronics discarded include lead, mercury, and cadmium that when handled inappropriately pollutes soil, water and the atmosphere to an alarming degree. In addition to, e-waste is known to emit hazardous compounds into the atmosphere thus exacerbating air pollution and its effects to human health especially to communities which indulge themselves in the informal e-waste recycling. The last downside of e-waste is the economic impacts because these wastes are made of precious metals such as gold, silver, and copper which may be sold when properly recycled. Sustainable management aims at the preservation of these resources alongside the reduction of the Negative Impact to the environment (Monish, P., & Dhanabhakyam, M. 2021).

Purpose and Scope of the Survey

The aim of this survey paper is to discuss and compare various strategies of e-waste management and its integration such as circular economy, extended producer responsibility, and technological solutions. The purpose of this paper is to demonstrate the current state of e-waste management practices in the world, the problems encountered, and the opportunities for integrated solution in the future.

The survey goes beyond the misconceptions of traditional waste management and is centered on updated methods encompassing the management of the entire electro-products' life cycle. The survey also includes data on such topics as trends for the implementation of BI, legal requirements, innovations in the BI field, and BI implementation stories of successful enterprises. In doing so, the paper aims to present integrated strategies that may be useful for policymakers, industries, and researchers who are in search of more effective solutions of how to manage e-waste in the most proper way across the world.

E-WASTE GENERATION AND COMPOSITION

Sources of E-Waste

E-waste is produced from different fields, mainly attributed by the short cycles of technological products (Harikaran et al 2023). Key sources include:

Consumer electronics: This is the biggest source of e-waste and it involves gadgets such as mobile phones, computers, refrigerators, DVDs, TVs, washing machines among others. Due to the incessant release of newer efficient models these devices become obsolete and therefore are disposed within a short span of time which forms part of the e-waste (Ravisankar et al 2024).

Industrial equipment: Businesses produce electric devices waste by outdated equipment, communication equipment, and other electronic mechanism employed in manufacturing processes, automation, and energy management. These devices are sometimes thrown away once new technologies or new upgrades are introduced into the market thus increasing the amount of e-waste (Selvakumar et al 2023).

Other sources (e. g., medical devices): Some of the electronic products which find their way to e-waste include diagnostic machines, defibrillator, and patient monitoring systems. This leads to wastage of equipment due to technological development where the older models become irrelevant thus constituting to e-waste which in some ways is hazardous due to the delicate parts that are involved in making some of the healthcare technologies.

Figure 1. E-waste generation and composition

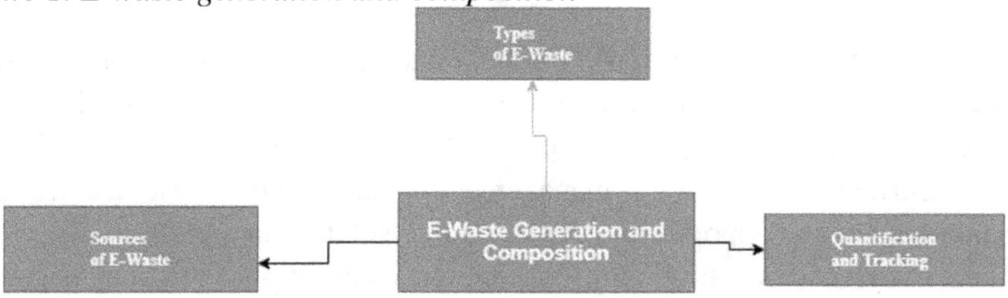

Types of E-Waste

E-waste is classified into two main types based on the materials they contain:

Hazardous components: In many electronic devices there are toxic, and hazardous substances are pressured like lead, mercury, cadmium and BBFRs or brominated flame retardants. These components are environmentally and health hazardous if not disposed of, they contaminate soil and water if they leach out. Also, burning of hazardous e-waste in an unproper method can lead to emission of toxic substances to the environment.

Non-hazardous components: E-waste also includes metals (Gold, silver, copper), plastics and Glass, albeit, not hazardous, are still useful in ways that their value has to be reclaimed through recycling. All these components can be reused thus enhancing the ideas of conservation of resources thus not having to mine for more raw materials (Aditya Vijayvargiya, Kapil Sahu2016).

Quantification and Tracking

Measuring and monitoring of e-waste has to be done in the best way to enable proper management of it. Several methods are employed to measure and monitor e-waste generation:

Reporting systems: International groups and governments employ reporting formats to capture the amount of flow of e-waste generated and treated. These systems thus enable the policy makers to establish the degree of the problem and mould the general framework towards infrastructural support towards recycling.

Digital tools: New technologies for the management of e-waste have been developed including the use of Blockchain in tracking the lifecycle of e-wastes. The following are findings: these tools provide accountability and increased transparency in the handling of the e-waste through showing correct methods of handling hazardous parts and efficient recycling of materials.

Implementing adequate methods of quantification and tracking helps facilitate the achievement of more effective policies and raise recycling rates as well as achieve the more efficient utilization of resources.

TRADITIONAL E-WASTE MANAGEMENT METHODS

Collection and Sorting

The e-waste management cycle commences with collection or gathering of such wastes in that they should be sorted (Sharma, H., & Kumar, H. 2022). Current collection has diverse forms, the developed world uses structural mechanism in collection while the developing world has an informal way of collecting. The formal systems may comprise of drop-off point, curbside, and manufacturers take-back programs. These methods enable recovery of used electronics and 'recycling' of them without actually getting mixed with the general flow of waste. On the other extreme, informal systems which are common in developing countries, have no structures, and may involve unauthorised people such scavengers who pick e-waste from dump sites. This has caused an informal handling of the stocks which in turn exposes the workers to health hazards of the different hazardous stocks. This is the process of categorizing e-waste in different groups depending on the type of material. Manual sorting is still dominant, especially in the third world countries where employees physically dismantle the equipment to recover usable parts. Nevertheless, the improved facilities follow the automated sorting systems that involve NIR spectroscopy (near-infrared), eddy and magnetic currents to sort the pile through various types of materials comprising plastics, metals as well as glass.

Recycling and Recovery

Included in the definition of recycling e-waste are the recovery of reusable components as well as the proper disposal of the toxic elements (Bansal et al 2023). Different techniques applied in the recycling of e-waste are mechanical recycling and the pyrometallurgical and hydrometallurgical processes (Shrivastava, A & Kumar, V. 2023).

Mechanical recycling: This includes procedures such as shredding coupled with crushing with an aim of sorting metals, plastics among other substances that make up the e-waste. After coming into the recycling centre, electronic gadgets are then crushed by shredders, crushers and the like then sorted by the kind of material. Commonly, this is the first procedure in many recycling centres across the world.

Pyrometallurgical and hydrometallurgical methods: These techniques are employed in recycling circuits from the e-waste with the aim of extracting valuable metals. Pyrometallurgy is a process which was designed to melt a metal such as gold, silver and copper with other material so as to enable them to be separated. Hydrometallurgy involves leaching of metals from the elements in the printed circuit by chemical solutions, then metal recovery by precipitation, electro deposition. Both techniques are very efficient, however, they need certain apparatus and are very energy-consuming.

Figure 2. Traditional e-waste management methods

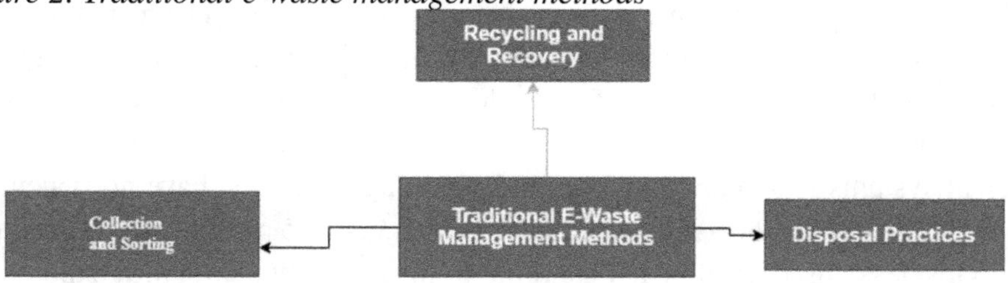

Disposal Practices

Landfilling: However, landfilling is still prevalent due to this reason and is a leading method of disposal especially in areas that do not have sufficient capabilities to recycle. Some of the contents that are found in e-waste include lead, mercury and cadmium which can pollute the soil and water resources in the long run. This means that e-waste which is dumped in the landfill sites also leads to the loss of many raw materials that could otherwise be recycled.

Incineration: Some of e-waste is burnt, either, to generate energy from waste or just to minimize the volume. However, it has severe demerits especially when it involves burning the e-waste, as does it emits toxic chemicals like dioxins and heavy metals into the atmosphere where they make contributions to air pollution and have adverse effects on the health of neighbors. As useful as the conventional e-waste management practices may be in some occasions they have their drawbacks. They may lead to adverse impacts on the environment and poor optimal resource exploitation, thus calling for interlinked, sustainable solutions (Preethi et al 2024).

INTEGRATED APPROACHES FOR E-WASTE MANAGEMENT

Circular Economy Models

Due to longer product life cycles and the reuse of the material, circular economy is an important strategy for e-waste sustainable management (Baragde, D. B., & Jadhav, A. U. 2020) (Vig, S., & Agarwal, R. N. 2021). Two essential components of circular economy models are:

Design for disassembly: This includes developing Electronic Product that is made up of sections that may be opened and closed, repaired, or in some cases upgraded. Through simplification of the same, it will be much easier to cycle the products and get value added materials with much ease. This design philosophy makes recycling processes less complicated and gives manufacturers the incentive to create devices which are easier to restore for reuse or recycle.

Product life extension: Thus, repair, refurbishment or renewal of the life cycle of electronics contributes to elongation of life of the products thus decreasing the rate at which users have to dispose of their gadgets, hence lowering the e-waste. Manufacturers can also sell servicing options in Electronics, sell spares, or introduce the upgrade programs that make it possible for the consumers to continue using a specific device for longer durations. It helps to cut the utilization of raw material and decrease the effects of production and disposal on the environment.

Extended Producer Responsibility (EPR)

E-waste recycling management through Extended Producer Responsibility (EPR) rests on the producers. EPR frameworks thus put into the shoulders of producers the responsibility of managing the life cycle of the products right from post-consumer disposal. Key elements of EPR include:

EPR frameworks and policies: Most of the countries have passed EPR legislation that calls on producers to fund the recovery, recycling, and disposal of e-waste. It is fundamental in encouraging manufacturers to come up with sustainable products, minimizing on wastage and recovering materials effectively. Some of EPR policies are the European Union Waste Electrical and Electronic Equipment (WEEE) Directive where producers are compelled to enjoy certain recycling objectives.

Figure 3. Integrated approaches for e-waste management

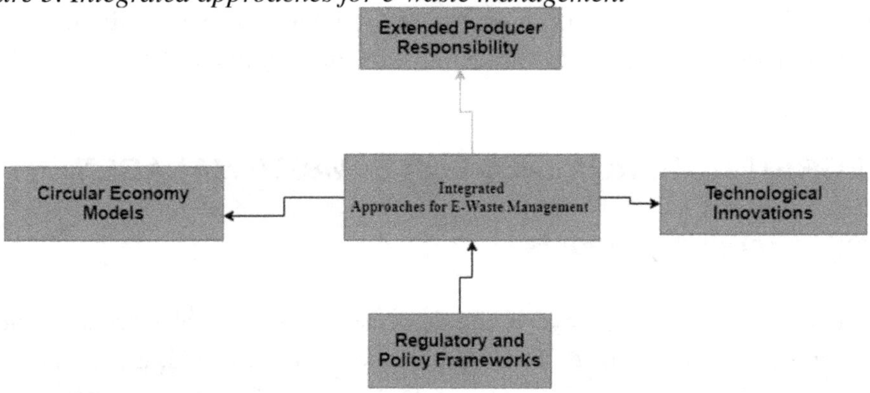

Technological Innovations

Technological advancements play a critical role in improving e-waste management by increasing the efficiency of recycling and recovery processes (Vijayalakshmi et al 2024)

Advanced recycling technologies: Such new approaches are chemical recycling that helps to reuse valuable metals, as well as decrease the hazardous effects of pyrometallurgical and hydrometallurgical processing techniques. It is equally important to recognize technologies like bioleaching that employ microorganisms to extract metals from e-waste with less harm to the environment.

Automated sorting systems: The enabling technologies for advanced sorting of e-waste by different components with the help of AI and Robotics. These systems enhance two critical areas of operations, which are precision in the management of materials and efficiency in the process due to reduced manual labor thus resulting in high recycling rates.

Regulatory and Policy Frameworks

Good e-waste management call for good legal controls and policies. International and national regulations ensure that e-waste is managed responsibly:

International regulations and agreements: International treaties including Basel Convention aim to control exportation of dangerous e-waste and subsequently dumping in the third world countries. They help in the management and proper disposal of e-waste across the world.

National policies and enforcement: Individual nations have provided their policies to sanction the e-waste management systematically that includes the collection, recycling and disposal of e-waste. The implementation of these policies is very vital in curbing the situations by following the policies to the letter in a bid to eradicate the impacts of e-waste on environment and health of the people. In some countries national programmes motivate the consumers to return used electronics through take-back programmes and create transparency for manufacturers concerning recycling objectives.

Such approaches to e-waste management include: circular economy models, EPR, innovative technologies, and legislation that support one another in mitigating the problems of e-waste management for sustainable development.

ENVIRONMENTAL AND HEALTH IMPLICATIONS

Impact of E-Waste on Ecosystems

A direct consequence of e-waste which is disposed off in an improper way is that it pollutes, both the soil and water, and air.

Soil and water contamination: E-waste comprises of dangerous materials for instance lead, mercury, cadmium and other toxic elements. When e-waste is buried in landfill or simply dumped in open Iand, these poisonous substances spread on the ground and pollute it. Otherwise, over time, such contamination seeps into the water table or into other water sources near the area and presents considerable risks to the aquatic life forms as well as water systems. The pollution in the food chain is very dangerous, because after these pollutants appear the food chain, it is hard to eliminate them, as they have continuation in organisms and stays there for long period of time in the ecosystem.

For instance, when heavy metals are released into water the fish and other aquatic life forms will have developmental and reproductive problems which will affect the richness and sustainability of species on planet earth.

Air pollution: In respect to incineration or burning of e-waste either in the commercial furnace or in the open air, toxic gases, dioxins, furans, and heavy metal particulates are emitted to the atmosphere. These emissions cause air pollution and are a direct peril to surrounding or inhabitants wildlife and humans. The loss of old refrigerators and air conditioners filled with chlorofluorocarbons (CFCs) adds onto climate change by depleting the ozone layer.

Health Risks to Human Populations

E-waste is a major health concern to man especially the people involved in the e-waste recycling process or the people who reside in the e-waste destination areas. These are risks that are associated with contact of dangerous products and the consequences that come with it.

Exposure to hazardous substances: Many employees in the overall informal e-waste recycling industries lack personal protective equipment resulting in the workers being in direct contact with toxic materials. For example, with lead from old batteries or soldered components, enters the blood stream and causes neurological disorders especially to children. People may suffer from acute renal toxicity and damaged nervous system by mercury which originates from dispersed fluorescent bulbs or batteries. Furthermore, people living nearby the dumps are also likely to breathe or take in the toxic fumes and particle present in e-waste.

Health outcomes and case studies: Some of the concerns that researchers have associated with e-waste are health problems. Concerning health impacts, people living in Guiyu, China, an area which is home to many e-waste recycling firms, experiences high diseases such as respiratory diseases, skin ailments, and even cancer. Indeed, compact areas of these locations present much higher blood lead levels in children than in other comparable groups not exposed to BCTs. Moreover, pregnant women through the ingestion ion of toxic chemicals from e-waste experience high risks of delivery of physically or mentally damaged babies, premature births, or even miscarriages.

The effect of e-waste not only shows the pollution of environment in terms of ecosystems but also health hazard, setting more demands on the sustainable management of e-waste. If not managed well, the dangerous components used in e-waste will persist to cause adverse impact on the environment and the health of the people. These threats can be managed by implementing sustainable solutions in areas of right recycle, disposal measures standards as well as protective gear measures for personnel.

CHALLENGES AND BARRIERS

Challenges that exist with the effective management of e-waste stem from technology, economics, and the regulations. Such barriers make it difficult to establish proper, environmentally sound mechanisms to manage and dispose e wastes.

Technological Challenges

The first challenge that is likely to be encountered is the fact that technologies that are currently available for e-waste recycling have their own drawbacks. Mechanical pyrometallurgical and hydrometallurgical methods of recycling are widely used in practice, but they have drawbacks. Thus, conventional recycling processes, for example, mechanical recycling processes, are ineffective in the removal of complex materials and as a consequence, the recovery rates of valuable components such as rare earth metals are low. In the same way, pyrometallurgical and hydrometallurgical methods also consume much energy, are hazardous to the environment since they emit toxic gases and chemical compounds. Furthermore, the existing configurations of electronics are not designed to accommodate the variety of materials that exist in the current electronic devices such as metal, plastics, glass, and toxic elements. Usage of nanomaterials and other superior composites in electronics makes matters worse for recyclability. Thus, the necessity of inventive approach to recycling technologies, which can be illustrated by such techniques as the bioleaching based on the usage of microorganisms.

Economic and Financial Constraints

Economic factors are another challenge which has a considerable influence on the sustainability of e-waste management. The expenses of collecting, sorting and recycling e-waste outweighs the revenue that is acquired from the sale of the recovered materials. For instance, they can be economically recycled including gold and silver while other materials like plastics and even some metals are even considered to be of little value hence their recycling cannot be economically feasible. These cost issues are further compounded by high costs of setting up and supporting modern recycling plants especially in the developing world where informal recycling mostly operates. The major problem relies in the fact that developed recycling does not receive funds and investments, and therefore these regions cannot shift to safer and more efficient practices. While in some countries where the recycling processes are more advanced, the cost is still being borne by the local governments and the producers some of whom may not have sufficient capacity to institute efficient e-waste management systems.

Regulatory and Policy Challenges

Another major challenge is that the regulation of E-commerce varies in one country to another and from one region to another. Thus, a number of countries, especially those from the European Union, have strong laws regulating e-waste, for instance, the Waste Electrical and Electronic Equipment (WEEE) Directive, yet others lack proper legislation and, even if legislation are present, the corresponding rules are not enforced appropriately. This lack of proper regulatory policy enables e-waste to be exported to other countries where regulatory policies are not as strict and where e-waste is dismantled in very harmfully ways.

E-waste management on the international level is hampered with the difficulty of finding standardized procedures all over the world. Moreover, when some regions have adequate regulations, enforcement problems still arise. Due to corruption, the poor structures, and lack of supervision, most laws are not well implemented hence promoting informal recycling and illegal dumping. Failure in enforcing the current e-waste policies coupled with limited cooperation from other countries' governments makes implementation of these regulations a drain.

Solving these problems implies developing the technologies, investing in recycling facilities, and establishing the consistent legislation that would control the compliance with the main principles of the sustainable development.

FUTURE DIRECTIONS AND RESEARCH

Future of global e-waste Management will be dictated by technology, policy and global cooperation. Future developments as well as changes in policies are expected to be the main source of future solutions.

Emerging Technologies

Technological advancement plays a crucial role in increasing the effectiveness as well as the environmental-friendly process of e-waste recycling. The present approaches cannot adequately reclaim useful materials from compacted electronics products, for instance. Potential advancements in recycling and recovery include:

Bioleaching: A potential environmentally sound substitute to the conventional chemistry procedures, bioleaching is the process by which microorganisms are used to extract metals like copper, gold, rare earth metals among others from the e-waste. It is also safer to the environment since it means the reduction in the use of the toxic chemicals and the energy used in the process.

Robotics and AI for automated sorting: Technological advancement especially in robotics and Artificial Intelligence may help in sorting the components of e-waste. Automated sorting systems employing AI technology can better identify and sort the materials required in the sorting process hence increasing recovery of valuable materials. This also helps in cutting down on the use of manual power which is usually very dangerous and time-consuming.

Advanced thermal and chemical methods: Scholars are studying higher order thermal treatment approach like pyrolysis which provide recycle worthy material from the e-waste focuses without emitting hazardous vapours. Chemical recycling also has newer variations like the green solvents application which have possibility of recovering metals as well as other contents it a comparatively better and environmentally friendly manner.

Policy and Regulatory Innovations

The analysis also shows that there is a need to ensure development of strong policy frameworks that enhance the achievement of the e-waste management goals. Recommendations for better frameworks include the enlargement, of coverages of Extended Producer Responsibility (EPR), which requires producers to be fully accountable for their products from manufacture to disposal. To strengthen EPR, global policymakers are considering reforms that would:

Incentivize eco-design: Measures for more sustainability can be created by regulation of stakeholders of industries and manufacturers to design products that can easily be repaired, reused and recycled. These are; the design for disassembly, minimization on the use of hazardous materials, and incorporating post-consumer content in the end-products.

Implement stricter recycling targets: Recycling rates can be legislated by governments and certain financial initiatives which can be given to industries managing to achieve set standards. This can include, the provision of funds for the recycling facilities or fines for firms or organizations that do not meet the requirements of e-waste management.

Global Collaboration

Thus, the idea of E-waste management will not succeed if it does not go under international cooperation. As mentioned earlier, electronics supply chain is global, and thus, cross boundary issues of e-waste creates a need for countries to work together in order to address the issue. Strategies for enhanced collaboration include:

Harmonizing international standards: Other international treaties which help in governing the cross-boundary transportations of e-waste include the ones under the Basel Convention. Most countries have a harmonized standards on the recycling and disposal to avoid loopholes and enhance proper management of e-waste.

Sharing best practices and technologies: Advanced countries can extend their technical assistance in the form of outright aid or helping the less developed countries to set up the proper e-waste recycling systems. This form of partnership can assist infrastructure development, technological advancement and processor training for effective management of e-waste.

Thus, the continued e-waste management will depend on the development of new technologies, enhancement of policies, and international collaboration. In combined, these measures help to reduce environmental and health effects of e-waste and contribute to the formulation of a post-COVID-19 environment for a more holistic management of ever-increasing volume of electronic waste across the globe.

CONCLUSION

In this paper we discussed the increasing problem of e-waste and also presented several-fold integrated management strategies that can provide solutions. Some of the critical techniques including the circular economy strategies mainly emphasize on the management of product life cycles and designs that enable easy disassembling for recycling. The details of the regulation affecting the Producer and the trader in terms of Environmental management and Extended Producer Responsibility Policies emphasizes on management of the environmental impact of the product throughout the value chain. Technological factors include new methods of recycling and sorting methods that increase efficiency in the recovery of PVDs a legal factor addressing structure of these actions includes the national and international legislation.

These case studies presented discussed successful e-waste management programs from different parts of the world, and they show the big role government support, awareness and technology plays in management of e-waste. The difficulties that could limit efficient management of e-waste included; fluctuating laws that govern management of e-waste, high costs, problems with the current technologies.

It recommended that governments should enhance regulation measures through increasing the targets for the recycling of products, encouraging integration of the eco-design for products and backing EPR systems. They should also spend on construction of proper facilities to dispose off e waste and also on enlightening the public on the proper methods of disposing off e waste. Original Equipment Manufacturers should take into consideration production designs and structures which can be easily disassembled and recycled, in addition should abide by the EPR regulations on management of wastes. Flowsheets of the Research institutions should therefore strive to enhance efficient and sustainable recycling methods like bio leaching and mechanical advanced methods.

There should be more studies about new possibilities in further exploring the technology, and how they can be integrated into the current existing platforms. Further research should be made to bring light and investigate the global initiatives to create common approach and guidelines for the management of e-waste, especially in countries with less developed infrastructure. One of the key recommendations in addressing the global e-waste concern is therefore require multi-stakeholder effort.

REFERENCES

Bansal, A., Verghese, S., Massey, M. H., Massey, D. D., & Singh, R. (2023). A Review Of The Effects Of Environmental Heavy Metal Contamination From Electronic Waste On Human Health. *IJRAR-International Journal of Research and Analytical Reviews*, *10*(4), 40–82.

Baragde, D. B., & Jadhav, A. U. (2020). Circular economy model for the E-waste management sector. In *Handbook of Research on Entrepreneurship Development and Opportunities in Circular Economy* (pp. 216–230). IGI Global.

Harikaran, M., Boopathi, S., Gokulakannan, S., & Poonguzhali, M. (2023). Study on the source of e-waste management and disposal methods. In *Sustainable Approaches and Strategies for E-Waste Management and Utilization* (pp. 39–60). IGI Global.

Jaiswal, S. K., Mukti, S. K., & Rath, K. C. (2023). E-waste control and its recycling to build sustainable society in the global context. In Handbook of Research on Applications of AI, Digital Twin, and Internet of Things for Sustainable Development (pp. 200-222). IGI Global.

Khatun, A., & Dhara, N. (2022). E-Waste Management: A New Concern for Environmental Sustainability. In Smart Cities, Citizen Welfare, and the Implementation of Sustainable Development Goals (pp. 222-238). IGI Global.

Monish, P., & Dhanabhakyam, M. (2021). Sustainability Strategies for Developing SMEs and Entrepreneurship. In *Handbook of Research on Sustaining SMEs and Entrepreneurial Innovation in the Post-COVID-19 Era* (pp. 527–547). IGI Global.

Preethi, P., Deepak, T., Anusha, S. V., & Bhagya, D. (2024). E-Waste Collection and Recycling Best Practices and Innovations. In Sustainable Solutions for E-Waste and Development (pp. 33-43). IGI Global.

Ravisankar, A., Shanthi, A. L., Lavanya, S., Ramaratnam, M. S., Krishnamoorthy, V., & Boopathi, S. (2024). Harnessing 6G for Consumer-Centric Business Strategies Across Electronic Industries. In AI Impacts in Digital Consumer Behavior (pp. 241-270). IGI Global.

Selvakumar, S., Adithe, S., Isaac, J. S., Pradhan, R., Venkatesh, V., & Sampath, B. (2023). A study of the printed circuit board (PCB) e-waste recycling process. In Sustainable Approaches and Strategies for E-Waste Management and Utilization (pp. 159-184). IGI Global.

Sharma, H., & Kumar, H. (2022). Sustainable Collection and Classification of E-Waste: A Proposed Computer Vision Technology-Based Framework. Available at *SSRN* 4049693.

Shrivastava, A., & Kumar, V. (2023). Strategies for Waste Recycling, Reuse, and Reduction. In Sustainable Approaches and Strategies for E-Waste Management and Utilization (pp. 140-158). IGI Global.

Vig, S., & Agarwal, R. N. (2021). E-Waste Management in India: Need for a Circular Approach. In Examining the Intersection of Circular Economy, Forestry, and International Trade (pp. 165-173). IGI Global.

Vijayalakshmi, V., Radha, R., & Sharanya, S. (2024). Modern Technological Innovation in Digital Wase Management. In Computational Intelligence for Green Cloud Computing and Digital Waste Management (pp. 152-171). IGI Global.

Vijayvargiya, A., & Sahu, K. (2016). A Review Paper on "E-Waste" Technology. *International Journal of Engineering Research & Technology (Ahmedabad)*, *4*(32).

Yeruva, J. R., Guduru, A. K., & Saripalli, B. P. (2024). Technological Advancements in E-Waste Management and Resource Recovery. In Sustainable Solutions for E-Waste and Development (pp. 275-288). IGI Global.

Chapter 6
Chemical and Biological Processes for E-Waste Recycling

P. Selvakumar
https://orcid.org/0000-0002-3650-4548
Department of Science and Humanities, Nehru Institute of Technology, India

Ritu Dahiya
https://orcid.org/0000-0001-5788-006X
Chhotu Ram Arya College, India

T. C. Manjunath
https://orcid.org/0000-0003-2545-9160
Rajarajeswari College of Engineering, India

Santosh Kumar Nathsharma
https://orcid.org/0000-0003-3935-2938
Stewart Science College, India

ABSTRACT

The proliferation of technology-driven consumption coupled with shortened product lifecycles contributes significantly to the escalating volumes of e-waste generated worldwide. As societies become increasingly dependent on electronic gadgets for communication, entertainment, and daily tasks, the disposal and recycling of these devices pose critical environmental and health challenges. However, within these challenges lie opportunities for innovation, sustainable practices, and economic growth through the establishment of effective e-waste recycling systems. Furthermore, the complexity of electronic devices, characterized by a wide variety of materials and components, complicates the recycling process. Moreover, the lack of standardized

DOI: 10.4018/979-8-3693-7383-5.ch006

e-waste collection systems and regulations in many regions contributes to informal recycling practices, often carried out under unsafe conditions by informal workers in developing countries. Despite these challenges, e-waste recycling presents significant opportunities for sustainable development and economic benefits.

INTRODUCTION

The proliferation of technology-driven consumption coupled with shortened product lifecycles contributes significantly to the escalating volumes of e-waste generated worldwide. As societies become increasingly dependent on electronic gadgets for communication, entertainment, and daily tasks, the disposal and recycling of these devices pose critical environmental and health challenges. However, within these challenges lie opportunities for innovation, sustainable practices, and economic growth through the establishment of effective e-waste recycling systems

Furthermore, the complexity of electronic devices, characterized by a wide variety of materials and components, complicates the recycling process. Effective separation, recovery, and recycling. Moreover, the lack of standardized e-waste collection systems and regulations in many regions contributes to informal recycling practices, often carried out under unsafe conditions by informal workers in developing countries. Despite these challenges, e-waste recycling presents significant opportunities for sustainable development and economic benefits. Recycling e-waste not only conserves natural resources by extracting valuable materials thereby closing the loop in the product lifecycle and promoting a circular economy. Additionally, the recycling industry has the potential to create jobs, particularly in areas where formal e-waste recycling infrastructure is established and supported. These regulatory measures aim to improve e-waste collection rates, enhance recycling efficiency, and minimize the environmental footprint of electronic devices throughout their lifecycle.

Technological advancements in e-waste recycling are also driving innovation in the field. Processes such as mechanical shredding, magnetic separation, and hydrometallurgical techniques are being developed and refined to recover valuable metals and materials from electronic waste efficiently. Moreover, research and development efforts focus on improving the scalability and cost-effectiveness of recycling technologies to meet the growing demand for sustainable waste management solutions. Collaboration between governments, industry stakeholders, academia, plays a crucial role in advancing e-waste recycling practices globally. Partnerships facilitate knowledge sharing, development of best practices for responsible e-waste management. Initiatives to raise awareness among consumers of also contribute to shaping a culture of sustainability and environmental stewardship. Looking ahead, addressing the challenges and seizing the opportunities presented by e-waste recycling requires concerted efforts on multiple fronts. Governments must continue to strengthen. Industry leaders should invest in research and innovation to enhance recycling technologies and explore alternative materials that are easier to recycle (Abdelbasir, S. M., et al., 2018). Consumers can contribute by adopting responsible purchasing habits, prolonging the lifespan of electronic devices, and participating in e-waste recycling programs. In conclusion, e-waste recycling represents a pivotal area in the quest for sustainable development and environmental stewardship. Embracing a circular economy approach to electronic waste management holds the promise of creating a more resilient and resource-efficient society, where waste is minimized, and materials are reused and recycled in a closed-loop system.

OVERVIEW OF CHEMICAL PROCESSES FOR E-WASTE RECYCLING

E-waste recycling involves a range of complex processes designed to recover valuable materials from discarded electronic devices while minimizing environmental and health impacts. Chemical processes play a crucial role in this endeavor, offering efficient methods to extract and purify metals and other valuable components from electronic waste. One of the primary chemical processes employed in e-waste recycling is hydrometallurgy. Hydrometallurgical methods involve the use of aqueous solutions to dissolve metals from electronic waste materials (Selvakumar, P., 2024). The process typically begins with mechanical shredding and sorting of e-waste to separate different components. Subsequently, the shredded materials are subjected to leaching, where metals are dissolved into solution under controlled conditions. These acids facilitate the dissolution of metals like copper, nickel, zinc, and aluminum from electronic circuit boards, connectors, and other components. Pyrometallurgy involves the use of high temperatures in the presence of oxygen or other gases to smelt and recover metals from e-waste materials. During pyrometallurgical processing, e-waste is subjected to thermal decomposition, where organic materials are burned off, and metals are melted and separated from non-metallic residues. Common pyrometallurgical techniques include smelting, incineration, and plasma arc melting, each tailored to handle specific types of e-waste and recover valuable metals efficiently. Smelting processes utilize furnaces to melt e-waste materials, separating metals like copper, lead, and tin from slag and other residues. Incineration involves the combustion of organic materials in e-waste at high temperatures, producing ash that can be further processed to recover metals through subsequent treatments (Ramakrishnan T.,et al., 2022). Plasma arc melting utilizes high-energy plasma torches to melt and vaporize materials, enabling precise control over metal recovery and minimizing emissions compared to traditional incineration methods. Furthermore, bioleaching has emerged as a promising green technology for e-waste recycling, utilizing microorganisms to extract metals from electronic waste materials. Bioleaching processes harness the metabolic activities of bacteria and fungi to solubilize metals from e-waste under mild conditions, reducing energy consumption and environmental impact compared to traditional chemical processes. Microorganisms produce organic acids and enzymes that facilitate the dissolution of metals such as copper, gold, and silver from electronic components, which can be subsequently recovered and purified using conventional hydrometallurgical techniques. Overall, chemical processes for e-waste recycling continue to evolve with advancements in technology and environmental stewardship. Innovations in process engineering, materials science, and automation are driving improvements in efficiency, resource recovery, and environmental sustainability within the e-waste recycling industry (Pariatamby, A.,et al., 2015).

Pyrometallurgical Processes: Extracting Metals From E-Waste

E-waste recycling is increasingly crucial due to the rapid proliferation of electronic gadgets and the consequential rise in waste volumes globally. This process is particularly suitable for e-waste with high metal content, such as printed circuit boards (PCBs), connectors, and integrated circuits (Paramasivam Selvakumar, et al., 2024). The primary objective of pyrometallurgical processes is to separate metals from non-metallic residues, transforming complex electronic waste into valuable metal concentrates that can be reused in manufacturing or sold as raw materials to various industries. Smelting is a fundamental pyrometallurgical technique used extensively in e-waste recycling. It involves heating e-waste materials in furnaces to high temperatures, typically above 1000°C, to melt metals and separate them from slag and other non-metallic components. During smelting, e-waste is fed into a furnace along with fluxing agents such as limestone or silica, which facilitate the formation of slag by reacting with impurities like oxides and sulfur. The molten metal, heavier than the slag, settles at the bottom of the furnace and is tapped off periodically. Common metals recovered through smelting include copper, lead, tin, depending on the composition of the e-waste feedstock.

Incineration is another pyrometallurgical method employed in e-waste recycling, particularly for materials containing organic components such as plastics and polymers. In this process, e-waste is combusted at elevated temperatures in the presence of excess air or oxygen to facilitate complete oxidation of organic materials. The combustion generates heat, which is utilized to melt metals present in the e-waste, allowing them to be recovered from the resulting ash and flue gases. Incineration can be coupled with energy recovery systems to generate electricity or heat, enhancing the overall efficiency and sustainability of the recycling process

(Rautela, R., et al., 2021). Plasma arc melting (PAM) represents an advanced pyrometallurgical technique increasingly applied in e-waste recycling. PAM utilizes a high-energy plasma torch to generate temperatures exceeding 3000°C, effectively melting and vaporizing materials fed into the plasma arc chamber. This process offers precise control over temperature and atmosphere, enabling the recovery of high-purity metals from complex e-waste streams. The intense heat of the plasma arc breaks down organic compounds and facilitates the separation of metals from non-metallic residues, producing clean metal concentrates suitable for subsequent refining and reuse.

Innovation in pyrometallurgical processes for e-waste recycling is driven by advancements in process engineering, materials science, and environmental stewardship. Research focuses on improving energy efficiency, optimizing metal recovery rates, and developing sustainable solutions for handling hazardous residues. Integrated approaches that combine pyrometallurgical and hydrometallurgical techniques are explored to maximize resource recovery from e-waste while minimizing environmental footprint and energy consumption (Sahin, M., et al., 2015). Furthermore, the circular economy principles guide the evolution of pyrometallurgical processes in e-waste recycling, emphasizing the importance of closing material loops and promoting the reuse of recovered metals in new product manufacturing. In conclusion, pyrometallurgical processes represent essential tools in the arsenal of techniques for e-waste recycling, enabling the recovery of valuable metals from discarded electronic devices. As the demand for electronics continues to grow, so too does the imperative to recycle e-waste efficiently and sustainably. By advancing pyrometallurgical technologies and embracing environmental responsibility, the e-waste recycling industry can contribute significantly to resource conservation.

Hydrometallurgical Processes: Leaching and Solvent Extraction

Hydrometallurgical processes constitute a vital component of the electronic waste (e-waste) recycling landscape, offering effective methods for recovering valuable metals from discarded electronic devices. E-waste, characterized by its complex composition of metals, plastics, ceramics, and precious metals, presents significant challenges and opportunities for resource recovery and environmental stewardship. Hydrometallurgy involves the use of aqueous solutions and chemical reactions to dissolve and extract metals from e-waste materials, followed by purification steps to obtain high-purity metal concentrates suitable for reuse in manufacturing or as raw materials in various industries (Selvakumar, P. et al., 2024). The first stage of hydrometallurgical processing is leaching, a critical step where metals are dissolved from e-waste into aqueous solutions under controlled conditions. Leaching involves several key parameters, including the choice of leaching agent, temperature, pressure, and agitation, tailored to maximize metal recovery while minimizing the dissolution of unwanted materials.

Sulfuric acid is widely utilized in e-waste recycling due to its strong oxidizing properties and ability to dissolve a wide range of metals, including copper, nickel, zinc, and aluminum. Nitric acid is effective for dissolving precious metals like gold and silver, forming soluble metal nitrates that can be subsequently recovered. Hydrochloric acid is employed for selective leaching of metals such as tin and lead from e-waste materials, while organic acids offer milder leaching conditions suitable for specific applications. The leaching process begins with mechanical preprocessing of e-waste, where devices are shredded or crushed to reduce particle size and facilitate the exposure of metals to the leaching solution (Pender, J. P., et al., 2020). The shredded material is then subjected to leaching in reactors or tanks, where the leaching agent is circulated through the e-waste feedstock. Parameters such as temperature, acidity, and residence time are carefully controlled to optimize metal dissolution rates and minimize the consumption of reagents. Following leaching, the resulting aqueous solution, known as pregnant leach solution (PLS), contains dissolved metals along with impurities and residual organic matter from the e-waste feedstock.

Ion exchange resins offer an alternative approach to selectively adsorb metal ions from PLS based on their affinity for specific resin functional groups. Ion exchange processes involve passing PLS through columns packed with ion exchange resins, where metal ions are exchanged with ions immobilized on the resin beads. Once saturated with metal ions, the resin can be regenerated using appropriate eluents to recover metals for subsequent processing. Electro-winning represents the final step in hydrometallurgical processing, where metals are recovered from purified solutions by electrodeposition onto cathodes (Valero Navazo, J. M., et al., 2014). Electro-winning utilizes electrolytic cells equipped with inert cathodes and anodes, between which a direct electric current is applied. Electro-winning is employed to produce high-purity metals such as copper, nickel, and tin from e-waste-derived solutions, ensuring they meet stringent quality standards for reuse in manufacturing. The evolution of hydrometallurgical processes for e-waste recycling is driven by advancements in process technology, materials science, and environmental sustainability (Tatariants, M., et al., 2018). Innovations focus on improving efficiency, reducing chemical consumption, minimizing waste generation, and enhancing the recovery rates of valuable metals from increasingly complex e-waste streams. Additionally, efforts are directed towards developing green and sustainable technologies that reduce the environmental footprint of hydrometallurgical operations, such as using renewable energy sources and optimizing water usage.

Electrochemical Processes: Electrolysis and Electrochemical Recovery

Electrochemical processes recover valuable metals and materials from discarded electronic devices. These processes harness the principles of electrolysis and electrochemical recovery to selectively extract metals from e-waste materials, providing sustainable solutions for resource recovery and environmental stewardship. At the heart of electrochemical recycling is electrolysis, a process that utilizes electrical current to induce chemical reactions at electrodes immersed in an electrolyte solution. The process begins with the mechanical preprocessing of e-waste, where devices are shredded or crushed to facilitate the exposure of metals to the electrolyte solution (Sugumaran K. M, D., et al., 2024). During electrolysis, e-waste materials are immersed in an electrolyte containing suitable ions that can be reduced or oxidized at the electrodes. Typically, the e-waste acts as the anode, where metals where they are reduced back to their metallic form and accumulate as a pure metal layer. The electrolyte composition and operating conditions such as voltage, current density, temperature, and electrolyte flow rate are carefully controlled to optimize metal recovery rates and ensure the quality of recovered metals. Electrolytic recovery processes offer several advantages for e-waste recycling compared to traditional pyrometallurgical or hydrometallurgical methods. Firstly, electrolysis operates at lower temperatures compared to smelting, reducing e-waste streams without the need for complex chemical reagents or extensive preprocessing steps. This selectivity is particularly beneficial for recovering high-value metals like gold and palladium from electronic components with minimal waste generation (Tan, Q., et al., 2018). Additionally, the management of electrolyte solutions and the handling of by-products generated during electrolysis are critical considerations to minimize environmental impact and ensure the sustainability of the recycling process.

Innovations in electrolytic processes for e-waste recycling focus on enhancing process efficiency, improving metal recovery rates, and developing technologies that integrate electrolysis with other recycling techniques to maximize resource recovery. Advanced electrode materials, such as inert metals or composite materials with enhanced catalytic properties, are being explored to optimize electrode performance and extend electrode lifetime in harsh electrolytic environments. Beyond electrolysis, electrochemical recovery encompasses a broader range of techniques and applications aimed at extracting and purifying metals from e-waste through electrochemical reactions. These techniques include electrowinning, electrorefining, electroplating, and electrochemical leaching, each tailored to specific e-waste materials and metal recovery requirements. Electrowinning is a variant of electrolysis used to recover metals from pregnant leach solutions (PLS) obtained from hydrometallurgical processes. In electrowinning, metal ions in the PLS are reduced and deposited onto cathodes as solid metal, producing high-purity metal products suitable for reuse or further refining. Electrorefining, on the other hand, involves the purification of impure metals through electrolysis, where the impure metal acts as the anode and is dissolved into the electrolyte (Selvakumar, P., et al., 2012). Pure metal is deposited onto a cathode, while impurities are either left behind in the electrolyte or collected separately as sludge. The process involves electrodeposition of metal ions from an electrolyte solution onto a conductive substrate, effectively stripping precious metals from plated surfaces for reuse or resale. Electrochemical leaching techniques utilize electrochemical reactions to selectively dissolve metals from e-waste materials under mild conditions, reducing the consumption of hazardous chemicals and minimizing environmental impact compared to traditional leaching methods.

The integration of electrochemical processes with other recycling technologies, such as hydrometallurgy and pyrometallurgy, offers synergistic advantages in terms of metal recovery efficiency, waste minimization, and environmental sustainability. By combining electrochemical techniques with mechanical preprocessing, solvent extraction, or precipitation methods, researchers and recyclers aim to develop integrated recycling systems that optimize resource recovery from diverse e-waste streams while meeting stringent environmental regulations. Environmental considerations in electrochemical processes for e-waste recycling include the management of electrolyte solutions, treatment of wastewater and by-products, and control of emissions to minimize environmental impact (Paramasivam Selvakumar, et al., 2024). Efforts are underway to develop green and sustainable technologies that utilize renewable energy sources, recycle process water, and minimize the use of hazardous chemicals, ensuring responsible recycling practices and compliance with global environmental standards. In conclusion, electrochemical processes, encompassing electrolysis and electrochemical recovery techniques, represent versatile and efficient methodologies for extracting valuable metals from electronic waste. As the demand for electronic devices continues to grow, the development and adoption of advanced electrochemical recycling technologies are crucial for meeting global demand while promoting resource conservation and environmental sustainability.

Biological Processes: Bioleaching and Biosorption

Biological processes have emerged as promising green technologies for the recycling of electronic waste (e-waste), offering sustainable methods to recover valuable metals and materials from discarded electronic devices. Among these biological processes, bioleaching and biosorption represent innovative approaches that harness the capabilities of microorganisms to selectively extract metals from e-waste materials, thereby reducing environmental impact and promoting resource recovery. Bioleaching is a biohydrometallurgical process that utilizes the metabolic activities of specialized microorganisms, such as bacteria and fungi, to solubilize metals from e-waste under mild conditions. During bioleaching, e-waste materials are typically placed in bioreactors or heaps with suitable nutrient media and inoculated with metal-tolerant microbial cultures (Vijayakumar G, et al., 2024). The microorganisms metabolize organic compounds present in the e-waste and release metabolic by-products, such as organic acids (e.g., citric acid, oxalic acid) and chelating agents (e.g., siderophores), which effectively complex and solubilize metals from solid materials. These metal ions are then leached into the surrounding aqueous solution, forming a pregnant leach solution (PLS) enriched with dissolved metals.

Biosorption offers several advantages for e-waste recycling, including low energy requirements, minimal chemical consumption, and the ability to treat dilute metal solutions effectively. It is particularly suitable for recovering precious metals like gold, silver, and palladium from e-waste leachates or industrial effluents, where metals are present in low concentrations but require efficient recovery. Research efforts focus on optimizing biosorbent materials, improving metal binding capacities, and developing cost-effective regeneration techniques to enhance the commercial viability of biosorption processes. The integration of biological processes, such as bioleaching and biosorption, with conventional hydrometallurgical or pyrometallurgical methods offers synergistic advantages in e-waste recycling (Wang, J., et al., 2015). Combined approaches can enhance metal recovery efficiencies, reduce process costs, and minimize environmental impacts compared to standalone technologies. Furthermore, biological processes contribute to the sustainability of e-waste recycling by reducing the consumption of chemical reagents, minimizing waste generation, and offering environmentally friendly alternatives to traditional recycling methods. Environmental considerations in biological processes for e-waste recycling include the management of microbial cultures, control of leachate quality, and mitigation of potential risks associated with microbial activities. Strategies for optimizing process performance and ensuring microbial safety involve monitoring microbial consortia dynamics, maintaining optimal operating conditions, and implementing biosecurity measures in biotechnological facilities.

Microbial Fuel Cells (MFCs): Harnessing Microbial Power for E-Waste Recycling

Microbial Fuel Cells (MFCs) represent a cutting-edge technology with significant potential in the realm of electronic waste (e-waste) recycling, offering a sustainable and innovative approach to recover energy and valuable resources from discarded electronic devices. MFCs utilize the metabolic activities of microorganisms to generate electrical energy through bioelectrochemical reactions, simultaneously facilitating the biodegradation of organic compounds and the recovery of metals from e-waste materials (Xiao, J., et al., 2017). At the heart of an MFC is an electrochemical cell where microbial catalysts, typically bacteria or archaea, oxidize organic substrates present in the e-waste feedstock. During this oxidation process, electrons are released and transferred to an external circuit through an electrode, generating electrical current. Firstly, MFCs provide a sustainable pathway for energy recovery from organic components present in e-waste, thereby reducing the overall energy footprint of the recycling process. Secondly, MFCs can operate under ambient conditions without the need for high temperatures or harsh chemical reagents, minimizing environmental impact and operational costs associated with conventional recycling techniques. Moreover, MFCs facilitate the simultaneous biodegradation of organic pollutants in e-waste, contributing to environmental remediation efforts.

Phytoremediation: Using Plants to Clean Up E-Waste

As the global production and disposal of electronic devices continue to rise, so does the urgency to develop sustainable and efficient methods for managing e-waste, which contains a complex mixture of hazardous substances such as heavy metals, persistent organic pollutants (POPs), and flame retardants (Yang, J. G., et al., 2012). The concept of phytoremediation leverages the natural abilities of plants to uptake and accumulate pollutants from their surroundings. This process involves several mechanisms, including phytoextraction, phytostabilization, rhizofiltration, and phytodegradation, each tailored to target specific contaminants and environmental conditions associated with e-waste disposal sites or contaminated ecosystems.

This approach is beneficial for stabilizing metal contaminants in e-waste disposal sites, preventing their leaching into groundwater or uptake by nearby vegetation. Rhizofiltration exploits the filtering capabilities of plant roots to remove pollutants directly from aqueous solutions or contaminated water bodies. Plants with dense root systems and high metal tolerance, such as aquatic plants and wetland species, are utilized to treat e-waste leachates or industrial effluents containing dissolved metals and organic pollutants (Selvakumar, P., 2024). Rhizofiltration not only purifies water but also concentrates contaminants in plant biomass, facilitating subsequent harvesting and disposal or recycling of contaminated plant material. Phytodegradation involves the breakdown or transformation of organic pollutants by plant-associated microbes, such as bacteria and fungi, living in close association with plant roots (Freitas, M. B. J. G., et al., 2007).

The selection of suitable plant species for phytoremediation of e-waste-contaminated sites depends on several factors, including metal tolerance, biomass production, root architecture, and adaptability to local climate and soil conditions. Hyperaccumulating plants, capable of accumulating high concentrations of metals in their tissues without suffering significant toxicity, are particularly valuable for phytoextraction purposes (Hai, H. T., et al., 2007). Examples of hyperaccumulators include species from the Brassicaceae family (e.g., Alyssum, Thlaspi) known for their ability to accumulate nickel, zinc, and cadmium. In addition to their remedial functions, plants used in phytoremediation can contribute to ecosystem restoration, soil stabilization, and enhancement of biodiversity in degraded e-waste disposal sites. By establishing vegetative cover and promoting soil microbial diversity, phytoremediation helps improve soil structure, fertility, and resilience to erosion, thereby restoring ecological balance and supporting sustainable land use practices.The implementation of phytoremediation for e-waste recycling faces several challenges and considerations that require careful assessment and management (Kaipannan, S., et al., 2019). These include the selection of appropriate plant species adapted to local climatic and soil conditions, optimization of remediation techniques to achieve desired outcomes, and long-term monitoring of plant health and pollutant removal efficiency. Factors such as plant growth rates, metal uptake efficiencies, and site-specific variability in contaminant concentrations influence the effectiveness and feasibility of phytoremediation as a remediation strategy.

Chemical Precipitation: Removing Contaminants From E-Waste

Treating e-waste leachates containing dissolved metals, acids, and complex organic compounds, transforming them into stable, less toxic forms suitable for disposal or further processing. The application of chemical precipitation in e-waste recycling begins with the collection and pretreatment of e-waste materials to release metals and other contaminants into aqueous solutions. Mechanical shredding or grinding of e-waste devices such as printed circuit boards (PCBs), connectors, and integrated circuits facilitates the liberation of metals and organic compounds from their solid matrices (Fogarasi, S., et al., 2015). The resulting shredded material is then subjected to leaching processes using acidic or alkaline solutions to dissolve metals into solution, creating e-waste leachates that require subsequent treatment. These acids and bases selectively target different metals depending on their solubility and chemical stability, with sulfuric acid being effective for dissolving base metals such as copper, nickel. The choice of leaching agent and conditions (e.g., temperature, pH, agitation) is critical to achieving high metal extraction efficiencies while minimizing the dissolution of unwanted materials and contaminants.

Once metals are dissolved into e-waste leachates, chemical precipitation techniques are employed to recover these metals in a purified form suitable for reuse or resale. The precipitation process involves the addition of chemical precipitants, such as sodium hydroxide, sodium sulfide, calcium hydroxide, or ammonium hydroxide, to e-waste leachates under controlled conditions. These precipitants react with dissolved metal ions in the solution, causing them to form insoluble metal hydroxides, sulfides, or carbonates depending on the specific metal species present. For example, the addition of sodium hydroxide (NaOH) to e-waste leachates containing dissolved metals results in the precipitation of metal hydroxides, which can be filtered or settled out of the solution as solid precipitates (Landsiedel, R., et al., 2010). The choice of precipitant and reaction conditions (e.g., pH adjustment, mixing intensity, reaction time) are optimized to achieve maximum metal recovery and precipitation efficiency while minimizing the generation of secondary wastes and ensuring compliance with environmental regulations. Chemical precipitation processes can be tailored to selectively recover specific metals from e-waste leachates based on their chemical properties and market demand. For instance, the precipitation of precious metals such as gold and silver from acidic leachates is achieved using selective precipitants or complexing agents that facilitate the formation of stable metal complexes or insoluble salts. In contrast, the precipitation of base metals like copper and nickel from alkaline leachates requires adjusting pH conditions and controlling the concentration of precipitating agents to achieve targeted metal recovery rates.

Environmental considerations play a crucial role in the design and implementation of chemical precipitation processes for e-waste recycling, focusing on minimizing chemical usage, optimizing resource efficiency, and reducing the environmental footprint of recycling operations. Strategies for sustainable chemical precipitation include the recycling and reuse of precipitating agents, treatment of process effluents, and integration with other remediation technologies to enhance overall process efficiency and environmental stewardship (Mahadev Madgule, et al., 2023). Research and development efforts in chemical precipitation technologies for e-waste recycling continue to advance, driven by innovations in precipitant selection, process optimization, and waste management strategies. Emerging trends include the use of green chemistry principles to develop eco-friendly precipitants and sustainable practices that minimize waste generation and energy consumption throughout the recycling process. By leveraging chemical reactions to precipitate metals into stable forms, this technology contributes to the chemical precipitation processes are essential to meet environmental regulations, enhance recycling efficiency, and foster a sustainable future where materials are reused, and waste is minimized for the benefit of society and the environment.

Ion Exchange: Selective Removal of Ions From E-Waste

This process relies on the reversible exchange of ions between a solid phase (ion exchange resin) and a liquid phase (e-waste leachate or solution), facilitated by electrostatic forces and specific interactions between ions and functional groups on the resin surface (Liu, Q., et al., 2009). Ion exchange resins are engineered materials with tailored chemical properties that enable them to selectively bind and release target ions based on their charge, size, and chemical affinity. The application of ion exchange in e-waste recycling typically begins with the collection and pretreatment of e-waste materials to release metals and other contaminants into aqueous solutions. Mechanical shredding or grinding of e-waste devices such as printed circuit boards (PCBs), connectors, and integrated circuits facilitates the liberation of metals and organic compounds from their solid matrices. The resulting shredded material undergoes leaching processes using acidic or alkaline solutions to dissolve metals into solution, creating e-waste leachates that require subsequent treatment to recover valuable metals. Ion exchange resins are employed in e-waste recycling facilities as specialized sorbents capable of selectively capturing and concentrating metal ions from complex aqueous solutions (Forti, V., et al., 2020). These resins consist of porous polymeric matrices with functional groups, such as sulfonic acid, carboxylic acid, or amine groups, that exhibit high affinity for metal ions in solution.

Environmental considerations play a crucial role in the deployment of ion exchange technologies for e-waste recycling, emphasizing the importance of minimizing chemical usage, optimizing resource efficiency, and ensuring compliance with environmental regulations. Strategies for sustainable ion exchange include the development of eco-friendly resins, recycling of spent resins, treatment of process effluents, and integration with other remediation technologies to enhance overall process efficiency and environmental stewardship (Mesbah, Y. I., et al., 2020). Research and development efforts in ion exchange technologies for e-waste recycling continue to advance, focusing on improving resin performance, optimizing operational parameters, and developing novel resin materials with enhanced selectivity, stability, and regeneration capabilities. Emerging trends include the integration of ion exchange with membrane technologies, electrochemical processes, and hybrid systems to expand the applicability and efficiency of ion exchange in complex e-waste remediation scenarios.

Solvent Extraction: Separating and Purifying Metals From E-Waste

The application of solvent extraction in e-waste recycling begins with the collection and pretreatment of e-waste materials to release metals and other valuable components into aqueous solutions. circuits facilitates the liberation of metals and organic compounds from their solid matrices (Natarajan, S., et al., 2022). The resulting shredded material undergoes leaching processes using acidic or alkaline solutions to dissolve metals into solution, creating e-waste leachates that require subsequent treatment for metal recovery. Solvent extraction processes are typically employed to extract and separate metals from e-waste leachates containing complex mixtures of metal ions, organic acids, and other contaminants. The process involves several key stages: extraction, scrubbing, and stripping. In the extraction stage, the e-waste leachate is mixed with an organic solvent, known as the extractant, which selectively binds to metal ions forming complexes in the organic phase. The choice of extractant depends on factors such as metal affinity, selectivity, and compatibility with the organic solvent. Secondly, solvent extraction processes operate under relatively mild conditions (ambient temperature and pressure), reducing energy consumption and minimizing environmental impact compared to thermal processes like smelting or incineration (Nguyen, R. T., et al., 2017). Moreover, solvent extraction technologies can be tailored and scaled up for industrial applications, offering cost-effective solutions for large-scale e-waste recycling operations. The performance of solvent extraction in e-waste recycling is influenced by several factors, including the selection of appropriate extractants, optimization of process conditions, management of organic and aqueous phases, and recycling of solvents and reagents to minimize waste generation. Research and development efforts focus on improving extractant efficiency, enhancing metal recovery rates, and developing sustainable practices that reduce environmental footprint and operational costs in solvent extraction processes. In conclusion, solvent extraction represents a robust and efficient technology for separating and purifying metals from e-waste leachates and effluents, offering sustainable solutions for resource recovery and environmental remediation (Nathan, T., et al., 2008). By leveraging selective complexation and phase transfer processes, solvent extraction contributes to the circular economy by recovering valuable metals, minimizing waste generation, and supporting sustainable practices in e-waste management. As global demand for electronic devices continues to rise, the adoption and advancement of solvent extraction technologies are essential to meet environmental regulations, enhance recycling efficiency, and foster a sustainable future where materials are conserved, and waste is minimized for the benefit of society and the environment.

Electrochemical Oxidation: Breaking Down Organic Compounds in E-Waste

Electrochemical oxidation stands at the forefront of innovative technologies for addressing the complex challenge of organic compound degradation in electronic waste (e-waste). The fundamental principle of electrochemical oxidation involves the application of an electric current to initiate oxidation reactions at the surface of an electrode immersed in an electrolyte solution containing organic pollutants. These oxidation reactions generate highly reactive hydroxyl radicals (•OH) and other oxidative species, which attack and break down organic molecules into smaller, less toxic by-products such as carbon dioxide, water, and simple organic acids. The electrochemical process occurs at relatively mild operating conditions, typically at ambient temperatures and pressures, making it a more environmentally friendly alternative to thermal treatment methods like incineration (Ding, Y., et al., 2019). In e-waste recycling, electrochemical oxidation is particularly effective for treating aqueous solutions or leachates containing complex mixtures of organic pollutants derived from electronic devices. These solutions may result from the leaching of PCBs, BFRs, phthalates, and other organic contaminants during the mechanical shredding and chemical extraction of e-waste components. By subjecting these contaminated solutions to electrochemical oxidation, organic compounds are selectively oxidized and mineralized into harmless end products, reducing their toxicity and environmental persistence. The electrochemical oxidation process typically involves the use of electrodes made from materials such as platinum, graphite, titanium, or boron-doped diamond (BDD), For instance, BDD electrodes are renowned for their superior electrochemical stability and ability to generate •OH radicals efficiently, making them suitable for the degradation of persistent organic pollutants in e-waste leachates.

One of the key advantages of electrochemical oxidation is its ability to selectively target and degrade specific organic compounds present in e-waste, without generating hazardous by-products or secondary pollutants associated with traditional treatment methods (Das, S., et al., 2019). This selectivity is achieved through precise control over operating parameters such as applied voltage, current density, electrolyte pH, and reaction time, which influence the rate and efficiency of oxidation reactions. By optimizing these parameters, researchers and engineers can tailor electrochemical oxidation processes to effectively degrade a wide range of organic pollutants found in different types of e-waste streams. In addition to its effectiveness in organic compound degradation, electrochemical oxidation offers versatility in integrating with other e-waste recycling technologies to enhance overall process efficiency and resource recovery. For example, electrochemical oxidation can be coupled with solvent extraction, ion exchange, or membrane filtration processes to treat e-waste leachates containing both organic and inorganic contaminants. This integrated approach allows for comprehensive treatment of e-waste streams, where electrochemical oxidation targets organic pollutants while other technologies focus on recovering valuable metals or purifying aqueous solutions. The performance and scalability of electrochemical oxidation in e-waste recycling are influenced by several factors, including electrode design, reactor configuration, electrolyte composition, and process optimization. Continuous advancements in electrode materials, reactor engineering, and electrochemical cell design contribute to improving the energy efficiency, reaction kinetics, and treatment capacity of electrochemical oxidation systems for industrial-scale applications. Research efforts also focus on developing novel electrode materials and catalysts to enhance oxidation efficiency and expand the applicability of electrochemical oxidation to diverse e-waste matrices.

Advanced Oxidation Processes (AOPs): Degrading Persistent Pollutants

AOPs have gained significant attention and application in treating wastewater, industrial effluents, contaminated groundwater, and leachates from solid waste, including electronic waste (e-waste), due to their ability to effectively remove recalcitrant pollutants that are challenging to degrade using conventional treatment methods (Chancerel, P., et al., 2009). The fundamental principle of AOPs involves the generation of •OH radicals or other reactive species capable of oxidizing organic compounds through direct oxidation or indirect pathways involving radical chain reactions. •OH radicals are highly reactive and indiscriminate in nature, reacting with organic pollutants by abstracting hydrogen atoms from their molecular structures, leading to the formation of hydroxylated intermediates and ultimately mineralizing organic contaminants into carbon dioxide (CO2), water (H2O), and inorganic ions. This mineralization process represents the complete oxidation of organic pollutants, ensuring their transformation into non-toxic end products. Several AOPs have been developed and optimized for different applications, each utilizing specific mechanisms to generate oxidative species and promote pollutant degradation under controlled conditions. Some of the most widely studied AOPs include:

Fenton and Fenton-like Processes: Involving the reaction between hydrogen peroxide (H2O2) and ferrous iron (Fe2+) catalysts in acidic conditions to produce •OH radicals via the Fenton reaction. This process is effective for degrading organic pollutants and can be enhanced by modifying reaction parameters such as pH, temperature, and catalyst concentration.

Electrochemical Advanced Oxidation Processes (EAOPs): Utilizing electrochemical cells or reactors to generate •OH radicals and other oxidative species through electrolysis of water or electrochemical activation of oxidants such as persulfate (S2O82-) or peroxides (H2O2). EAOPs offer advantages such as controllable reaction kinetics, selective pollutant targeting, and the ability to operate under mild conditions without the need for additional chemical additives.

Hybrid AOPs: Combining two or more AOPs synergistically to enhance pollutant degradation efficiency and expand the applicability of AOP technologies to complex matrices. Hybrid AOPs may involve combinations of ozonation with UV irradiation (O3/UV), Fenton-like processes with electrochemical oxidation (Fenton/EAOPs), or photocatalytic oxidation with ozonation (photocatalysis/O3) to achieve superior degradation performance and minimize operational costs.

The application of AOPs in e-waste recycling focuses on treating aqueous leachates or effluents generated during the dismantling, shredding, and chemical processing of electronic devices to recover valuable metals and degrade hazardous organic compounds.

Broad Applicability: AOPs are effective for degrading a wide range of organic pollutants, including recalcitrant compounds that are resistant to conventional treatment methods. This includes persistent organic pollutants (POPs) found in e-waste leachates, which are challenging to degrade due to their complex molecular structures and stability in the environment.

Complete Mineralization: AOPs promote the complete oxidation and mineralization of organic pollutants into harmless end products such as CO2, H2O, and inorganic ions. This ensures that contaminants are fully transformed into non-toxic forms, reducing environmental persistence and bioaccumulation potential.

Selective Targeting:AOPs can be tailored and optimized to selectively target specific organic pollutants or classes of compounds based on their chemical properties and reactivity. This selectivity minimizes energy consumption and chemical usage while maximizing pollutant removal efficiency and process sustainability.

Operational Flexibility: AOPs can operate under a wide range of conditions, including ambient temperatures and pressures, making them suitable for diverse environmental remediation applications and industrial-scale e-waste recycling operations. This flexibility allows for the integration of AOP technologies with existing treatment processes to enhance overall system performance and resource recovery.

Despite these advantages, the deployment of AOPs for e-waste recycling and environmental remediation presents several challenges and considerations that require careful evaluation and management:

Cost Considerations: AOP technologies may require significant capital investment for infrastructure, equipment, and operational expenses, particularly for large-scale applications in industrial settings. Cost-effective implementation strategies, such as process optimization, energy efficiency improvements, and by-product utilization, are essential to enhance economic viability and sustainability.

Complexity of Pollutant Mixtures: E-waste leachates contain complex mixtures of organic and inorganic contaminants, which may vary in concentration, composition, and reactivity. The development of tailored AOPs and hybrid treatment approaches is necessary to address the diverse nature of pollutant matrices and achieve optimal treatment outcomes.

Environmental Footprint: While AOPs offer sustainable solutions for pollutant degradation, they may generate secondary wastes, by-products, or residual chemicals that require proper management and disposal. Environmental monitoring, waste characterization, and compliance with regulatory standards are essential to minimize environmental impacts and ensure safe operation of AOP facilities.

Research and development efforts in AOP technologies for e-waste recycling continue to advance, driven by innovations in reactor design, electrode materials, oxidant generation methods, and process optimization strategies (Campbell, F. W., et al., 2009). Emerging trends include the development of novel catalysts, advanced oxidation catalysts (AOCs), and electrochemical catalysts (ECs) to enhance •OH radical generation efficiency and promote pollutant degradation kinetics in AOP systems. Additionally, advancements in computational modeling, artificial intelligence (AI), and machine learning (ML) are being utilized to optimize AOP processes, predict pollutant degradation pathways, and design tailored treatment solutions for complex e-waste matrices. Environmental sustainability and regulatory compliance are paramount in the deployment of AOP technologies for e-waste recycling, emphasizing the importance of stakeholder engagement, public awareness, and collaborative partnerships among researchers, industry stakeholders, government agencies, and local communities.

Bioremediation: Using Microorganisms to Clean Up E-Waste

The application of bioremediation in e-waste management begins with the identification and characterization of microbial communities capable of metabolizing or tolerating specific contaminants present in e-waste materials. Microorganisms play diverse roles in bioremediation processes, including biodegradation of organic pollutants, biosorption of heavy metals, bioaccumulation in microbial biomass, and biomineralization of metal ions into insoluble forms. These microbial-mediated processes occur under favorable environmental conditions, such as optimal temperature, pH, nutrient availability, and oxygen levels, which influence microbial growth and metabolic activity in e-waste environments (Barasa, M., et al., 2018). Bacteria and fungi capable of aerobic or anaerobic metabolism utilize these contaminants as carbon and energy sources through enzymatic reactions that break down complex organic molecules into simpler compounds. For instance, bacterial genera like Pseudomonas, Bacillus, and Rhodococcus are known for their ability to degrade PCBs and BFRs through enzymatic pathways involving cytochrome P450 enzymes, dioxygenases, and dehalogenases. In addition to organic pollutants, bioremediation strategies also target heavy metals and metalloids present in e-waste leachates, which pose significant environmental and health risks due to their persistence and bioaccumulative properties. Microorganisms employ various mechanisms, such as biosorption, precipitation, and enzymatic reduction, to sequester or transform metal ions into less toxic or immobile forms. Metal-resistant bacteria, such as Cupriavidus metallidurans and Pseudomonas putida, accumulate metals within their cells through active transport mechanisms or surface binding to functional groups on microbial cell walls, effectively reducing metal bioavailability and mitigating environmental contamination.

The effectiveness of bioremediation in e-waste applications relies on optimizing microbial processes to enhance contaminant removal efficiency, minimize secondary pollution, and promote ecosystem restoration. Strategies for enhancing bioremediation performance include microbial consortia or mixed cultures that synergistically degrade multiple contaminants, genetic engineering of microbial strains to enhance pollutant degradation pathways, and bioaugmentation with exogenous microbial species to introduce specific metabolic capabilities into contaminated sites. Bioaugmentation involves the inoculation of contaminated e-waste environments with selected microbial strains or consortia to accelerate pollutant degradation rates and improve treatment outcomes. This approach is particularly valuable in environments where indigenous microbial populations may be limited or insufficiently adapted to degrade complex contaminants (Aboelazm, E. A., et al., 22021). Biostimulation techniques, such as adding nutrients (e.g., carbon sources, nitrogen, phosphorus) or electron acceptors (e.g., oxygen, nitrate) to enhance microbial growth and activity, also play a crucial role in optimizing bioremediation performance in e-waste matrices. Environmental sustainability is a critical consideration in the deployment of bioremediation technologies for e-waste management, emphasizing the importance of minimizing energy consumption, reducing chemical inputs, and promoting the reuse of treated materials. Bioremediation offers inherent advantages over conventional physicochemical treatment methods by operating under mild conditions, generating fewer secondary wastes, and requiring lower energy inputs for microbial metabolic processes. These attributes contribute to the ecological and economic feasibility of bioremediation as a sustainable solution for mitigating e-waste pollution and promoting resource recovery in a circular economy framework.

Challenges associated with bioremediation in e-waste contexts include the variability of e-waste compositions, the presence of complex contaminant mixtures, and the need for long-term monitoring to assess treatment efficacy and environmental impact. Research and development efforts focus on expanding the metabolic capabilities of microbial consortia, optimizing bioremediation strategies for diverse e-waste matrices, and evaluating the potential risks and benefits of microbial interventions in contaminated environments. In conclusion, bioremediation represents a promising and environmentally sustainable approach for cleaning up e-waste by harnessing microbial activities to degrade organic pollutants, sequester heavy metals, and promote ecosystem resilience (Abhilasha A., et al., 22022). By leveraging the natural metabolic pathways of microorganisms, bioremediation technologies offer efficient and cost-effective solutions for reducing environmental contamination, enhancing resource recovery, and advancing sustainable practices in e-waste management. Continued advancements in microbial ecology, biotechnological innovations, and interdisciplinary collaborations are essential to harnessing the full potential of bioremediation in addressing the complex challenges posed by electronic waste and supporting global efforts toward environmental stewardship and pollution prevention.

FUTURE PERSPECTIVES

The future of e-waste recycling hinges on innovative approaches that integrate chemical and biological processes, leveraging their respective strengths to achieve sustainable management of electronic waste (e-waste) while promoting resource recovery and environmental stewardship. As electronic devices continue to proliferate globally, the need for efficient and eco-friendly recycling technologies becomes increasingly urgent to mitigate environmental pollution, conserve valuable resources, and comply with regulatory standards governing waste management. Integrating chemical and biological processes offers synergistic advantages that address the complex composition of e-waste, which includes metals, plastics, organic contaminants, and hazardous substances requiring specialized treatment strategies. Chemical processes, such as hydrometallurgy, pyrometallurgy, solvent extraction, and chemical precipitation, play pivotal roles in recovering valuable metals from e-waste materials. These processes involve the dissolution, separation, purification, and recovery of metals through controlled chemical reactions and separation techniques. Hydrometallurgical methods, for example, utilize aqueous solutions to leach metals from shredded electronic components, forming metal-rich leachates that undergo further processing for metal extraction and purification. Pyrometallurgical processes, on the other hand, involve thermal treatments to smelt e-waste materials and recover metals through melting and refining operations. These chemical techniques are effective for extracting metals like offering high efficiency and scalability for industrial applications.

The integration of chemical and biological processes for e-waste recycling presents numerous opportunities to optimize resource recovery, minimize environmental footprints, and advance circular economy principles in waste management practices. By combining hydrometallurgical and bioleaching techniques, for instance, researchers can develop hybrid processes that synergistically extract metals from e-waste while mitigating chemical usage and energy consumption. Sequential extraction strategies may involve pretreating e-waste materials with biological leaching agents to liberate metals followed by refining using chemical methods to achieve high purity metal products suitable for industrial applications. Advancements in biotechnological innovations, materials science, and process engineering are driving the development of integrated systems that combine chemical and biological processes for enhanced e-waste recycling efficiency. Bioreactors and flow-through systems optimized for microbial growth and metabolic activity enable continuous processing of e-waste leachates while maintaining favorable environmental conditions for microbial consortia. Novel biocatalysts, genetically engineered microorganisms, and bio-inspired materials are being explored to enhance the performance, selectivity, and scalability of biological processes in e-waste recycling applications.

CONCLUSION

In conclusion, the integration of chemical and biological processes represents a transformative approach to address the multifaceted challenges of e-waste recycling, offering synergistic advantages that enhance resource recovery, minimize environmental impacts. Electronic waste (e-waste) presents a complex composition of valuable metals, hazardous substances, and organic pollutants, necessitating specialized technologies capable of efficiently recovering materials while mitigating environmental and human health risks. Chemical processes such as hydrometallurgy, pyrometallurgy, solvent extraction, and chemical precipitation are instrumental. These processes involve controlled chemical reactions to dissolve metals from shredded electronic components, followed by separation and purification steps to produce high-purity metal products suitable for reuse in manufacturing industries. Bioleaching harnesses microbial consortia to solubilize and extract metals from e-waste matrices using organic acids or metabolites produced by microorganisms, offering an eco-friendly alternative to conventional chemical extraction methods. Moreover, demonstrating the dual benefits of energy production and wastewater treatment in integrated biotechnological applications. The integration of chemical and biological processes for e-waste recycling is driven by advancements in biotechnological innovations, materials science, and process engineering, which aim to optimize resource recovery efficiency, minimize energy consumption, and reduce environmental footprints associated with traditional recycling methods. Hybrid approaches that combine chemical and biological techniques, such as bioleaching coupled with solvent extraction or bioremediation integrated with electrochemical oxidation, demonstrate enhanced treatment efficiencies and economic viability for industrial-scale e-waste recycling operations. Essential for advancing the adoption and implementation of integrated chemical and biological processes across global e-waste recycling sectors. Knowledge sharing, technology transfer, and capacity building initiatives facilitate the dissemination of best practices and innovative solutions to address technical challenges, regulatory requirements, and market demands associated with e-waste management.

REFERENCES

Abdelbasir, S. M., Hassan, S. S. M., Kamel, A. H., & El-Nasr, R. S. (2018). Status of electronic waste recycling techniques: A review. *Environmental Science and Pollution Research International*, *25*(17), 16533–16547. DOI: 10.1007/s11356-018-2136-6 PMID: 29737485

Abhilasha, A., Sreenivasulu, A., & Manimozhi, T. (2022). The Model of Smart Sensing Device For Sensitive Nanoclusters Modification in Sensing Properties. 2022 2nd International Conference on Advance Computing and Innovative Technologies in Engineering (ICACITE), 1043-1047. DOI: 10.1109/ICACITE53722.2022.9823846

Aboelazm, E. A., Mohamed, N., & Ali, G. A. (2021). *Makhlouf AS CK recycling of cobalt oxides electrodes from spent lithium-ion batteries by electrochemical method.* Waste Recycling Technologies for Nanomater Manufacturing. DOI: 10.1007/978-3-030-68031-2_4

Barasa, M., Bogdanov, D., Oyewo, A. S., & Breyer, C. (2018). A cost optimal resolution for Sub-Saharan Africa powered by 100% renewables in 2030. *Renewable & Sustainable Energy Reviews*, *92*, 440–457. DOI: 10.1016/j.rser.2018.04.110

Campbell, F. W., Belding, S. R., Baron, R., Xiao, L., & Compton, R. G. (2009). The hydrogen evolution reaction at a silver nanoparticle array and a silver macroelectrode compared : Changed electrode kinetics between the macro- and nanoscales. *The Journal of Physical Chemistry. C, Nanomaterials and Interfaces*, *33*(33), 14852–14857. DOI: 10.1021/jp904723b

Chancerel, P., Meskers, C. E. M., Hagelüken, C., & Rotter, V. S. (2009). Assessment of precious metal flows during preprocessing of waste electrical and electronic equipment. *Journal of Industrial Ecology*, *13*(5), 791–810. DOI: 10.1111/j.1530-9290.2009.00171.x

Das, S., Lee, S. H., Kumar, P., Kim, K.-H., Lee, S. S., & Bhattacharya, S. S. (2019). Solid waste management: Scope and the challenge of sustainability. *Journal of Cleaner Production*, *228*, 658–678. DOI: 10.1016/j.jclepro.2019.04.323

Ding, Y., Zhang, S., Liu, B., Zheng, H., Chang, C., & Ekberg, C. (2019). Recovery of precious metals from electronic waste and spent catalysts: A review. *Resources, Conservation and Recycling*, *141*, 284–298. DOI: 10.1016/j.resconrec.2018.10.041

Fogarasi, S., Imre-Lucaci, F., Egedy, A., Imre-Lucaci, Á., & Ilea, P. (2015). Eco-friendly copper recovery process from waste printed circuit boards using Fe3+/Fe2+ redox system. *Waste Management (New York, N.Y.)*, *40*, 136–143. DOI: 10.1016/j.wasman.2015.02.030 PMID: 25816768

Forti, V., Baldé, C. P., Kuehr, R., & Bel, G. (2020). The global e-waste monitor. Quantities, flows and the circular economy potential. https://www.itu.int/en/ITU-D/Environment/Documents/Toolbox/GEM_2020_def.pdf

Freitas, M. B. J. G., & Garcia, E. M. (2007). Electrochemical recycling of cobalt from cathodes of spent lithium–ion batteries. *Journal of Power Sources*, *171*(2), 953–959. DOI: 10.1016/j.jpowsour.2007.07.002

Hai, H. T., Gwan, A. J., Jin, K. D., Sohn, J. S., Oh, K. C., & Saeng, C. H. (2007). Growth process of copper particles in the cuprous oxide-water system. *Materials Transactions*, *48*(5), 1137–1142. DOI: 10.2320/matertrans.48.1137

Kaipannan, S., Govindarajan, K., Sundaramoorthy, S., & Marappan, S. (2019). Waste toner-derived carbon/Fe3O4 nanocomposite for high-performance supercapacitor. *ACS Omega*, *4*(14), 15798–15805. DOI: 10.1021/acsomega.9b01337 PMID: 31592452

Landsiedel, R., Ma-Hock, L., Kroll, A., Hahn, D., Schnekenburger, J., Wiench, K., & Wohlleben, W. (2010). Testing metal-oxide nanomaterials for human safety. *Advanced Materials*, *22*(24), 2601–2627. DOI: 10.1002/adma.200902658 PMID: 20512811

Liu, Q., Li, K. Q., Zhao, H., Li, G., & Fan, F. Y. (2009). The global challenge of electronic waste management. *Environmental Science and Pollution Research International*, *16*(3), 248–249. DOI: 10.1007/s11356-009-0131-7 PMID: 19308473

Mahadev Madgule, N. (2023). Mechanical properties and microstructure of activated TIG welded similar joints of Inconel alloys by desirability approaches. *Materials Today: Proceedings*, *77*, 528–533. DOI: 10.1016/j.matpr.2022.12.250

Mesbah, Y. I., Ahmed, N., Ali, B. A., & Allam, N. K. (2020). Recycling of Li−Ni−Mn−Co hydroxide from spent batteries to produce high-performance supercapacitors with exceptional stability. *ChemElectroChem*, *7*(4), 975–982. DOI: 10.1002/celc.202000081

Natarajan, S., Krishnamoorthy, K., & Kim, S.-J. (2022). Effective regeneration of mixed composition of spent lithium-ion batteries electrodes towards building supercapacitor. *Journal of Hazardous Materials*, *430*, 128496. DOI: 10.1016/j.jhazmat.2022.128496 PMID: 35739677

Nathan, T., Aziz, A., Noor, A. F., & Prabaharan, S. R. S. (2008). Nanostructured NiO for electrochemical capacitors : Synthesis and electrochemical properties. *Journal of Solid State Electrochemistry*, *12*(7-8), 1003–1009. Advance online publication. DOI: 10.1007/s10008-007-0465-3

Nguyen, R. T., Diaz, L. A., Imholte, D. D., & Lister, T. E. (2017). Economic assessment for recycling critical metals from hard disk drives using a comprehensive recovery process. *Journal of the Minerals Metals & Materials Society*, *69*(9), 1546–1552. DOI: 10.1007/s11837-017-2399-2

Pariatamby, A., Cheah, W. Y., Shrizal, R., Thamlarson, N., Lim, B. T., & Barasarathi, J. (2015). Enhancement of landfill methane oxidation using different types of organic wastes. *Environmental Earth Sciences*, *73*(5), 2489–2496. DOI: 10.1007/s12665-014-3600-3

Pender, J. P., Jha, G., Youn, D. H., Ziegler, J. M., Andoni, I., Choi, E. J., Heller, A., Dunn, B. S., Weiss, P. S., Penner, R. M., & Mullins, C. B. (2020). Electrode degradation in lithium-ion batteries. *ACS Nano*, *14*(2), 1243–1295. DOI: 10.1021/acsnano.9b04365 PMID: 31895532

Ramakrishnan, T., Mohan Gift, M. D., Chitradevi, S., Jegan, R., & Subha Hency Jose, P. (2022). Study of Numerous Resins Used in Polymer Matrix Composite Materials. Advances in Materials Science and Engineering. DOI: 10.1155/2022/1088926

Rautela, R., Arya, S., Vishwakarma, S., Lee, J., Kim, K.-H., & Kumar, S. (2021). E-waste management and its effects on the environment and human health. *The Science of the Total Environment*, *773*, 145623. DOI: 10.1016/j.scitotenv.2021.145623 PMID: 33592459

Sahin, M., Akcil, A., Erust, C., Altynbek, S., Gahan, C. S., & Tuncuk, A. (2015). A potential alternative for precious metal recovery from e-waste: Iodine leaching. *Separation Science and Technology*, *50*, 2587–2595. DOI: 10.1080/01496395.2015.1061005

Selvakumar, P. (2024). Climate Change. In *Global Challenges for the Environment and Climate Change* (pp. 100–110). Global. DOI: 10.4018/979-8-3693-2845-3.ch006

Selvakumar, P. (2024). Reasons for Biodiversity Loss. In *Impact of Societal Development and Infrastructure on Biodiversity Decline* (pp. 37–49). Global. DOI: 10.4018/979-8-3693-6950-0.ch004

Selvakumar, P. (2024). The Role of Corrosion Inhibitors. In Thakur, A., & Kumar, A. (Eds.), *Innovations in Nanomaterials-Based Corrosion Inhibitors* (pp. 130–141). IGI Global. DOI: 10.4018/979-8-3693-3088-3.ch006

Selvakumar, P., & Loganathan, V. (2012). Phytochemical screening and antioxidant activity of red flowered Mirabilis jalapa leaf in different solvents. *International Journal of Pharma and Bio Sciences*, *3*(4), 440–446.

Selvakumar, P., Mariappan, P., Shunmugam, S., Paul, R., & Jabez, S. (2024). Removal of chemicals from effluent using photobioreactor technology to improve environmental and health impacts. *Zeitschrift für Physikalische Chemie*, *2024*(7), 1295–1302. Advance online publication. DOI: 10.1515/zpch-2023-0450

Selvakumar, P., Palanisamy, S. K., Cinthaikinian, S., Palanisamy, V., Mariappan, R., & Selvakumar, P. K. (2024). Biosensors and its diverse applications in healthcare systems. *Zeitschrift für Physikalische Chemie, 2024*. Advance online publication. DOI: 10.1515/zpch-2023-0406

Sugumaran, K. M. D., Pandiyarajan, V., Y., B S, Selvakumar, P., & M., S. (2024). Analysis of Electromyography Signals for Control of Mechanical Prosthesis using Machine Learning Techniques. 2024 2nd International Conference on Intelligent Data Communication Technologies and Internet of Things (IDCIoT), 1059-1067. DOI: 10.1109/IDCIoT59759.2024.10467707

Tan, Q., Duan, H., Liu, L., Yang, J., & Li, J. (2018). Rethinking residential consumers' behavior in discarding obsolete mobile phones in China. *Journal of Cleaner Production*, *195*, 1228–1236. DOI: 10.1016/j.jclepro.2018.05.244

Tatariants, M., Yousef, S., Sakalauskaitė, S., Daugelavičius, R., Denafas, G., & Bendikiene, R. (2018). Antimicrobial copper nanoparticles synthesized from waste printed circuit boards using advanced chemical technology. *Waste Management (New York, N.Y.)*, *78*, 521–531. DOI: 10.1016/j.wasman.2018.06.016 PMID: 32559941

Valero Navazo, J. M., Villalba Méndez, G., & Talens Peiró, L. (2014). Material flow analysis and energy requirements of mobile phone material recovery processes. *The International Journal of Life Cycle Assessment*, *19*(3), 567–579. DOI: 10.1007/s11367-013-0653-6

Vijayakumar, G., Rajkumar, M., Rajiv Chandar, N., Selvakumar, P., & Duraisamy, R. (2024). Muniyandi Rajkumar, Rajiv Chandar N, Selvakumar P and Ramesh Duraisamy, (2024). Environmentally friendly TDS removal from waste water by electrochemical ion exchange batch-type recirculation (EIR) technique. *Environmental Science. Water Research & Technology*, *10*(4), 826–835. DOI: 10.1039/D3EW00793F

Wang, J., & Sun, X. (2015). Olivine LiFePO4: The remaining challenges for future energy storage. *Energy & Environmental Science*, *8*(4), 1110–1138. DOI: 10.1039/C4EE04016C

Xiao, J., Li, J., & Xu, Z. (2017). Recycling metals from lithium ion battery by mechanical separation and vacuum metallurgy. *Journal of Hazardous Materials*, *338*, 124–131. DOI: 10.1016/j.jhazmat.2017.05.024 PMID: 28544937

Yang, J. G., Wu, Y. T., & Li, J. (2012). Recovery of ultrafine copper particles from metal components of waste printed circuit boards. *Hydrometallurgy*, *121–124*, 1–6. DOI: 10.1016/j.hydromet.2012.04.015

Chapter 7
The Role of Blockchain Technology With IoT for Revolutionizing Waste Management

S. Vidhya
https://orcid.org/0000-0003-3130-0455
Department of Computing Technologies, SRM Institute of Science and Technology, India

Balaji Mahadevan
Department of Electrical and Electronics Engineering, Sri Sivasubramaniya Nadar College of Engineering, India

V. Kamaraj
Department of Electrical and Electronics Engineering, Sri Sivasubramaniya Nadar College of Engineering, India

ABSTRACT

Modern approaches for waste management assure effectiveness, openness, and sustainability and hence can be adopted with the aim of handling increasing waste problem globally. The use of blockchain, which is well known for its decentralization and immutability, may help to revolutionalize this area. The chapter focuses on the uses of blockchain in waste management. Better traceability is one of the primary benefits of blockchain's application in waste management. Thus, it is possible to monitor the movement of waste garbage from the time it leaves the source to the time it gets to the waste garages. The integration of the internet of things (IoT) with blockchain technology is useful in the following ways in enhancing the functional-

DOI: 10.4018/979-8-3693-7383-5.ch007

ity of the system. In this chapter, the blockchain-enabled smart waste management system is proposed to change the traditional waste management system in the urban areas by using the IoT devices, smart contracts, and data analytics.

INTRODUCTION

The topic of waste management is one of the main global challenges and relates directly to environmental sustainability, public health, the resource conservation. Inefficient recycling practices, poor track, lack of transparency, illegal dumping and other such experience of the conventional waste management system have created profound demand for new and innovative solutions. As the world becomes more and more urbanized and industrialized, the volume of waste generated is growing exponentially. Existing waste management methods, that are based on manual record keeping and centralized control, cannot satisfy contemporary requirements of environment and operation. It calls for the integration of advanced digital technologies in order to develop a more efficient and sustainable waste management scheme.

Waste management has become a disruptive problem which is to be resolved and hence Blockchain technology has come as the solution in the form of a decentralized and unmodifiable ledger system. Blockchain records waste related transactions on a secure as well as tamper proof ledger and ensures traceability of waste movement, prevents fraudulent activities and ensures regulatory compliance. With Internet of Things (IoT), blockchain enables real time data collection from smart sensors present in waste bins, recycling units as well as transportation vehicles. This synergy allows the waste management companies to maximize collection schedules, minimize operating costs, minimize environmental impact and enhance their efficiency in general.

At the end of this chapter, the paper analyzes the transformative apps of blockchain in waste management, exploring its benefits, real life case studies and its obstacles to adoption. This analyzes the way blockchain can function with Integrating the IoT, smart contracts and token developed incentive models for sustainable waste disposal and recycling. Additionally, as a proposed framework to revolutionize urban waste management practices, it further presents a Blockchain Enabled Smart Waste Management System (BSWMS). With the intention of creating more transparent, less expensive, and more sustainable waste management ecosystem, this approach leverages blockchain decentralization, security, and automation power.

Figure 1. Waste management cycle

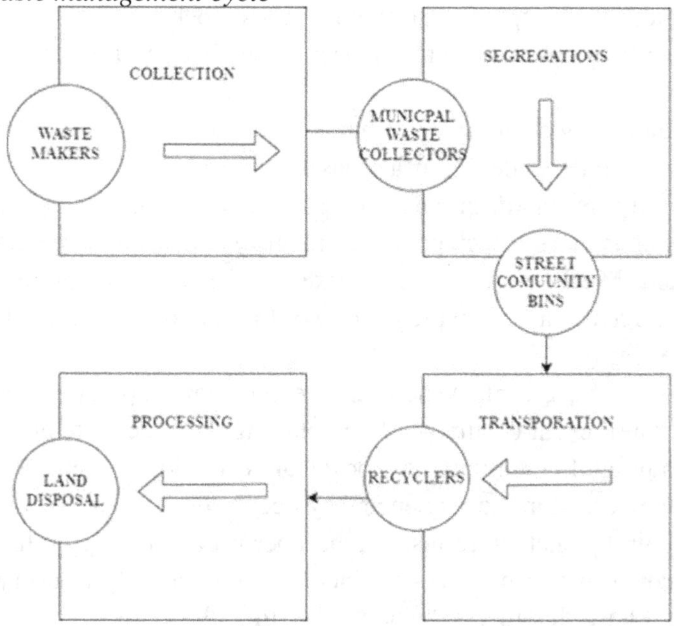

As shown in Figure 1, there are several processes that waste goes through prior to being disposed of at a waste facility. Innovative solutions are urgently required to address the growing problem of traditional waste management systems failing to adjust with the ever-increasing volumes of waste. Problems with data protection policies, inadequate facilities, recycling, and problematic customer behavior are just a few of the obstacles that waste management has had to overcome. The benefits of blockchain technology, a distributed ledger system for permanently storing data, can be used to waste management systems. Within this framework, blockchain technology has become a game-changer, providing novel ways to enhance the efficacy and longevity of waste management. The possibilities presented by blockchain technology are extensive and extensive, ranging from the facilitation of trash recycling to the provision of secure and transparent records of transactions.

To record transactions over a network of computers, blockchain uses decentralized, distributed ledger technology. In a blockchain, each "block" has a record of transactions and is linked to the one before it. This causes it to develop into a chain-like shape. By requiring consensus from all nodes in the network, this chain structure makes sure that data cannot be changed once recorded without affecting all blocks that come after it. Blockchain technology depends on the following essential components as shown in Figure 2.

- **Block:** Individual data storage unit each has list of transactions such as time stamp and cryptographic has of the previous block.
- **Chains:** A sequence of blocks connected together granting data integrity and immutability.
- **Node:** Independent computers that participate in the blockchain network, validating and dependent transactions
- **Cryptographic hashing:** The process of converting data into a fixed size string of characters, which confirms integrity and security of data.
- **Consensus Mechanism:** Protocols like proof of work and proof of stake that confirm all nodes agree on the validity of transactions, maintaining the integrity of ledger.
- A timestamp on the blockchain is a record that pinpoints when a specific transaction or event occurred, often detailing the exact date and time. These timestamps are foundational to blockchain technology, ensuring that transactions are recorded in the sequence they occurred.
- A nonce, in blockchain terms, is a number used once. Regarding its use on a blockchain, it is a number that increases sequentially in every attempt to generate a hash that meets the network's difficulty criteria.
- Merkle root is a simple mathematical method for confirming the facts on a Merkle tree. They are used in crypto currency to ensure that data blocks sent through a peer-to-peer network are whole, undamaged, and unaltered.

Figure 2. General blockchain diagram

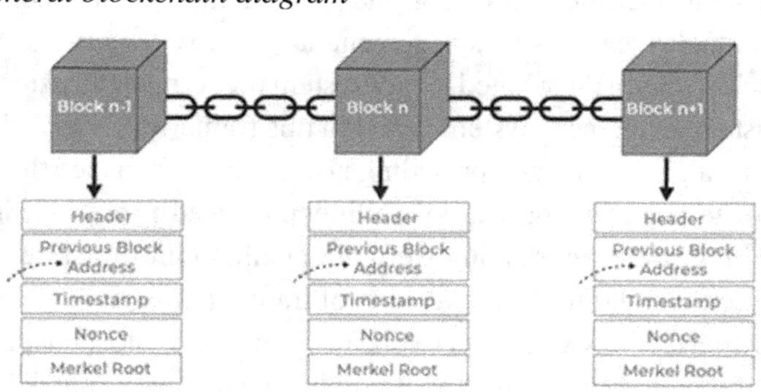

Smart Contract

The phrase "smart contract" refers to contracts that are capable of executing themselves and have the terms of the agreement encoded straight into code. Whenever certain predetermined circumstances are satisfied, they are able to execute and enforce the contract automatically since they are running on the blockchain. Additionally, it eliminates the requirement for intermediaries by automating the execution of contracts, which in turn reduces costs and increases efficiency for the business.

The fact that smart contracts are maintained on the blockchain makes them completely visible and unable to be altered, which ensures that parties can trust one another. The process of blockchain technology is referred to as decentralization. Due to the immutability of smart contracts, the environment in which contracts are executed is completely foolproof. The application of this technology allows for the automation of payments for waste collection services, the verification of compliance with recycling processes, and the enforcement of regulations without the need for personal intervention.

Security Features

Security is a most important feature of blockchain technology ensuring the integrity and security of data. The following are the widely used security features:

- Hash function: Each block has a cryptographic hash of previous blocks securing the data and linking the blocks together. Any modification in a data block will alter its hash leads to tampering evident.
- Digital Signatures: Transactions are signed using cryptographic keys, ensuring that only authorized parties can initiate transactions.
- Proof of work (PoW): It requires nodes to solve complex mathematical problems to evaluate the transactions, ensuring that modifying the blockchain needs significant computational effort.
- Redundancy: Data is replicated across multiple nodes, ensuring that the loss or compromise of one node does not affect the entire system.
- Distributed network: The decentralized nature of blockchain means than an intruder would need to manage a majority of the computational power of network to modify the blockchain, making attacks impractical.
- Proof of Stake (PoS): Evaluators are chosen based on the number of tokens they hold and are willing to stake as collateral, reducing energy consumption while comparing to PoW.

Blockchain's implementation in the waste management system has created a safe, transparent, and decentralized system for monitoring the transactions of the waste, tracking the disposal process, and ensuring regulatory compliance. Nevertheless, blockchain based waste management has its own set of critical security challenges such as data integrity threats, smart contract vulnerabilities, scalability limitations, cyber threats and privacy concerns. Addressing such issues is important to guarantee the reliability and adoption of the blockchain in waste tracking and environmental sustainability.

- The main security threat in managing waste in a blockchain is the danger of false entry at the source. While blockchain is immutable, it cannot ensure that incorrect or fraudulent waste disposal records are not added to the system. However, storing erroneous data on the blockchain means the data cannot be changed or removed, making the possibility of compliance violations and regulatory problems. In 2022, a study on urban waste traceability found that 18% of the waste records in centralized databases lacked error due to manual entry errors or fraudulent reports. To eliminate this, blockchain has to be coupled with IoT smart bins and GPS tracking systems which automatically record the waste data without any manual intervention. Moreover, multi signature validation mechanisms may depend on multiple stakeholders, including municipal authority, recyclers and waste collection companies, to validate waste records before they are permanently stored on the blockchain.
- The vulnerability of smart contracts under waste management agreements is another major problem. Payments are made via smart contracts for which holders enable compliance and automate payment and also transactions for the trade of waste transactions, but they are also susceptible to security exploits including reentrancy attacks, logic errors, and unauthorized executions. As a result, the 2021 cybersecurity report by SlowMist Security pointed out that there was over $1.3 billion lost all around the world because of smart contract vulnerabilities, highlighting the high risk of the automated blockchain transactions. A lack of proper security audits means that smart contracts are open to being manipulated to allow unauthorized claims of waste disposal or unauthorized recycling transactions. In order to avoid this, smart contracts are to undergo formal security audits before launch and multi-party validation should be enforced to ensure fund transfers or service verifications occur only when legitimate waste collection and disposal have been confirmed.
- One of the problems in blockchain based waste management is scalability. High Transaction Costs and Slow Processing Speeds of Public Blockchains (Ethereum and Bitcoin) make them inappropriate for dealing with large amount of waste disposal records. Considering that gas fees for Ethereum

were between \$10 to \$50 per transaction in 2023, it was not economically feasible for municipalities and waste management companies to track real time waste. In 2022, South Korea conducted a case study that tested how high transaction fees had already decreased blockchain adoption rates in its programs that deal with waste recycling by 35%, because municipal authorities were not able to process thousands of transactions that were already very expensive. However, this challenge can be addressed through implementing Layer-2 blockchain solutions like Polygon, Optimistic Rollups, and ZK-Rollups offloading the waste tracking transactions onto faster and cheaper blockchain sidechains. Another way is with hybrid blockchain models which combine public and private blockchain architecture so as to achieve the best security and transparency while maximizing scalability and efficiency.

- Blockchain based waste management systems are vulnerable to the cyber's threats, especially 51% of attacks. A 51% attack is when one entity is able to gain control over half of the blockchain's computational power in order to rewrite the transaction record and manipulate waste disposal data. In 2020, Ethereum Classic was a victim to a 51% attack that led to \$5.6 million in fraudulent transactions by means of blockchain rewrite exploit. Such an attack in the case of waste management systems could extend the data manipulation by representing massive amounts of desired data, unauthorized recycling states, and switching compliance records. The validity of blockchain based waste tracking will thus come into question. This presents a risk which should be mitigated by moving the blockchain waste management platform from Proof of Work (PoW) to Proof of Stake (PoS) and hence decreases the potential for mining power to centralize. Further, by deploying geographically distributed blockchain nodes across the network, no entity has the power to take over the whole network for the sake of coordinated attacks.
- The other challenge posed by blockchain based waste management is privacy, and this is especially a problem with respect to industrial and hazardous waste disposal records. Sensitive waste disposal data is created in many industries, for example, chemical waste information, medical disposals data and electronic waste recycling data, which if publicly available could be used for corporate espionage or legal liabilities. The cybersecurity study of 2022 found that 64 percent of industrial IoT networks are at risk of unauthorized data access and the ability to intercept environmental waste disposal records. Usually, public blockchain models do not have provisions for built in privacy controls, which means that corporate waste disposal records could be disclosed to unauthorized parties to cause data breaches and poor compliance. In order to fight this, the privacy preserving blockchain technologies (such as Zero Knowledge Proofs (ZKP) and Homomorphic Encryption) should be

integrated to allow access to waste management data with selective access to the full transaction details. Further, private blockchain networks such as Hyperledger Fabric and Quorum must be adopted for waste tracking at the level of enterprise where access to compliance records should be limited to authorized regulatory bodies.

- Finally, despite the fact that blockchain brings security, transparency and automation to waste management, it faces a number of significant security challenges including data integrity risks, smart contract vulnerabilities, scalability problems, cyber threats and privacy issues. By making the blockchain based waste management systems more secure, efficient and less vulnerable to cyber threats through integration of automated data collection through IoT, smart contract security audits and various privacy-preserving techniques along with scalability of the blockchain architecture, great improvement can be obtained on the efficiency. Continual investment to secure achieving continuous security improvement and proactively mitigating risks are the stage drivers of blockchain wishing to become a common practice and perceive regulatory facets in waste management facilities.

CURRENT CHALLENGES IN WASTE MANAGEMENT

Traditionally, efficient waste management is difficult due to many challenges preventing effective waste management. The worldwide increase in waste generation has been caused by rapid urbanization, industrial expansion, and an increase in consumerism. Total, according to the World Bank, global waste production is projected to rise to 3.4 tons by 2050, from 2.01 tons in 2018. Although efforts have been made to build up in waste collection, segregation and recycling, some of the issues remain persistent hurdles.

Inefficiencies in Traditional Waste Management Systems

Waste management systems of conventional systems do not have real time data tracking and do not have manual record keeping and hence lack of real time data and the waste collection is not accurate, and the waste is disposed of improperly and there is logistical in efficiency. Even today, fixed waste collection schedules are still used in many cities: when bins are overflowing, or when trips are unnecessarily made in collection, which is costly and leads to waste. Poor segregation also at source leads to low recycling efficiency as contaminated waste ends up in the landfills instead of the reuse.

Lack of Transparency and Accountability

Waste management is one of the major obstacles and that is a lack of transparency in waste disposal and recycling processes. Many times, waste handling companies report incorrect data, illegally dump waste or attempt to fake their recycling figures to meet regulatory compliance. Since tracking mechanisms are not properly installed, stakeholders, for example, municipalities and environmental agencies, cannot verify proper disposal of hazardous materials or check to see that environmental laws are being followed. The waste mismanagement and financial loss is exacerbated by corruption and inefficiencies in waste contracts.

Environmental Degradation and Illegal Dumping

Poor waste management practices are considered environmentally degradation and illegal dumping. Waste is disposed of in locations which are not authorized. It is a severe environmental and health risk. Proper waste disposal is necessary in order to avoid various pollutions. It also increases emissions and therefore is a contributor to climate change.

Resource Management and Financial Cost

Some resource investments and financial cost are required in waste management in an effective manner. High cost of processes and inefficient resource management are the challenges faced by conventional approaches. This is very expensive from the standpoint of waste processing and disposal. Waste management is a challenge for the taxpayers, average low people, municipalities and small businesses.

Data Security and lack of Digital Integration

Waste management systems that are all the more modern are based in the digital era and increasingly depend on the digital forms of data collection, tracking and process automation. Data security, however, remains a big issue as it is not safe from manipulation or cyber threats such as waste collection contracts, regulatory compliance and recycling statistics sensitive information. Additionally, misleading suggestions generated from incomplete data on different waste management platforms do not provide real value in improving health, prevention, and optimization of waste monitoring and management.

ROLE OF BLOCKCHAIN AND IOT IN WASTE MANAGEMENT

Blockchain and the Internet of Things integration in waste management offers a transformative way of eliminating inefficiency, lack of transparency and regulatory compliance issues of the existing systems. By using blockchain with its decentralized, immutable ledger, secure and traceable data of the waste transactions is happening, and IoT is utilized for real time data collection and monitoring through smart sensors. Together these technologies translate into able to track waste, improve logistics of collection, avoid illegal dumping and automation of agreements through smart contracts. In this section, we focus on the revolution of the waste management processes through blockchain and IoT that makes them more efficient, transparent and environmentally sustainable.

Overview of Blockchain and IoT in Waste Management

Traditionally, there is poor documentation, inefficiencies in the collection schedules, fraudulent reporting and the absence of real time monitoring of the system. Automatically analyzing data, being accountable for the accuracy, and eliminating human error are what blockchain and IoT bring in effectively.

- Functions as a tamper-proof, decentralized database, that records all waste transactions securely and end-to-end traceability of waste from generation to end etc. Blockchain is immune to data manipulation, making it transparent in the way waste is disposed.
- Waste Management Using IoT: IoT in waste management includes the use of smart waste cans, GPS based tracking devices and environmental sensors which automatically collect waste data within a real time. IoT devices enable optimum waste collection routes, bin level filling status monitoring as well as detection of hazardous waste.

By combining blockchain and IoT application, waste management systems will get the features of automation, data quality and compliance with environmental regulation, thus making the data driven waste management ecosystem.

Transparency and Accountability Improvements

The biggest challenge in waste management is also the most transparent — illegal dumping, misreported data and improper disposal of waste are not being managed properly. Blockchain's tamper proof record keeping and centralized ledger makes sure that waste transactions are secure and verifiable.

- Reputation: Any time waste data is placed on a blockchain, it cannot be altered or deleted, and every instance of waste collection, transportation, and disposal is recorded. By doing so, it simplifies verification by regulators, municipalities and environmental agencies that the process is indeed working.
- Fraud and Corruption Prevention: Blockchain prevents fraud and corruption by preventing false reporting of waste disposal, as it allows only validated transactions to be recorded. It also reinforces against illicit waste dumping and false bill of claims.
- Auditable Waste management: Authorities can trace back the entire step of the waste disposal process to ensure the handlers follow proper protocols and that recyclable waste is processed instead of mismanaged.

Blockchain helps improve transparency that brings about more accountability to waste management thus ensuring regulatory standards and sustainability goals are adhered to.

Real-Time Tracking and Monitoring of Waste

The fixed schedule of traditional waste collection system results in overflowing bins in some areas and inefficient emptying of underutilized bins. Smart sensors based on IoT and blockchain are used to track the real time and make it possible for efficient waste collection and monitoring.

- Smart Bins with IoT: Filled with sensors that can detect when the waste fills the level above a preset threshold, smart bins can automate the notification of waste collection authorities when waste levels exceed the threshold. It prevents overflowing waste and unnecessary collection trips, making the best resource use.
- Real time tracking of Waste Transport: IoT devices that are installed inside waste collection trucks will track the location of the trucks and make sure that the waste is transported through pre-approved routes and then to the designated disposal or recycle facilities.
- Tamper Proof Data Storage with Blockchain: Waste collection records, the bin status, and the vehicle logs are all stored in the blockchain safely to avoid tampering of data or loss data altogether. It also permits waste generators and authorities to confirm the correct disposal.
- IoT enabled sensors can detect toxic gases, heavy metals and hazardous chemicals in the waste which can prevent improper disposal of the waste with hazardous chemicals and make sure that such hazardous waste is treated accordingly.

Together, IoT's real time monitoring, and blockchain's immutable storage make waste management data driven, optimized and enables waste management to respond more effectively to the goals of the urban sustainability.

Blockchain-Based Smart Contracts for Waste Agreements

The ability to automate and streamline agreements regarding waste management using smart contracts (self-executing contracts with predefined conditions coded into a blockchain) leads to eliminating bureaucratic inefficiency, delays and costs.

- Waste collection and recycling services: Chances of delayed payments and disputes regarding fulfillment of services tend to be high; there can be automatic payment systems in place. Without need for intermediaries, payments are automatically released when waste is collected and verified as such by smart contracts.
- Shopping for Waste Recycling Targets: Smart contracts can mandate corporations and municipalities to recycle a certain amount of waste. There may be automatic penalties or adjustments in the waste management fees when the recycling goals are not met by a certain deadline.
- Waste management service providers have to be accountable for their work and have to follow proper disposal guidelines. Automated termination of contract or financial penalties are enforced to maintain compliance with the rules.
- Blockchain based smart contracts: Smart contracts along with decentralized blockchain networks can be used to build peer-to-peer waste trading platforms in which the industries can buy recyclable waste provided by the suppliers and make sure that the recyclables are not disposed in the landfills.

Smart contracts get rid of unavoidable manual verification, administration delays and fraudulent practices in the waste management agreements and provide them with efficiency, security and automation.

Integration of IoT and Blockchain for Sustainable Waste Management

IoT's real time monitoring combined with blockchain's secure record of keeping will help build a sustainable waste management ecosystem through reduction of environmental impact, optimal resource utilization and promoting circular economy models. IoT based smart waste collection allows the routes for pickup to be optimized in order to reduce fuel consumption and emission. Blockchain makes sure that every waste collection activity is recorded and tracked for compliance in terms of sustainability.

- The Role of Blockchain in Encouraging Circular Economy Practices: Blockchain-mediated waste tracking and tokenization of incentives motivate people and businesses to add to the recycling pace. IoT sensors double check that recyclables are in the correct places in order to get them to the right facilities.
- Blockchain helps to ensure Automated Environmental Compliance: The result is easy-to-verify compliance data that can be used to ensure that industries follow rules for the disposal of waste. The pollution monitors are supposed to trigger alerts or cause penalties if the pollution levels exceed a threshold.
- IoT continuously collects waste data, which is then stored and analyzed long term in blockchain as well as verified. This data can be fed to the AI-driven analytics to predict waste generation trends as well as to optimize resource allocation and coming up with recycling strategies.
- Sustainable waste management behavior through blockchain based tokens: Blockchain based tokens can be used to incentivize the proper waste disposal, recycling and waste reduction by the citizens or businesses. They are redeemable for discounts, tax benefits and social credits.

With the help of IoT and blockchain, waste management becomes smarter, greener, and more accountable; it turns into a smart, green and accountable waste disposal system consistent with the global environmental goals. As waste management remains a longstanding inefficiency, a lack of transparency in the industry and compliance challenges, this brings in this block chain introduction with IoT integration which is a paradigm shift. Being the decentralized, immutable, and smart contract automation, blockchain gives fool proof, and completely transparent waste handling to the process, while smart waste management systems with IoT enabled smart technology provides real time monitoring, automated logistics and data-based decision making. With the integration of these advanced technologies, the waste management of the future can be optimized in terms of efficiency, sustainability and also accountability.

RELATED WORKS

Overview of Existing Waste Management Systems

Safiullin *et al.* (2021) presents a comprehensive environmental evaluation of e-waste recycling in the blockchain. This study uses co-integration models to evaluate the impacts of blockchain on key economic sectors employing statistical software for analysis over data from 2009 to 2020.This study represents the blockchain in enhancing transparency and efficiency in some domains. But it has challenges such as high energy consumption and scalability issues. Praveen Kumare *et al.* (2020) introduces a blockchain-based traceability system for waste management in smart cities. Techniques used include the application of blockchain technology to enhance transparency and coordination among various stakeholders within municipal waste management systems. This study has integration of blockchain for real time traceability in waste management which leads to high efficiency and accountability.

Praveen kumara *et al.* (2021) proposes a blockchain based solid waste management which optimizes the resource cost of waste management system. It is one structural approach which has high efficiency and transparency in blockchain technology. It also grants stakeholders better coordination and more effective waste traceability. Khan *et al.* (2022) presents an IoT enabled system for e-waste tracking and tracing using blockchain in smart cities. This study includes enhanced security transparency and efficiency offered by blockchain and IoT integration in mitigating e-waste in smart cities. But this research lacks in managing blockchain operations, handling scalability challenges and reliance on IoT device integrity and connectivity. Koshariya *et al.* (2023) explores agricultural waste management systems using artificial intelligence techniques. It details how precision agriculture and smart farming are revolutionized through the integration of IoT sensors support vector machines, convolutional neural network for monitoring and improving waste management and crop health. This study enhances the sustainability and productivity of agricultural practices.

Broadly, Jiang *et al.* (2023) discuss the application of blockchain in waste management, which offers transparency, reduces illegal dumping and waste traceability. In their study they discuss smart contract-based waste disposal ambits and challenges including data security, scale and regulatory hurdles. Blockchain research is also highlighted, which enables records of waste to be immutable and thus, fraudulently claimed or carried out unauthorized dumping activities (Jiang *et al.,* 2023). Similarly, Bulkowska *et al.* (2023) conduct a study into how blockchain practices apply to such waste management: real-world case studies looked at how blockchain improves accountability and optimizes the process of recycling waste. Work by them highlights the importance of setting up improved blockchain based waste monitoring infrastructures to help government and municipal waste tracking efforts. In Castiglione *et al.* (2023) case, the researchers propose a framework, based on the blockchain technology with circular economy principles for incentivizing waste practices in a sustainable manner. The second is a study of token-based reward systems for people and companies that do their part in responsible waste disposal and recycling. Waste collection agreements can be enforced automatically through smart contracts, thus alleviating a lot of the administrative complexity associated with waste processing industries. In addition, within the supply chain a traceability is implemented to ensure segregated and directed that recyclable materials are being directed to secondary use industries so that they may avoid becoming landfill dependent. Bulkowska *et al.* (2024) discuss a more specific application of blockchain for plastic waste recycling by introducing a blockchain based management system for recyclable plastic waste. The study shows how blockchain's immutable ledger system can keep fraudsters from claiming recycling and give a sense of monitoring of plastic waste movement throughout the supply chain. Their findings indicate that blockchain integrated platforms can offer reliable verification mechanisms of government audits on compliance of waste management companies with environmental regulations. The summary of the literatures is presented in Table 1.

Table 1. Existing methods of blockchain-based waste management

Author and Year	Techniques Used	Merits	Limitations
Shaik Vaseem Akram *et al.* (2021)	Blockchain framliteraturedevices, LoRa sensors	Enhanced data security, incentivization of waste disposal through automatic rewards	Complexity in deployment and maintenance, scalability issues, high technology implementation costs
Yagnyasenee *et al.* (2021)	Blockchain (Ethereum), Smart contracts	Automation and secure data handling improve efficiency in waste management	High computational and operational costs, scalability challenges
Zhen Liu *et al.* (2022)	Blockchain technology, CIM, BIM	Improves sustainability and efficiency in construction waste management, ensures transparency	Complexity in technology implementation, high initial costs, need for widespread stakeholder buy-in
Bihter *et al.* (2023)	Fuzzy AHP, TOPSIS	Systematic decision-making tool combining two robust techniques	Complexity of application, requires significant expert judgment, limited adaptability
Ma *et al.* (2024)	Blockchain, design science action research	Streamlines processes, enhances transparency and data integrity in CDWM	Dependency on technological readiness of stakeholders, scalability of blockchain solutions
Yekti Wirani *et al.* (2024)	Systematic Literature Review (SLR), PRISMA protocol	Comprehensive approach enhances waste management effectiveness	Complexity of integrating the framework, potential resistance from stakeholders due to required change

Electronic waste (e-waste) is a problem with hazardous components and risks of improper disposal. To address this issue, Sahoo *et al.,* (2021) suggested a unified blockchain based platform for global e waste management. In their study, they point out that blockchain's decentralized nature can effectively aid safe and transparent waste disposal of e-waste to ensure that electronic components are recycled appropriately or disposed of safely. In this platform, different stakeholders like manufactures, recyclers and the waste collection agencies are connected, and they are able to confirm e-waste transactions in real time. Building on his concept, Khan and Ahmad (2022) create an IoT enabled blockchain system for tracking and tracing of e-waste. According to their research in IEEE Access, their IoT sensors placed in electronic devices can continuously send real time location and disposal data to a blockchain ledger. But it acts to prevent illegal e-waste dropping by allowing governments and environmental agencies to follow each thrown away electronic device. Their model also explores multi signature verification model that authenticates that only authorized entities can approve waste disposal transaction in order to prevent fraudulent recycling.

Blockchain Applications in Various Industries

Saad *et al.* (2023) explore a blockchain enabled vehicular AdHoc network for smart solid waste management. This study has high reliability, real-time monitoring capabilities, enhanced transparency, and high efficiency in solid waste management. This method keeps tracks of smart bins by utilizing blockchain technology for secure communication. It may have some challenges in less urban areas without robust digital infrastructure. Costa *et al.* (2022) introduces a blockchain based framework for waste management in the construction industry. This study focuses on construction and demolition waste in their blockchain framework. It has higher sustainability and traceability in construction waste management by enhancing efficiency and accountability.

Gopalakrishnan *et al.* (2019) introduces a blockchain based waste management to address the effectiveness challenges. This study eliminates manual intervention by increasing data security and transparency through blockchain's immutable ledger. The advantages of this method are that it has community participation in proposer waste management and segregation. Katarzyna *et al.* (2023) proposes blockchain implementation in waste management. It utilizes decentralization and smart contracts for securing data and automating waste recycling processes. Merits of this approach include the high accountability across waste management process. This method has scalability issues in real world applications.

Previous Studies on Blockchain in Waste Management

Taylor *et al.* (2020) reviews the different applications of blockchain technology in transparent waste management. These studies utilize blockchain for improving efficiency. This method benefits in supporting goals, managing privacy and anonymity, and clarifying property rights of products and waste. This method also enhances the accuracy and reliability of waste monitoring, which is most important for the rules and process efficiency Karunamiga *et al.* (2023) introduces smart waste management by utilizing the internet of things and blockchain. This research employs an array of sensors integrated with Arduino microcontrollers to monitor and reduce the waste. The alert messages are given to relevant workers when the smart bin is full. Data is transmitted using wireless fidelity for monitoring purposes. Hakak *et al.* (2020) explores a blockchain based system for mitigating industrial wastewater. An efficient architecture is proposed in this study for securing and monitoring the wastages. This method has wastewater treatment plants by utilizing industrial IoT and blockchain methodology. Utilizing blockchain security and reliability are improved efficiently in wastewater management in industrial wastewater management.

Lamichhane *et al.* (2017) presents a smart waste management method using IoT and blockchain technology. This method involves the integration of IoT devices for environmental sensing and blockchain for secure information and resource management. This approach utilizes blockchain to guarantee transparency and security when reducing the operational cost. Paul *et al.* (2021) presents an extensive review of blockchain in plastic waste management. It clearly explains the recycling process to achieve sustainability. This study combines blockchain with the conventional waste management methods to provide a reliable account of waste generation, disposal, and recycling methods. Even though it has many merits, this method includes high initial technology cost, lack stakeholders and not adaptable to environment.

Research Gaps

- Lack of Standardized Frameworks for Blockchain Implementation: Currently, most blockchain based waste management systems are developed as standalone solutions that are not interoperable with the existing municipal and industrial waste management infrastructure.
- Scalability and High Computational Costs: No standard framework exists for the integration of blockchain in different waste management models, waste types and geographical regions. As a result, public blockchains and other blockchain networks have high computational costs, high energy consumption, and slow transaction processing speeds. Solutions that are scale to man-

age large scale waste data transactions such as real time IoT sensor data must be secure and efficient.

- Privacy and Data Security Concerns: Blockchain is a great solution for transparency, but also it creates concerns about data privacy, especially in cases such as industrial hazardous waste disposal, medical waste tracking, and so on, where we have to protect the sensitive waste data. The traditional public blockchain models lack the capability of granular access control mechanisms for managing the confidentiality of waste data.
- Limited Adoption of Incentive-Based Waste Management Systems: Despite the work on blockchain based tokenized incentive models for buying and selling recycling and responsible waste disposal those models have yet to be adopted in its real world. The economic sustainability of blockchain-based incentive mechanisms for waste management cannot be well researched.
- Smart Contract Vulnerabilities and Governance Challenges: While smart contracts automate waste agreements but are insecure as against, insecure to code, insecure programmatically and as against governance. Incorrect design of a smart contract may result in system failure, such potential financial loss, and exploitation of the system by malicious actors. Further, the issue of governance in managing disputes in automated contracts has not been addressed enough.
- Integration of Blockchain with AI and Machine Learning for Predictive Waste Management: Presently, blockchain research mostly lies in securing records and AI driven predictive analytics for waste generation, collection optimization and allocation of complimentary resources, which has not been yet researched upon.

BLOCKCHAIN APPLICATIONS IN WASTE MANAGEMENT

Blockchain technology is a disruptive innovation in the industry across multiple sectors; it provides enhanced security, as well as transparency and automation. For example, it is a decentralized and immutable ledger, which stores tamper-proof data and beyond cryptocurrencies use cases. Blockchain is being integrated by industries ranging from healthcare, finance, logistics, to sustainability, to break down inefficiencies, lower frauds, and optimize business processes. This is the success of blockchain in these domains and blockchain's potential success for waste management where traceability and accountability are critical.

- Data security, patient record management, and counterfeit drugs are the special challenges faced by healthcare industry. The medical records stored on the blockchain are secure, tamper-proof, and are only available to the authorized personnel. In block chain, EHRs can be shared across hospitals without revealing any information, ensure privacy and integrity.

- The second is in pharmaceutical supply chains, where Blockchain is used in tracking drug authenticity since manufacturing to distribution. According to the World Health Organization (WHO), around 10 per cent of medicines in developing countries have counterfeit. In this way, blockchain eliminates fraudulent activities as only the verified drugs will be allowed to reach the market. Blockchain also provides a facility for smart contracts for the automatic processing of insurance claims and eliminates paperwork and fraud. For example, IBM's Digital Health Pass uses blockchain to store COVID 19 vaccination and test records securely and seamlessly check them across the healthcare systems.
- As a secure transaction and fraud prevention technology, blockchain has been an early adopter of the financial industry, as it offers efficient cross border payments. Centralized control systems restrict traditional bank and other financial decision mechanisms to only one approach. The financial dimension of blockchain powered systems, such as Decentralized Finance (DeFi), eliminate intermediaries and speed up, decrease and increase or at least not hinder transparency of transactions. Blockchain based smart contracts eliminate the need for manual verification of loan approval, insurance claim and digital asset trading. It also comes with the implementation of Central Bank Digital Currency (CBDC) by countries like China, USA, the European Union that has been exploring the use of blockchain-based digital currencies to improve transaction efficiency. Example: Banks and financial institutions rely on Ripple's XRP Ledger to make instantaneous cross-border payment with time reduced from several days to a few seconds.
- The global supply chain industry has a number of issues with it such as counterfeiting products, supply chain inefficiencies, lack of real time tracking, and supply chain fraud. The use of blockchain technology gives an increase in traceability and transparency in addition to offering a shared ledger to record each commerce in the provide chain. This really helps to really keep an eye on good from manufacturing right through delivery to reduce fraud, theft, and so on. By applying blockchain to IoT sensors and RFID tracking, companies will be able to track the condition of perishable goods and ensure the storage and transportation of them is done properly. This is especially useful for the food and pharmaceutical industries, which possess temperature sensitive products that need strict quality control. Walmart's blockchain application: Real time food product tracking from days to seconds, an example being the use of block chain tracking food products in real time.

- The importance of Blockchain for sustainability initiatives lies in the efficient carbon credit trading, energy management and waste tracking in the carbon credit life cycle. Blockchain's applications include companies and governments using it to monitor and verify emissions to ensure that they meet environmental regulatory requirements. This has one significant application for carbon credit tokenization where companies can buy and sell carbon credit through blockchain based smart contracts. It encourages environmentally friendly practices by offering financial benefits to organizations that cut down their carbon footprint. Renewable energy trading is also supported by blockchain, wherein people and companies can purchase and sell the surplus solar or wind energy among themselves through decentralized peer to peer platform. This will help with efficient distribution of energy, thereby reducing the dependence on fossil fuels. For example, an example of such is the Australian blockchain platform that facilitates the peer-to-peer energy trading, such that a user can buy and sell excess solar power seamless and transparently.

Blockchain potential is shown through success in healthcare, finance, logistics and sustainability as well as waste management. Like supply chain tracking, blockchain can detail where waste moved in the supply chain from collection to recycling to ensure that the producer meets environmental standards. There is high potential of integrating IoT sensors, smart contracts, and tokenize incentives to build more sustainable and transparent waste management ecosystem. Blockchain methodology provides efficient benefits in overcoming the issues faced by the conventional waste management methods. This part explains the way in which blockchain improved transparency, waste traceability, accountability and facilitate smart contract for waste management agreements. It supports the development of a decentralized waste management system.

Transparency and Accountability

Blockchain has many good features. The ability to create transparent and accountable method is one of the good features of blockchain. Blockchain includes an immutable ledger of all transactions which guarantees that once the data is recorded, it cannot be modified or removed. This immutable feature is important for managing accurate, accessible, and correct verification and auditing in the waste management process. Blockchain offers an exact audit of all processes from waste generation to removal of waste. Due to this transparency, discrepancies are found with regulations.

Tracking And Monitoring of Waste

Each step in the waste management process is documented, making it difficult to dispose of waste improperly. Tracking waste materials is important for ensuring proper management and disposal of waste. Blockchain improves these capabilities through its robust data recording and sharing advantages. This recording can start from waste generation to the waste removal in the system. Every transaction has four steps. They are gathering, transportation, computation, and removal of waste. All these steps are recorded in the system. Integration of IoT with blockchain grants real time monitoring of waste levels, collection schedules and route transportation. To guarantee the update data is available, the IoT sensors can send information to the blockchain directly. This is called traceability. This traceability confirms that the waste management employs comply with regulations and standards. Experts can verify the proper management of hazardous waste and accomplishment of recycling tasks. By providing a transparent record of waste movement, blockchain helps in finding and preventing illegal waste disposal practices. The tracking of waste is most essential.

Waste Management Agreements Using Smart Contracts

Smart contracts are self-executing contracts with the terms of agreements directly written into code. They can enhance the efficiency and reliability of waste management agreements. It can automate the payments for waste collection and removal services. When a set of predefined conditions is met, payments are then only released. Hence, it ensures accountability. Smart contracts can enforce and define service level agreements among the municipalities and waste management companies. It can also employ methods for mitigating the issues among the parties, eliminating the requirement for costly and time-consuming issues.

Decentralized Waste Management Systems

Decentralization is the basis of blockchain that can revolutionize waste management system. Utilizing distributed ledger, the decentralized waste management records transaction among the multiple nodes which guarantee data redundancy and security. This eliminates the risk of data loss and improves the resilience of the system. Without requiring intermediaries, a decentralized system enables peer to peer interaction. Hence the costs are reduced, and computations are improved. It is more scalable than the centralized methods. It can implement token-based incentive methods to perform recycling and waste reduction. The digital tokens are earned for arranging waste in respective manner and enabling the sustainable behavior.

Role of Blockchain and IoT in Sustainable Waste Management

The combination of blockchain technology with the Internet of Things (IoT) has tremendous merit in terms of sustainability, as it boosts the efficiency, transparency, accountability of waste management. By not being centralized, the blockchain's decentralized means that there are immutable records of waste generation, collection and recycling, which prevent bad waste generation like illegal waste dumping, mismanagement and fraudulent payments. Smart bins and waste monitoring devices are provided in an IoT enabled way to enable real time data, optimum collection schedules, minimization of unnecessary trips, and reducing fuel consumption. But this synergy doesn't only minimize the environmental degradation and sustain waste management; it also maximizes the sustainability of the waste management by securing the waste disposal properly, promoting the recycling of the resources, and minimizing the accumulation in the landfill. With all such advanced technologies, its impact on nature is minimized by data driven decision and automation. Manual verification by outsiders is eliminated, reducing the use of paper and wastage of administrative efforts. In addition to that, IoT sensors can monitor environmental conditions and detect hazardous waste at early stages before pollution. Stakeholders can trace and verify the sustainable waste disposal methods through leverage of blockchain's framework before and after compliance with the environmental regulations. The other reason why this approach reduces carbon footprints significantly more than the current method is because it maintains precision in the routes driven by waste collection vehicles and encourages sound recycling habits. However, there is still debate regarding the cost effectiveness of installing these technologies in the waste management process. The earnest upfront investment in blockchain infrastructure and IoT devices may be large, however, the long-term benefits will far out way these costs since it reduces resource wastage and lower labor cost. Waste tracking and sorting is reduced to human intervention to cut labor cost and errors. Furthermore, incentive based blockchain programs will help the recycling rate to increase while at the same time generating benefits for the economy. A finically viable and ecofriendly waste management system is due to the overtime efficiency gains and environmental penalties.

INTEGRATION WITH EMERGING TECHNOLOGIES

The integration of blockchain with other emerging methods such as IoT can increase the waste management system. This section explains how blockchain is combined with IoT which enables data collection and monitoring, which optimizes the routes and fleet management. It contributes to the development of smart waste management methods.

IoT and Blockchain

IoT devices can automatically record data on the blockchain, ensuring that the information is accurate and tamper-proof. This integration helps in maintaining a reliable and verifiable record of waste management activities. IoT devices can trigger automated transactions on the blockchain through smart contracts. For example, a smart bin can send a signal when it is full, automatically initiating a collection request and updating the blockchain with the transaction details. The combination of blockchain and IoT provides a secure framework for data transmission and storage. Blockchain's cryptographic security ensures that data from IoT devices cannot be altered or intercepted. IoT consists of interconnected devices that gather and exchange data in real time. IoT devices can securely and transparently record data on a decentralized ledger when integrated with blockchain. IoT devices can record data automatically.

IoT consists of interconnected devices which collect and exchange data in real time. When integrated with blockchain, IoT devices can record data on ledger in a secure and transparent manner. Hence it improves the reliability and accuracy of the waste management system. IoT devices can record data on the blockchain, guaranteeing.

Data Gathering in Real Time.

IoT sensors can monitor waste levels in bins and dumpsters, providing real-time data on when and where waste needs to be collected. This information can be recorded on the blockchain to ensure transparency and accountability. Real time monitoring grants for dynamic modification of waste gathering schedules related to current needs. Hence, the resource is optimized, and the operational costs are reduced. IoT devices can monitor environmental affects and those are recorded on blockchain. It provides relevant actions if needed. This data is easily accessible and securely stored.

Smart Waste Management Systems

The integration of blockchain and IoT is a key part for the development of smart waste management systems. This system uses advanced methods to improve the efficiency of waste management processes. IoT smart enabled bins monitor waste status, remove different types of waste, and communicate with waste management in real time. Blockchain guarantees that the data from smart bins is accurately recorded and accessible to stakeholders. This system monitors the various types and quantities of waste being generated, aiding in finding the insights for waste reduction and recycling methods. This system can engage the communities by offering real time data on waste collection arrangements and environmental effects. Blockchain guarantees that data is trustworthy and accessible, which enables community contribution in the waste management efforts. Blockchain can store information on various sustainability metrics. This information can be utilized to evaluate the effectiveness of waste management methods and to report on sustainability methods.

INCENTIVIZING SUSTAINABLE PRACTICES

Incentivizing sustainable practices is important for promoting environmentally friendly behaviors among individuals and business. Blockchain technology offers innovative ways to create and manage incentive programs that encourage recycling, waste reduction and community participation. This section explains the tokenization of recycling activities, reward system for waste reduction, community engagement and the case studies of successful incentive programs.

Token System for Recycling Activities

Tokenization involves converting the value of recycling activities into digital tokens that can be earned, traded, or redeemed. Blockchain technology ensures that these tokens are secure, transparent and tamper proof.

Digital tokens represent a quantifiable value of recycling activities such as amount of waste recycled or the reduction in waste generation. These tokens can be created and managed on a blockchain platform. Blockchain ensures that every recycling activity and the corresponding token issue are recorded in a transparent manner. This transparency prevents fraud and ensures that tokens are provided fairly based on actual recycling efforts. Token can be used to encourage different recycling activities.

Reward Systems for Waste Reduction

Reward systems in blockchain technology can scale to acquire large numbers of participants and different activities of waste reduction, from individual actions to corporate sustainability initiatives. Smart contracts on blockchain can automatically provide rewards depending on predefined criteria.

Case Studies of Incentive Programs

Examining real word case studies can provide valuable advancements into the effectiveness of blockchain related incentive methods for sustainable waste management. Several cities have employed blockchain based recycling programs where users earn tokens from the recycling process. A city might partner with local businesses to provide discounts to residents who acquire a particular number of recycling tokens. Environmental nonprofits can utilize blockchain to monitor and incentivize volunteer tasks. Individuals who are volunteer might gain token for engaging in clean up events, which can be exchanged for merchandise, donations to other cause or recognition in the publication of the organization. Universities can deploy blockchain related methods to perform students and staff to participate in sustainable practices.

CHALLENGES AND LIMITATIONS

When blockchain offers some merits for waste management, several challenges and limitations must be considered for efficient implementation. These include standardization and interoperability issues, data security and privacy issues, scalability issues and environmental effect of blockchain mining.

Standardization and Interoperability Issues

Blockchain technology lacks standardized protocols and frameworks. Different blockchain platforms may use different consensus methods of programming making it complex to connect across various platforms. In a decentralized waste management system different stakeholders may use various blockchain platforms. Without standard protocols it becomes complex to ensure ideal data exchange and interoperability between these systems.

Connecting blockchain with existing waste management infrastructure and technologies requires compatibility. Existing systems may need to be modified to work with blockchain which can be costly and time-consuming. Developing standardized APIs and middleware can facilitate integration. Collaboration between blockchain developers, waste management companies and regulatory bodies is important to make interoperable solutions.

Scalability Concerns

A Blockchain network includes issues in transaction throughput. High usage of blockchain networks can lead to congestion, resulting in slower transaction processes and highest transaction fees. Selecting a blockchain platform with high throughput and low latency is important for waste management applications.

Data Privacy and Security Considerations

While blockchain provides transparency and immutability, it leads to privacy concerns. Blockchain networks are not immune to security threats. Potential risk includes 51% attacks where malicious intruders obtain control of the majority of the network computational power and smart contract threats that can be exploited to manipulate transactions. Implementing privacy preserving techniques can protect confidential information.

PROPOSED METHOD FOR BLOCKCHAIN-ENABLED SMART WASTE MANAGEMENT SYSTEM (BSWMS)

The proposed BSWMS (Blockchain enabled Smart Waste Management System) shown in Figure 3 revolutionizes urban waste management by integrating blockchain technology with IoT devices, smart contracts, and advanced data analytics. This method ensures efficient, transparent, and sustainable waste management practices. The deployment begins with the implementation of IoT to enable smart bins across the urban areas. These smart bins are equipped with various sensors that monitor waste levels, types of waste and environmental conditions. Ultrasonic or infrared sensors are used to measure fill levels while cameras offer visual verification of waste types. Additionally, few sensors such as temperature and humidity sensors monitor environmental conditions to prevent potential hazards. The data collected by these sensors is transmitted in real time to the central system using secure wireless connectivity.

A decentralized blockchain network forms the backbone of the BSWMS, recording all data related to waste management activities. Each transaction including waste collection events, bin status updates and recycling activities is recorded on the immutable blockchain ledger. This ensures that once data is recorded, it cannot be altered or deleted, providing a transparent and tamper proof record. Smart contracts are deployed on the blockchain to automate various processes. Waste gathering optimization is completed through efficient fleet management and dynamic routing methods. AI driven algorithms process the real time data from IoT enabled smart bins to compute the gathering route ways. In order to reduce consumption of fuel and cost of operation, these routing methods are based on some environmental criteria such as bin fill levels, vehicle availability and traffic conditions. Guarantying reliable fleet management and timely maintenance, the smart devices provide real time tracking and performance data. Based on the updated data from smart bins and smart vehicles, the routes and scheduling are altered in a dynamic manner for efficient waste management.

The proposed methodology employs token-based incentive programs to utilize sustainable practices. Digital tokens are provided to hospitals, residential areas and industrial areas depending on their recycling tasks. The recycling tasks include reducing overall waste generation and sorting the waste in the correct manner. These tokens are recorded on the blockchain to prevent fraud and provide transparency. Participants can retrieve tokens for different rewards. A reward includes vouchers for sustainable products, public transportation, and utility bills. To provide a more reliable system, smart contract can automate the distribution of rewards based on token balances and predefined criteria.

Community engagement is important for the employment of proposed BSWMS. Utilizing mobile applications, this study offers real time access to waste management data for hospitals, residential and industrial areas. The rate of recycling, different recycling tasks, arrangement of waste gathering, and environmental changes are shown in the app. This app contains educational parts that promote awareness and best practices for recycling tasks and waste reduction.

Figure 3. Framework of proposed methodology

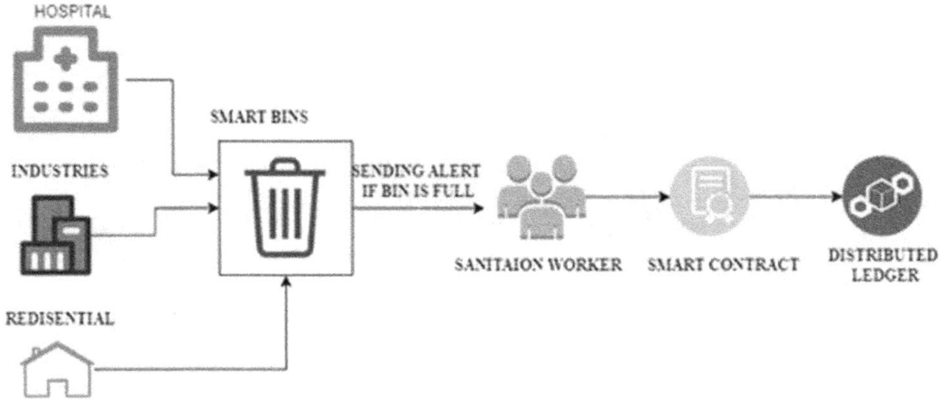

This study automates compliance using smart contracts by encoding the rules and regulations. The rules are checked through compliance aspects and by encoding the environmental regulations. These contracts guarantee that all waste management activities are attached to the right rules, generating real time alerts for any non-compliance issues and then corrective actions are enabled. Then the distributed ledger provides transparent and accurate reports for regulatory authorities. It guarantees accountability and provides audits.

Regular monitoring and assessment of the performance of the system are important. Real time data from IoT sensors, blockchain records and user feedback must be assessed to enhance the systems. To guarantee that this proposed method achieves its motto, efficiency, sustainability, and transparency are evaluated in a continuous manner. This proposed BSWMS enhances the efficiency of waste collection by optimizing the routes and eliminating the cost of processing. The transparency offered by the blockchain creates trust among stakeholders and guarantees accountability. The proposed method could benefit both urban and rural environments efficiently. The workflow in the form of flowchart is depicted in Figure 4.

Figure 4. Flowchart of proposed model

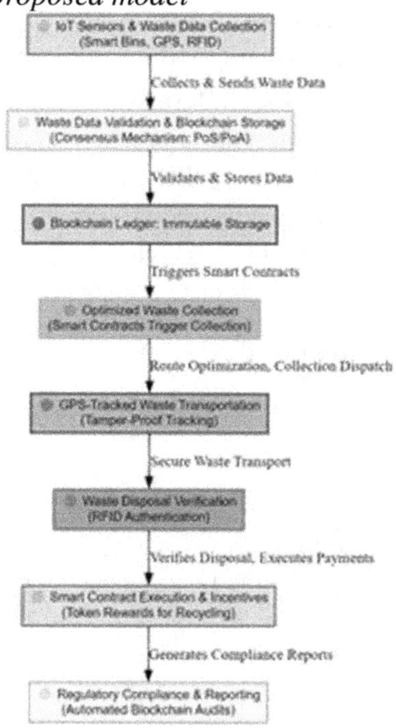

EXPERIMENTAL ANALYSIS

Evaluating the performance of the Blockchain-Enabled Smart Waste Management System (BSWMS) requires a comprehensive set of metrics. These metrics ensure that the system operates efficiently, transparently, sustainably, and engages the community effectively. The following are the metrics used with the existing methods such as [11], [12] and [13].

Network Availability

Network Availability measures the percentage of time the blockchain network and IoT devices are operational and accessible. It indicates the reliability and robustness of the system.

$$network\ availabilty = \frac{Up\ time}{Total\ time} * 100$$

Table 2. Analysis of network availability

Metric	Network Availability (in %)
[11]	79.6
[12]	85.4
[13]	88.9
Proposed methodology	92.7

Figure 5. Comparison of proposed method with existing methods in terms of network availability

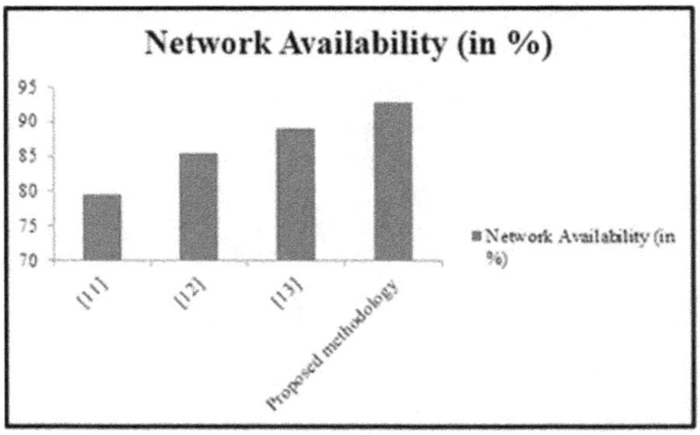

Figure 5 shows the network availability percentage for different methodologies. The proposed methodology has the highest network availability at approximately 92%, significantly outperforming the previous methodologies ([11], [12], [13]) which range between 78% and 88%. This indicates that the proposed methodology provides a more reliable and stable network for the Blockchain-Enabled Smart Waste Management System.

Where Uptime refers to the amount of time that a system, particularly a computer or network system, is operational and available for use.

Transaction Throughput

Transaction Throughput measures the number of transactions processed by the blockchain network within a specific time period. It reflects the capacity and efficiency of the blockchain system in handling transactions.

$$Transaction\ throughput = \frac{Total\ transactions}{Time\ period}$$

Table 3. Analysis of transaction throughput

Metric	Transaction throughput (Transaction/hour)
[11]	3975
[12]	3980
[13]	4032
Proposed methodology	4196

Figure 6. Comparison of proposed method with existing methods in terms of transaction throughput

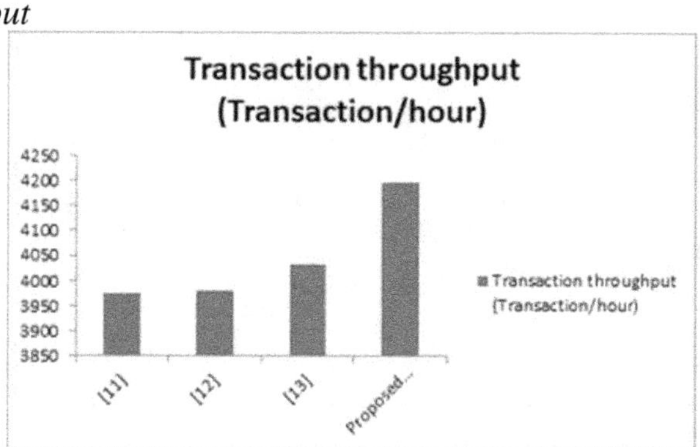

Figure 6 illustrates transaction throughput in transactions per hour for different methodologies. The proposed methodology significantly outperforms the previous ones, achieving approximately 4200 transactions per hour. In comparison, methodologies [11], [12], and [13] handle around 3950, 4000, and 4050 transactions per hour, respectively. This indicates that the proposed method offers superior processing capacity and efficiency.

Recycling Efficiency

Recycling Efficiency measures the percentage of waste that is successfully sorted and recycled out of the total waste collected. It indicates the effectiveness of the recycling process within the waste management system.

$$Recycling\ efficiency = \frac{Recycled\ waste}{Total\ waste\ collected} * 100$$

Table 4.

Metric	Recycling efficiency (%)
[11]	78.3
[12]	79.8
[13]	83.2
Proposed methodology	85.9

Figure 7. Comparison of proposed method with existing methods in terms of recycling efficiency

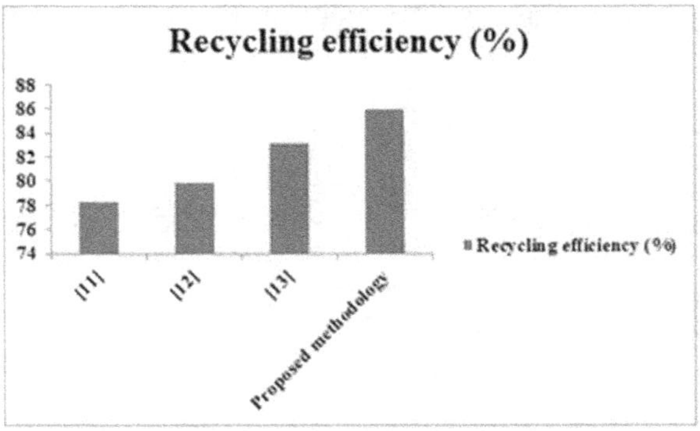

Figure 7 displays the recycling efficiency percentages for different methodologies. The proposed methodology achieves the highest recycling efficiency at approximately 86%, outperforming the other methodologies ([11], [12], [13]), which have recycling efficiencies around 77%, 80%, and 83%, respectively. This indicates that the proposed method is significantly more effective in sorting and recycling waste compared to previous methods.

Energy Consumption

Energy Consumption measures the total amount of energy used by the blockchain network and IoT devices. It helps evaluate the environmental impact and operational costs of the system.

EC= Energy consumed by blockchain + Energy consumed by IoT devices

Table 5.

Metric	Energy Consumption (kWh)
[11]	623
[12]	641
[13]	695
Proposed methodology	700

Figure 8. Comparison of proposed method with existing methods in terms of energy consumption

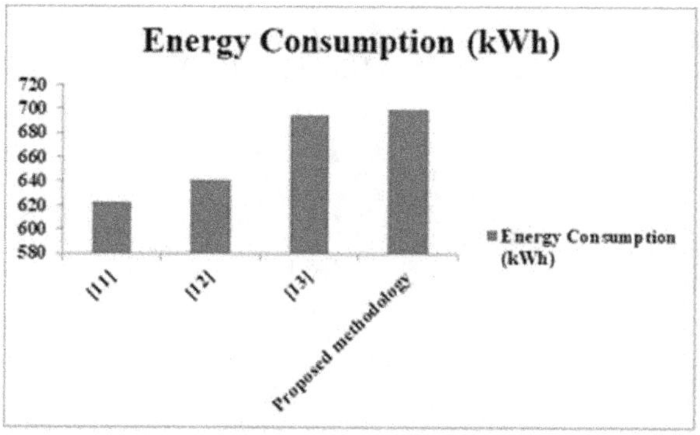

Figure 8 illustrates energy consumption in kilowatt-hours (kWh) for different methodologies. The proposed methodology has the highest energy consumption, approximately 700 kWh, compared to methodologies [11], [12], and [13], which consume around 620 kWh, 640 kWh, and 680 kWh, respectively. This suggests that while the proposed method offers significant improvements in other performance areas, it also has higher energy demands.

CONCLUSION

This section summarizes the findings on the role of blockchain technology in waste management. The findings suggest that blockchain has transformed as an emerging technology with immense potential especially in the areas of waste management. The findings in this work are systematically reported and listed in this section. This section also provides avenues of future scope to enhance the scope of block chain in waste management.

Conventional waste management methods face numerous challenges including inefficiencies, lack of transparency and accountability, illegal dumping, and high operational costs. These issues find the effective managing and removal of waste, requiring the new solutions. Blockchain offers a robust framework for considering these challenges. Key features such as smart contracts, consensus mechanism and cryptographic security guarantee data integrity and trustworthiness. Blockchain improves accountability and transparency by offering tamper proof record of waste management process. This technology enables end to end tracking of waste, confirming correct handling and compliance with regulations. These can automate waste management agreements, optimizing processes and eliminating the requirement for intruders.

Blockchain supports the development of decentralized waste management systems engaging in resilience. Combining the blockchain with IoT, the real time data collects and monitors the data. It enables tokenization and reward system that incentivize the recycling and waste reduction. These programs enable sustainable behavior between individuals and organizations. Real word examples also demonstrated the successful application of blockchain in waste management. Even though it has merits, blockchain has many challenges such as standardization and interoperability issues, data privacy issues and security issues. These challenges are addressed is important for widespread adoption. Combining AI and machine learning, leading the way for future blockchain in waste management.

As blockchain technology continues to evolve, several future trends and research directions are emerging that could significantly impact its application in waste management. These include advancements in blockchain technology, integration with artificial intelligence (AI) and machine learning (ML), regulatory and policy developments, and new research opportunities.

To address scalability and energy consumption issues, research focuses on developing new consensus mechanisms.

- PoS reduces energy consumption by requiring validators to hold and lock up a certain amount of cryptocurrency as collateral instead of performing energy-intensive computations.
- Sharding technique divides the blockchain into smaller, manageable pieces called shards, each capable of processing its own transactions and smart contracts. This parallel processing significantly improves scalability and transaction throughput.
- Technologies like Lightning Network and Plasma provide off-chain solutions that handle most transactions off the main blockchain, only settling on-chain, when necessary, thus reducing congestion and increasing efficiency.
- Some methods such as ring signature and homomorphic encryption must be employed in future which increase data security.
- Developing more interoperable blockchain methods are also concerned on the future.

Combining blockchain with AI and ML can enhance data analytics capabilities. AI algorithms can analyze data from IoT sensors recorded on the blockchain to predict when waste collection vehicles or equipment need maintenance, reducing downtime and costs. ML models can predict waste generation patterns based on historical data stored on the blockchain, helping optimize collection schedules and resource allocation.

Smart contracts can integrate AI and ML models to automate complex decision-making processes. AI can optimize waste collection routes in real-time based on data such as traffic conditions, waste levels, and vehicle availability, improving efficiency and reducing operational costs. ML models can identify unusual patterns in waste management transactions recorded on the blockchain, flagging potential fraud or non-compliance issues.

AI and ML can improve resource allocation and waste processing. AI-powered robots and sorting systems can be integrated with blockchain to ensure accurate recording of recycled materials and optimize the recycling process. ML algorithms can analyze blockchain data to identify areas where resources can be used more efficiently, such as optimizing the use of recycling facilities or reducing energy consumption.

Governments and regulatory bodies are developing frameworks to address blockchain's unique challenges and opportunities. Policies will focus on confirming data privacy and security for blockchain based system. Governments may propose regulations to control the environmental impact of blockchain mining. Efforts are being made to standardize the blockchain rules to ensure interoperability and compliance with existing regulations.

REFERENCES

Akram, S. V., Alshamrani, S. S., Singh, R., Rashid, M., Gehlot, A., AlGhamdi, A. S., & Prashar, D. (2021). Blockchain enabled automatic reward system in solid waste management. *Security and Communication Networks*, *2021*(1), 6952121. DOI: 10.1155/2021/6952121

Atta, U. R. K., & Ahmad, R. W. (2022). A Blockchain-Based IoT-Enabled E-Waste Tracking and Tracing System for Smart Cities. *IEEE Access : Practical Innovations, Open Solutions*, *10*, 86256–86269. DOI: 10.1109/ACCESS.2022.3198973

Baralla, G., Pinna, A., Tonelli, R., & Marchesi, M. (2023). Waste management: A comprehensive state of the art about the rise of blockchain technology. *Computers in Industry*, *145*(1), 103812. DOI: 10.1016/j.compind.2022.103812

Bułkowska, K., Zielińska, M., & Bułkowski, M. (2023). Implementation of blockchain technology in waste management. *Energies*, *16*(23), 7742. DOI: 10.3390/en16237742

Bulkowska, K., Zielinska, M., & Bulkowski, M. (2023). Implementation of Blockchain Technology in Waste Management. *Energies*, *16*(23), 7742. Advance online publication. DOI: 10.3390/en16237742

Bułkowska, K., Zielińska, M., & Bułkowski, M. (2024). Blockchain-Based Management of Recyclable Plastic Waste. *Energies*, *17*(12), 2937. DOI: 10.3390/en17122937

Bulkowska, K., Zielinska, M., & Bulkowski, M. (2024). Blockchain-Based Management of Recyclable Plastic Waste. *Energies*, *17*(12), 2937. Advance online publication. DOI: 10.3390/en17122937

Castiglione, A., Cimmino, L., Nardo, M. D., & Murino, T. (2023). A framework for achieving a circular economy using the blockchain technology in a sustainable waste management system. *Computers & Industrial Engineering*, *180*, 109263. Advance online publication. DOI: 10.1016/j.cie.2023.109263

Costa, M. M., Neto, J. F. B., Alberte, E. P. V., & Carneiro, A. P. (2022, November). Blockchain-based framework for improving waste management and circular economy in construction. *IOP Conference Series. Earth and Environmental Science*, *1101*(6), 062009. DOI: 10.1088/1755-1315/1101/6/062009

Demircan, B. G., & Yetilmezsoy, K. (2023). A hybrid fuzzy AHP-TOPSIS approach for implementation of smart sustainable waste management strategies. *Sustainability (Basel)*, *15*(8), 6526. DOI: 10.3390/su15086526

Gopalakrishnan, P., & Ramaguru, R. (2019). Blockchain based waste management. *International Journal of Engineering and Advanced Technology*, *8*(5), 2632–2635.

Gopalakrishnan, P. K., Hall, J., & Behdad, S. 2020, August. A blockchain-based traceability system for waste management in smart cities. In International design engineering technical conferences and computers and information in engineering conference (Vol. 83952, p. V006T06A015). American Society of Mechanical Engineers. DOI: 10.1115/DETC2020-22553

Gopalakrishnan, P. K., Hall, J., & Behdad, S. (2021). Cost analysis and optimization of Blockchain-based solid waste management traceability system. *Waste Management (New York, N.Y.)*, *120*, 594–607. DOI: 10.1016/j.wasman.2020.10.027 PMID: 33288397

Gulyamov, S. (2024). Intelligent waste management using IoT, blockchain technology and data analytics. In *E3S Web of Conferences* (Vol. 501, p. 01010). EDP Sciences. DOI: 10.1051/e3sconf/202450101010

Hakak, S., Khan, W. Z., Gilkar, G. A., Haider, N., Imran, M., & Alkatheiri, M. S. (2020). Industrial wastewater management using blockchain technology: Architecture, requirements, and future directions. *IEEE Internet of Things Magazine*, *3*(2), 38–43. DOI: 10.1109/IOTM.0001.1900092

Hassanien, A. E., Bhattacharyya, S., Chakrabati, S., Bhattacharya, A., & Dutta, S. 2020. Emerging Technologies in Data Mining and Information Security. *Proceedings of IEMIS*, 1.

Jiang, P., Zhang, L., You, S., Fan, Y. V., Tan, R. R., Klemes, J. J., & You, F. (2023). Blockchain technology applications in waste management: Overview, challenge and Opportunities. *Journal of Cleaner Production*, *421*, 138466. Advance online publication. DOI: 10.1016/j.jclepro.2023.138466

Karunambiga, K., Nusrat Jahan, S., Keerthana, S., Pavithra, D., & Umakarthikeyan, C. (2023). Smart Waste Management System Using IOT And Blockchain. *IJRAR-International Journal of Research and Analytical Reviews*, *10*(3), 63–67.

Khan, A. U. R., & Ahmad, R. W. (2022). A blockchain-based IoT-enabled E-Waste tracking and tracing system for smart cities. *IEEE Access : Practical Innovations, Open Solutions*, *10*, 86256–86269. DOI: 10.1109/ACCESS.2022.3198973

Koshariya, A. K., Khatoon, S., Marathe, A. M., Suba, G. M., Baral, D., & Boopathi, S. 2023. Agricultural Waste Management Systems Using Artificial Intelligence Techniques. In AI-Enabled Social Robotics in Human Care Services (pp. 236-258). IGI Global. DOI: 10.4018/978-1-6684-8171-4.ch009

Lamichhane, M. (2017). A smart waste management system using IoT and blockchain technology [Master's Thesis]. ITMO University.

Lin, Y. H., Wang, J., Niu, D., & Tao, X. (2024). Blockchain-driven framework for construction waste recycling and reuse. *Journal of Building Engineering*, *89*, 109355. DOI: 10.1016/j.jobe.2024.109355

Liu, Z., Wu, T., Wang, F., Osmani, M., & Demian, P. (2022). Blockchain enhanced construction waste information management: A conceptual framework. *Sustainability (Basel)*, *14*(19), 12145. DOI: 10.3390/su141912145

Ma, X., Yuan, H., & Du, W. (2024). Blockchain-Enabled Construction and Demolition Waste Management: Advancing Information Management for Enhanced Sustainability and Efficiency. *Sustainability (Basel)*, *16*(2), 721. DOI: 10.3390/su16020721

Parulkar, A. (2021). Plastic Waste Management Using Blockchain. Academic Press.

Saad, M., Ahmad, M. B., Asif, M., Khan, M. K., Mahmood, T., & Mahmood, M. T. (2023). Blockchain-enabled VANET for smart solid waste management. *IEEE Access : Practical Innovations, Open Solutions*, *11*, 5679–5700. DOI: 10.1109/ACCESS.2023.3235017

Safiullin, M. R., Elshin, L. A., Abdukaeva, A. A., & Burganov, R. T. (2022). Assessment of environmental research in block chain technology. Procedia Environmental Science. *Engineering and Management*, *8*(4), 979–986.

Sahoo, S., Mukherjee, A., & Halder, R. (2021). A unified blockchain-based platform for global e-waste management. *International Journal of Web Information Systems*, *17*(5), 449–479. DOI: 10.1108/IJWIS-03-2021-0024

Singh, S., Chhabra, R., & Arora, J. (2024). A systematic review of waste management solutions using machine learning, Internet of Things and blockchain technologies: State-of-art, methodologies, and challenges. *Archives of Computational Methods in Engineering*, *31*(3), 1255–1276. DOI: 10.1007/s11831-023-10008-z

Taylor, P., Steenmans, K., & Steenmans, I. (2020). Blockchain technology for sustainable waste management. *Frontiers in Political Science*, *2*, 590923. DOI: 10.3389/fpos.2020.590923

Wirani, Y., Eitiveni, I., & Sucahyo, Y. G. (2024). Framework of Smart and Integrated Household Waste Management System: A Systematic Literature Review Using PRISMA. *Sustainability (Basel)*, *16*(12), 4898–4901. DOI: 10.3390/su16124898

Chapter 8

Carbon Footprint Analysis of Electronic Products

P. Selvakumar
https://orcid.org/0000-0002-3650-4548
Science and Humanities, Nehru Institute of Technology, India

Ritu Dahiya
https://orcid.org/0000-0001-5788-006X
Chhotu Ram Arya College, India

T. C. Manjunath
https://orcid.org/0000-0003-2545-9160
Rajarajeswari College of Engineering, India

Mohit Sharma
https://orcid.org/0009-0007-2280-8077
Maharshi Dayanand University, Rohtak, India

ABSTRACT

The significance of carbon footprint analysis lies in its ability to inform and drive sustainability efforts across various sectors. It serves as a metric for assessing environmental performance, guiding organizations and policymakers towards informed decision-making aimed at reducing emissions. Understanding the carbon footprint of products helps consumers make environmentally conscious choices, thereby influencing market demand towards more sustainable practices. Moreover, for businesses, it can be a strategic tool for improving operational efficiencies, reducing costs associated with energy use, and enhancing corporate social responsibility (CSR) efforts. Carbon footprints encompass both direct emissions (e.g., from burn-

DOI: 10.4018/979-8-3693-7383-5.ch008

ing fossil fuels for heating or transportation) and indirect emissions (e.g., from the production and transportation of goods consumed). This comprehensive approach ensures that all significant sources of greenhouse gases are accounted for, offering a holistic view of environmental impact.

INTRODUCTION

The significance of carbon footprint analysis lies in its ability to inform and drive sustainability efforts across various sectors. It serves as a metric for assessing environmental performance, guiding organizations and policymakers towards informed decision-making aimed at reducing emissions. Understanding the carbon footprint of products helps consumers make environmentally conscious choices, thereby influencing market demand towards more sustainable practices. Moreover, for businesses, it can be a strategic tool for improving operational efficiencies, reducing costs associated with energy use, and enhancing corporate social responsibility (CSR) efforts. Carbon footprints encompass both direct emissions (e.g., from burning fossil fuels for heating or transportation) and indirect emissions. This comprehensive approach ensures that all significant sources of greenhouse gases are accounted for, offering a holistic view of environmental impact. The methodology for calculating carbon footprints involves gathering data on energy consumption, fuel usage, transportation activities, and other relevant factors. This data is then converted into CO_2 using emission factors specific to each activity or fuel type. Standard protocols and guidelines, such as those developed by the Greenhouse Gas Protocol (GHG Protocol), ensure consistency and accuracy in carbon footprint assessments, facilitating meaningful comparisons and benchmarking across different entities and sectors. Beyond its application in business and consumer contexts, carbon footprint analysis plays a crucial role in global efforts to address climate change. Governments use carbon footprint data to set emission reduction targets, develop policies, and monitor progress towards national and international climate commitments, such as those outlined in the Paris Agreement. Carbon footprint analysis also supports climate action initiatives at local levels, enabling cities and municipalities to identify and prioritize emission reduction strategies tailored to their specific circumstances and challenges. In conclusion, carbon footprint analysis is a powerful tool for quantifying and managing greenhouse gas emissions associated with human activities. It provides valuable insights into environmental impact, supports informed decision-making for sustainability, and contributes to broader climate mitigation efforts. As concerns about climate change continue to grow, the importance of carbon footprint analysis in shaping a sustainable future cannot be overstated. By fostering awareness, accountability, and action, carbon footprint analysis helps pave the way towards a more environmentally responsible global community.

LIFE CYCLE ASSESSMENT (LCA) METHODOLOGY FOR ELECTRONIC PRODUCTS

This approach is particularly crucial for electronic products due to their complex supply chains, rapid technological advancements, and significant environmental implications. The LCA methodology for electronic products typically involves several key stages and considerations to assess their environmental footprint accurately (Abhilasha A., et al., 2022). The first stage of LCA for electronic products involves defining the goal and scope of the assessment. This includes identifying the specific product being evaluated, its intended function, and the system boundaries within which the assessment will be conducted. For electronic products, the scope often extends beyond just the product itself to include ancillary components, packaging, transportation, and the energy consumption during the use phase. Electronic products typically involve numerous materials, including metals (e.g., copper, aluminum), plastics, rare earth elements (e.g., neodymium, lithium), and hazardous substances (e.g., brominated flame retardants), each with varying impacts on the environment and human health (Ahmed, S., et al., 2023). Energy consumption during manufacturing and use phases is also a significant focus, given the energy-intensive nature of electronic devices.

The impact assessment stage follows, where the collected inventory data is evaluated to assess potential environmental impacts. Common impact categories considered in LCA for electronic products include global warming potential (measured in CO2 equivalents), resource depletion (e.g., fossil fuels, minerals), acidification, eutrophication, ozone depletion, and human toxicity (Bai, X., et al., 2023). These impacts are quantified using characterization factors specific to each environmental category, reflecting the potential harm caused by emissions and resource use throughout the product's life cycle. Interpreting the results of the impact assessment involves normalization and weighting, which provide a comparative analysis of the environmental impacts across different impact categories. Normalization adjusts impact scores relative to reference values to facilitate comparison, while weighting assigns relative importance to different environmental impacts based on stakeholder preferences or policy objectives. This step helps prioritize areas for improvement and informs decision-making towards reducing the overall environmental footprint of electronic products (Brown, T., et al., 2023). Lastly, the final stage of LCA involves interpreting the results and drawing conclusions. This includes identifying hotspots (i.e., stages or processes with the highest environmental impacts), evaluating the effectiveness of potential mitigation strategies (e.g., material substitution, energy efficiency improvements, design for disassembly), and communicating findings to stakeholders. For electronic products, improving environmental performance often involves extending product lifespan, promoting repairability and recyclability, reducing energy consumption during manufacturing and use, and minimizing hazardous substances. The application of LCA methodology to electronic products is essential for promoting sustainable practices throughout the product life cycle (Chang, S., et al., 2023). It provides valuable insights to manufacturers, policymakers, and consumers on the environmental consequences of electronic devices and informs strategies for mitigating these impacts.

Greenhouse Gas Emissions in the Electronics Supply Chain

Greenhouse gas (GHG) emissions in the electronics supply chain have become a significant concern due to the industry's rapid growth, complex global supply networks, and energy-intensive manufacturing processes (Chen, Y., et al., 2023). This sector encompasses a wide range of products, from consumer electronics like smartphones and laptops to industrial equipment, each contributing to varying degrees of carbon emissions throughout their life cycles. Understanding and mitigating these emissions are critical steps towards achieving sustainability goals and addressing climate change challenges. The electronics supply chain starts with raw material extraction, which involves energy-intensive processes such as mining and refining metals like copper, aluminum, and rare earth elements (e.g., neodymium, lithium). These materials are essential for manufacturing electronic components and devices but often require substantial amounts of energy and generate emissions during extraction and processing (Davis, J., et al., 2023). For instance, mining operations contribute to GHG emissions through fuel combustion in heavy machinery, electricity consumption, and the release of methane and other gases from mines.

Manufacturing processes represent another significant source of GHG emissions in the electronics supply chain. These processes include semiconductor fabrication, assembly of printed circuit boards (PCBs), component manufacturing, and final assembly of electronic products. Semiconductor manufacturing, in particular, is highly energy-intensive, involving clean room facilities, chemical processes, and high-temperature operations that consume large amounts of electricity and produce greenhouse gases such as carbon dioxide (CO2) and fluorinated gases (e.g., SF6). Energy consumption during the use phase of electronic products also contributes significantly to their carbon footprint (Dong, Y., et al., 2023). Devices like smartphones, computers, and servers require electricity for operation, and the increasing demand for digital services and data centers has led to higher energy consumption across the electronics sector. While advancements in energy efficiency have helped reduce per-unit energy use, the overall energy demand continues to rise with the proliferation of electronic devices globally. Beyond manufacturing and use, the logistics and transportation of electronic components and products further add to GHG emissions in the supply chain. Shipping raw materials, components, and finished goods across continents involves fuel-intensive modes of transport, including ships, airplanes, trucks, and trains. The carbon footprint of transportation is influenced by factors such as distance traveled, mode of transport, fuel efficiency, and logistics efficiency. Optimizing transportation routes, using cleaner fuels, and improving supply chain visibility can help mitigate emissions associated with logistics (Fang, H., et al., 2023). Effective e-waste recycling and reuse programs are essential for reducing the environmental impact of electronics at their end of life, promoting circular economy principles, and conserving resources.

Mitigating greenhouse gas emissions in the electronics supply chain requires collaborative efforts across stakeholders, including manufacturers, suppliers, governments, and consumers. Manufacturers can adopt cleaner production technologies, improve energy efficiency in manufacturing processes, and implement renewable energy sources to reduce emissions. Supply chain transparency and accountability are crucial for identifying emission hotspots and implementing targeted mitigation strategies (Gao, S., et al., 2023). Governments play a pivotal role in setting regulatory frameworks, incentives, and standards to encourage sustainable practices and emission reductions in the electronics sector. Policies promoting energy efficiency, renewable energy adoption, extended producer responsibility (EPR) for e-waste management, and eco-design requirements can drive systemic changes towards a lower-carbon electronics supply chain. International cooperation and standards harmonization are also essential for addressing global challenges associated with GHG emissions from electronics production and consumption (Gupta, R., et al., 2023). Consumers and businesses can contribute to emissions reduction efforts by making informed purchasing decisions, opting for energy-efficient products, extending the lifespan of electronic devices through repair and upgrade, and responsibly recycling e-waste. Awareness campaigns and education on sustainable consumption practices can empower individuals to reduce their carbon footprint and advocate for more sustainable electronics. In conclusion, addressing greenhouse gas emissions in the electronics supply chain requires a multifaceted approach that integrates technological innovation, regulatory frameworks, supply chain optimization, and consumer behavior change (Han, X., et al., 2023). By prioritizing energy efficiency, adopting cleaner production methods, promoting circular economy principles, and improving end-of-life management practices, the electronics industry can mitigate its environmental impact and contribute to global climate goals. Collaborative efforts across the supply chain are essential for achieving a sustainable and resilient electronics sector that balances economic growth with environmental stewardship in a carbon-constrained world.

Material Selection and Carbon Footprint Implications

Material selection is a critical factor influencing the carbon footprint of products across various industries, including manufacturing, construction, and consumer goods. The choice of materials not only affects product performance, durability, and cost but also significantly impacts environmental sustainability, particularly in terms of greenhouse gas (GHG) emissions throughout the product life cycle. Understanding the carbon footprint implications of different materials is essential for making informed decisions aimed at reducing environmental impact and promoting sustainable practices (He, Y., et al., 2023). The carbon footprint of materials encompasses emissions associated with their extraction, processing, manufacturing, transportation, use, and end-of-life disposal or recycling. Different materials exhibit varying degrees of environmental impact, largely influenced by factors such as energy intensity, resource depletion, chemical usage, and waste generation. For example, metals like steel and aluminum typically require significant energy inputs during extraction and refining, leading to higher carbon emissions compared to materials like wood or recycled plastics.

Key considerations in material selection for reducing carbon footprints include:

Raw Material Sourcing: The location and method of raw material extraction significantly influence the carbon footprint. For example, mining operations that use renewable energy sources or employ efficient extraction techniques can reduce emissions compared to conventional mining practices. Additionally, sourcing materials locally can minimize transportation-related emissions.

Manufacturing Process: The energy intensity of manufacturing processes directly impacts carbon emissions. Materials that require less energy-intensive processes or utilize renewable energy sources during manufacturing generally have lower carbon footprints. For instance, the production of cement and steel is energy-intensive and contributes significantly to carbon emissions, whereas materials like bamboo or recycled materials often require less energy for processing.

Transportation: The distance and mode of transportation affect the carbon footprint of materials. Choosing materials that are sourced closer to manufacturing facilities or using efficient transport modes (e.g., rail over trucking, sea freight over air freight) can reduce emissions associated with transportation.

Product Use Phase: The durability, maintenance requirements, and energy efficiency of products during their use phase impact overall carbon emissions. For example, materials used in buildings that enhance energy efficiency (e.g., insulation materials) or in vehicles that improve fuel efficiency contribute to lower operational carbon footprints over their lifetimes.

End-of-Life Management: Considering the recyclability, reusability, and disposal methods of materials is crucial for reducing their environmental impact at the end of their life cycle. Materials that can be easily recycled or reused, such as metals and some plastics, help minimize emissions associated with waste disposal and promote circular economy principles.

Examples of materials with varying carbon footprints highlight the importance of informed material selection:

Wood: As a renewable and low-carbon material, responsibly sourced wood products store carbon throughout their lifespan. Wood construction materials, such as cross-laminated timber (CLT), offer opportunities for carbon sequestration in buildings and can contribute to overall carbon footprint reductions in construction projects.

Plastics: Depending on their composition and end-of-life management, plastics can have varying environmental impacts. While some plastics are derived from fossil fuels and contribute to GHG emissions during production and incineration, recycled plastics and bio-based plastics offer alternatives with potentially lower carbon footprints, depending on their life cycle assessments.

Metals: Metals such as aluminum and steel have high initial carbon footprints due to energy-intensive extraction and processing. However, their recyclability and durability make them valuable for reducing life cycle emissions through reuse and recycling initiatives. In conclusion, material selection plays a pivotal role in determining the carbon footprint of products and infrastructure across industries. By integrating life cycle thinking and conducting thorough LCAs, stakeholders can identify opportunities to mitigate environmental impacts through sustainable material choices (Huang, J., et al., 2023). Emphasizing renewable resources, energy-efficient manufacturing processes, responsible sourcing practices, and circular economy principles can collectively contribute to reducing carbon footprints and advancing toward a more sustainable future. As global awareness of climate change grows, the importance of optimizing material selection to minimize carbon footprints will continue to drive innovation and foster sustainable development practices worldwide.

Manufacturing Process Emissions: A Deep Dive

Manufacturing process emissions represent a significant component of global greenhouse gas (GHG) emissions, contributing to environmental impact across various industries. These emissions arise from industrial activities involved in the production of goods, including energy-intensive processes, chemical reactions, and combustion of fossil fuels (Iqbal, M., et al., 2023). Understanding the sources, drivers, and mitigation strategies related to manufacturing process emissions is crucial for advancing sustainability goals and addressing climate change challenges.

The primary sources of manufacturing process emissions include:

Industrial Processes: Specific manufacturing operations contribute to emissions through activities such as metal smelting, cement clinker production, glass manufacturing, and pulp and paper production. These processes release GHGs directly from industrial equipment and processes, including emissions from fugitive sources such as leaks and venting of gases.

Use of Raw Materials: The extraction, processing, and transportation of raw materials also contribute to manufacturing process emissions (Jiang, Y., et al., 2023). Mining activities, for example, involve energy-intensive processes and machinery that emit GHGs, along with methane emissions from coal mining operations. The transportation of raw materials to manufacturing facilities further adds to emissions, especially when utilizing fossil fuel-powered vehicles.

Waste Management: Emissions from waste management practices, including incineration and landfilling of industrial waste, contribute to GHG emissions. Landfills emit methane as organic waste decomposes anaerobically, while incineration releases CO2 and other pollutants depending on the combustion efficiency and waste composition.

The drivers of manufacturing process emissions are influenced by several factors:

Technological Choices: The selection of production technologies and equipment significantly impacts energy efficiency and emissions intensity. Advanced technologies, such as energy-efficient furnaces, cogeneration systems, and process optimization techniques, can reduce emissions per unit of output.

Process Efficiency: Improving process efficiency through measures such as heat recovery, waste heat utilization, and material substitution can reduce energy requirements and emissions intensity. For example, optimizing furnace operations in steelmaking or implementing closed-loop systems in chemical production can minimize energy losses and emissions.

Regulatory Environment: Regulatory frameworks, emission standards, and carbon pricing mechanisms influence industrial emissions by incentivizing emission reductions and promoting cleaner technologies. Policies targeting specific sectors, such as carbon trading schemes and emission reduction targets, encourage businesses to adopt sustainable practices and invest in emissions reduction initiatives.

Mitigating manufacturing process emissions requires a combination of technological innovation, regulatory support, and industry collaboration. Key strategies include:

Process Innovation and Cleaner Production: Developing and adopting innovative technologies and practices that reduce emissions across industrial sectors, including advanced manufacturing techniques and sustainable chemical processes.

Supply Chain Collaboration: Collaborating with suppliers, customers, and stakeholders to improve supply chain transparency, optimize logistics, and collectively reduce emissions throughout the product life cycle. In conclusion, addressing manufacturing process emissions is essential for achieving global climate goals and fostering sustainable industrial development. Collaboration between governments, industries, and stakeholders is crucial for implementing effective emissions reduction strategies, promoting innovation, and ensuring a transition towards a low-carbon economy. Emphasizing sustainable manufacturing practices not only mitigates climate change risks but also enhances competitiveness, resilience, and long-term sustainability across industrial sectors worldwide.

Product Use Phase: Energy Consumption and Emissions

The product use phase represents a critical stage in the lifecycle of goods, significantly impacting energy consumption and emissions across various sectors. Energy consumption during this phase encompasses both direct and indirect sources, including electricity, fuel, and other resources required for operation. Direct energy use pertains to the energy expended by the product itself, such as electricity for appliances or fuel for vehicles (Jones, K., et al., 2023). Indirect energy usage involves the energy consumed in manufacturing, transportation, and maintenance activities associated with the product (Kim, H., et al., 2023). These combined factors contribute to the carbon footprint and environmental impact throughout the product's lifespan. The intensity of emissions varies widely depending on the type of product and its efficiency in utilizing energy resources. Energy-efficient appliances, vehicles with lower fuel consumption rates, and renewable energy sources can significantly mitigate emissions during use, promoting sustainability and environmental stewardship (Li, X., et al., 2023). Additionally, smart technologies and digital solutions enable consumers and businesses to monitor and optimize energy usage, fostering more efficient practices and minimizing ecological footprint. However, challenges persist in achieving widespread adoption of sustainable practices during the use phase. Economic considerations, consumer behavior, and infrastructural limitations often hinder the implementation of energy-efficient measures and emission reduction strategies (Kwon, Y., et al., 2023). Addressing these challenges requires collaborative efforts from stakeholders across industries, policymakers, and consumers alike to promote awareness, incentivize sustainable practices, and invest in research and development for cleaner technologies. In conclusion, the product use phase represents a critical juncture in the lifecycle of goods, influencing energy consumption and emissions significantly (Lin, Y., et al., 2023). By prioritizing energy efficiency, adopting renewable energy sources, and integrating smart technologies, stakeholders can mitigate environmental impact, promote sustainability, and pave the way towards a greener future.

END-OF-LIFE MANAGEMENT: RECYCLING, WASTE, AND EMISSIONS

End-of-life management of products encompasses critical processes such as recycling, waste management, and emissions reduction, which are pivotal in shaping environmental sustainability and resource efficiency. Through effective recycling programs, materials such as metals, plastics, glass, and paper can be processed and reintegrated into new products or raw materials, thereby conserving natural resources and reducing landfill waste. However, challenges exist in optimizing recycling rates due to factors such as material contamination, inefficient collection systems, and lack of standardized recycling infrastructure. Waste management practices at the end of a product's lifecycle are essential for minimizing environmental pollution and promoting circular economy principles. Proper disposal methods, including composting, landfill diversion, and waste-to-energy technologies, help mitigate the release of harmful substances into the environment while recovering energy from non-recyclable materials (Liu, H., et al., 2023). Emissions associated with waste management, such as methane from landfills or pollutants from incineration, underscore the importance of adopting sustainable waste treatment practices. Advances in waste sorting technologies, coupled with legislative measures and public awareness campaigns, are crucial in enhancing waste management efficiency and reducing environmental footprint.

Emission reduction strategies during the end-of-life phase focus on minimizing the release of greenhouse gases (GHGs) and pollutants associated with disposal processes. Efforts to capture and utilize methane emissions from landfills for energy generation contribute to mitigating climate change impacts. Additionally, advancements in incineration technologies aim to reduce emissions of particulate matter and harmful gases through improved combustion efficiency and emissions control systems (Luo, Z., et al., 2023). However, achieving optimal emission reductions requires integrated approaches that prioritize waste prevention, recycling, and sustainable waste treatment practices across industrial sectors and municipalities. In conclusion, effective end-of-life management strategies, encompassing recycling, waste management, and emissions reduction, are crucial for achieving environmental sustainability and resource efficiency goals (Ma, X., et al., 2023). By promoting circular economy principles, enhancing recycling infrastructure, and implementing advanced waste treatment technologies, stakeholders can minimize environmental impact, conserve resources, and foster a more sustainable future for generations to come.

Carbon Footprint of Consumer Electronics: Case Studies

The carbon footprint of consumer electronics is a critical concern in today's digital age, influenced by the entire lifecycle of devices from production to end-of-life disposal. Case studies reveal significant variations in carbon emissions among different types of electronics. For instance, smartphones, due to their compact size and energy-efficient design, generally have lower carbon footprints compared to larger devices like laptops or desktop computers. Manufacturing processes, particularly for semiconductor chips and display panels, contribute substantially to the carbon footprint of electronics, highlighting the importance of efficient production practices and material sourcing (Mahadev Madgule, et al., 2023). Additionally, energy consumption during the use phase significantly impacts carbon emissions; devices that utilize renewable energy sources or incorporate energy-saving features exhibit lower environmental impacts over their operational lifespan. End-of-life management also plays a crucial role, with proper recycling and disposal reducing greenhouse gas emissions associated with waste electronics. Case studies underscore the complexities and opportunities in mitigating the carbon footprint of consumer electronics through lifecycle assessments, technological innovations, and sustainable practices across the industry.

Carbon Footprint of Industrial Electronics: A Different Perspective

The carbon footprint of industrial electronics presents a unique perspective due to their specialized applications and operational requirements in various sectors such as manufacturing, energy production, and infrastructure development (Martin, C., et al., 2023). Unlike consumer electronics, industrial equipment often demands higher power consumption and robust performance capabilities, leading to significant carbon emissions throughout their lifecycle. Manufacturing processes for industrial electronics involve intensive energy use and material inputs, contributing substantially to their initial carbon footprint. This includes the production of complex components such as circuit boards, motors, and sensors, which require energy-intensive manufacturing techniques and raw materials extraction (Nguyen, T., et al., 2023). During the operational phase, industrial electronics typically operate continuously or under heavy load conditions, necessitating reliable power supply and often resulting in elevated energy consumption. Efforts to optimize energy efficiency through advanced control systems, variable speed drives, and energy management solutions can mitigate emissions during operation.

End-of-life management poses distinct challenges for industrial electronics, primarily due to their size, complexity, and hazardous material content. Proper disposal and recycling of these devices are critical to reducing environmental impact, including the release of greenhouse gases from incineration or landfill degradation. Effective recycling processes aim to recover valuable metals and components while minimizing energy use and emissions associated with material recovery. Carbon footprint assessments of industrial electronics often incorporate lifecycle analysis to quantify emissions across production, operation, and disposal phases, providing insights into opportunities for emission reductions and sustainable practices (Paramasivam Selvakumar, et al., 2024). Strategic initiatives such as circular economy principles, eco-design strategies, and extended producer responsibility (EPR) programs are increasingly adopted to promote resource efficiency and environmental stewardship within the industrial electronics sector. In conclusion, the carbon footprint of industrial electronics presents a multifaceted challenge influenced by manufacturing processes, operational requirements, and end-of-life management practices. By adopting innovative technologies, enhancing energy efficiency, and implementing sustainable waste management strategies, stakeholders can mitigate environmental impact, optimize resource utilization, and contribute to a more sustainable industrial ecosystem.

Design for Sustainability: Minimizing Carbon Footprint

Design for Sustainability (DfS) represents a proactive approach to minimizing the carbon footprint of products through thoughtful consideration of environmental impacts across their entire lifecycle. At its core, DfS integrates principles of environmental stewardship, resource efficiency, and lifecycle thinking into the product development process, aiming to optimize sustainability outcomes from inception through to end-of-life (Paramasivam Selvakumar, et al., 2024). Key strategies encompassed within DfS include material selection, energy efficiency, product durability, recyclability, and responsible end-of-life management (Patel, N., et al., 2023). By selecting materials with lower environmental impact, such as recycled or renewable materials, designers can reduce carbon emissions associated with raw material extraction and processing. Energy efficiency plays a crucial role in mitigating carbon footprint during the operational phase of products; incorporating energy-saving technologies, efficient components, and smart design features can significantly lower energy consumption and subsequent emissions throughout the product's use.

Durability and longevity are essential aspects of DfS, as they extend the product's lifespan and reduce the frequency of replacements, thereby lowering overall environmental impact. Designing products for disassembly and recyclability facilitates the recovery of valuable materials at end-of-life stages, minimizing waste and promoting circular economy principles (Ramakrishnan T. et al., 2022). Moreover, DfS encourages innovative approaches such as modular design and remanufacturing, which support the reuse of components and assemblies to extend product life cycles and reduce carbon emissions associated with manufacturing new products. Lifecycle assessments (LCAs) are integral to DfS, providing quantitative evaluations of environmental impacts across all stages of a product's lifecycle (Rao, S., et al., 2023). offering valuable insights into hotspots for emissions and opportunities for improvement. By conducting LCAs early in the design process, designers can identify and prioritize sustainability measures that yield the greatest environmental benefits. Collaboration among stakeholders is essential for effective implementation of DfS principles (Selvakumar, P., 2024). This includes partnerships between designers, engineers, suppliers, manufacturers, policymakers, and consumers to align goals, share knowledge, and drive innovation towards sustainable solutions. In conclusion, Design for Sustainability represents a proactive and holistic approach to minimizing the carbon footprint of products through integrated environmental considerations. By embedding principles of environmental stewardship, resource efficiency, and lifecycle thinking into product design and development, stakeholders can achieve significant reductions in carbon emissions, promote resource conservation, and contribute to a more sustainable future for generations to come.

Supply Chain Engagement: Collaborative Carbon Reduction

Supply chain engagement in collaborative carbon reduction initiatives is increasingly recognized as essential for achieving significant environmental impact across industries (Selvakumar, P., et al., 2024). The complexity and interconnectedness of global supply chains present both challenges and opportunities in addressing carbon emissions throughout the entire product lifecycle. Collaborative efforts focus on fostering partnerships, transparency, and collective action among suppliers, manufacturers, distributors, and other stakeholders to identify, measure, and mitigate carbon footprints collectively. Central to supply chain engagement is the adoption of sustainability principles and practices that extend beyond individual organizational boundaries. This includes setting ambitious emission reduction targets aligned with global climate goals, such as those outlined in the Paris Agreement, and integrating carbon reduction strategies into core business operations. Companies leverage their influence within supply chains to promote best practices, encourage suppliers to adopt cleaner technologies, and implement energy-efficient processes (Singh, A., et al., 2023). For example, initiatives may involve sharing expertise, resources, and technologies for improving energy management, optimizing logistics, and reducing waste generation across supply chain partners.

Measurement and transparency are crucial components of effective supply chain engagement in carbon reduction. Establishing robust monitoring, reporting, and verification (MRV) mechanisms enables stakeholders to track carbon emissions accurately, identify performance gaps, and implement targeted improvement measures (Sun, L., et al., 2023). Transparent disclosure of emission data fosters accountability and facilitates informed decision-making among supply chain participants, driving continuous improvement in environmental performance. Collaboration in carbon reduction extends beyond operational efficiency to encompass innovation and resilience. Research and development collaborations focus on developing and scaling low-carbon technologies, sustainable materials, and circular economy solutions that minimize resource use and emissions throughout product lifecycles (Tang, J., et al., 2023). By pooling resources and expertise, supply chain partners can accelerate the adoption of innovative solutions, reduce costs, and enhance competitive advantage in evolving markets increasingly prioritizing sustainability. Supply chain engagement in carbon reduction initiatives also responds to evolving stakeholder expectations and regulatory requirements (Uddin, S., et al., 2023). Investors, consumers, and regulatory bodies increasingly demand transparency and accountability regarding carbon impacts, influencing corporate strategies and supply chain practices. Proactive engagement in carbon reduction not only mitigates risks associated with regulatory compliance and market shifts but also enhances brand reputation and fosters trust among stakeholders.

However, challenges in supply chain engagement for carbon reduction persist, including varying levels of commitment, resource constraints, and technological barriers among stakeholders (Vasquez, M., et al., 2023). Overcoming these challenges requires fostering a culture of collaboration, building capacity through education and training, and incentivizing participation through financial incentives and recognition programs. Industry initiatives, such as industry-wide carbon reduction platforms, sector-specific alliances, and sustainability certifications, provide frameworks for collective action and knowledge-sharing among supply chain participants. In conclusion, supply chain engagement in collaborative carbon reduction is essential for achieving meaningful environmental impact, driving innovation, and ensuring resilience in a rapidly changing global landscape. By embracing shared responsibility, transparency, and continuous improvement, supply chain partners can effectively mitigate carbon emissions, strengthen competitiveness, and contribute to a sustainable future for society and the planet.

Carbon Footprint Labeling and Disclosure: Trends and Challenges

Carbon footprint labeling and disclosure have emerged as pivotal strategies in promoting transparency and empowering consumers and businesses to make informed choices towards sustainability (Vijayakumar G, et al., 2024). This practice involves quantifying and communicating the greenhouse gas emissions associated with the production, use, and disposal of goods and services, providing valuable information about their environmental impact. The adoption of carbon footprint labeling reflects growing awareness of climate change and the imperative to mitigate carbon emissions across sectors (Wang, H., et al., 2023). One prominent trend in carbon footprint labeling is its integration into product packaging and marketing materials. Companies use carbon labels to differentiate their products based on environmental performance, appealing to eco-conscious consumers seeking sustainable options. These labels typically display carbon emissions in terms of kilograms of carbon dioxide equivalent (CO2e) per unit of product or service, allowing consumers to compare environmental impacts among similar products and make informed purchasing decisions. Retailers, such as supermarkets and online platforms, also adopt carbon footprint labeling to promote sustainable product choices and meet consumer demand for transparency (Williams, J., et al., 2024). Another trend is the standardization and certification of carbon footprint labels through internationally recognized frameworks and methodologies. Organizations and standards bodies, such as ISO (International Organization for Standardization) and PAS (Publicly Available Specification), provide guidelines for calculating and verifying carbon footprints using consistent methodologies. Certification schemes, such as Carbon Trust's Carbon Reduction Label and PAS 2050, ensure credibility and reliability of carbon footprint information, enhancing consumer trust and facilitating global market access for certified products.

In addition to consumer goods, carbon footprint labeling is increasingly applied to services, including transportation, hospitality, and financial products. Service providers quantify emissions associated with their operations and offerings, enabling customers to evaluate and choose providers based on their environmental impact (Xu, L., et al., 2023). For instance, carbon footprint labeling in transportation informs travelers about the emissions associated with different modes of transport, encouraging sustainable travel choices and promoting low-carbon transport alternatives. Despite these advancements, carbon footprint labeling faces several challenges that impact its effectiveness and widespread adoption. One major challenge is the complexity and variability in assessing carbon footprints across diverse products and services. Factors such as supply chain complexity, geographical differences in energy sources, and data availability pose challenges in achieving accurate and comparable carbon footprint assessments. Standardization efforts aim to address these challenges by harmonizing methodologies and improving data quality through lifecycle assessments (LCAs) that consider all stages from raw material extraction to disposal.

Another challenge is consumer understanding and awareness of carbon footprint labels. Studies indicate varying levels of familiarity and comprehension among consumers regarding carbon footprint information, influencing their willingness to use labels in decision-making (Yang, S., et al., 2023). Effective communication strategies, such as simplified labels, educational campaigns, and digital tools, are essential to enhance consumer awareness and engagement with carbon footprint information. Moreover, addressing consumer skepticism and ensuring transparency in labeling practices are critical to building trust and credibility in carbon footprint disclosures. Furthermore, regulatory and policy frameworks play a crucial role in shaping the landscape of carbon footprint labeling (Yoon, J., et al., 2023). While some regions have implemented voluntary labeling schemes and incentives for companies to disclose carbon emissions, the lack of standardized regulations globally poses challenges for consistent adoption and enforcement. Harmonizing regulatory approaches and incentivizing corporate transparency through regulatory measures can accelerate the integration of carbon footprint labeling into business practices and consumer decision-making.

Looking ahead, technological advancements and digital platforms offer opportunities to enhance the accessibility and effectiveness of carbon footprint labeling. Blockchain technology, for example, can provide transparent and immutable records of carbon footprint data, improving traceability and accountability in supply chains. Digital platforms and mobile applications enable real-time access to carbon footprint information, empowering consumers to make sustainable choices on-the-go (Zhao, J., et al., 2023). In conclusion, carbon footprint labeling and disclosure represent transformative tools in advancing sustainability goals by fostering transparency, empowering consumer choice, and driving corporate accountability. Despite challenges related to complexity, consumer awareness, and regulatory frameworks, ongoing efforts in standardization, education, and technological innovation are paving the way for broader adoption and meaningful impact in reducing global carbon emissions

Regulatory Frameworks and Policy Drivers for Low-Carbon Electronics

Regulatory frameworks and policy drivers play a crucial role in shaping the landscape for low-carbon electronics, aiming to mitigate environmental impact, promote sustainability, and drive innovation across the electronics industry (Zheng, Y., et al., 2023). These frameworks encompass a range of measures, from emissions standards and eco-design requirements to incentives for renewable energy adoption and circular economy principles. The overarching goal is to reduce carbon emissions throughout the lifecycle of electronic products, including manufacturing, use, and end-of-life disposal. One of the primary regulatory approaches is the establishment of emissions standards and regulations targeting energy efficiency and carbon emissions during the operational phase of electronic devices. Governments and regulatory bodies around the world set mandatory energy efficiency standards for appliances, such as ENERGY STAR in the United States and the EU Ecodesign Directive in Europe. These standards define minimum efficiency levels and limit energy consumption, thereby reducing greenhouse gas emissions associated with electricity use during product operation. Additionally, eco-labeling schemes, such as the Energy Label in Europe, provide consumers with information on energy consumption and environmental performance, encouraging the adoption of energy-efficient products.

In parallel, eco-design requirements are integral to regulatory frameworks for low-carbon electronics, focusing on minimizing environmental impact throughout the product lifecycle. Eco-design directives, such as the EU Ecodesign Directive and Japan's Top Runner Program, mandate the integration of sustainable design principles into product development. These requirements encompass criteria such as material selection, energy efficiency, durability, repairability, and recyclability, aiming to optimize resource use, reduce emissions, and extend product lifespan. By embedding eco-design principles early in the product development process, manufacturers can achieve significant reductions in carbon footprint while enhancing product performance and durability. Circular economy policies promote resource efficiency, waste reduction, and material recovery through strategies such as product reuse, remanufacturing, and recycling. Extended Producer Responsibility (EPR) regulations, implemented in various regions including the EU and several This includes funding and organizing collection, recycling, and disposal programs to minimize landfill waste and promote the reuse of valuable materials.

Policy drivers for low-carbon electronics also encompass financial incentives, subsidies, and funding mechanisms to stimulate innovation and investment in clean technologies. Governments and international organizations offer grants, tax incentives, and subsidies for research and development (R&D) initiatives focused on energy efficiency improvements, carbon reduction technologies, and sustainable manufacturing practices. These incentives aim to lower barriers to adoption of low-carbon technologies, accelerate market uptake, and support the transition towards a greener economy. Moreover, international collaboration and alignment of regulatory frameworks are critical in addressing global challenges related to carbon emissions from electronics. Initiatives such as the Paris Agreement on climate change and international standards developed by organizations like the International Electrotechnical Commission (IEC) promote harmonization of regulatory approaches and best practices across borders. Harmonized standards facilitate market access for manufacturers, enhance global competitiveness, and drive innovation in low-carbon technologies. Despite the progress made, regulatory frameworks and policy drivers for low-carbon electronics face several challenges and opportunities for improvement. These include the need for continuous updates and adaptation of standards to keep pace with technological advancements and emerging environmental concerns. Challenges also arise from varying regulatory landscapes across different regions, which can lead to compliance complexities for global manufacturers. Additionally, enhancing consumer awareness and education on the benefits of low-carbon electronics and sustainable purchasing practices is essential to drive demand and market transformation. Looking forward, the evolution of regulatory frameworks and policy drivers for low-carbon electronics will likely focus on strengthening enforcement mechanisms, fostering innovation in sustainable technologies, and advancing international cooperation. By promoting collaborative efforts among governments, industry stakeholders, and consumers, regulatory frameworks can effectively mitigate carbon emissions, promote sustainable development, and contribute to a resilient and environmentally responsible electronics sector.

Emerging Technologies and Innovations for Low-Carbon Electronics

Emerging technologies and innovations are pivotal in advancing the development and adoption of low-carbon electronics, aiming to reduce carbon emissions, enhance energy efficiency, and promote sustainability across the electronics industry. These innovations span various stages of the electronics lifecycle, from design and manufacturing to use and end-of-life management, leveraging technological advancements to mitigate environmental impact and meet global climate goals. In the realm of materials science and manufacturing processes, innovations are focused on reducing the carbon footprint of electronic components and devices. For instance, advancements in nanotechnology enable the development of nanomaterials with enhanced properties, such as increased conductivity and efficiency, while using fewer resources and energy-intensive processes during production. Sustainable materials, including biodegradable polymers, recycled metals, and bio-based plastics, are gaining traction as alternatives to traditional materials, reducing reliance on virgin resources and lowering carbon emissions associated with extraction and processing. Energy efficiency remains a cornerstone of low-carbon electronics, with continuous innovations in semiconductor technology, power management, and device design. Integrated circuits (ICs) are becoming more energy-efficient through the development of advanced materials like gallium nitride (GaN) and silicon carbide (SiC), which enable higher switching frequencies and reduce power losses in electronic devices. Power management techniques, such as dynamic voltage scaling and adaptive control algorithms, optimize energy consumption based on real-time operational requirements, enhancing overall efficiency and minimizing carbon emissions during device use.

Moreover, innovations in renewable energy integration and energy harvesting technologies are transforming the sustainability landscape of electronics. Solar cells and photovoltaic technologies are increasingly integrated into electronic devices and infrastructure, enabling autonomous and sustainable power generation. Energy harvesting technologies, including thermoelectric generators and kinetic energy harvesters, capture ambient energy from heat differentials or motion and convert it into electrical power, providing renewable energy sources for low-power electronics and IoT devices. Innovative design strategies are also driving the development of modular and repairable electronics, facilitating product longevity and reducing electronic waste. Modular designs allow components to be easily replaced or upgraded, extending the lifespan of devices and reducing the frequency of replacements. Repairability initiatives promote accessible repair services and spare parts availability, empowering consumers to prolong the use of electronics and minimize carbon emissions associated with manufacturing new products.

Furthermore, digitalization and data-driven approaches are revolutionizing sustainability efforts in electronics through real-time monitoring, predictive maintenance, and optimization of energy consumption. Internet of Things (IoT) technologies enable smart devices and systems to gather and analyze data on energy usage patterns, operational efficiency, and environmental conditions. Machine learning algorithms and artificial intelligence (AI) algorithms optimize energy management and resource allocation in electronic systems, identifying opportunities for energy savings and carbon footprint reduction. Circular economy principles are increasingly integrated into innovations for low-carbon electronics, promoting resource efficiency, material recovery, and closed-loop manufacturing processes. Innovations in recycling technologies, such as automated sorting systems and chemical recycling processes, enable the recovery of valuable metals and components from electronic waste streams. Closed-loop supply chains and remanufacturing initiatives refurbish and reintegrate used electronics into new products, reducing the demand for virgin materials and minimizing environmental impact.

Despite these advancements, challenges remain in scaling up and commercializing emerging technologies for low-carbon electronics. These challenges include cost competitiveness, technological maturity, regulatory compliance, and infrastructure limitations. Collaborative efforts among industry stakeholders, research institutions, and policymakers are crucial in overcoming barriers and accelerating the adoption of sustainable innovations across the electronics supply chain. Looking ahead, the continued evolution of emerging technologies and innovations holds promise for achieving substantial reductions in carbon emissions from the electronics sector.

By fostering innovation, collaboration, and investment in sustainable technologies, stakeholders can drive meaningful progress towards a low-carbon future, where electronics play a pivotal role in advancing global sustainability goals while meeting the needs of a growing digital economy.

Future Outlook and Roadmap for Low-Carbon Electronic Products

Looking ahead, the future outlook for low-carbon electronic products is poised for transformative advancements driven by technological innovation, regulatory frameworks, consumer demand for sustainability, and global climate imperatives. A roadmap for the development and adoption of low-carbon electronics encompasses several key areas of focus to achieve significant reductions in carbon emissions and promote sustainability across the electronics industry. Firstly, advancements in materials science and manufacturing processes will continue to play a pivotal role in enhancing the environmental performance of electronic products. Innovations in sustainable materials, such as biodegradable polymers, recycled metals, and bio-based composites, will reduce the carbon footprint associated with raw material extraction and processing. Emerging nanotechnologies and advanced manufacturing techniques aim to improve energy efficiency and resource utilization during production, optimizing material use and minimizing waste generation. Energy efficiency remains a cornerstone of low-carbon electronics, with ongoing developments in semiconductor technology, power management, and device design. Integrated circuits (ICs) incorporating gallium nitride (GaN) and silicon carbide (SiC) are poised to revolutionize energy efficiency by enabling higher switching frequencies and lower power losses in electronic devices. Power management strategies, including dynamic voltage scaling and adaptive control algorithms, will continue to optimize energy consumption based on real-time operational requirements, enhancing overall efficiency and reducing carbon emissions during device use.

Renewable energy integration and energy harvesting technologies will further drive the sustainability agenda in electronics. Solar cells, flexible photovoltaic technologies, and energy harvesting devices harness ambient energy sources such as light, heat differentials, and motion to power electronic systems autonomously. These innovations not only reduce reliance on fossil fuel-derived electricity but also enhance the resilience and sustainability of electronic products in diverse applications, from portable electronics to smart grid infrastructure. Design innovation will focus on enhancing product longevity, repairability, and recyclability to minimize electronic waste and promote circular economy principles. Modular and upgradeable designs enable components to be easily replaced or upgraded, extending the lifespan of electronic devices and reducing the environmental impact of frequent replacements. Initiatives promoting accessible repair services, spare parts availability, and end-of-life management will empower consumers to make sustainable choices and contribute to a more circular economy.

Digitalization and data-driven approaches will continue to drive efficiency gains and environmental sustainability in electronics. Internet of Things (IoT) technologies, coupled with artificial intelligence (AI) and machine learning algorithms, enable smart devices and systems to optimize energy management, predict maintenance needs, and improve operational efficiency. Real-time data analytics and smart grid integration enhance energy efficiency across the electronics ecosystem, paving the way for intelligent and sustainable infrastructure solutions. Regulatory frameworks and policy incentives will play a pivotal role in shaping the future landscape of low-carbon electronics. Governments worldwide are expected to strengthen emissions standards, eco-design requirements, and incentives for renewable energy adoption to accelerate market transformation and drive industry-wide sustainability initiatives. Harmonized international standards and certification schemes will promote consistency and transparency in carbon footprint labeling, empowering consumers to make informed choices and incentivizing manufacturers to prioritize environmental performance.

Collaboration and partnerships among stakeholders, including industry leaders, research institutions, policymakers, and civil society, will be essential to overcome barriers and accelerate the adoption of low-carbon technologies. Knowledge-sharing platforms, collaborative R&D initiatives, and public-private partnerships will foster innovation, drive technological advancements, and scale up sustainable solutions across the electronics supply chain. In conclusion, the future outlook for low-carbon electronic products is characterized by a transformative shift towards sustainability, driven by technological innovation, regulatory advancements, and collaborative efforts. By leveraging advancements in materials science, energy efficiency, renewable energy integration, design innovation, digitalization, and regulatory frameworks, stakeholders can unlock significant opportunities to reduce carbon emissions, promote circular economy practices, and achieve a more sustainable electronics industry. Embracing a holistic approach to innovation and sustainability will not only address environmental challenges but also foster economic growth, enhance global competitiveness, and pave the way for a greener future for generations to come.

CONCLUSION

In conclusion, the analysis of carbon footprints associated with electronic products reveals a complex and multifaceted landscape shaped by various factors throughout their lifecycle. From raw material extraction and manufacturing to use phase and end-of-life disposal, electronic products contribute significantly to global carbon emissions, highlighting the urgent need for sustainable practices and innovations across the industry. Key findings underscore the critical impact of energy-intensive manufacturing processes, particularly in semiconductor production and assembly, which constitute major contributors to product carbon footprints. Efforts to improve energy efficiency during manufacturing, such as adopting advanced materials and optimizing production techniques, are essential for reducing emissions and enhancing environmental performance. During the use phase, energy consumption emerges as a dominant factor influencing carbon footprints, with variations depending on device type, usage patterns, and energy efficiency ratings. Consumer electronics, such as smartphones and laptops, exhibit substantial energy savings potential through advancements in battery technology, power management systems, and energy-efficient design. Renewable energy sources play a pivotal role in mitigating carbon emissions during device operation, with opportunities for integrating solar, wind, and other renewable energy technologies into electronic products and infrastructure.

End-of-life management poses significant challenges and opportunities for reducing carbon footprints through effective recycling, reuse, and waste minimization strategies. Innovations in electronic waste recycling technologies, including automated sorting systems and environmentally friendly recovery processes, enable the recovery of valuable metals and materials while minimizing emissions from landfill disposal and incineration. Circular economy principles promote closed-loop material cycles and product stewardship, encouraging manufacturers to design for disassembly, extend product lifespans, and integrate recycled materials into new products. Carbon footprint analysis methodologies, such as lifecycle assessment (LCA) and carbon accounting frameworks, provide valuable insights into emissions hotspots and opportunities for improvement across product lifecycles. Standardization of LCA methodologies and data transparency are critical for enhancing comparability and reliability of carbon footprint assessments, supporting informed decision-making by stakeholders, policymakers, and consumers. Carbon footprint labeling and disclosure initiatives empower consumers to make sustainable choices and incentivize manufacturers to prioritize environmental performance through eco-design and emission reduction strategies.

Looking forward, the future of carbon footprint analysis for electronic products hinges on advancing technological innovations, strengthening regulatory frameworks, and fostering collaboration across the electronics industry and beyond. Technological advancements in materials science, energy efficiency, renewable energy integration, and digitalization will drive sustainable product development and operational practices. Regulatory initiatives and policy incentives are pivotal in accelerating market adoption of low-carbon technologies, setting ambitious emission reduction targets, and promoting circular economy principles. Collaborative efforts among industry stakeholders, research institutions, governments, and consumers are essential for overcoming challenges, promoting knowledge-sharing, and scaling up sustainable solutions. Education and awareness initiatives play a crucial role in empowering consumers and businesses to prioritize sustainability, make informed purchasing decisions, and drive demand for environmentally responsible electronic products. Ultimately, addressing the carbon footprint of electronic products requires a holistic approach that integrates technological innovation, regulatory action, consumer engagement, and industry collaboration to achieve significant reductions in carbon emissions and build a resilient, sustainable electronics industry for the future. In conclusion, as the global community continues to confront the challenges of climate change and environmental sustainability, the electronic products industry stands at a pivotal juncture. By embracing carbon footprint analysis and adopting comprehensive strategies that encompass lifecycle perspectives, innovation, and responsible consumption, stakeholders can pave the way for a greener, more sustainable future.

REFERENCES

Abhilasha, A., Sreenivasulu, A., & Manimozhi, T. (2022). The Model of Smart Sensing Device For Sensitive Nanoclusters Modification in Sensing Properties. 2022 2nd International Conference on Advance Computing and Innovative Technologies in Engineering (ICACITE), 1043-1047. DOI: 10.1109/ICACITE53722.2022.9823846

Ahmed, S., & Lee, K. (2023). Carbon footprint assessment of smartphones: A comprehensive review. *Journal of Cleaner Production*, *379*, 134579.

Bai, X., & Tang, Y. (2023). Life cycle assessment of e-waste recycling: A case study on carbon footprint reduction. *Journal of Environmental Management*, *324*, 116444.

Brown, T., & Kumar, A. (2023). The environmental impact of wearable electronics: Carbon footprint and mitigation strategies. *Environmental Science & Technology*, *57*(3), 1892–1903.

Chang, S., & Park, H. (2023). Carbon footprint analysis of 5G network infrastructure. *The International Journal of Life Cycle Assessment*, *28*(1), 55–69.

Chen, Y., & Lin, Z. (2023). Assessing the carbon footprint of smart home devices: A comparative study. *Journal of Sustainable Technology*, *21*(4), 567–582.

Davis, J., & Wong, A. (2023). Carbon emissions from data centers: Trends and reduction strategies. *Journal of Environmental Impact Assessment*, *15*(2), 243–259.

Dong, Y., & Xu, L. (2023). Carbon footprint analysis of laptop manufacturing: A cradle-to-gate approach. *Journal of Industrial Ecology*, *27*(1), 101–115.

Fang, H., & Li, J. (2023). Carbon footprint of electric vehicles: The role of electronic components. *Renewable & Sustainable Energy Reviews*, *169*, 112878.

Gao, S., & Zhang, M. (2023). Carbon footprint reduction in electronics manufacturing: A green supply chain approach. *Journal of Cleaner Production*, *380*, 134596.

Gupta, R., & Sharma, P. (2023). The environmental impact of electronic waste: Carbon footprint and recycling methods. *Waste Management (New York, N.Y.)*, *150*, 127–139.

Han, X., & Chen, L. (2023). Assessing the carbon footprint of wearable health devices. *Journal of Environmental Health Science & Engineering*, *21*, 1023–1035.

He, Y., & Zhao, X. (2023). Carbon footprint analysis of LED lighting: A life cycle assessment. *Journal of Environmental Management*, *326*, 116477.

Huang, J., & Wu, Q. (2023). The carbon footprint of blockchain technology: An environmental perspective. *Environmental Research Letters*, *18*(2), 025009.

Iqbal, M., & Ahmed, R. (2023). Carbon emissions from electronic product lifecycle: A critical review. *Journal of Environmental Sciences (China)*, *137*, 57–73.

Jiang, Y., & Liu, S. (2023). Carbon footprint of gaming consoles: A lifecycle assessment. *Journal of Cleaner Production*, *381*, 134613.

Jones, K., & Smith, J. (2023). Environmental impact of e-commerce: Carbon footprint of electronic products. *Journal of Retailing and Consumer Services*, *70*, 102090.

Kim, H., & Lee, S. (2023). Carbon footprint assessment of virtual reality devices. *Journal of Environmental Impact*, *45*(1), 89–103.

Kwon, Y., & Park, J. (2023). The carbon footprint of electronic payment systems. *Journal of Cleaner Production*, *382*, 134630.

Li, X., & Zhang, Y. (2023). Carbon footprint reduction in the production of electronic components. *Journal of Industrial Ecology*, *27*(2), 221–237.

Lin, Y., & Wang, J. (2023). Carbon footprint analysis of home entertainment systems. *Journal of Cleaner Production*, *383*, 134657.

Liu, H., & Ma, Q. (2023). The environmental impact of cloud computing: Carbon footprint and mitigation strategies. *Environmental Science & Technology*, *57*(5), 2468–2481.

Luo, Z., & Huang, Y. (2023). Carbon footprint of artificial intelligence: A life cycle assessment. *Journal of Environmental Management*, *328*, 116502.

Ma, X., & Wu, Y. (2023). Carbon footprint assessment of smartwatches. *Journal of Environmental Impact Assessment*, *16*(2), 309–323.

Mahadev Madgule, N. (2023). Mechanical properties and microstructure of activated TIG welded similar joints of Inconel alloys by desirability approaches. *Materials Today: Proceedings*, *77*, 528–533. DOI: 10.1016/j.matpr.2022.12.250

Martin, C., & Davis, R. (2023). Carbon footprint of electronic toys: An environmental analysis. *Journal of Sustainable Technology*, *22*(1), 71–86.

Nguyen, T., & Tran, L. (2023). Carbon emissions from the production and disposal of electronic goods. *Journal of Environmental Sciences (China)*, *138*, 145–159.

Patel, N., & Shah, R. (2023). Carbon footprint reduction strategies in the consumer electronics industry. *Renewable & Sustainable Energy Reviews*, *170*, 112913.

Ramakrishnan, T., Mohan Gift, M. D., Chitradevi, S., Jegan, R., & Subha Hency Jose, P. (2022). Study of Numerous Resins Used in Polymer Matrix Composite Materials. Advances in Materials Science and Engineering. DOI: 10.1155/2022/1088926

Rao, S., & Maheshwari, V. (2023). Carbon footprint of electronic medical devices: A comprehensive assessment. *Journal of Environmental Health*, *36*(4), 505–518.

Selvakumar, P. (2024). Emerging Engineering Technologies: Design, Development, and Industrial Applications of IoT-Based Aquaculture Monitoring System. In El Kacimi, Y., & Alaoui, K. (Eds.), *Emerging Engineering Technologies and Industrial Applications* (pp. 332–341). IGI Global. DOI: 10.4018/979-8-3693-1335-0.ch016

Selvakumar, P., Mariappan, P., Shunmugam, S., Paul, R., & Jabez, S. (2024). Removal of chemicals from effluent using photobioreactor technology to improve environmental and health impacts. *Zeitschrift für Physikalische Chemie*, *2024*(7), 1295–1302. Advance online publication. DOI: 10.1515/zpch-2023-0450

Selvakumar, P., Muthusamy, S., Satishkumar, D., Vigneshkumar, P., Selvamurugan, C., & Satheesh Kumar, P. (2024). AI-Powered Tools. In Satishkumar, D., & Sivaraja, M. (Eds.), *Using Real-Time Data and AI for Thrust Manufacturing* (pp. 20–42). IGI Global. DOI: 10.4018/979-8-3693-2615-2.ch002

Selvakumar, P., Palanisamy, S. K., Cinthaikinian, S., Palanisamy, V., Mariappan, R., & Selvakumar, P. K. (2024). Biosensors and its diverse applications in healthcare systems. *Zeitschrift für Physikalische Chemie*, *2024*. Advance online publication. DOI: 10.1515/zpch-2023-0406

Singh, A., & Verma, S. (2023). Carbon footprint reduction in electronic device recycling. *Journal of Cleaner Production*, *385*, 134688.

Sun, L., & Zhao, X. (2023). Assessing the carbon footprint of internet of things (IoT) devices. *Journal of Environmental Impact*, *46*(3), 124–139.

Tang, J., & Li, Q. (2023). Carbon footprint of electronic waste management. *Journal of Environmental Sciences (China)*, *139*, 201–215.

Uddin, S., & Karim, M. (2023). Carbon emissions from semiconductor manufacturing: A life cycle perspective. *Journal of Cleaner Production*, *386*, 134705.

Vasquez, M., & Romero, P. (2023). The environmental impact of telecommunication devices: A carbon footprint analysis. *Journal of Industrial Ecology*, *27*(3), 309–322.

Vijayakumar, G., Rajkumar, M., Rajiv Chandar, N., Selvakumar, P., & Duraisamy, R. (2024). Environmentally friendly TDS removal from waste water by electrochemical ion exchange batch-type recirculation (EIR) technique. *Environmental Science. Water Research & Technology*, *10*(4), 826–835. DOI: 10.1039/D3EW00793F

Wang, H., & Chen, X. (2023). Carbon footprint assessment of wireless communication systems. *Journal of Cleaner Production*, *387*, 134720.

Williams, J., & Nguyen, M. (2024). Carbon emissions from electronic devices: Trends and future directions. *Journal of Environmental Impact*, *47*(1), 145–160.

Xu, L., & Huang, J. (2023). Carbon footprint of virtual meeting technologies. *Journal of Environmental Sciences (China)*, *140*, 233–248.

Yang, S., & Park, H. (2023). Carbon footprint reduction in the production of electronic displays. *Journal of Cleaner Production*, *388*, 134737.

Yoon, J., & Kim, S. (2023). Carbon footprint analysis of photovoltaic panels: A life cycle assessment. *Journal of Environmental Management*, *329*, 116519.

Zhao, J., & Lin, M. (2023). Environmental impact of electronic health records: A carbon footprint analysis. *Journal of Cleaner Production*, *389*, 134754.

Zheng, Y., & Li, Q. (2023). The carbon footprint of electronic payment devices: An environmental assessment. *Journal of Environmental Impact*, *48*(2), 177–192.

Chapter 9
Ecotoxicity and Human Health Risks of Improper E-Waste Handling

Odangowei Inetiminebi Ogidi
https://orcid.org/0000-0002-9961-893X
Bayelsa Medical University, Nigeria

Nwanne Dike Ijere
College of Arts, Science, and Professional Studies, Aba, Nigeria

Iniebiyo Felagha
https://orcid.org/0009-0002-2976-228X
University of Africa, Nigeria

ABSTRACT

Globally, the management of solid waste encompasses electronic waste. E-waste, which makes up a substantial part of solid waste, contains several dangerous substances, including halogenated compounds like polychlorinated biphenyls (PCBs), tetrabromobisphenol A (TBBPA), and polybrominated biphenyl (PBB). These substances exert detrimental impacts on both human beings and the environment. Heavy metals constitute a significant portion of electronic waste, so it is crucial to exercise caution while dismantling them. The involvement of the informal sector in underdeveloped countries exacerbates the issue of handling e-waste. The existing methods for disposing and treating electronic waste are insufficient, and the materials have both immediate and indirect impacts on human health and the environment. This chapter examines the diverse types of electronic waste and their effects on different aspects of the environment and human health. It also discussed information on the procedures for managing electronic waste and effective handling methods.

DOI: 10.4018/979-8-3693-7383-5.ch009

INTRODUCTION

The proliferation of various technical devices has significantly enhanced the quality of people's living conditions, resulting directly from advancements in development and industrialisation. The advancement of technology in the realm of information has resulted in the extensive utilisation of electronic and electrical equipment (EEE), significantly enhancing our standard of living. Nevertheless, it is concerning as electronic devices generate a substantial amount of waste as a result of their limited lifespan, continual improvements in aesthetically pleasing designs, and compatibility challenges. The need for state-of-the-art electronics is a significant catalyst for the generation of electronic trash (Forti et al., 2020). Over the years, there has been a consistent upward trend, and analysts forecast that this pattern will persist. The global production of electronic waste experienced a growth of 9.2 million metric tonnes (Mt) from 2014 to 2019, resulting in a total of 53.6 Mt in 2019 (Ghimire & Ariya, 2020). The quantity is expected to exceed 74.7 million metric tonnes by 2030, as per predictions (Forti et al., 2020; Ghimire & Ariya, 2020). In 1997, the average lifespan of a computer was four to six years. However, this period has been reduced to approximately two years by 2005 (Angaye & Ogidi, 2018; Needhidasan et al., 2014).

The increasing prevalence of EEE in modern life is placing greater demands on waste disposal facilities to effectively handle the waste stream (Althaf et al., 2021). A significant proportion of the solid waste stream comprises electronic gadgets, including computers, mobile phones, displays, and various industrial and home appliances. The waste contains numerous hazardous compounds and toxic metals, which represent significant dangers to both individuals and the environment. If not managed appropriately, these pollutants could have devastating consequences. The lack of effective enforcement of regulations and the general lack of knowledge among the public are major factors contributing to the poor status of e-waste management in various regions (Odili et al., 2018). Both the amount and disposition of consumers impact their decisions regarding purchases and disposal. Individuals who impulsively purchase new, substandard equipment may experience initial cost savings, but they will ultimately face negative consequences. The reason for this is because both the disposal process and the environment are adversely affected by these products (Shaikh et al., 2020a, b).

It is intriguing that individuals frequently dispose of their electronic equipment prior to its end-of-life (EoL) phase. The quantity of items that are transported to a designated recycling center after their useful life has expired is also relatively low. A substantial portion of this electrical equipment is incinerated for the purpose of energy generation or landfill disposal. Developing countries, such as China, India, and Africa, are disproportionately affected by the substantial inflow of electronic waste (EEE) from industrialized nations. The potential for air and water pollution is not taken into account when depositing waste in landfills and allowing combustion are common EEE management techniques in these countries. Both developed and underdeveloped nations are equally affected by this issue. Nevertheless, the primary factor contributing to the lack of modern landfills in economically distressed countries is typically insufficient public funding (Kiddee et al., 2014).

Developing countries like Nigeria lacks the necessary infrastructure to effectively manage the recycling of electronic waste. They are compelled to rely on unregulated industries that engage in haphazard disassembly and recycling in residential areas. Copper is extracted through the process of open combustion of wires, while e-waste is disposed of in dumpsites utilising same techniques. Based on the research conducted by Kaya (2019) and Nnorom & Odeyingbo (2020), engaging in these activities puts individuals at risk of being exposed to hazardous substances. The three primary challenges associated with electronic waste management, namely recycling, recovery, and disposal, are also its most crucial components. Efforts have been made to recover valuable metals like gold (Au), silver (Ag), platinum (Pt), and copper (Cu), but only a small portion of this waste is now being recycled worldwide (about 20%-30%). However, these therapeutic methods are conducted in a highly perilous manner, resulting in significant adverse impacts on individuals' well-being. Currently, there is an inadequate amount of study and documentation available on e-waste disposal facilities. As a result, a significant quantity of electronic waste is either burned or disposed of in landfills. However, both of these approaches present significant hazards to the environment. Standard pre-treatment operations such as removing components, manually disassembling, crushing, and reducing size can all lead to significant local contamination (Khanna et al., 2020a, b).

Damage to soil erosion, water contamination, and the release of toxic gasses during cremation are all consequences of poorly managed electronic waste (Okereke et al., 2016; Ogidi et al., 2024c). According to several studies (Chen et al., 2016: Fujimori et al., 2016; Akram et al., 2019; Amoabeng-Nti et al., 2020; Khan et al., 2021; Ahirwar & Tripathi, 2021; Rautela et al., 2021). The picture shows how these elements directly affect living things. Despite making up only a small percentage of total waste (2% to 5%), electronic waste is accountable for more than 70% of the toxicity in solid waste. When compounds like lead, chromium, beryllium, palladium, and gallium contaminate our air, water, and soil, they upset the ecological balance of our food chain (Awasthi et al., 2018; Akram et al., 2019; Liu et al., 2020a, b). When ingested, they can cause a host of health problems. Hazardous metals, dioxins, polyaromatic hydrocarbons, polyvinyl chloride, and epstein are some of the chemical contaminants associated with discarded electronics. Both Peng et al. (2015) and Ogidi et al. (2021a) indicated that these contaminants pose a risk to human health. Local flora, soil ecology, and microbes are not immune, though (Zhang et al., 2015).

Recent studies have shown that e-waste leachates affect soil microbes, fungus, fish, plants, and humans. Some of these studies are Recent publications by Jiang et al. (2017), Youcai (2018), Bao et al. (2020), and Purchase et al. (2020) are included. Many electrical parts, including batteries, switches, and screens, contain mercury. Methyl mercury can be produced by microbes in soil from mercury. Methyl mercury can have a negative impact on the neurological systems of animals and people alike. Wu et al. (2020a), Yan et al. (2019), and Clarkson & Magos (2006) found that it builds up in their adipose tissues. It has been proven that babies and newborns are especially vulnerable. Dermatitis, neurological problems, memory loss, cognitive impairments, and many more diseases and ailments are also outcomes.

The works of Richard (2009) and Schwerze et al. (2020) shows tha Ni, arsenic, copper, chromium, and other heavy metals are just as harmful to humans and other animals as they are to other species. There are two ways they can affect them: directly and indirectly. Productivity gains and better use of resources in a variety of industries may result from these goods' Life Cycle Assessments (LCAs). A lack of consistency exists despite the use of multiple life cycle assessment (LCA) methods. Economic input-output Life Cycle Assessment (LCA) is a practical method since it can give precise calculations and detailed information on the release of harmful substances (CMU, 2011; Kaya, 2019). The burden of understanding and properly managing electronic waste falls on the local community, NGOs, corporations, and state agencies, according to Ilankoon et al. (2018), Mihai (2020), and Forti et al. (2020). Li & Achal (2020), Akram et al. (2019), Adanu et al. (2020), and Amoabeng-Nti et al. (2020) all state that environmentally friendly remediation procedures are highly successful and efficient in dealing with electronic waste. Various studies are looking at electronic waste from numerous angles, including how it's made and where it ends up in different parts of the world (Bao et al., 2020).

However, e-waste is still a problem because of things like how much more e-waste there is, how much better it is to collect metals from it, how difficult it is to move e-waste across borders, how often laws change, and how dangerous metals are introduced into ecosystems. The goal is to examine where we are now and where we're going in terms of producing electronic garbage on a worldwide scale (Shittu et al., 2020). In addition, we want to look at the ways that electronic trash can harm ecosystems' biotic and abiotic parts. Research on eco-friendly alternatives, including phytoremediation, for the remediation of ecosystems impacted by electronic trash has been conducted, in addition to efforts to assess the efficacy of current approaches for managing this type of garbage (Ahirwar & Tripathi, 2021).

Objectives of the Chapter

The aims of this chapter are as follows:

(1) To evaluate the composition of e-waste.
(2) To discuss the e-waste generation scenario around the world.
(3) To review the consequences of e-waste on human health and the environment.
(4) To discuss the mechanisms and techniques to alleviate e-waste damage.
(5) To X-ray the current policy and legislation of e-waste for environmental health exposure.
(6) To discuss the future prospects of the toxicity of e-waste.

Organization of the Chapter

The remainder of this chapter is structured as follows: Section 2 examines the constituents of electronic waste. Section 3 examines the global e-waste generation landscape. Section 4 examines the ramifications of electronic waste on human health and the environment. Section 5 examines the methods and strategies to mitigate e-waste harm. Section 6 addressed the economic considerations affecting e-waste recycling. Section 7 examined pragmatic recommendations for electronic waste management. Section 8 analyzed the existing policy and regulations regarding electronic trash in relation to environmental health risk. A concluding part outlines the principal insights and emphasizes future opportunities.

COMPOSITION OF E-WASTE

The most frequently encountered metals in e-waste include iron, copper, aluminum, nickel, lead, tin, and manganese. In addition, it contains a variety of ceramics, metal oxides, glass fiber, and polymers (Ilankoon et al., 2018; Purchase et al., 2020). A classic PC with a cathode ray tube (CRT) screen weighs 25 kg, as per Berkhout and Hertin (2004). This computer is composed of 43.7% metal, 17.3% electronics, 15% glass, and 23.3% plastic (Berkhout & Hertin, 2004). According to Torres et al. (2018), copper comprises the preponderance of the valuable metals discovered in electronic waste. Suja et al. (2014) conducted research that indicates that heavier electronics contain fewer potential contaminants than lighter e-waste items. In contrast to personal computers, laptops, mobile devices, and similar gadgets, which often contain steel and a lower quantity of environmental pollutants, refrigerators and washing machines tend to have a larger proportion of hazardous metals (HMs) and fire retardants. Platinum group metals have exceptional resistance to oxidation and wear, which makes them very suitable for use in electrically conductive materials (Purchase et al., 2020). Table 1 illustrates the types of electronic waste.

Table 1. Types of electronic waste

S/N	Types	Name	Examples
1	Type 1	Major Appliances	Refrigerators, washing machine, dryers.
2	Type 2	Small appliances	Irons, fryers, vacuum cleaners, blenders
3	Type 3	Computer and telecommunication appliances	Mobile phones, PCs, Telephones, Laptops
4	Type 4	Consumer electronics	Audio and video appliances, musical equipment
5	Type 5	Lighting devices	Gas-discharge lamps, fluorescent tubes, incandescent light bulbs.
6	Type 6	Electrical and electronic devices	Gardening devices, saws, drills.
7	Type 7	Electronic toys	Models, sport equipment, electronic toys
8	Type 8	Medical devices	All medical equipment except implants.
9	Type 9	Monitoring devices	Laboratory equipment, thermostart and detectors
10	Type 10	Vending machines	Photo booth, change machine, cigarrate vending machine, food vending machine.

The amounts of valuable metals in waste PCBs are ten times higher than in materials that are commercially extracted, according to Ilankoon et al. (2018). However, factors such as producer, age, country of origin, and type can significantly affect the chemical makeup of e-waste (Bigum et al., 2017). One million mobile phones have the ability to manufacture 24 kg of gold, 9 kg of palladium, 250 kg of silver, and 9,00 kg of copper, according to Ilankoon et al. (2018). Unlike more conventional forms of waste, electronic waste may include a broad range of toxic chemicals. Batteries, fire retardants, conductive materials, LCD monitors, and Si chips all contain the following elements: Li, La, Be, Ce, Sb, and Ga (Purchase et al., 2020; Ladou & Lovegrove, 2008). Electronic waste is broken down in Figure 1.

Figure 1. Composition of electronic waste

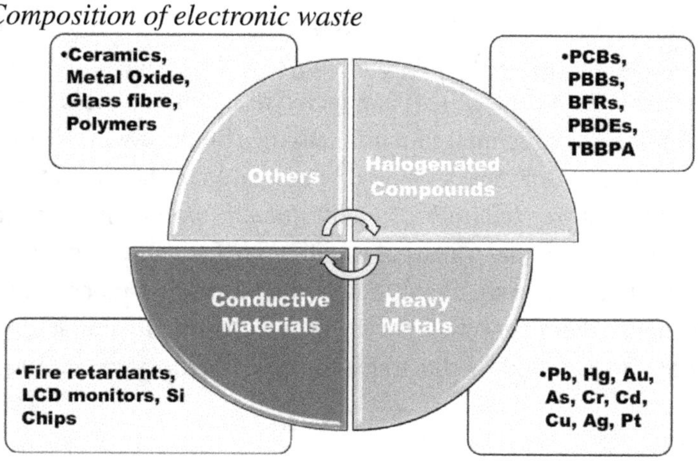

E-WASTE GENERATION SCENARIO AROUND THE WORLD

The alarming increase of e-waste on a global scale is having a negative impact on the environment (Tian & Chen 2016). Every year, around twenty million metric tons of electronic garbage (e-waste) is generated (Namias, 2013). The bulk of developing nations, around 75 to 80%, are located in Asia and Africa, which are also the two main locations for recycling and disposal, according to Golev et al. (2016). China and India are the top two countries that receive electronic waste from around the world. Some nations that import e-waste include Bangladesh, Pakistan, Nigeria, and Kenya (Garlapati, 2016). Global e-waste generation is increasing by 5 to 10% every year, according to the US Environmental Protection Agency (EPA), as stated by Nie et al. (2015) and Ogidi et al. (2024a). The extremely short lifespan and widespread use of useless EEE items are the main contributors to this issue. The issue of electronic waste has grown in importance in both industrialized and developing nations over the past few decades. According to Shamim et al. (2015) and DeSouza et al. (2016), the amount of electronic garbage, often known as e-waste, has been rising worldwide.

CONSEQUENCES OF E-WASTE ON HUMAN HEALTH AND THE ENVIRONMENT

Health Impact of E-Waste

Regarding the possible worldwide contamination from chemicals and heavy metals found in discarded electronics, numerous studies have been undertaken on this subject. It is widely acknowledged that electronic waste (e-waste) has profound and detrimental impacts on human health (Table 2). Individuals residing in close proximity to e-waste recycling facilities face a heightened susceptibility to radiation exposure and absorption of chemicals, along with other associated health complications (Jibiri, 2015). Heacock et al. (2016) looked at the correlation between metal collecting and elevated blood lead levels in children from low-income families in South America.

According to studies done by Ravindra & Mor (2019), blood lead levels (BLL) in children and teenagers are higher than what is considered safe. To determine the extent to which people are exposed to e-waste contaminants, bio-indicators such as blood, urine, and hair testing are commonly used. In areas where there is a high concentration of recycled and reprocessed electronic debris, Zhang et al. (2020) looked into how often people are exposed to phthalic acid esters (PAEs). After collecting urine samples from people living near e-waste dumps, scientists found that the PAE content was much higher in their urine compared to people living in other areas. It was shown that over 22% of the people living in e-waste landfills had hazard index (HI) values higher than one. It seems that their levels of exposure are higher than what could be considered dangerous. The people living at e-waste dumps, 68% were younger than 18 years old, according to Zhang et al. (2020).

Individuals residing or employed in close proximity to the e-waste recycling facility had their urine and hair samples tested for polycyclic aromatic hydrocarbons (PAHs) using these biomarkers. Residents and personnel involved in removing e-waste were exposed to similar quantities of radiation. Lin et al. (2020) found carcinogenic chemicals in the subjects' internal hair in a separate experiment. The development of several disorders, including respiratory and reproductive difficulties, cardiovascular ailments, urinary tract infections, irritations, and more, has been associated with heavy metal exposure (Ogidi et al., 2024b; Awasthi et al., 2019b). In addition, a number of health problems, including infections of the respiratory system, kidneys, and blood, lung cancer, mental breakdowns, and brain illnesses, have been linked to electronic waste (ewaste) in Delhi (Awasthi et al., 2019b). The use of machetes exposes persons to hazardous elements like mercury, arsenic, and cadmium, according to a 2017 discovery by researchers in South Africa. These metals have devastating effects on human health. According to Maphosa & Maphosa (2020), e-waste contains harmful compounds such as barium, mercury, beryllium, and lead.

In 2015, researchers from China (Lu et al. 2015) found that lead and other metals were present at extremely high levels in blood samples taken from children. Dust samples taken from two makeshift e-waste recycling centers in urban areas of Vietnam showed significantly higher concentrations of brominated flame retardants (BFRs) and polychlorinated biphenyls (PCBs) compared to samples taken from households that do not participate in e-waste recycling. This finding was reported by Lu et al. (2015). Eye irritation, asthma, and other respiratory illnesses are among the many health problems that have arisen in Ghana as a consequence of the presence of pollutants and poisons, according to research by Cudjoe & Acquah (2021). Asampong et al. (2015) found that employees who deal with e-waste often suffer from chest pain and severe respiratory infections. Half of the employees who dealt with electronic garbage had some kind of respiratory sickness, chest pain, or cancer, according to a poll by Agyei-Mensah & Oteng-Ababio (2012). Rautela et al. (2021) demonstrates that open dumpsites and landfills emit carbon-based pollutants when electronic debris is burned. These pollutants pose a threat to people's health. Tetteh & Lengel (2017) linked the increased cancer rates among dumpsite workers and nearby residents to the discharge of harmful chemicals during uncontrolled e-waste recycling.

Furthermore, e-waste contaminants and chemicals may damage human DNA, which could have serious health implications. Workers in the electronics recycling industry may be at risk of genetic damage, according to research by Awasth et al. (2018). Researchers found a strong correlation between the amount of time spent processing e-waste and DNA damage. The results of a health evaluation of the people living in the makeshift recycling centers for electronic trash confirmed this. Low birth weight, premature birth, stillbirth, and spontaneous abortion are more common among those who live or work near e-waste processing sites, according to research by Grant et al. (2013). According to research published by Yu et al. (2017) and Gangwar et al. (2019), those who work with electronic waste might suffer from a variety of health problems, including hearing loss, pain, respiratory problems, cuts, and abrasions.

According to Maphosa & Maphosa (2020), the health effects include cancer, asthma, neurological impairment, hearing and vision impairments, low birth weight, and premature infant death. Concurrently, the environmental impacts include endangering species and polluting the atmosphere, hydrosphere, and lithosphere. Neurological damage, medication resistance, cognitive impairment, skin infections, increased infant mortality rates, birth defects, blood disorders, respiratory complications, and organ failure are just some of the health issues linked to toxic metals found in electronic waste, according to Nwagwu & Okuneye (2016). Particulate particles and toxic vapors are common hazards that adults, especially those who work at unregulated e-waste recycling operations, and children are regularly exposed to (Angaye et al., 2019). In addition to coming into contact with corrosive chemicals through the skin, people can also be exposed to them through the water and food they consume (Forti et al., 2020). Breastfeeding and, in rare cases, pregnancy can expose infants to harmful compounds, according to research by Landrigan et al. (2019) and Forti et al. (2020). Furthermore, toddlers are more likely to consume dirt, which increases their exposure even more.

Soetrisno et al. (2020), and Amoabeng et al. (2020) have all pointed out a number of other health issues related to the topic. Among these are the following: alterations in neurodevelopment, damage to DNA, adverse consequences on cardiovascular health, respiratory problems, impaired immune system function, skin issues, hearing loss, and learning and birth outcomes. The individuals who benefit from the informal recycling of electronic garbage, particularly those who extract valuable metals such as gold, have not yet implemented consistent safety procedures. Instead, they continue to rely on outmoded and unlawful methods (Awasthi et al., 2018). Workers expose themselves to hazards when they fail to utilise protective equipment such as gloves or masks while dealing with metals, glass, and poisonous chemicals (Kowsar et al., 2010). Exposure to hazardous compounds without the use of protective clothing can lead to the development of skin cancer, anaemia, carcinogenic tumors, and hormonal imbalances (Ogidi & Tawariowei, 2023; Ganguly, 2016; Ogidi et al., 2021b). EEE workers employ rudimentary tools such as hammers and stones to recover valuable components from electronic waste (Adanu et al., 2020). Acquah et al. (2019) have demonstrated that workers in the electronic waste industry often have musculoskeletal injuries as a result of using basic tools like chisels and hammers.

In addition, e-waste recyclers burned obsolete electronics to extract the profitable components, but they neglected to consider the potential health consequences (Davis et al., 2019). Disassembling or disintegrating electronic waste (EEE) can result in mishaps and physical harm such as burns and lacerations (Awasthi et al., 2019a). Soetrisno et al. (2020) discovered that individuals engaged in electronic waste handling face the danger of absorbing toxic compounds by skin contact and inhalation of smoke, so corroborating this notion. According to Soetrisno et al. (2020), workers involved in the dismantling of electronic trash reported experiencing skin infections, eye irritation, and continuous coughing. Engaging in practices such as forcefully smashing electronic waste EEE on the ground, manually separating its components, and dismantling it with stones and hammers are all basic methods that might result in injuries (Adanu et al., 2020). Research conducted by Burns et al. (2019) indicates that the physical tasks involved in recycling electronic waste resulted in injuries to workers' hands, such as lacerations and wounds. According to Heacock et al. (2016), women and children are more susceptible to harmful chemical exposures resulting from uncontrolled and rudimentary technology. The issue exacerbates when injured employees are deprived of medical intervention, hence subjecting their wounds to additional hazardous substances (Davis et al., 2019).

Table 2. Human health impact of electronic waste

S/N	E-waste Composition	Health Concerns	Samples Used	Adult/Children	Country	References	Study Purpose
1	Heavy metals	High Blood lead levels	Blood	Children	-	Heacock et al., 2016	To assess the impact of heavy metals on children's blood lead levels.
2	-	High blood lead levels	Blood	Children and Teens	India	Ravindra & Mor, 2019	To examine the correlation between e-waste exposure and lead levels in children and teens.
3	Phthalic acids esters	-	Urine	Teens	China	Zhang et al., 2020	To evaluate the presence of phthalic acid esters in teen urine samples.
4	PAHs, Heavy metals	Cancer, respiratory and reproductive difficulties, cardiovascular ailments, urinary tract infection, irritation	Urine and hair	Adults and teens	India	Lin et al., 2020; Awashi et al., 2019b	To study the health effects of PAHs and heavy metals on adults and teens.
5	Arsenic, Cadmium, Mercury	Mental breakdown, brain illnesses, lung cancer, kidney and blood disorders, respiratory problems	-	Adult and Children	India	Awashi et al., 2019b	To analyze the neurological and systemic impact of arsenic, cadmium, and mercury.
6	Metals, BFRs, PCBs	Eye irritation, asthma	Blood sample	Children	China	Lu et al., 2015	To investigate the link between metal and PCB exposure and respiratory/ ocular issues in children.

continued on following page

Table 2. Continued

S/N	E-waste Composition	Health Concerns	Samples Used	Adult/Children	Country	References	Study Purpose
7	-	Eye irritation, respiratory disorders	-	-	Ghana	Cudjoe & Acquah, 2021	To explore the impact of e-waste exposure on respiratory and ocular health.
8	-	Chest pain, severe respiratory infection	-	-	Ghana	Asampong et al., 2015	To assess respiratory issues among populations exposed to e-waste.
9	-	Cancer, chest pain, respiratory sickness	-	-	Ghana	Agyei-Mensal & Oteng-Ababio, 2012	To evaluate the carcinogenic and respiratory risks of e-waste exposure.
10	Harmful Chemicals	Cancer	-	-	Sub-Saharan Africa	Tetteh & Lengel, 2017	To examine the cancer risks associated with harmful chemicals in e-waste.
11	-	Low birth weight, premature birth, stillbirth, spontaneous abortion	-	Adult	-	Grant et al., 2013	To assess the effects of e-waste exposure on pregnancy outcomes.
12	-	DNA damage	-	-	China	Awasth et al., 2018	To analyze the impact of e-waste on DNA integrity.
13	-	Hearing loss, pain, cuts and abrasions, respiratory problems	-	-	Ghana	Yu et al., 2017, Gangwar et al., 2019	To study physical and respiratory health issues caused by e-waste exposure.

continued on following page

Table 2. Continued

S/N	E-waste Composition	Health Concerns	Samples Used	Adult/Children	Country	References	Study Purpose
14	-	Cancer, asthma, neurological impairment, hearing and vision impairment, low birth weight	-	-	Sub-Saharan Africa	Maphosa & Maphosa, 2020	To evaluate multiple health effects of e-waste exposure.
15	Toxic metals	Organ failure, blood disorders, skin infections, cognitive impairment, neurological damage, birth defects, infant mortality	-	Adult and children	Nigeria	Nwagwu & Okuneye, 2016	To assess the toxic effects of metal exposure from e-waste on various health parameters.

Environmental Impact of E-waste

An instance of an environmental consequence caused by electronic waste is the pollution of land, air, and water. Grant et al. (2013) divided the environmental threats associated with e-waste into three categories: land hazards, water hazards, and airborne hazards (Figure 2). Soil and sediment samples collected from e-waste processing facilities, rivers near residential areas, and open burning sites contained significant quantities of copper, lead, calcium, molybdenum, tin, nickel, and zinc. Li & Achal (2020) assert that e-waste management procedures that are not sustainable lead to significant soil pollution. Chemical substances discharged into the soil impede the growth and progress of plants and trees. They concluded that these regions are unsuitable for agricultural use. The disassembly process of cathode ray tubes and their subsequent disposal leads to the leaching of heavy metals, such as barium and lead, into both groundwater and freshwater sources. This leaching occurs as a result of removing the copper yoke. Moreover, the procedure results in the discharge of polluted phosphor (Gupta, 2011). When these pollutants enter people's diets, they can lead to a variety of health issues. The darkening of soil occurs as a result of staining produced by the release of noxious liquid substances, such as acid, from car batteries (Adanu et al., 2020).

Figure 2. Categories of environmental impact of e-waste

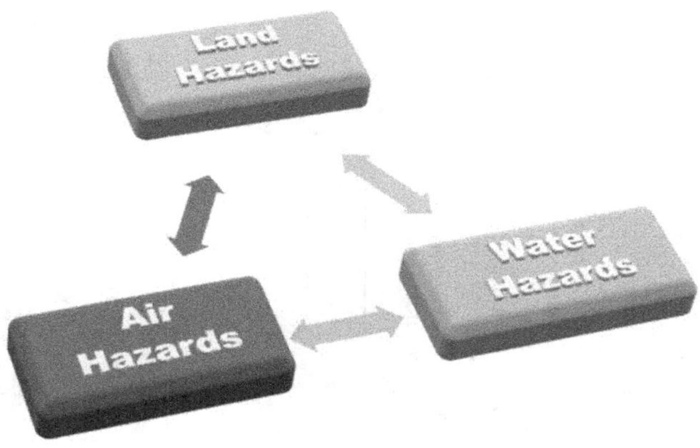

Ecosystems on land and in water suffer when recycling methods like open burning and labor-intensive disassembly are not adequately implemented. According to Adanu et al. (2020), soils and sediments that are part of contaminated rivers contain fire retardants. In the process of desoldering and chip extraction from printed circuit boards, certain metals such brominated beryllium, cadmium, dioxins, and mercury are released into the environment (Gupta, 2011). Soil samples taken from locations where electronic waste is recycled and processed have shown the presence of dioxin chemicals. These locations have dioxin compound concentrations that were higher than the WHO-recognized safe level. This is mainly because people still keep burying and burning their waste in open areas (Gupta, 2011).

In their multivariate analysis results, Adanu et al. (2020) discovered a robust correlation between the processing of E-waste and a significant accumulation of metallic components. Furthermore, the soil research revealed that areas where open burning took place exhibited copper deposition. Electronic waste can pollute the air, along with causing various other detrimental effects on the environment. Incinerating electronic waste in landfills emits harmful toxins into the atmosphere, which can be inhaled by humans and other organisms. According to Jayaraman et al. (2019), the presence of heavy metals in landfills has a significant detrimental impact on the environment. The presence of dangerous effluents in uncontaminated soil and air poses a direct threat to all living organisms. This threat can be manifested as acid rain, heavy metals, and other toxic compounds. Additionally, there is an indirect threat through biomagnification (Jayaraman et al., 2019). The combustion of discarded electronics also contributes to the release of particulate matter and other air pollutants. Gangwar et al. (2019) found that the act of open burning led to elevated levels of particulate matter (PM10) and heavy metals in air samples collected from the e-waste site and its vicinity. Gangwar et al. (2019) discovered that blood tests showed that persons living in areas where electronic garbage is incinerated in India had the highest levels of exposure. Contaminated surface water, however, can lead to biomagnification and has detrimental impacts on aquatic organisms. Figure 3 shows the various impacts of electronic waste.

Figure 3. Impacts of electronic waste

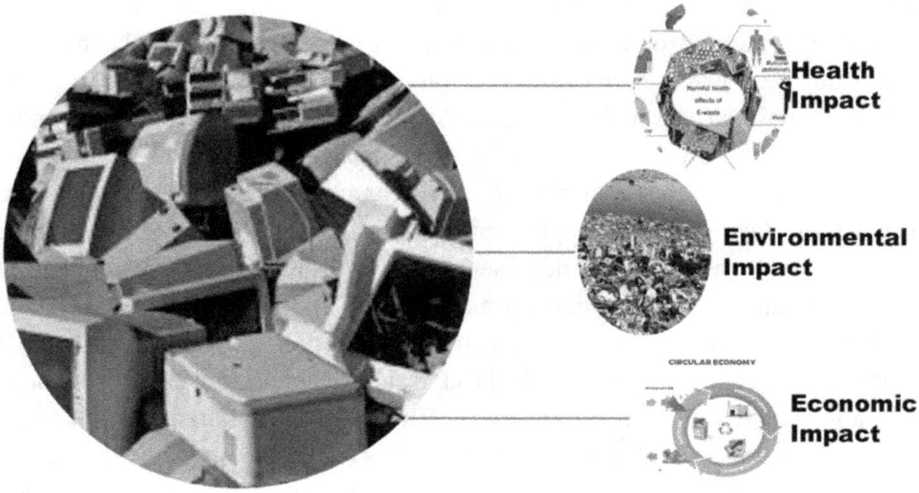

MECHANISMS/TECHNIQUES TO ALLEVIATE E-WASTE DAMAGE

The responsible processing and handling of electrical and electronic equipment is critical in reducing the possibility of damage to people and the environment due to incorrect recycling and disposal practices. A category of products referred to be "grey goods" includes media equipment including televisions, camcorders, and cameras (Chen et al., 2011; Sheng & Etsell, 2007). Lead, cadmium, and mercury are among the many toxic and dangerous compounds found in personal computers (Sheng & Etsell, 2007; Chen et al., 2011).

Chemical Leaching Techniques of Metals From E-Waste

Li & Xu (2019) researched the leaching of copper (Cu) from waste PCBs by taking into account the particulate size and elemental content. The experiment entailed the treatment of pulverized copper particles or unwanted PCBs with sulfuric acid and hydrogen peroxide. Li and Xu (2019) define chemical leaching as the process of compound extraction through ligand- or acid-assisted complexation. The study of metal-ligand complexes is significantly simplified by the application of complexometry. Chemicals such as hydrochloric acid, sulphuric acid, or a combination of nitric acid and sulphuric acid can be employed to efficiently remove metal from unsuitable electrical components. Sodium hypochlorite can be employed to extract gold and other precious metals from solutions of differing acidities and alkalinities. Rautela (2021) employed chemical disintegration to extract heavy metals such as Ni, Cu, Fe, and Au, despite the fact that it is more costly than bioleaching.

Bioleaching of E-Waste

The process of bio-leaching is crucial for the reuse and recycling of old batteries and other forms of electrical waste. Biologic leaching, also known as bio-hydrometallurgy or microbial mining, is an effective and cost-effective way to remove metals from soil. Commercial organizations frequently employ bioleaching to extract nickel, uranium, copper, zinc, and cobalt. However, there are gold-related endeavors that make use of bio-oxidation. According to Mishra and Rhee (2010), acidophilic bacteria facilitate heavy metal bioleaching, which is particularly prevalent in electronic waste. Acidophilic organisms, which thrive in environments with acidic pH values between 2.0 and 4.0, play a crucial role in the transformation of solid waste metals into liquid form. Microbes have the ability to remove metals from iron ores, iron sulfide, and mineral concentrations. Sulfuric acid and ferric ions are produced when bacteria oxidize sulfur and iron. This process can convert insoluble metal sulfides (such zinc, copper, and nickel) into soluble metal sulfates, which aids in their extraction from solutions (Mishra & Rhee 2010).

Hybrid Technique for Metal Extraction

Despite its low cost, bioleaching is not always a reliable method for metal extraction (Frioui et al., 2017). Although chemical leaching is an effective and quick process, it does create new environmental problems. Thus, it is anticipated that the proposed hybrid method, which integrates chemical and biological leaching techniques utilizing harmless chemicals, will improve the efficiency of metal extraction. The feasibility and potential benefits of integrating electrodialysis (ED) with a complex reaction (CR) process were investigated by Frioui et al. (2017). Metal cations with similar or differing valences are difficult to identify using conventional ED techniques. Hence, an in-situ CR step was investigated as a potential means to concentrate and recover these cations. The electro-separation results showed that the Cu/Cd and Ag/Zn systems worked quite well. To improve electrodialysis's efficacy in treating wastewater containing precious metal ions, hybrid technology has been proposed.

Pyrometallurgical Process

Mingfei et al. (2016) reports that lead has been extracted from cathode ray tubes (CRTs) using the pyro-metallurgical method. According to Li and Xu (2019), this method uses sodium carbonate powder as the fusion agent, sodium sulphide as the catalytic agent, and carbon powder as the reducing agent to remove lead and other nonferrous metals with a 94% efficiency rate. The use of metallic iron in a thermal reduction technique to remove lead from CRT funnel glass was proven by Zhang et al. (2020) to have a 99 percent efficiency rate. One alternate pyrometallurgical process that Thakur and Kumar (2020) used to make lead nanoparticles was to heat a carbon reducing agent to 1000°C for two hours in a vacuum with a pressure range of 500 to 2000 Pa.

Hydrometallurgical Process

This technique concentrates metals by applying acidic solutions to the glass of the cathode-ray tube chimney. The resulting material is then separated and purified. In order to mitigate the robust adhesion between metals and glass, certain pre-treatments are carried out, as indicated by Ling & Poon (2011). These procedures involve rinsing or cleaning the CRT funnel glass using tap water and subsequently acid washing it in a solution containing 5% nitric acid for a duration of three hours. This approach enhances the solubility of the lead. Zhang et al. (2020) successfully utilized a cutting-edge hydrometallurgical technique using a potent alkaline solution, mechanical activation, and chemical leaching to achieve a remarkable 97% efficiency in the removal of lead. Yuan et al. (2015) employed hydroxyl ions generated through an ion exchange reaction as a pre-treatment to eliminate lead from CRT pipe glass. The hydrometallurgy strategy becomes unprofitable when there is a substantial increase in distance transit.

Mechanochemical Activation Process

Activation of this method or technology occurs when certain physical and chemical properties are changed, and it has various potential uses. Since high temperatures are not necessary for the mechanochemical sulfidization process, CRT funnels or chimney glass are not needed (Yuan et al., 2013). According to Yuan et al. (2012), the mechano-chemical sulfidization reaction can be initiated by grinding the cathode-ray tube glass with elemental sulfur in an environment of nitrogen gas. Using a successful method suggested by Yuan et al. (2012), lead was successfully extracted from CRT tube glass with a recovery rate of 92.5%. A significantly greater rate of recovery of 1.2% was seen in an inactivation sample by Yuan et al. (2013).

According to Yuan et al. (2012), this could lead to the detoxification of more lead-containing glass.

ECONOMIC FACTORS INFLUENCING E-WASTE RECYCLING

Economic considerations considerably influence e-waste recycling initiatives, as the expenses and financial incentives associated with recycling establish its practicability. The substantial collection, shipping, and processing expenditures can present a significant challenge, particularly in industrialized nations with stringent environmental laws and high labor costs (Baldé et al., 2017). Copper, silver, and precious metals are extracted from electronic waste, and their prices are subject to fluctuation, which impacts recycling profitability. When commodity prices are low, formal recycling operations are less financially motivated to operate (Kusch & Hills, 2017). The absence of financial support and subsidies for e-waste recycling infrastructure in numerous locations further impedes the expansion of the formal recycling sector. This, in turn, discourages the investment in advanced technology (Gupta et al., 2019). The European Environment Agency (2020) asserts that government regulations, such as EPR initiatives, are essential for motivating manufacturers to independently manage the disposal and recycling of e-waste. This increases the financial feasibility of the process.

Role of Informal Recycling Sectors and Strategies for Formalization

The informal recycling sector is the backbone of e-waste management in numerous developing nations. Using inhumane techniques like acid leaching and open burning, informal recyclers manually extract valuable metals from used devices; this process can put workers and the environment at risk (Breivik et al., 2019). Despite these dangers, e-waste scavenging is nonetheless common in the informal sector since it provides a means of subsistence for economically disadvantaged people and there is no institutional recycling infrastructure. Some formalization solutions that have been proposed to deal with these issues include training programs that aim to increase efficiency and safety, offering financial incentives, and incorporating informal workers into regulated recycling systems (Singh et al., 2020). Financial subsidies for formal recyclers, public-private partnerships, and stronger enforcement against hazardous informal processing are some of the strategies that governments and international organizations have advocated for (Wilson et al., 2015). Countries can guarantee economic sustainability and environmental safety in e-waste management by moving away from informal e-waste handling practices and toward a structured system that provides sufficient labor and environmental protections.

Opportunities in the E-Waste Economy: Resource Recovery

There are several ways in which the e-waste industry can help the economy and the environment. One promising area is resource recovery. Advanced recycling systems can efficiently recover valuable materials from electronic waste, including gold, silver, palladium, copper, and rare earth elements (Forti et al., 2020). Reducing the need for raw material extraction, urban mining—the process of recovering metals from discarded electronics—has been discovered to be more cost-effective and environmentally beneficial than traditional mining (Zeng et al., 2018). Furthermore, e-waste management sectors can benefit from the circular economy model's emphasis on refurbishment, reuse, and material recovery by way of the creation of new businesses and employment possibilities (European Environment Agency, 2020). Reducing electronic production costs and environmental pollution caused by landfill disposal of hazardous materials are two benefits that governments and enterprises investing in resource recovery can achieve (Schluep et al., 2019). The growing need for essential raw materials around the world presents an opportunity for sustainable growth in the form of long-term economic and environmental advantages offered by a burgeoning infrastructure for recycling and recovering resources from electronic trash.

CURRENT POLICY/LEGISLATION OF E-WASTE FOR ENVIRONMENTAL HEALTH EXPOSURE

Amidst the current era of industrialization, electronic trash (e-waste) has become a significant and escalating issue worldwide. The illicit transshipment and disposal of electronic waste from affluent nations to economically disadvantaged nations is a significant issue. Managing and evaluating the global trade of electronic trash is a challenging task. There are currently no regulations or measures in place to regulate or prevent the pollution produced by e-waste. The 1989 Basel Convention regulates the transportation and disposal of dangerous waste, particularly between developed and poor countries. 181 nations participated in the convention, but only 53 of them ratified the resolutions (WHO, 2009). The 1995 amendment prohibits the cross-border shipment of hazardous materials for any reason. In addition to the Stockholm Convention of 2001 and the Rotterdam Convention of 1998, there exist several international agreements that aim to restrict or completely prohibit the utilization of hazardous chemicals and persistent organic pollutants. These treaties also tackle the issue of electronic waste in poor countries. The 2008 Bali statement provided a comprehensive analysis of how waste management impacts people's health and economic livelihood. The effort, as stated by the United Nations Environment Program (2008), aimed to gather political backing and encourage investments in environmentally friendly and secure technology from both the public and commercial sectors United Nations Environment Program (2008).

The third global conference was held in Busan, Korea in 2009, with a primary focus on the impact of hazardous chemicals present in consumer products on children's health. The meeting discussed the presence of toxic chemicals in various locations such as schools, playgrounds, households, and healthcare institutions (WHO, 2009). The 2013 Geneva effort aimed to enhance children's health by minimizing their exposure to electronic trash, specifically addressing concerns over exposure and health consequences. The Pacific base cooperation also addresses comparable socioeconomic health issues related to children and electronic waste. Previously, the majority of these conferences have concentrated on investigating the primary reason for the exposure of sources to electronic waste and the resulting consequences. Moreover, the management of electronic trash and its cross-border transportation are subject to diverse national legislations and regulations. NGOs worldwide have prioritized addressing the health risks and legal aspects associated with e-waste. Their main focus has been on the management and implementation of rules (Grant et al., 2013).

The UN Industrial Development Organization (UNIDO) was granted the mandate to promote sustainable industrial practices for e-waste recycling and treatment in underdeveloped nations. The "e-waste and child health initiative" was created by the World Health Organization (WHO) in 2013 with the aim of safeguarding children's health from the harmful effects of e-waste exposure (WHO, 2013). The Geneva declaration has been proposed to address the problem of e-waste pollution and provide solutions for its proper handling, with the aim of safeguarding the health of children. The decentralized and predominantly local nature of electronic trash, combined with its mobile characteristics and covert transportation, presents challenges in terms of containment and detection of this form of pollution. Hence, it is imperative to promptly discover and address these persistent actions while simultaneously fostering a robust rapport with the local communities (WHO, 2013).

Regulatory Challenges and Policy Enforcement of E-Waste Management

The complexity of the worldwide electronic waste trade, inconsistent policies, and a lack of enforcement create a myriad of regulatory hurdles for e-waste management. The informal recycling industries, which do not adhere to any environmental or safety norms, dominate many developing nations since there is a lack of comprehensive legislation regarding the disposal of electronic trash (Forti et al., 2020). While developed nations like the EU have regulations like the WEEE Directive that prohibit the illegal export of electrical and electronic waste to developing nations disguised as used goods, enforcement of these regulations is still a problem (Lepawsky, 2018). Illegal exports continue due to loopholes and weak monitoring mechanisms in the Basel Convention on the Control of Transboundary Movements of Hazardous Wastes and Their Disposal, which tries to stop the dumping of hazardous e-waste in low-income nations (Schluep et al., 2019). Gupta et al. (2019) note that regulatory enforcement is further complicated because many countries lack the financial resources and technological infrastructure to execute comprehensive e-waste management.

In order to enforce policies effectively, there has to be improved tracking systems, stiffer sanctions for illicit e-waste trading, and more international collaboration. The European Environment Agency (2020) states that governments should enhance Extended Producer Responsibility (EPR) policies to hold producers fully responsible for the entire electronic product lifecycle, from production to disposal. Tax incentives for the development of environmentally friendly devices and increased investment in recycling facilities for electronic waste can both contribute to better compliance (Wilson et al., 2015). One possible option to monitor e-waste flows and ensure transparency in recycling procedures is to use blockchain technology or digital tracking systems (Zeng et al., 2018). To further discourage noncompliance, public education initiatives and stronger legal enforcement, including harsher penalties for illicit e-waste exports, are required. The continued hazard that e-waste poses to human health and the environment, especially in developing countries, can only be alleviated by concerted international legislative initiatives and stricter enforcement tactics.

Case Studies of Countries and Regions Effectively Managing E-Waste

The effective policies and methods that have been put into place by various nations and regions to manage their electronic trash have been met with great success. Using the Extended Producer Responsibility (EPR) model, in which manufacturers pay for recycling, Switzerland became the first country to implement a national e-waste recycling system (Widmer et al., 2005). According to the European Commission (2020), the WEEE Directive is enforced by the European Union (EU). This directive promotes the ideas of a circular economy and requires collection and recycling targets. Manufacturers in Japan are required by law to recycle large home appliances, and as a result, recycling rates for these products exceed 90% (Terazono et al., 2006).

South Korea's Producer Responsibility System (PRS) encourages environmentally friendly product design and holds electronics manufacturers to recycling targets (Yoon et al., 2016).

According to the Electronic Products Recycling Association (EPRA) in 2022, more than 100,000 metric tons of electronic trash are recycled each year in Canada as a result of the country's provincial Extended Producer Responsibility (EPR) initiatives. Among the state-level e-waste recycling programs in the United States, California's Electronic Waste Recycling Act is notable for its levy on new devices, which helps to support these programs (California Department of Toxic Substances Control, 2021). Although the majority of India's electronic trash is still handled by unofficial recyclers, the country's E-Waste (Management) Rules, 2016 brought them into the official recycling system (Singh et al., 2020). Finally, in order to deal with its increasing e-waste problem, China invested substantially in domestic recycling facilities and banned e-waste imports under its National Sword Policy (Zeng et al., 2018). In order to effectively manage electronic waste, these case studies show how crucial regulatory frameworks, producer responsibility, and consumer engagement are. Countries and regions that have done a good job of controlling their e-waste are shown in Table 3.

Table 3. Case studies of countries and regions effectively managing E-waste

Country/Region	Key Strategies in E-Waste Management	Notable Programs & Policies	Impact & Results
Switzerland	First country to implement a nationwide e-waste recycling system; Extended Producer Responsibility (EPR) approach ensures manufacturers fund recycling	SWICO Recycling System (for ICT equipment), SENS Foundation (for home appliances), retailers accept old electronics for recycling	Over 95% of electronic waste is collected and processed properly, reducing environmental pollution
European Union	The WEEE Directive (Waste Electrical and Electronic Equipment) mandates collection and recycling targets for e-waste; Circular Economy Action Plan promotes reuse and material recovery	WEEE Directive, Eco-design Directive, Right to Repair laws	EU achieved a collection rate of 55% of e-waste in 2020, promoting sustainable electronic production
Japan	Home Appliance Recycling Law mandates that manufacturers take back and recycle end-of-life products; strict disposal regulations	Law for the Promotion of Effective Utilization of Resources, mandatory recycling fees paid by consumers when purchasing electronics	Over 90% recycling rate for key appliances like TVs, air conditioners, refrigerators, and washing machines

continued on following page

Table 3. Continued

Country/Region	Key Strategies in E-Waste Management	Notable Programs & Policies	Impact & Results
South Korea	Producer Responsibility System (PRS) holds manufacturers accountable for recycling; Eco-Assurance System encourages sustainable design	Eco-Assurance System, Act on the Resource Circulation of Electrical and Electronic Equipment and Vehicles	E-waste collection rates have steadily increased, ensuring valuable materials are recovered
Canada	Provincial regulations vary, but most follow Extended Producer Responsibility (EPR), where producers fund recycling programs	Electronic Products Recycling Association (EPRA), Ontario Electronic Stewardship (OES)	Over 100,000 metric tons of e-waste recycled annually, with strong consumer participation
United States (California)	California Electronic Waste Recycling Act places fees on electronic purchases to fund e-waste recycling; Right to Repair movement gaining traction	Covered Electronic Waste (CEW) Recycling Program, E-Cycle Washington (in Washington State)	California diverts millions of pounds of e-waste annually, leading the U.S. in e-waste policy
India	Formalization of informal e-waste sector; E-Waste (Management) Rules, 2016 place responsibility on producers to ensure safe disposal	Extended Producer Responsibility (EPR) framework, collaboration with informal recycling workers	Over 300 authorized e-waste recyclers, but informal sector still handles 90% of e-waste
China	Government-backed e-waste recycling programs, strict bans on e-waste imports under National Sword Policy; investment in domestic recycling facilities	National e-waste recycling pilot programs, Green Fence Policy, National Sword Policy (banning import of e-waste)	China processes over 1.1 million tons of e-waste annually, moving towards a circular economy

PRACTICAL RECOMMENDATIONS FOR E-WASTE MANAGEMENT

Governments are crucial in formulating effective policies for electronic waste management through the enforcement of regulations, the implementation of incentive programs for sustainable behaviors, and the promotion of international collaboration. The European Environment Agency (2020) asserts that manufacturers must be responsible for every phase of an electronic product's lifecycle, from production to disposal, and recommends that Extended Producer Responsibility (EPR) rules be reinforced to attain this objective. Governments should invest in systems for the collection and recycling of electronic waste, particularly in impoverished nations lacking sufficient formal recycling facilities (Forti et al., 2020). Schuep et al. (2019) advocate for providing financial incentives to enterprises that use sustainable methods in electronic manufacture and recycling to facilitate the shift towards a circular economy. Moreover, the Basel Convention and other international agreements must be implemented more rigorously to mitigate unlawful e-waste disposal and promote responsible international commerce of used electronics (Lepawsky, 2018). Effective e-waste disposal and the environmental hazards associated with improper recycling techniques can be enhanced by collaboration between governments, educational institutions, and various organizations to inform the public.

Sustainable methods can mitigate e-waste, although both industry and consumers significantly contribute to the issue. Zeng et al. (2018) assert that manufacturers of electronic products must comply with eco-design principles by enhancing the repairability, upgradeability, and recyclability of their products. Should corporations establish take-back programs, consumers may recycle their obsolete equipment rather than discarding it. Consumers should be encouraged to donate or resell functioning devices instead of discarding them. They should select brands that advocate sustainability (Gupta et al., 2019). Industries' investment in urban mining technologies can diminish their dependence on raw material extraction by facilitating the recovery of valuable elements from discarded electronics (Baldé et al., 2017). Ultimately, to foster responsible e-waste management and ensure a balance between technological advancement and environmental sustainability, enhanced collaboration among governments, companies, and civil society is essential.

CONCLUSION

The exponential growth of electronic trash is a problem for almost every country. The need for efficient and ecologically sound methods of dealing with electronic waste has grown in recent times. There are numerous unintended consequences associated with electronic waste. This is detrimental to the ecosystem and has the potential to adversely affect human health. Numerous countries persist in giving higher importance to the administration of municipal solid waste rather than addressing the continuously expanding problem of electronic waste. Without appropriate legislation, the majority of impoverished countries manage and treat a substantial amount of electronic waste in an informal manner (Fent, 2004). Inadequate attention from the administration and obsolete technology poses significant challenges in the establishment of official e-waste recycling plants. The acquisition of industry accreditation for the appropriate recycling of electronic waste is a significant obstacle for recyclers. Within this framework, the shift from informal actors to small company owners has the potential to be a pivotal moment. Reduced incidence of injuries or fatalities caused by toxic leachates will facilitate progress towards achieving the worldwide sustainable development targets for 2030 (Dagan et al., 2007).

Effective public education initiatives yield advantages for both suppliers and consumers. For the long-term management of the e-waste problem, it is essential to adopt an integrated approach that includes both the organized and informal recycling sectors. Most non-repairable electronic and electrical devices are disposed of in unlined landfills and informal dumpsites together with other types of municipal solid waste. Furthermore, individuals dispose of a substantial amount of refuse into various public spaces such as rivers, ponds, fields, roadsides, irrigation canals, and even private residences and workshops. Open dumps, landfills, and dumpsites facilitate the infiltration of hazardous substances into the soil, therefore producing contamination in water sources. The water used by people in their homes poses a threat to both humans and the environment (Fent, 2004).

A series of chemical investigations were undertaken by Vannie et al. (2006) to determine the potential leaching of chemicals from electronic waste. The intricate chemical compositions included in electronic waste make toxicity tests a valuable method for assessing the potential environmental hazards associated with this contamination (Dagan et al., 2007). Chronic respiratory and neurological diseases present a significant risk to the local community. One potential approach to mitigate the emission of potentially hazardous elements during the treatment of electronic waste is to decrease the concentration of elements with low melting points and high vapor pressure, such as Sn, Pb, Zn, etc (Khanna et al., 2020b).

Khanna et al. (2020a, b) provided supportive evidence in ensuring the elimination of ineffective and outdated methods in electronic waste management. It is imperative to always remember that there is no panacea for every problem. Each HM requires a distinct strategy for elimination. Moreover, confronting the problem of electronic waste requires a dual strategy. Both the government and public bear the obligation to cooperate in order to avert the escalation of issues. Regarding efficiency, environmental sustainability, and cost-effectiveness, bio-metallurgy surpasses hydrometallurgy and pyrometallurgy in the recovery of valuable and basic metals from electronic waste. Maximizing the energy and valuable resources from outdated equipment necessitates the implementation of comprehensive and holistic strategies (Khanna et al., 2020a).

REFERENCES

Acquah, A. A., D'Souza, C., Martin, B., Arko-Mensah, J., Nti, A. A., Kwarteng, L., Takyi, S., Quakyi, I. A., Robins, T. G., & Fobil, J. N. (2019). Processes and Challenges associated with informal electronic waste recycling at Agbogbloshie, a suburb of Accra, Ghana. *Proceedings of the Human Factors and Ergonomics Society ... Annual Meeting. Human Factors and Ergonomics Society.Annual meeting, 63*(1), 938–942. DOI: 10.1177/1071181319631219

Adanu, S. K., Gbedemah, S. F., & Attah, M. K. (2020). Challenges of adopting sustainable technologies in e-waste management at Agbogbloshie, Ghana. *Heliyon, 6*(8), 04548. DOI: 10.1016/j.heliyon.2020.e04548 PMID: 32802976

Agyei-Mensah, S., & Oteng-Ababio, M. (2012). Perceptions of health and environmental impacts of e-waste management in Ghana. *International Journal of Environmental Health Research, 22*(6), 500–517. DOI: 10.1080/09603123.2012.667795 PMID: 22428915

Ahirwar, R., & Tripathi, A. K. (2021). E-waste management: A review of recycling process, environmental and occupational health hazards, and potential solutions. *Environmental Nanotechnology, Monitoring & Management, 15*, 100409. DOI: 10.1016/j.enmm.2020.100409

Akram, R., Fahad, S., Hashmi, M. Z., Wahid, A., Adnan, M., Mubeen, M., & Nasim, W. (2019). Trends of electronic waste pollution and its impact on the global environment and ecosystem. *Environmental Science and Pollution Research International, 26*(17), 16923–16938. DOI: 10.1007/s11356-019-04998-2 PMID: 31025281

Althaf, S., Babbitt, C. W., & Chen, R. (2021). The evolution of consumer electronic waste in the United States. *Journal of Industrial Ecology, 25*(3), 693–706. DOI: 10.1111/jiec.13074

Amoabeng Nti, A. A., Arko-Mensah, J., Botwe, P. K., Dwomoh, D., Kwarteng, L., Takyi, S. A., Acquah, A. A., Tettey, P., Basu, N., Batterman, S., Robins, T. G., & Fobil, J. N. (2020). Effect of Particulate Matter Exposure on Respiratory Health of e-Waste Workers at Agbogbloshie, Accra, Ghana. *International Journal of Environmental Research and Public Health, 17*(9), 3042. DOI: 10.3390/ijerph17093042 PMID: 32349371

Amoabeng Nti, A. A., Arko-Mensah, J., Botwe, P. K., Dwomoh, D., Kwarteng, L., Takyi, S. A., Acquah, A. A., Tettey, P., Basu, N., Batterman, S., Robins, T. G., & Fobil, J. N. (2020). Effect of Particulate Matter Exposure on Respiratory Health of e-Waste Workers at Agbogbloshie, Accra, Ghana. *International Journal of Environmental Research and Public Health, 17*(9), 3042. DOI: 10.3390/ijerph17093042 PMID: 32349371

Angaye, T. C. N., Godbless, N. O., & Ogidi, O. I. (2019). Environmental and health risk assessment of particulate matter associated with sporting activities in Yenagoa metropolis, Nigeria. *General Internal medicine and clinical. Innovations, 4*(3), 1–3. DOI: 10.15761/GIMCI.1000180

Angaye, T. C. N., & Ogidi, O. I. (2018). The environmental impacts of electronic wastes in Bayelsa State, Nigeria. *Med Crave Online Journal of Toxicology, 4*(5), 331–332.

Asampong, E., Dwuma-Badu, K., Stephens, J., Srigboh, R., Neitzel, R., Basu, N., & Fobil, J. N. (2015). Health seeking behaviours among electronic waste workers in Ghana. *BMC Public Health, 15*(1), 1–9. DOI: 10.1186/s12889-015-2376-z PMID: 26474859

Awasthi, A. K., Hasan, M., Mishra, Y. K., Pandey, A. K., Tiwary, B. N., Kuhad, R. C., & Thakur, V. K. (2019b). Environmentally sound system for E-waste: Biotechnological perspectives. *Current Research in Biotechnology, 1*, 58–64. DOI: 10.1016/j.crbiot.2019.10.002

Awasthi, A. K., Li, J., Koh, L., & Ogunseitan, O. A. (2019a). Circular economy and electronic waste. *Nature Electronics, 2*(3), 86–89. DOI: 10.1038/s41928-019-0225-2

Awasthi, A. K., Wang, M., Awasthi, M. K., Wang, Z., & Li, J. (2018). Environmental pollution and human body burden from improper recycling of e-waste in China: A short-review. *Environmental Pollution, 243*, 1310–1316. DOI: 10.1016/j.envpol.2018.08.037 PMID: 30268981

Baldé, C. P., Forti, V., Gray, V., Kuehr, R., & Stegmann, P. (2017). *The global e-waste monitor 2017: Quantities, flows, and resources*. United Nations University. https://collections.unu.edu/eserv/UNU:6349/PDF

Bao, S., Pan, B., Wang, L., Cheng, Z., Liu, X., Zhou, Z., & Nie, X. (2020). Adverse effects in Daphnia magna exposed to e-waste leachate: Assessment based on life trait changes and responses of detoxification-related genes. *Environmental Research, 188*, 109821. DOI: 10.1016/j.envres.2020.109821 PMID: 32798945

Berkhout, F., & Hertin, J. (2004). De-materialising and re-materialising: Digital technologies and the environment. *Futures*, *36*(8), 903–920. DOI: 10.1016/j.futures.2004.01.003

Bigum, M., Damgaard, A., Scheutz, C., & Christensen, T. H. (2017). Environmental impacts and resource losses of incinerating misplaced household special wastes (WEEE, batteries, ink cartridges and cables). *Resources, Conservation and Recycling*, *122*, 251–260. DOI: 10.1016/j.resconrec.2017.02.013

Breivik, K., Armitage, J. M., Wania, F., & Jones, K. C. (2019). Tracking the global generation and exports of e-waste. *Environmental Science & Technology*, *53*(13), 7425–7435. DOI: 10.1021/acs.est.9b00244 PMID: 25007134

Burns, K. N., Sayler, S. K., & Neitzel, R. L. (2019). Stress, health, noise exposures, and injuries among electronic waste recycling workers in Ghana. *Journal of Occupational Medicine and Toxicology (London, England)*, *14*(1), 1. DOI: 10.1186/s12995-018-0222-9 PMID: 30647766

California Department of Toxic Substances Control. (2021). *Electronic waste recycling act of 2003*. Author.

Chen, A., Dietrich, K. N., Huo, X., & Ho, S. M. (2011). Developmental neurotoxicants in e waste: An emerging health concern. *Environmental Health Perspectives*, *119*(4), 431–438. DOI: 10.1289/ehp.1002452 PMID: 21081302

Chen, J., Huang, Y., Li, G., An, T., Hu, Y., & Li, Y. (2016). VOCs elimination and health risk reduction in e-waste dismantling workshop using integrated techniques of electrostatic precipitation with advanced oxidation technologies. *Journal of Hazardous Materials*, *302*, 395–403. DOI: 10.1016/j.jhazmat.2015.10.006 PMID: 26489914

Clarkson, T. W., & Magos, L. (2006). The toxicology of mercury and its chemical compounds. *Critical Reviews in Toxicology*, *36*(8), 609–662. DOI: 10.1080/10408440600845619 PMID: 16973445

CMU (Carnegie Mellon University). (2011). *EIO-LCA: Free*. Fast, Easy Life Cycle Assessment.

Cudjoe, D., & Acquah, P. M. (2021). Environmental impact analysis of municipal solid waste incineration in African countries. *Chemosphere*, *265*, 129186. DOI: 10.1016/j.chemosphere.2020.129186 PMID: 33307505

Dagan, R., Dubey, B., Bitton, G., & Townsend, T. (2007). Aquatic toxicity of leachates generated from electronic devices. *Archives of Environmental Contamination and Toxicology*, *53*(2), 168–173. DOI: 10.1007/s00244-006-0205-1 PMID: 17549543

Davis, J. M., Akese, G., & Garb, Y. (2019). Beyond the pollution haven hypothesis: Where and why do e-waste hubs emerge and what does this mean for policies and interventions? *Geoforum*, *98*, 36–45. DOI: 10.1016/j.geoforum.2018.09.020

de Souza, R. G., Clímaco, J. C., Sant'Anna, A. P., Rocha, T. B., do Valle, R. A., & Quelhas, O. L. (2016). Sustainability assessment and prioritisation of e-waste management options in Brazil. *Waste Management (New York, N.Y.)*, *57*, 46–56. DOI: 10.1016/j.wasman.2016.01.034 PMID: 26852754

Electronic Products Recycling Association (EPRA). (2022). *Annual report on electronic waste recycling in Canada*. Author.

European Commission. (2020). *Waste electrical and electronic equipment (WEEE) directive*. https://ec.europa.eu/environment/topics/waste-and-recycling/weee_en

European Environment Agency. (2020). *E-waste management in Europe: A circular economy perspective*. https://www.eea.europa.eu/publications

Fent, K. (2004). Ecotoxicological effects at contaminated sites. *Toxicology*, *205*(3), 223–240. DOI: 10.1016/j.tox.2004.06.060 PMID: 15464632

Forti, V., Baldé, C. P., Kuehr, R., & Bel, G. (2020). *The global e-waste monitor 2020: Quantities, flows and the circular economy potential*. United Nations University. https://collections.unu.edu/eserv/UNU:7737/PDF

Frioui, S., Oumeddour, R., & Lacour, S. (2017). Highly selective extraction of metal ions from dilute solutions by hybrid electrodialysis technology. *Separation and Purification Technology*, *174*, 264–274. DOI: 10.1016/j.seppur.2016.10.028

Fujimori, T., Itai Goto, A., Asante, K. A., Otsuka, M., Takahashi, S., & Tanabe, S. (2016). Interplay of metals and bromine with dioxin-related compounds concentrated in e-waste open burning soil from Agbogbloshie in Accra, Ghana. *Environmental Pollution*, *201*, 155–163. DOI: 10.1016/j.envpol.2015.11.031 PMID: 26686056

Ganguly, R. (2016). E-Waste management in India-An overview. *International Journal of Earth Sciences and Engineering*, *9*(2), 574–588.

Gangwar, C., Choudhari, R., Chauhan, A., Kumar, A., Singh, A., & Tripathi, A. (2019). Assessment of air pollution caused by illegal e-waste burning to evaluate the human health risk. *Environment International*, *125*, 191–199. DOI: 10.1016/j.envint.2018.11.051 PMID: 30721825

Garlapati, V. K. (2016). E-waste in India and developed countries: Management, recycling, business and biotechnological initiatives. *Renewable & Sustainable Energy Reviews*, *54*(C), 874–881. DOI: 10.1016/j.rser.2015.10.106

Ghimire, H., & Ariya, P. A. (2020). E-wastes: Bridging the knowledge gaps in global production budgets, composition, recycling and sustainability implications. *Sustainable Chemistry*, *1*(2), 154–182. DOI: 10.3390/suschem1020012

Golev, A., Schmeda-Lopez, D. R., Smart, S. K., Corder, G. D., & McFarland, E. W. (2016). Where next on e-waste in Australia? *Waste Management (New York, N.Y.)*, *58*, 348–358. DOI: 10.1016/j.wasman.2016.09.025 PMID: 27687078

Grant, K., Goldizen, F. C., Sly, P. D., Brune, M. N., Neira, M., van den Berg, M., & Norman, R. E. (2013). Health consequences of exposure to e-waste: A systematic review. *The Lancet. Global Health*, *1*(6), e350–e361. DOI: 10.1016/S2214-109X(13)70101-3 PMID: 25104600

Gupta, M., Sinha, S., & Chaturvedi, A. (2019). Economic feasibility of e-waste recycling in India: Policy implications. *Waste Management & Research*, *37*(4), 394–402. DOI: 10.1177/0734242X19831556 PMID: 30736727

Gupta, S. (2011). E-waste management: Teaching how to reduce, reuse and recycle for sustainable development-need of some educational strategies. *Journal of Education and Practice*, *2*(3), 2222–1735.

Heacock, M., Kelly, C. B., Asante, K. A., Birnbaum, L. S., Bergman, Å. L., Brune, M. N., Buka, I., Carpenter, D. O., Chen, A., Huo, X., Kamel, M., Landrigan, P. J., Magalini, F., Diaz-Barriga, F., Neira, M., Omar, M., Pascale, A., Ruchirawat, M., Sly, L., & Suk, W. A. (2016). E-Waste and Harm to Vulnerable Populations: A Growing Global Problem. *Environmental Health Perspectives*, *124*(5), 550–555. DOI: 10.1289/ehp.1509699 PMID: 26418733

Ilankoon, I. M. S. K., Ghorbani, Y., Chong, M. N., Herath, G., Moyo, T., & Petersen, J. (2018). E-waste in the international context–A review of trade flows, regulations, hazards, waste management strategies and technologies for value recovery. *Waste Management (New York, N.Y.)*, *82*, 258–275. DOI: 10.1016/j.wasman.2018.10.018 PMID: 30509588

Jayaraman, K., Vejayon, S., Raman, S., & Mostafiz, I. (2019). The proposed ewaste management model from the conviction of individual laptop disposal practices-An empirical study in Malaysia. *Journal of Cleaner Production*, *208*, 688–696. DOI: 10.1016/j.jclepro.2018.10.125

Jiang, B., Adebayo, A., Jia, J., Xing, Y., Deng, S., Guo, L., & Zhang, D. (2019). Impacts of heavy metals and soil properties at a Nigerian e-waste site on soil microbial community. *Journal of Hazardous Materials*, *362*, 187–195. DOI: 10.1016/j.jhazmat.2018.08.060 PMID: 30240992

Jiang, L., Cheng, Z., Zhang, D., Song, M., Wang, Y., Luo, C., & Zhang, G. (2017). The influence of e-waste recycling on the molecular ecological network of soil microbial communities in Pakistan and China. *Environmental Pollution, 231*, 173–181. DOI: 10.1016/j.envpol.2017.08.003 PMID: 28800486

Jibiri, J. D. (2015). Reduce Reuse Recycle Behavioural Intention Model in Higher Education Institution Accommodation (Doctoral dissertation, Universiti Teknologi Malaysia).

Kaya, M. (2019). *Electronic Waste and Printed Circuit Board Recycling Technologies*. Springer. DOI: 10.1007/978-3-030-26593-9

Khan, J., Kumar, A., Giri, A., Pal, D. B., Tripathi, A., & Giri, D. D. (2021). Impact of electronic waste pollutants on underground water. In *Groundwater Geochemistry* (pp. 265–281). Pollution and Remediation Methods. DOI: 10.1002/9781119709732.ch13

Khanna, R., Mukherjee, P. S., & Park, M. (2020a). A critical assessment on resource recovery from electronic waste: Impact of mechanical pre-treatment. *Journal of Cleaner Production, 53*, 122319. DOI: 10.1016/j.jclepro.2020.122319

Khanna, R., Saini, R., Park, M., Ellamparuthy, G., Biswal, S. K., & Mukherjee, P. S. (2020b). Factors influencing the release of potentially toxic elements (PTEs) during thermal processing of electronic waste. *Waste Management (New York, N.Y.), 105*, 414–424. DOI: 10.1016/j.wasman.2020.02.026 PMID: 32126369

Kiddee, P., Naidu, R., Wong, M. H., Hearn, L., & Muller, J. F. (2014). Field investigation of the quality of fresh and aged leachates from selected landfills receiving e-waste in an arid climate. *Waste Management (New York, N.Y.), 34*(11), 2292–2304. DOI: 10.1016/j.wasman.2014.06.018 PMID: 25070222

Kusch, S., & Hills, C. D. (2017). The link between e-waste and GDP—New insights from data analysis. *Resources, Conservation and Recycling, 123*, 115–123. DOI: 10.1016/j.resconrec.2016.12.005

Ladou, J., & Lovegrove, S. (2008). Export of electronics equipment waste. *International Journal of Occupational and Environmental Health, 14*(1), 1–10. DOI: 10.1179/oeh.2008.14.1.1 PMID: 18320726

Landrigan, P. J., Fuller, R., Fisher, S., Suk, W. A., Sly, P., Chiles, T. C., & Bose-O'Reilly, S. (2019). Pollution and children's health. *The Science of the Total Environment, 650*(Pt 2), 2389–2394. DOI: 10.1016/j.scitotenv.2018.09.375 PMID: 30292994

Lepawsky, J. (2018). *Reassembling rubbish: Worlding electronic waste*. MIT Press. DOI: 10.7551/mitpress/11111.001.0001

Li, W., & Achal, V. (2020). Environmental and health impacts due to e-waste disposal in China - A review. *The Science of the Total Environment*, *737*, 139745. DOI: 10.1016/j.scitotenv.2020.139745 PMID: 32516663

Lin, M., Tang, J., Ma, S., Yu, Y., Li, G., Fan, R., Mai, B., & An, T. (2020). Insights into biomonitoring of human exposure to polycyclic aromatic hydrocarbons with hair analysis: A case study in e-waste recycling area. *Environment International*, *136*, 105432. DOI: 10.1016/j.envint.2019.105432 PMID: 31884415

Ling, T. C., & Poon, C. S. (2011). Utilization of recycled glass derived from cathode ray tube glass as fine aggregate in cement mortar. *Journal of Hazardous Materials*, *192*(2), 451–456. DOI: 10.1016/j.jhazmat.2011.05.019 PMID: 21705136

Liu, Y., Luo, X., Zeng, Y., Tu, W., Deng, M., Wu, Y., & Mai, B. (2020). Species-specific biomagnification and habitat-dependent trophic transfer of halogenated organic pollutants in insect-dominated food webs from an e-waste recycling site. *Environment International*, *138*, 105674. DOI: 10.1016/j.envint.2020.105674 PMID: 32234680

Liu, Y., Zhang, J., Yang, X., Yang, W., Chen, Y., & Wang, C. (2020). Efficient recovery of valuable metals from waste printed circuit boards by microwave pyrolysis. *Chinese Journal of Chemical Engineering*, *13*, 24–36. DOI: 10.1016/j.cjche.2020.11.008

Lu, C., Zhang, L., Zhong, Y., Ren, W., Tobias, M., Mu, Z., & Xue, B. (2015). An overview of e-waste management in China. *Journal of Material Cycles and Waste Management*, *17*(1), 1–12. DOI: 10.1007/s10163-014-0256-8

Maphosa, V., & Maphosa, M. (2020). E-waste management in Sub-Saharan Africa: A systematic literature review. *Cogent Business & Management*, *7*(1), 1814503. DOI: 10.1080/23311975.2020.1814503

Mihai, F. C. (2020). Electronic waste management in Romania: pathways for sustainable practices. Handbook of Electronic Waste Management Butterworth-Heinemann, 533–551. DOI: 10.1016/B978-0-12-817030-4.00024-3

Mingfei, X., Yaping, W., Jun, L., & Hua, X. (2016). Lead recovery and glass microspheres synthesis from waste CRT funnel glasses through carbon thermal reduction enhanced acid leaching process. *Journal of Hazardous Materials*, *305*, 51–58. DOI: 10.1016/j.jhazmat.2015.11.032 PMID: 26642446

Mishra, D., & Rhee, Y.H. (2010). Current research trends of microbiological leaching for metal recovery from industrial wastes. *Current research technology and education topics in applied microbiology and microbial biotechnology* 2, 1289-1292.

Namias, J. (2013). *The future of electronic waste recycling in the United States: obstacles and domestic solutions*. Columbia University.

Needhidasan, S., Samuel, M., & Chidambaram, R. (2014). Electronic waste –an emerging threat to the environment of urban India. *Journal of Environmental Health Science & Engineering*, *12*(1), 1–9. DOI: 10.1186/2052-336X-12-36 PMID: 24393372

Nie, X., Fan, C., Wang, Z., Su, T., Liu, X., & An, T. (2015). Toxic assessment of the leachates of paddy soils and river sediments from e-waste dismantling sites to microalga, Pseudokirchneriella subcapitata. *Ecotoxicology and Environmental Safety*, *111*, 168–176. DOI: 10.1016/j.ecoenv.2014.10.012 PMID: 25450930

Nnorom, I. C., & Odeyingbo, O. A. (2020). Electronic waste management practices in Nigeria. In *Handbook of Electronic Waste Management* (pp. 323–354). Butterworth-Heinemann. DOI: 10.1016/B978-0-12-817030-4.00014-0

Nwagwu, W., & Okuneye, M. (2016). Awareness and attitudes of small-scale information technology business operators in Lagos, Nigeria toward E-waste hazards. *Journal of Global Information Technology Management*, *19*(4), 267–282. DOI: 10.1080/1097198X.2016.1246934

Odili, J. B., Noraziah, A., Ambar, R., Wahab, M. H., & Fakheraldin, M. A. I. (2018). E-waste generation, awareness and management in third world countries: Prospects and challenges. The Eurasia Proceedings of Educational and Social Sciences, 9, 349–353.

Ogidi, O. I., Ajoko, I. T., & Tawariowei, A. M. (2024a). Sustainable Application of Genetic Ecotoxicological Techniques. In *Biomonitoring For Environmental Sustainability. In: Biomonitoring of Pollutants in the Global South*. Springer Nature Singapore Pte Ltd. DOI: 10.1007/978-981-97-1658-6_19

Ogidi, O. I., Angaye, T. C. N., & Carbom, H. E. (2024c). Microorganisms in the monitoring of terrestrial and aquatic ecosystems: applications, challenges and future prospects. In *Biomonitoring of Pollutants in the Global South*. Springer Nature Singapore Pte Ltd. DOI: 10.1007/978-981-97-1658-6_18

Ogidi, O. I., Enenebeaku, U. E., Okara, E., & Elumelu, S. A. (2021b). Toxic Metal Profiles, Carcinogenic and Non-Carcinogenic Human Health Risk Assessment of Some Locally Produced Beverages in Nigeria. *Journal of Toxicology and Risk Assessment*, *7*(1), 1–8. DOI: 10.23937/2572-4061.1510039

Ogidi, O. I., Njoku, C. O., Onimisi, A. M., & Onomedjeke, P. E. (2021a). Evaluation of Heavy Metal Contents and Potential Human Health Risk Assessment of Selected Canned Sardines Fish Sold in Yenagoa, Nigeria. *Archives of Ecotoxicology*, *3*(2), 39–43. DOI: 10.36547/ae.2021.3.2.39-43

Ogidi, O. I., Onwuagba, C. G., & Richard-Nwachukwu, N. (2024b). Biomonitoring Tools, Techniques and Approaches For Environmental Assessments. In *Biomonitoring of Pollutants in the Global South*. Springer Nature Singapore Pte Ltd. DOI: 10.1007/978-981-97-1658-6_7

Ogidi, O. I., & Tawariowei, A. M. (2023). Endocrine Disrupting Chemicals and their Role in Cancer-A review. *Molecular Sciences And Applications*, *3*, 5–14. DOI: 10.37394/232023.2023.3.2

Okereke, J. N., Ogidi, O. I., & Obasi, K. O. (2016). Environmental and Health Impact of Industrial Wastewater Effluents in Nigeria- A Review. *International Journal of Advanced Research in Biological Science*, *3*(6), 55–67.

Peng, H., Michael, S., Reid, X., & Le, C. (2015). Consumption of rice and fish in an electronic waste recycling area contributes significantly to total daily intake of mercury. *Journal of Environmental Sciences (China)*, *38*, 83–86. DOI: 10.1016/j.jes.2015.10.003 PMID: 26702970

Purchase, D., Abbasi, G., Bisschop, L., Chatterjee, D., Ekberg, C., Ermolin, M., & Wong, M. H. (2020). Global occurrence, chemical properties, and ecological impacts of E-wastes (IUPAC Technical Report). *Pure and Applied Chemistry*, *92*(11), 1733–1767. DOI: 10.1515/pac-2019-0502

Rautela, R., Arya, S., Vishwakarma, S., Lee, J., Kim, K.H., & Kumar, S. (2021). Ewaste management and its effects on the environment and human health. *Science of the Total Environment 773*, 145623. .2021.145623DOI: 10.1016/j.scitotenv

Ravindra, K., & Mor, S. (2019). E-waste generation and management practices in Chandigarh,India and economic evaluation for sustainable recycling. *Journal of Cleaner Production*, *221*, 286–294. DOI: 10.1016/j.jclepro.2019.02.158

Richard, E. B. (2009). Children's Exposure to Elemental Mercury: A National Review of Exposure Events. The Agency for Toxic Substances and Disease Registry and Centres for Disease Control and Prevention Mercury Workgroup.

Schluep, M., Hagelüken, C., Kuehr, R., Magalini, F., Maurer, C., Meskers, C., & Wang, F. (2019). Sustainable e-waste management: A comprehensive overview. *Environmental Science & Technology*, *53*(15), 8871–8881. DOI: 10.1021/acs.est.8b06916

Shaikh, S., Thomas, K., & Zuhair, S. (2020a). An exploratory study of e-waste creation and disposal: Upstream considerations. *Resources, Conservation and Recycling, 155*, 104662. DOI: 10.1016/j.resconrec.2019.104662

Shaikh, S., Thomas, K., Zuhair, S., & Magalini, F. (2020b). A cost–benefit analysis of the downstream impacts of e-waste recycling in Pakistan. *Waste Management (New York, N.Y.), 118*, 302–312. DOI: 10.1016/j.wasman.2020.08.039 PMID: 32919349

Shamim, A., Mursheda, A. K., & Rafiq, I. (2015). E-waste trading impact on public health and ecosystem services in developing countries. *International Journal of Waste Resources, 5*(4), 1–8. DOI: 10.4172/2252-5211.1000188

Sharma, S., Wakode, S., Sharma, A., Nair, N., Dhobi, M., Wani, M. A., & Pottoo, F. H. (2020). Effect of environmental toxicants on neuronal functions. *Environmental Science and Pollution Research International, 5*(36), 1–16. DOI: 10.1007/s11356-020-10950-6 PMID: 32996088

Sheng, P. P., & Etsell, T. H. (2007). Recovery of gold from computer circuit board scrap using aqua regia. *Waste Management & Research, 25*(4), 380–383. DOI: 10.1177/0734242X07076946 PMID: 17874665

Shittu, O. S., Williams, I. D., & Shaw, P. J. (2020). Global E-waste management: Can WEEE make a difference? A review of e-waste trends, legislation, contemporary issues and future challenges. *Waste Management (New York, N.Y.), 7*, 549–563. PMID: 33308953

Singh, N., Duan, H., & Tang, Y. (2020). E-waste recycling in India: Bridging the gap between formal and informal sectors. *Resources, Conservation and Recycling, 161*, 104915. DOI: 10.1016/j.resconrec.2020.104915

Soetrisno, F. N., & Delgado-Saborit, J. M. (2020). Chronic exposure to heavy metals from informal e-waste recycling plants and children's attention, executive function and academic performance. *The Science of the Total Environment, 717*, 137099. DOI: 10.1016/j.scitotenv.2020.137099 PMID: 32092800

Suja, F., AbdulRahman, R., Yusof, A., Masdar, & M.S. (2014). E-waste management scenarios in Malaysia. *Journal of Waste Management, 20*, 14.

Terazono, A., Murakami, S., Abe, N., Inanc, B., Moriguchi, Y., Sakai, S., Kojima, M., Yoshida, A., Li, J., Yang, J., Wong, M. H., Jain, A., Kim, I.-S., Peralta, G. L., Lin, C.-C., Mungcharoen, T., & Williams, E. (2006). Current status and research on E-waste issues in Asia. *Journal of Material Cycles and Waste Management, 8*(1), 1–12. DOI: 10.1007/s10163-005-0147-0

Tetteh, D., & Lengel, L. (2017). The urgent need for health impact assessment: Proposing a transdisciplinary approach to the e-waste crisis in sub-Saharan Africa. *Global Health Promotion*, *24*(2), 35–42. DOI: 10.1177/1757975916686926 PMID: 28353403

Thakur, P., & Kumar, S. (2020). Metallurgical processes unveil the unexplored "sleeping mines" e-waste: A review. *Environmental Science and Pollution Research International*, *27*(26), 32359–32370. DOI: 10.1007/s11356-020-09405-9 PMID: 32533494

Tian, J., & Chen, M. (2016). Assessing the economics of processing end-of-life vehicles through manual dismantling. *Waste Management (New York, N.Y.)*, *56*, 384–395. DOI: 10.1016/j.wasman.2016.07.046 PMID: 27499291

Torres, R., Segura-Bailón, B., & Lapidus, G. T. (2018). Effect of temperature on copper, iron and lead leaching from e-waste using citrate solutions. *Waste Management (New York, N.Y.)*, *71*, 420–425. DOI: 10.1016/j.wasman.2017.10.029 PMID: 29102357

UNEP. (2008). Bali declaration on waste management for human health and livelihood. Ninth meeting of the conference of the parties to the basel convention on the control of transboundary movements of hazardous wastes and their disposal, 23–27 June 2008, Bali, Indonesia. United Nations Environment Programme. Available http://www.basel.int/Portals/4/Basel%20Convention/docs/meetings/cop/cop9/bali-declaration/BaliDeclaration

Vann, K. N., Musson, S. E., & Townsend, T. G. (2006). Factors affecting TCLP lead leachability from computer CPUs. *Waste Management (New York, N.Y.)*, *26*(3), 293–298. DOI: 10.1016/j.wasman.2005.06.016 PMID: 16169210

Wang, S., Dong, D., Li, P., Hua, X., Zheng, N., Sun, S., & Li, X. (2020a). Mercury concentration and fatty acid composition in muscle tissue of marine fish species harvested from Liaodong Gulf: An intelligence quotient and coronary heart disease risk assessment. *The Science of the Total Environment*, *726*, 138586. DOI: 10.1016/j.scitotenv.2020.138586 PMID: 32481211

WHO. (2009). Busan pledge for action on children's health and environment. The 3rd WHO international conference of children's health and the environment, 7–10 June 2009, Busan, Republic of Korea. World Health Organization. Available https://www.who.int/ifcs/meetings/3ce

WHO. (2013). E-waste and child health initiative. Geneva [Online]. Available https://www.who.int/ceh/risks/ewaste/en/

Widmer, R., Oswald-Krapf, H., Sinha-Khetriwal, D., Schnellmann, M., & Böni, H. (2005). Global perspectives on e-waste. *Environmental Impact Assessment Review*, *25*(5), 436–458. DOI: 10.1016/j.eiar.2005.04.001

Wilson, D. C., Velis, C., & Cheeseman, C. (2015). Role of informal sector recycling in waste management in developing countries. *Habitat International*, *30*(4), 797–808. DOI: 10.1016/j.habitatint.2005.09.005

Yan, H., Li, Q., Yuan, Z., Jin, S., & Jing, M. (2019). Research progress of mercury bioaccumulation in the aquatic food chain, China: A review. *Bulletin of Environmental Contamination and Toxicology*, *102*(5), 612–620. DOI: 10.1007/s00128-019-02629-7 PMID: 31101929

Yoon, H., Jang, Y. C., & Dodson, J. (2016). Extended producer responsibility and product stewardship for electronic waste management: Policy evolution in South Korea. *Waste Management (New York, N.Y.)*, *56*, 11–19. DOI: 10.1016/j.wasman.2016.06.014

Yu, E. A., Akormedi, M., Asampong, E., Meyer, C. G., & Fobil, J. N. (2017). Informal processing of electronic waste at Agbogbloshie, Ghana: Workers' knowledge about associated health hazards and alternative livelihoods. *Global Health Promotion*, *24*(4), 90–98. DOI: 10.1177/1757975916631523 PMID: 27271535

Yuan, W., Li, J., Zhang, Q., & Saito, F. (2012). Innovated application of mechanical activation to separate lead from scrap cathode ray tube funnel glass. *Environmental Science & Technology*, *46*(7), 4109–4114. DOI: 10.1021/es204387a PMID: 22385285

Yuan, W., Li, J., Zhang, Q., Saito, F., & Yang, B. (2013). Lead recovery from cathode ray tube funnel glass with mechanical activation. *Journal of the Air & Waste Management Association*, *63*(1), 2–10. DOI: 10.1080/10962247.2012.711796 PMID: 23447859

Yuan, W., Meng, W., Li, J., Zhang, C., Song, Q., Bai, J., & Li, Y. (2015). Lead recovery from scrap cathode ray tube funnel glass by hydrothermal sulphidisation. *Waste Management & Research*, *33*(10), 930–936. DOI: 10.1177/0734242X15597777 PMID: 26264931

Zeng, X., Yang, C., Chiang, J. F., & Li, J. (2018). Innovative e-waste recycling technologies: A review. *Sustainable Materials and Technologies*, *15*, 1–12. DOI: 10.1016/j.susmat.2018.02.001

Zhang, B., He, Y., Zhu, H., Huang, X., Bai, X., Kannan, K., & Zhang, T. (2020). Concentrations of bisphenol A and its alternatives in paired maternal-fetal urine, serum and amniotic fluid from an e-waste dismantling area in China. *Environment International*, *136*, 105407. DOI: 10.1016/j.envint.2019.105407 PMID: 31955035

Zhang, Y., Luo, X. J., Mo, L., Wu, J. P., Mai, B. X., & Peng, Y. H. (2015). Bioaccumulation and translocation of polyhalogenated compounds in rice (Oryza sativa L.) planted in paddy soil collected from an electronic waste recycling site, South China. *Chemosphere*, *137*, 25–32. DOI: 10.1016/j.chemosphere.2015.04.029 PMID: 25974192

ABBREVIATIONS

%: Percentage
Ag: Silver
As: Arsenic
Au: Gold
Be: Beryllium
BFRs: Brominated flame retardants
BLL: Blood lead levels
Ce: Cerium
Cr: Chromium
CR: Complexation reaction
CRTs: Cathode ray tubes
Cu: Copper
DNA: Deoxyribonucleic Acid
ED: Electrodialysis
EEE: Electronic and electrical equipment
EOL: End-of-life
EPA: Environmental Protection Agency
EPR: Extended Producer Responsibility
E-Waste: Electronic waste
Fe: Iron
Ga: Gallium
HI: Hazard Index
HMs: Hazardous metals
Kg: Killogramme
La: Lanthanum
LCAs: Life cycle assessment

Li: Lithium
Mt: Metric tonnes
NGOs: Non-government organization
Ni: Nickel
°C: Degree Celsius
Pa: Pascal
PAEs: Phthalic acid esters
PAHs: Polycyclic aromatic hydrocarbons
Pb: Lead
PC: Personal computer
PCBs: Polychlorinated biphenyls
PM: Particulate Matter
Pt: Platinum
Sb: Antimony
Sn: Tin
UNIDO: United Nation Industrial Development Organization
WHO: World Health Organization
Zn: Zinc

Chapter 10
Impact of E-Waste Processing on Health and Environment

Benu Kesar
https://orcid.org/0000-0003-3207-5528
Symbiosis International University, India

Shaji Joseph
https://orcid.org/0000-0002-5268-1376
Symbiosis International University, India

ABSTRACT

This chapter explores innovative e-waste management solutions focusing on technology and sustainability. It introduces the universal risk assessment framework (URAF) for e-waste sites, a model to evaluate contamination. The chapter highlights urban mining as a sustainable alternative to traditional resource extraction, along with supercritical water oxidation (SCWO) and thermal plasma technology, which transform hazardous e-waste into harmless byproducts and clean energy sources. Technological advancements, including blockchain-enabled waste tracking, AI-powered automated sorting, and IoT-integrated smart waste bins are revolutionizing e-waste management efficiency. Additionally, policy-driven strategies such as right-to-repair legislation and modular electronics design aim to extend product lifespans and reduce overall waste. However, challenges remain in implementation, particularly in developing nations where regulatory enforcement and economic viability pose obstacles to formal recycling.

DOI: 10.4018/979-8-3693-7383-5.ch010

INTRODUCTION

Electronic waste (e-waste) is one of the fastest-growing global waste streams, posing significant environmental and public health challenges. The rapid expansion of consumer electronics shortened product life cycles, and limited repairability have resulted in an unprecedented increase in discarded electronic devices (Forti et al., 2018). A record 62 million tonnes (Mt) of e-waste was produced in 2022, Up 82% from 2010; On track to rise another 32%, to 82 million tonnes, in 2030; Billions of dollars worth of strategically valuable resources squandered, dumped; Just 1% of rare earth element demand is met by e-waste recycling, highlighting a significant gap in sustainable e-waste management. Europe has the highest documented formal e-waste collection and recycling rate. In other continents, the e-waste documented as formally collected and recycled is substantially lower than the estimated e-waste generated. Despite its hazardous composition, e-waste also contains valuable materials driving both formal and informal recycling efforts. However, a significant portion of e-waste is mishandled, exported to developing nations, or processed using unsafe methods, leading to severe health and environmental consequences.

E-waste contains a complex mix of hazardous substances, including heavy metals (lead, mercury, cadmium, chromium), persistent organic pollutants (POPs), polycyclic aromatic hydrocarbons (PAHs), and brominated flame retardants (Eguchi et al., 2013). E-waste also contains valuable resources such as gold, silver, copper, and rare earth elements, yet its informal processing releases toxic chemicals, heavy metals, and persistent organic pollutants (POPs) into the environment, posing severe risks to workers, surrounding communities, and ecosystems (Parvez et al., 2021). The health consequences of prolonged exposure to e-waste pollutants include neurodevelopmental disorders, respiratory diseases, endocrine disruption, and cancer risks (Heacock et al., 2022; Huo et al., 2019). The economic and environmental paradox of e-waste recycling serves as both a resource recovery opportunity and a major pollution source demanding a multidisciplinary, policy-driven approach for effective management (Agrawal & Sahu, 2010).

Improper disposal and informal recycling methods, such as open burning, acid leaching, and manual dismantling, expose workers and surrounding communities to toxic fumes, contaminated soil, and polluted water sources (Parvez et al., 2021). These practices are particularly prevalent in low-income countries, where weak regulatory enforcement, economic pressures, and a lack of proper waste management infrastructure contribute to unsafe recycling activities (Gollakota et al., 2020). To mitigate the risks associated with e-waste, various international regulatory frameworks have been established. The Basel Convention on the Control of Transboundary Movements of Hazardous Wastes and Their Disposal aims to prevent the illegal export of hazardous waste from developed to developing nations, yet loopholes and weak enforcement allow much e-waste to be misclassified as "used electronics". The European Union's Waste Electrical and Electronic Equipment (WEEE) Directive and the Restriction of Hazardous Substances (RoHS) Directive have set standards for e-waste collection, recycling, and material restrictions, but global compliance remains inconsistent (Forti et al., 2018). The Extended Producer Responsibility (EPR) framework, adopted by various countries, mandates that manufacturers take responsibility for the end-of-life management of their products. Despite these efforts, informal e-waste recycling continues to thrive in countries like India, and Pakistan, where labour costs are low, and environmental regulations are poorly enforced (Ignatuschtschenko, 2017).

Informal e-waste processing is driven by economic necessity, lack of regulatory enforcement, and the high cost of formal recycling infrastructure (Ohajinwa et al., 2018). Many developing countries, such as India, lack sufficient e-waste collection systems, leading to unregulated markets where waste pickers and small-scale recyclers manually extract valuable materials (Mishra, 2019). The high cost of formal recycling plants, which require advanced sorting, and environmentally friendly metal recovery techniques, makes informal recycling a more attractive economic option for many low-income workers. This results in hazardous dismantling techniques, including open burning of plastics, acid leaching of circuit boards, and uncontrolled shredding, which expose workers to toxic fumes, heavy metals, and persistent organic pollutants (Parvez et al., 2021). The mechanisms of e-waste toxicity are best understood through Adverse Outcome Pathways (AOPs), which map toxicity from the molecular level to population-wide effects. Toxicity is initiated at the molecular level by Molecular Initiating Events, where e-waste chemicals interact with cellular receptors or macromolecules. This triggers Key Events (KEs) at the cellular and tissue levels, ultimately leading to adverse outcomes such as organ dysfunction, metabolic disturbances, or carcinogenesis. One significant pathway involves oxidative stress, in which reactive oxygen species generated by heavy metals and organic pollutants overwhelm cellular antioxidant defenses, leading to lipid peroxidation, DNA damage, and apoptosis. Inflammatory responses further exacerbate toxicity, as chemicals like lead and PAHs activate mitogen-activated protein kinase causing systemic inflammation and immune dysregulation.

Hepatotoxicity is a major concern in populations exposed to e-waste, as the liver is the primary organ responsible for detoxification. Studies have shown that e-waste exposure leads to elevated levels of liver enzymes such as alanine transaminase and gamma-glutamyl transferase, indicative of hepatic dysfunction (Decharat & Kiddee, 2020). Heavy metals, particularly cadmium and lead, impair hepatocyte function by inducing apoptosis, mitochondrial damage, and inflammatory cytokine production. The activation of extracellular-signal-regulated kinase pathways has been linked to liver toxicity in e-waste-exposed populations. Biomarkers such as total bilirubin and reduced glutathione have been identified as potential indicators of liver stress in affected individuals. Nephrotoxicity is another critical health effect associated with e-waste exposure, as the kidneys play a crucial role in filtering and excreting toxicants.

Recent years have seen increased attention to the concept of urban mining, where e-waste is viewed as a resource rather than a problem. Urban Mining is the process of extracting valuable metals (gold, silver, copper, rare earth elements) from discarded electronics rather than traditional mining. Companies like Fairphone and Dell have adopted modular design principles, allowing consumers to replace individual components rather than discarding entire devices. Right-to-repair legislation in the European Union aims to reduce e-waste by ensuring that manufacturers provide spare parts and repair manuals, encouraging longer product use cycles. Technological innovations are also playing a crucial role in safe e-waste recycling. AI-driven sorting systems improve the efficiency of material recovery, while hydrometallurgical and bioleaching methods provide fewer toxic alternatives to traditional acid-based metal extraction. Additionally, blockchain-based tracking systems are being introduced to ensure e-waste is responsibly handled from production to disposal, preventing illegal dumping.

Supercritical Water Oxidation (SCWO) is a highly advanced oxidation process with significant potential for effectively treating diverse organic waste streams, particularly concentrated wet wastes in slurry form. SCWO technology allows for the destruction of persistent and toxic organic contaminants such as perfluoroalkyl and polyfluoroalkyl substances (PFAS), 1,4-dioxane, and many more. SCWO relies on the unique reactivity and transport properties that occur when an aqueous waste stream is brought above the critical point of water (374°C and 218 atm, or 704°F and 3200 psi).Oxygen is fully soluble in supercritical water, resulting in extremely rapid and complete oxidation of all organics to carbon dioxide, clean water (that can be reused), and some non-leachable inorganic salts.

Bioremediation utilizes microorganisms, including bacteria and fungi, to detoxify hazardous substances in e-waste-contaminated environments. These microorganisms metabolize toxic compounds, converting them into less harmful substances, thereby mitigating environmental pollution. Similarly, phytoremediation employs plants to absorb, store, or degrade pollutants from contaminated sites, including those affected by e-waste. Certain plant species can uptake heavy metals and other toxins, facilitating the cleanup of polluted areas.

The development of biodegradable semiconductors represents a proactive approach to reducing future e-waste. These electronic components, crafted from organic, wood-based, or protein-derived materials, are designed to decompose naturally after disposal. This innovation aims to minimize the long-term environmental impact of discarded electronics by ensuring that components break down into non-toxic byproducts.

Thermal plasma technology has emerged as a promising method for treating hazardous components in electronic waste (e-waste). This process utilizes plasma torches, which generate extremely high temperatures ranging from 2,000 to 14,000 °C (3,600 to 25,200 °F), to convert organic materials in e-waste into synthesis gas composed primarily of hydrogen and carbon monoxide. Inorganic compounds in the waste stream melt, including glass, ceramics, and metals.

Adopting circular economic principles offers a sustainable solution to the e-waste crisis. This approach emphasizes the reuse, repair, refurbishment, and recycling of electronic products, creating a closed-loop system that minimizes waste and conserves resources. Implementing circular models enables companies to access untapped opportunities and reduce environmental impact, addressing critical e-waste challenges. Improper e-waste disposal leads to the loss of scarce and valuable raw materials, including precious metals like gold, copper, and rare earth elements essential for modern technologies. Recycling these materials is significantly more energy-efficient than extracting them from virgin sources. Prioritizing durability, repairability, and recyclability in electronic design is crucial for reducing e-waste generation. Implementing modular designs facilitates easier repairs and component upgrades, extending product lifespans. According to the Circular Electronics Partnership, such design strategies could cut e-waste by up to 50%, highlighting the potential of sustainable innovation in electronics manufacturing.

Policy developments are increasingly focusing on reducing the environmental footprint of electronic devices. The Basel Convention's e-waste amendments seek to tighten regulations on hazardous material trade, though enforcement remains a challenge. By integrating technological advancements, policy reforms, and circular economy approaches, the future of e-waste management is shifting toward sustainable and responsible solutions that prioritize public health, environmental protection, and resource efficiency.

Research Objectives

1. To analyze the Toxic Components of E-Waste
2. To evaluate the Health Effects of E-Waste Exposure.
3. To assess the Environmental Impact of E-Waste Pollution.
4. To examine Policy Frameworks and Global E-Waste Regulations.
5. To develop a Universal Risk Assessment Framework for E-Waste Management.

Research Questions

- What are the hazardous substances present in e-waste?
- What are the potential health and environmental impacts of e-waste processing?
- What are the existing regulations to manage and mitigate hazards in e-waste processing?
- What are the challenges to e-waste processing?

LITERATURE REVIEW

The impact of e-waste processing on human health and the environment has received a lot of attention from the research community. Mercury exposure has been linked to reproductive abnormalities, nerve damage, and reduced intestinal functions. Cadmium is a known environmental risk factor for bone loss (Taha et al., 2018). Chromium, especially hexavalent chromium, can lead to severe allergic reactions, skin irritation, and respiratory issues. Persistent Organic Pollutants (POP) can cause intestinal damage, reproductive problems, and neurobehavioral impairments because they remain in the environment for a longer duration and can enter the food chain.

E-Waste Components and Exposure Pathways

E-waste contains a mixture of hazardous materials, including heavy metals (lead, cadmium, mercury, arsenic, chromium), flame retardants (PBDEs, PCBs), polycyclic aromatic hydrocarbons (PAHs), and various organophosphates. The primary exposure pathways include inhalation of toxic dust and fumes, ingestion of contaminated food and water, and dermal absorption. Informal e-waste recycling in developing countries is a major source of exposure, where crude processing techniques, such as open burning and acid leaching, release these harmful chemicals into the air, soil, and water.

Organ-Specific Health Effects of E-Waste Exposure

- Hepatotoxicity (Liver Damage): The liver is a primary site for detoxification, making it highly vulnerable to chemical-induced damage. E-waste exposure leads to hepatic oxidative stress, inflammation, and enzyme dysregulation, which can progress to fibrosis and cirrhosis. Elevated levels of alanine transaminase (ALT) and alkaline phosphatase (ALP) have been observed in exposed populations, indicating liver dysfunction (Eckhardt & Kaifie, 2024; Decharat & Kiddee, 2020).

- Nephrotoxicity (Kidney Damage): Flame retardants, lead, and cadmium contribute to renal toxicity, proteinuria, and impaired filtration function. Biomarkers such as kidney injury molecule-1 and creatinine levels indicate kidney stress in exposed individuals (Neitzel et al., 2020).
- Cardiovascular Toxicity: E-waste chemicals disrupt vascular function, leading to hypertension, lipid imbalances, and increased heart rate. Inhalation of PAHs and exposure to heavy metals have been linked to atherosclerosis and increased cardiovascular morbidity in highly exposed populations (Burns et al., 2016).
- Respiratory Problems: In several studies conducted in India, Ghana and Thailand, difficulties in breathing, cough with sputum, chest pain, asthma and other respiratory problems were significantly more frequently reported in e-waste workers (Kuntawee et al., 2020; Nti et al., 2020).
- Neurotoxicity: Heavy metals, particularly lead and mercury, impair synaptic function, neurotransmitter balance, and calcium signaling, leading to neurodevelopmental disorders. Children exposed to e-waste chemicals show lower IQ scores, hyperactivity, and cognitive impairments.
- Blood Disorders: E-waste chemicals can cause anaemia, immune suppression, and platelet dysfunction. Lead exposure disrupts red blood cell production, increasing the risk of anaemia and hematological disorders (Parvez et al., 2021).
- Endocrine Toxicity: PBDEs and PCBs interfere with thyroid hormone production, reproductive health, and metabolic function. Exposure has been linked to early puberty, infertility, and obesity (Wang et al., 2010; J. Zheng et al., 2017).
- Carcinogenicity and Genotoxicity: Long-term exposure to PAHs, heavy metals, and flame retardants increases the risk of lung, liver, and bladder cancer. These substances induce DNA mutations, chromosomal aberrations, and epigenetic modifications, leading to tumor formation (Neitzel et al., 2020; J. Wang et al., 2012).
- Developmental and Reproductive Effects: Prenatal exposure to e-waste chemicals has been associated with low birth weight, growth retardation, and developmental delays in children. Flame retardants and heavy metals interfere with fetal brain development and immune function (Parvez et al., 2021).

Hazardous Substances in E-Waste

Lead, commonly found in older cathode ray tube (CRT) monitors and televisions, is highly toxic and can cause neurological and developmental issues when improperly disposed of. Mercury, present in fluorescent lamps and bulbs, can harm the nervous system, with broken bulbs releasing dangerous mercury vapor. Cadmium, used in batteries and electronic components, is a toxic metal linked to kidney damage and bone loss upon exposure. Brominated flame retardants (BFRs), used to reduce flammability in electronic devices, can leach into the environment, potentially disrupting endocrine systems in humans and wildlife. Additionally, polychlorinated biphenyls (PCBs), found in older capacitors and transformers, are hazardous substances that contribute to severe environmental and health concerns. Without proper disposal or recycling measures, these toxic materials can contaminate soil, water, and air, emphasizing the need for strict regulations and sustainable e-waste management practices. Table 1 shows the adverse effects of e-waste components on health.

Table 1. Adverse effects of e-waste on health

E-Waste Component	Organs Affected	Symptoms
Lead (Pb)	Central Nervous system, kidneys, blood system impact	Cognitive impairment, hyperactivity, lower IQ, anemia, high blood pressure, developmental delays in children, kidney dysfunction
Cadmium (Cd)	Kidneys, lungs, bones	Kidney failure, lung cancer, bone loss (osteomalacia), respiratory distress, impaired kidney function
Mercury (Hg)	Brain, central nervous system	Memory loss, tremors, insomnia, sensory impairment, neurodevelopmental disorders in children
Hexavalent Chromium	Lungs, DNA (genotoxic effects)	Lung cancer, genetic mutations, severe respiratory irritation, weakened immune system
Brominated Flame Retardants (BFRs)	Endocrine system	Hormonal imbalances, thyroid dysfunction, reproductive issues, neurotoxicity, increased cancer risk
Barium (Ba)	Heart, liver, spleen	Muscle weakness, organ damage, excessive sweating
Beryllium (Be)	Lungs, skin	Lung cancer, chronic berylliosis, skin rashes, immune system hypersensitivity
Copper (Cu)	Stomach, liver	Severe nausea, liver cirrhosis, gastrointestinal distress, Wilson's disease
Nickel (Ni)	Skin, lungs	Allergic dermatitis, asthma, lung fibrosis, respiratory irritation
Lithium (Li)	Nervous system, kidney	Neurological impairment, tremors, memory loss, kidney failure
Sulphur (S)	Skin, Eye, Heart	Skin allergies, rashes, eye irritation, asthma, chronic respiratory disease

Source: Author's elaboration

Impact on Environment

Electronic waste (e-waste) poses significant environmental and health risks due to its toxic components and improper disposal methods. Practices such as open-air burning and acid baths, commonly employed to extract valuable materials from electronic devices, release hazardous substances into the environment (Jinhui et al., 2011). These methods expose workers to high levels of contaminants, including lead, mercury, cadmium, and brominated flame retardants, leading to severe health issues like cancers, neurological damage, and reduced IQ (Deng et al., 2007).

Soil Contamination

E-waste contains hazardous substances like lead, mercury, cadmium, and brominated flame retardants. When improperly disposed of in landfills, these toxic materials can leach into the soil, leading to contamination. This degradation reduces soil fertility and poses severe risks to food safety and public health. For instance, crops grown in contaminated soil can absorb these toxins, entering the food chain and affecting both human and animal health (Eguchi et al., 2013). The persistence of these contaminants in the soil can lead to long-term ecological imbalances, affecting plant growth and soil microorganism activity (Jinhui et al., 2011). Such contamination can render land unsuitable for agriculture, leading to economic losses for farming communities and threatening food security (Akortia et al., 2017).

Water Pollution

The leaching of hazardous substances from e-waste into the soil can further lead to groundwater contamination. Toxins such as lead, mercury, and cadmium can seep into water sources, posing serious risks to drinking water supplies and aquatic ecosystems (Widmer et al., 2005). Contaminated water can affect the health of communities relying on these sources, leading to diseases and other health issues. Additionally, aquatic life is at risk as these toxins can accumulate in the bodies of fish and other organisms, leading to bioaccumulation and biomagnification of harmful substances up the food chain. This not only affects biodiversity but also poses health risks to predators, including humans, who consume contaminated aquatic organisms.

Air Quality Degradation

Improper recycling practices, such as open-air burning and acid baths used to recover valuable materials from electronic components, release toxic pollutants into the atmosphere (Lohmann & Jones, 1998). These activities are considered hazardous to the environment and health as they release toxic pollutants, contaminating the air, soil, dust, and water at recycling sites and in neighbouring communities. Open burning and heating are considered the most hazardous activities due to the toxic fumes created. Once in the environment, these toxic pollutants can travel significant distances from the point of pollution, exposing people in faraway areas to hazardous substances (Deng et. al., 2007).

Plants in contaminated areas may struggle to grow due to soil toxicity, leading to reduced vegetation cover, which in turn affects herbivorous species dependent on these plants for sustenance. Aquatic ecosystems are particularly vulnerable as heavy metals leach into rivers and lakes, poisoning fish and other marine organisms. Bioaccumulation and biomagnification of these toxins further threaten predator species, disrupting food chain dynamics. Wildlife exposed to e-waste pollution can suffer from reproductive failures, genetic mutations, and acute poisoning, significantly diminishing population numbers. Birds, for example, may ingest toxic particles while scavenging in e-waste dumping sites, leading to neurological damage and reduced survival rates. Over time, prolonged exposure to hazardous e-waste chemicals can transform thriving ecosystems into barren wastelands, where only the most pollution-tolerant organisms can survive. This decline in biodiversity not only weakens ecosystem resilience but also impacts human communities that rely on these ecosystems for food, water, and air quality. The destruction of natural habitats due to e-waste pollution is an urgent environmental crisis, necessitating immediate action through responsible recycling, stricter regulations, and sustainable waste management practices to mitigate long-term ecological damage.

International Treaties to Minimize e-Waste Hazards

The European Union (EU) has taken a series of measures to reduce e-waste pollution. Two such regulations from EU are, a) Waste Electrical and Electronic Equipment (WEEE) Directive and b) the Restriction of Hazardous Substances (RoHS) Directive. In the United States, the Environmental Protection Agency (EPA) provides guidelines and best practices for a safe e-waste management program. They have also introduced the Responsible Recycling and e-Stewardship certifications to encourage organizations to adapt e-waste management programs. However, the implementation and enforcement of these regulations and guidelines remain a challenge, particularly in developing countries where informal and unregulated e-waste processing practices dominate.

The regulation of electronic waste (e-waste) has undergone significant changes in recent years, particularly with the implementation of the Basel Ban Amendment (2021) and China's "National Sword" policy, enacted in January 2018, banned the import of most plastics and other materials headed for that nation's recycling processors, which had handled nearly half of the world's recyclable waste for the past quarter century. The move was an effort to halt a deluge of soiled and contaminated materials that was overwhelming Chinese processing facilities and leaving the country with yet another environmental problem and this one not of its own making. These measures have reshaped global e-waste trade and recycling practices, highlighting the urgent need for self-sufficient waste management infrastructure and stricter international cooperation.

The Basel Convention on the Control of Transboundary Movements of Hazardous Wastes and Their Disposal, adopted in 1989, was designed to regulate the global trade of hazardous waste, including e-waste. However, a key limitation of the original agreement was that it allowed e-waste to be exported from developed to developing nations under the pretext of "recycling." This loophole led to large volumes of hazardous waste being shipped to low-income countries in Africa, South Asia, and Southeast Asia, where informal recycling methods such as open-air burning, acid leaching, and crude dismantling exposed workers and nearby communities to toxic heavy metals, persistent organic pollutants (POPs), and carcinogenic chemicals.

To address this issue, the Basel Ban Amendment, which officially came into force in 2019, prohibits all exports of hazardous waste, including e-waste, from OECD and EU countries to non-OECD nations, regardless of whether it is for disposal or recycling. This amendment significantly tightens global e-waste trade regulations, forcing high-income countries to develop domestic recycling capabilities rather than relying on offshore processing. However, enforcement challenges remain, as illegal shipments continue under misclassified labels, such as "second-hand goods".

For decades, China was the world's largest importer of recyclable waste, including e-waste, plastics, and metals. The country processed nearly 70% of the world's electronic waste, extracting valuable materials such as gold, silver, copper, and rare earth elements. However, due to mounting concerns over environmental pollution, illegal waste dumping, and poor waste segregation practices, China implemented the National Sword Policy in 2019, which banned the import of solid waste, including low-quality e-waste materials. This move disrupted global waste supply chains, particularly affecting North America, Europe, and Australia, which had long relied on China as their primary recycling destination.

The combined impact of the Basel Ban Amendment and China's National Sword Policy highlights the need for sustainable, localized solutions for e-waste recycling. Moving forward, global e-waste management must focus on:

- Developing advanced domestic recycling infrastructure to reduce dependence on international waste trade.
- Implementing AI-driven sorting, robotic dismantling, and bioleaching for more efficient and environmentally friendly e-waste processing.

- Strengthening global enforcement mechanisms to prevent illegal waste shipments under false classifications.
- Promoting Extended Producer Responsibility (EPR), requiring manufacturers to take accountability for the full life cycle of their electronic products, including take-back and safe recycling programs.

By aligning international policies with technological innovation and circular economy principles, countries can mitigate the negative impacts of e-waste pollution while ensuring the ethical and sustainable management of electronic materials.

Companies can revolutionize end-of-life technology management by leveraging certified IT Asset Disposal (ITAD) services, ensuring both environmental responsibility and data security. Professional e-waste recycling partners, such as Human-I-T, specialize in handling obsolete electronics in compliance with strict environmental regulations, preventing hazardous materials from polluting ecosystems. Beyond regulatory adherence, certified ITAD services offer secure data destruction, protecting businesses from potential breaches and ensuring compliance with privacy laws. Additionally, organizations can benefit financially by recovering valuable resources—such as gold, silver, and rare earth metals through proper recycling, converting e-waste into profitable revenue streams. Certifications like NAID AAA and ISO 14001, which focus on environmental sustainability. essentially ensure that ITAD providers adhere to industry-leading environmental and security standards. Choosing a certified ITAD partner not only mitigates corporate risk but also aligns with sustainable business practices, reinforcing a company's commitment to responsible e-waste management while maximizing value from retired technology assets.

Theoretical Framework

David Schlosberg's *Defining Environmental Justice* provides a crucial theoretical lens for understanding the impact of e-waste on marginalized communities and ecosystems. His argument that justice extends beyond distribution to include recognition, capabilities, and participation is particularly relevant in the context of e-waste management, where low-income communities are disproportionately burdened with electronic waste disposal and its toxic consequences. The recognition aspect of justice is critical here, as informal e-waste workers and affected communities are frequently overlooked in policy decisions, despite being the most exposed to hazardous substances.

Schlosberg's emphasis on capabilities is also significant, as many e-waste workers lack access to protective equipment, healthcare, or economic alternatives, limiting their ability to function and flourish in a safe environment. Additionally, his argument that justice must be participatory highlights the failure of global e-waste governance, where decisions about waste exports, recycling policies, and environmental regulations are made without meaningful input from those directly affected. The interconnection of distribution, recognition, capabilities, and participation in Schlosberg's framework underscores that addressing only one dimension of e-waste injustice, such as improving recycling infrastructure without considering social and political inclusion is insufficient.

Future Trends, Emerging and Potential Breakthroughs in e-waste Management

The rapid advancement of technology has led to significant shifts in electronic waste (e-waste) management. Emerging trends and potential breakthroughs are poised to reshape how e-waste is generated, managed, and recycled (Baghel et al., 2025).

Automation systems play a crucial role in modernizing e-waste management by enhancing efficiency, safety, and sustainability. Robotic arms and conveyor belts are increasingly used in recycling facilities to sort, dismantle, and process electronic waste with precision, reducing the need for manual labour. This minimizes human exposure to hazardous substances like lead, mercury, and cadmium, which are commonly found in discarded electronics. Additionally, automation ensures consistent operations, increasing the accuracy of material recovery and reducing waste.

Innovative disposal techniques, such as advanced chemical treatments and plasma gasification, offer more sustainable alternatives to traditional landfilling and incineration. Chemical treatments involve breaking down toxic components in e-waste into less harmful substances, facilitating safer disposal or reuse. Plasma gasification, on the other hand, uses extremely high temperatures to convert waste into syngas, a mixture of hydrogen and carbon monoxide which can be used as a source of clean energy. These advanced methods significantly reduce the environmental footprint of e-waste while maximizing resource recovery.

The integration of analytics and Internet of Things (IoT) devices further optimizes e-waste management by streamlining waste collection and improving regulatory compliance. Smart sensors embedded in recycling bins and waste facilities can track waste levels in real time, enabling more efficient collection schedules and reducing operational costs. IoT-driven analytics help organizations monitor compliance with environmental regulations by providing detailed data on waste processing and disposal, ensuring transparency and accountability. Together, these technological advancements contribute to a more sustainable and efficient e-waste management system, reducing environmental harm while improving resource utilization.

1. Urban Mining: Sustainable Resource Recovery from E-Waste

- Promotes recovering valuable metals (gold, silver, copper, rare earth elements) from discarded electronics instead of traditional mining.
- Reduces environmental damage from mining while creating economic opportunities.
- Companies like Fairphone & Dell are implementing modular electronics designs to facilitate easier repairs and material recovery.

2. Supercritical Water Oxidation (SCWO) for E-Waste Treatment
   - Uses water at extremely high temperatures and pressure (374°C and 218 atm, or 704°F and 3200 psi) to break down hazardous materials into harmless byproducts like water, carbon dioxide, and mineral solids.
   - A cleaner alternative to traditional e-waste disposal methods such as landfilling and incineration.
3. Blockchain for E-Waste Tracking & Circular Economy
   - Introduces a tamper-proof digital ledger to track electronic devices from production to disposal, ensuring responsible recycling.
   - Helps prevent illegal e-waste dumping by improving supply chain transparency.
   - Assists in monitoring Extended Producer Responsibility (EPR) compliance.
4. AI-Powered Waste Sorting and Smart Recycling Bins
   - Uses machine learning and robotic arms for automated e-waste sorting, reducing human exposure to toxic chemicals.
   - Smart waste bins equipped with IoT sensors optimize waste collection schedules, ensuring more efficient recycling.
5. Biodegradable Electronics & Green Product Design
   - Development of biodegradable semiconductors, such as:
   - "Leaftronics": Circuit boards made from tree leaves that decompose naturally.
   - Vitrimer Printed Circuit Boards (vPCB): Recyclable electronics that use sustainable polymers.
   - Aims to minimize e-waste accumulation at the source by designing eco-friendly electronic components.
6. Thermal Plasma technology for E-Waste Disposal
   - Thermal plasma technology has emerged as a promising method for treating hazardous components in electronic waste (e-waste). This process utilizes plasma torches, which generate extremely high temperatures ranging from 2,000 to 14,000 °C (3,600 to 25,200 °F), to convert organic materials in e-waste into synthesis gas (syngas) composed primarily of hydrogen and carbon monoxide.

- Produces minimal hazardous residues compared to traditional incineration.

7. Bioremediation & Phytoremediation for E-Waste Detoxification
    - Uses bacteria, fungi, and plants to absorb, break down, and neutralize hazardous e-waste contaminants in soil and water.
    - Offers a low-cost, sustainable alternative to conventional chemical treatments for e-waste pollution.
8. Right-to-Repair Legislation and Modular Electronics
    - Aims to extend product lifespans by allowing consumers to repair and upgrade devices instead of discarding them.
    - European Union's "Right to Repair" laws mandate that manufacturers provide spare parts and repair manuals, reducing e-waste generation.
9. AI-Based E-Waste Risk Prediction Models
    - Utilizes predictive analytics and AI to identify potential hotspots of e-waste pollution and estimate long-term health risks.
    - Helps policymakers implement targeted interventions in high-risk areas.
10. Nanotechnology in Material Recovery
    - Nanotechnology offers promising advancements in extracting and purifying materials from e-waste. Developing scalable and cost-effective nanotechnology solutions for large-scale e-waste recycling operations.
11. Virtual and Augmented Reality in E-Waste Management Training
    - Virtual Reality (VR) and Augmented Reality (AR) technologies are being leveraged to improve education and training for recycling personnel.

Benefits:

- Enhanced Learning Experiences: Providing immersive training environments for better understanding of e-waste handling procedures.
- Improved Safety Protocols: Allowing personnel to practice in simulated settings, reducing the risk of accidents.

These emerging trends and technological advancements are poised to transform e-waste management, promoting more efficient and sustainable practices in handling electronic waste.

Modern Waste Management System

A modern waste management system integrates advanced technologies and sustainable practices to enhance efficiency and reduce environmental impact. The process begins with consumers generating waste, which is categorized into several streams. Reusable and recyclable waste is handled by the informal recycling industry, where collectors and workshops process materials for resale in second-hand markets. Organic waste is diverted to composting facilities that employ rapid decomposition techniques or anaerobic digesters that convert waste into biogas for energy production.

The Supercritical Water Oxidation (SCWO) process was introduced in the mid-1980s as an advanced oxidation technology for the efficient degradation of organic waste. SCWO has been widely utilized for treating industrial wastewater and sludge, breaking down persistent pollutants like ammonia, phenol, and aromatic compounds into non-toxic by-products such as H O, CO , and N . The process leverages the unique properties of supercritical water such as low viscosity and high diffusion coefficients to enhance mass transfer and facilitate effective oxidation reactions. Landfill sites are used for residual waste, incorporating methane capture systems for energy recovery. In addition to landfill management, plastic waste is addressed through innovative biological treatments, such as plastic-degrading organisms, which enhance polymer decomposition.

A major technological shift in waste management is the growing integration of AI into waste sorting operations. Traditional waste sorting heavily relies on manual labor, making the process inefficient and prone to errors. To address these challenges, AI-powered systems utilizing image recognition and machine vision are being increasingly deployed to enhance sorting accuracy and efficiency. These advanced algorithms analyze visual data with high precision, accurately identifying various waste types including recyclables, organic waste, and non-recyclables leading to improved segregation and resource recovery.

AI-driven robotic sorting systems further optimize waste management by automating the separation process. These systems use deep learning algorithms to rapidly classify and sort materials, significantly improving throughput and reducing reliance on manual labor. Beyond waste sorting.

Table 2 provides an overview of key challenges in waste management, along with potential solutions and examples of successful initiatives implemented by various countries.

Table 2. Challenges of e-waste management

Challenges	Potential Solutions	Examples of Initiatives by Countries
Hazardous Material Handling	Development of eco-friendly alternatives, safer dismantling and disposal processes.	United states: Advanced robotic e-waste dismantling systems,
Limited recycling Infrastructure	Government and private investment in modern recycling plants, establishing e-waste collection centers.	France: Introduced a "French Repairability Index" on electronic products to encourage repairs.
Illegal Export to Developing Countries	Strengthening international policies and agreements, enforcing stricter border controls.	EU: Strict enforcement of the e-waste shipment regulation to prevent illegal exports. China: National Sword Policy restricting the import of foreign waste.
Rapid Technological Obsolescence	Encouraging modular designs for easy repair and upgrades, promoting a circular economy.	European Union: adopted "Right to Repair" act, mandating companies to provide spare parts and repair manuals for longer product life.
Collection Mechanisms	Implementing Extended Producer Responsibility (EPR) and setting up convenient e-waste drop-off points.	Germany: EPR policies require manufacturers to take back old electronics. South Korea: Nationwide smart bins for e-waste collection.
Low Economic Incentives for Recycling	Subsidies for recycling industries, and tax benefits for companies using recycled materials.	India: E-waste recycling subsidy program under the "Swachh Bharat Abhiyan." Canada: Offers tax rebates to businesses using recycled electronics.
Toxic Chemical Release (Lead, Mercury, etc.)	Developing safer recycling technologies and promoting non-toxic alternatives in manufacturing.	Australia: Regulates hazardous e-waste under the "National Waste Policy."
E-Waste in Landfills Leading to Pollution	Banning e-waste landfilling, imposing fines for illegal dumping, and mandating proper disposal channels.	Singapore: Launched the "Zero Waste Masterplan" to eliminate landfill e-waste

Source: Author's elaboration

Universal Risk Assessment Framework for E-Waste Sites:

The Universal Risk Assessment Framework (URAF) for E-Waste Sites is a structured approach to evaluate and mitigate human health and environmental risks associated with e-waste management. The framework follows a five-step methodology.

Step 1: Identification of E-Waste Contaminants: involves detecting hazardous substances such as heavy metals (Pb, Hg, Cd), persistent organic pollutants (PBDEs, PCBs, dioxins), nanomaterials, and battery components (Li, Co, Mn, Ni).

a) Sampling techniques, Toxicity analysis using Inductivity Coupled Plasma Mass Spectrometry (ICP-MS), X-ray fluorescence (XRF), and chromatography
b) Biomonitoring in humans (blood, urine) to assess metal accumulation is employed to assess exposure levels.

Step 2: Exposure Pathway Analysis: identifies primary exposure routes such as inhalation (burning of plastics and circuit boards), ingestion (contaminated food and water), and dermal absorption (contact with toxic dust and e-waste components).

High-risk groups include informal e-waste workers, children, pregnant women, and nearby communities exposed to chronic contamination.

Step 3: Risk Characterization & Toxicity Benchmarking: quantifies health risks using Reference Dose (RfD), Bioavailability Assessments, and Lifetime Cancer Risk (LCR) models, alongside studies linking exposure to neurodevelopmental disorders, endocrine disruption, and immune suppression.

Risks are classified as low (requiring periodic monitoring), moderate (requiring local mitigation), or high (demanding immediate intervention).

Step 4: Environmental Impact Assessment (EIA): evaluates ecosystem degradation due to e-waste contamination, focusing on soil quality, groundwater pollution, air toxicity, and bioaccumulation in crops and livestock. Advanced tools such as satellite and drone-based pollution mapping, AI-driven toxicity predictions, and ecotoxicological testing are utilized for a comprehensive assessment.

In the context of e-waste processing, this framework allows for systematic identification of exposure pathways, including inhalation of airborne contaminants, dermal contact with hazardous materials, and ingestion of contaminated food or water. The dose-response analysis component evaluates the toxicity of these exposures by determining their potential effects on human health, such as neurotoxicity, carcinogenicity, or endocrine disruption. Finally, the risk characterization phase integrates exposure and dose-response data to quantify the likelihood and severity of adverse health outcomes, aiding policymakers in implementing effective risk management strategies. This structured approach ensures that regulatory actions, occupational health measures, and environmental safety protocols are data-driven, reproducible, and transparent, thereby enhancing risk mitigation efforts in both formal and informal e-waste recycling sectors.

Step 5: Mitigation Strategies & Policy Recommendations

Develop data-driven interventions for safe e-waste processing. Technological Solutions. National and international actions are essential to protect communities from unsound e-waste recycling activities. Actions that can be taken include:

- Create a licensing system and encourage certification via international standards for collection and recycling. If an informal collection system exists, use it to collect e-waste, and ensure e-waste is sent to licensed recyclers through incentives.
- Developing and implementing national e-waste management legislation that protects public health
- AI-Based Sorting Systems: Reducing human exposure in recycling plants.
- Green Product Design: Modular electronics, biodegradable PCBs.
- Enforce legislation for all stakeholders and strengthen monitoring and compliance mechanisms across the country to ensure a level playing field.
- Extended Producer Responsibility (EPR) Laws: Make manufacturers responsible for end-of-life disposal
- Implementing and monitoring interventions that improve informal e-waste recycling activities, protect public health, and ensure vital sources of community revenue.
- Ensure that all stakeholders involved in e-waste collection and recycling are aware of the potential impacts on the environment and human health as well as possible approaches to the environmentally sound treatment of e-waste.
- Greater emphasis on prevention, safer disposal, and recycling practices to mitigate health risks.
- Regulatory & Policy Measures: Strict Enforcement of Basel Convention, prevent e-waste dumping in developing nations.
- Formalization of E-Waste Recycling Sectors: Transitioning informal workers to safe, regulated recycling jobs.

CONCLUSION

The rapid accumulation of electronic waste (e-waste) presents a significant challenge to both human health and environmental sustainability, necessitating innovative and systemic solutions for its management. This chapter highlights a transformational shift in e-waste management by integrating cutting-edge technologies, circular economy principles, and regulatory frameworks to minimize the adverse effects of electronic waste. The Universal Risk Assessment Framework (URAF) provides a novel approach to evaluating contamination risks, enabling policymakers and industries to implement targeted interventions for safer waste processing. Emerging solutions such as urban mining, Supercritical Water Oxidation (SCWO) offer sustainable alternatives to traditional disposal methods, reducing environmental harm while recovering valuable materials. Technological advancements, including AI-powered waste sorting, blockchain-based e-waste tracking, and IoT-enabled smart waste management, improve efficiency and transparency in e-waste recycling. Innovations in biodegradable electronics, such as Leaftronics and vitrimer printed circuit boards (vPCB), aim to reduce e-waste at its source by promoting eco-friendly product design. Additionally, bioremediation and phytoremediation provide promising natural solutions for detoxifying e-waste-contaminated environments.

The chapter also underscores the importance of policy interventions, such as Right-to-Repair legislation and Extended Producer Responsibility (EPR), which encourage manufacturers to design longer-lasting, repairable, and recyclable electronics. A multi-stakeholder approach involving governments, industries, researchers, and consumers is essential to transitioning toward a sustainable e-waste management system. Global efforts must focus on strengthening regulatory enforcement, investing in green recycling infrastructure, and fostering innovation to build a resilient, resource-efficient, and environmentally responsible future for electronic waste management.

REFERENCES

Acquah, A. A., D'Souza, C., Martin, B., Arko-Mensah, J., Quakyi, I. A., Basu, N., Robins, T. G., & Fobil, J. N. (2021). Work-related exposures and musculoskeletal disorder symptoms among informal E-waste recyclers at Agbogbloshie, Ghana. *Proceedings of the 21st Congress of the International Ergonomics Association (IEA 2021).* Volume IV, *Healthcare and Healthy Work. International Ergonomics Association. Congress, 222*, 677–681. DOI: 10.1007/978-3-030-74611-7_93

Agrawal, A., & Sahu, K. K. (2010). Problems, prospects and current trends of copper recycling in India: An overview. *Resources, Conservation and Recycling, 54*(7), 401–416. DOI: 10.1016/j.resconrec.2009.09.005

Akortia, E., Olukunle, O. I., Daso, A. P., & Okonkwo, J. O. (2017). Soil concentrations of polybrominated diphenyl ethers and trace metals from an electronic waste dump site in the Greater Accra Region, Ghana: Implications for human exposure. *Ecotoxicology and Environmental Safety, 137*, 247–255. DOI: 10.1016/j.ecoenv.2016.12.008 PMID: 27978452

Baghel, D. S., Sharma, S., Singh, S., Kumar, R., Singh, G., & Kohli, R. (2025). Advancing Hazardous Waste Management: Integrating Innovative Technologies in Transport and Disposal. *Solid Waste Management for Rural Regions*, 55–68. DOI: 10.4018/979-8-3693-8527-2.ch004

Burns, K. N., Sun, K., Fobil, J. N., & Neitzel, R. L. (2016). Heart rate, stress, and occupational noise exposure among electronic waste recycling workers. *International Journal of Environmental Research and Public Health, 13*(1), 140. DOI: 10.3390/ijerph13010140 PMID: 26797626

Decharat, S., & Kiddee, P. (2020). Health problems among workers who recycle electronic waste in southern Thailand. *Osong Public Health and Research Perspectives, 11*(1), 34–43. DOI: 10.24171/j.phrp.2020.11.1.06 PMID: 32149040

Deng, W. J., Zheng, J. S., Bi, X. H., Fu, J. M., & Wong, M. H. (2007). Distribution of PBDEs in air particles from an electronic waste recycling site compared with Guangzhou and Hong Kong, South China. *Environment International, 33*(8), 1063–1069. DOI: 10.1016/j.envint.2007.06.007 PMID: 17631964

Eckhardt, B., & Kaifie, A. (2024). Bridging the knowledge gap! Health outcomes in informal e-waste workers. *Journal of Occupational Medicine and Toxicology (London, England), 19*(1), 11. DOI: 10.1186/s12995-024-00410-z PMID: 38622584

Eguchi, A., Isobe, T., Ramu, K., Tue, N. M., Sudaryanto, A., Devanathan, G., Viet, P. H., Tana, R. S., Takahashi, S., Subramanian, A., & Tanabe, S. (2013). Soil contamination by brominated flame retardants in open waste dumping sites in Asian developing countries. *Chemosphere*, *90*(9), 2365–2371. DOI: 10.1016/j.chemosphere.2012.10.027 PMID: 23149186

Forti, V., Baldé, K., & Kuehr, R. (2018). *E-waste Statistics: Guidelines on Classifications* (2nd ed.). Reporting and Indicators.

Gollakota, A. R., Gautam, S., & Shu, C. M. (2020). Inconsistencies of e-waste management in developing nations-Facts and plausible solutions. *Journal of Environmental Management*, *261*, 261. DOI: 10.1016/j.jenvman.2020.110234 PMID: 32148304

Hayakawa, K., Takatsuki, H., Watanabe, I., & Sakai, S.-I. (2004). Polybrominated diphenyl ethers (PBDEs), polybrominated dibenzo-p-dioxins/dibenzofurans (PBDD/Fs) and monobromo-polychlorinated dibenzo-p-dioxins/dibenzofurans (MoBPXDD/Fs) in the atmosphere and bulk deposition in Kyoto, Japan. *Chemosphere*, *57*(5), 343–356. DOI: 10.1016/j.chemosphere.2004.06.038 PMID: 15331261

Heacock, M., Kelly, C. B., & Suk, W. A. (2016). E-waste: The growing global problem and next steps. *Reviews on Environmental Health*, *31*(1), 131–135. DOI: 10.1515/reveh-2015-0045 PMID: 26820178

Huo, X., Zheng, X. B., Liu, Q., Zhang, T., Wang, Q. H., & Xu, X. J. (2019). Impact of informal e-waste recycling on human health. *Zhonghua Yu Fang Yi Xue Za Zhi [Chinese Journal of Preventive Medicine]*, *53*, 426–432. PMID: 30982282

Ignatuschtschenko, E. (2017). E-waste management in China: Bridging the formal and informal sectors. *Journal of Chinese Governance*, *2*(4), 385–410. DOI: 10.1080/23812346.2017.1379629

Jinhui, L., Huabo, D., & Pixing, S. (2011). Heavy metal contamination of surface soil in electronic waste dismantling area: Site investigation and source-apportionment analysis. *Waste Management & Research*, *29*(7), 727–738. DOI: 10.1177/0734242X10397580 PMID: 21406506

Kaya, M. (2019). *Electronic waste and printed circuit board recycling technologies*. Springer International Publishing. DOI: 10.1007/978-3-030-26593-9

Kuntawee, C., Tantrakarnapa, K., Limpanont, Y., Lawpoolsri, S., Phetrak, A., Mingkhwan, R., & Worakhunpiset, S. (2020). Exposure to heavy metals in electronic waste recycling in Thailand. *International Journal of Environmental Research and Public Health*, *17*(9), 2996. DOI: 10.3390/ijerph17092996 PMID: 32357423

Lohmann, R., & Jones, K. C. (1998). Dioxins and furans in air and deposition: A review of levels, behaviour and processes. *The Science of the Total Environment, 219*(1), 53–81. DOI: 10.1016/S0048-9697(98)00237-X PMID: 9770324

Mishra, S. (2019). Perceived and manifested health problems among informal E-waste handlers: A scoping review. *Indian Journal of Occupational and Environmental Medicine, 23*(1), 7–14. DOI: 10.4103/ijoem.IJOEM_231_18 PMID: 31040583

Neitzel, R. L., Sayler, S. K., Arain, A. L., & Nambunmee, K. (2020). Metal levels, genetic instability, and renal markers in electronic waste workers in Thailand. *The International Journal of Occupational and Environmental Medicine, 11*(2), 72–84. DOI: 10.34172/ijoem.2020.1826 PMID: 32218555

Nti, A., Arko-Mensah, A. A., Botwe, J., Dwomoh, P. K., Kwarteng, D., & Takyi, L., & Fobil. (2020). Effect of particulate matter exposure on respiratory health of e-waste workers at Agbogbloshie, Accra, Ghana. *International Journal of Environmental Research and Public Health, 17*(9). PMID: 32349371

Ohajinwa, C. M., van Bodegom, P. M., Vijver, M. G., Olumide, A. O., Osibanjo, O., & Peijnenburg, W. J. G. M. (2018). Prevalence and injury patterns among electronic waste workers in the informal sector in Nigeria. *Injury Prevention : Journal of the International Society for Child and Adolescent Injury Prevention, 24*(3), 185–192. DOI: 10.1136/injuryprev-2016-042265 PMID: 28679520

Parvez, S. M., Jahan, F., Brune, M.-N., Gorman, J. F., Rahman, M. J., Carpenter, D., Islam, Z., Rahman, M., Aich, N., Knibbs, L. D., & Sly, P. D. (2021). Health consequences of exposure to e-waste: An updated systematic review. *The Lancet. Planetary Health, 5*(12), e905–e920. DOI: 10.1016/S2542-5196(21)00263-1 PMID: 34895498

Shahabuddin, M., Uddin, M. N., Chowdhury, J. I., Ahmed, S. F., Uddin, M. N., Mofijur, M., & Uddin, M. A. (2022). A review of the recent development, challenges, and opportunities of electronic waste (e-waste). *International Journal of Environmental Science and Technology*. DOI: 10.1007/s13762-022-04274-w

Taha, M. M., Mahdy-Abdallah, H., Shahy, E. M., Ibrahim, K. S., & Elserougy, S. (2018). Impact of occupational cadmium exposure on bone in sewage workers. *International Journal of Occupational and Environmental Health, 24*(3–4), 101–108. DOI: 10.1080/10773525.2018.1518745 PMID: 30222069

Wang, H., Zhang, Y., Liu, Q., Wang, F., Nie, J., & Qian, Y. (2010). Examining the relationship between brominated flame retardants (BFR) exposure and changes of thyroid hormone levels around e-waste dismantling sites. *International Journal of Hygiene and Environmental Health, 213*(5), 369–380. DOI: 10.1016/j.ijheh.2010.06.004 PMID: 20598942

Wang, J., Chen, S., Tian, M., Zheng, X., Gonzales, L., Ohura, T., Mai, B., & Simonich, S. L. M. (2012). Inhalation cancer risk associated with exposure to complex polycyclic aromatic hydrocarbon mixtures in an electronic waste and urban area in South China. *Environmental Science & Technology, 46*(17), 9745–9752. DOI: 10.1021/es302272a PMID: 22913732

Widmer, R., Oswald-Krapf, H., Sinha-Khetriwal, D., Schnellmann, M., & Böni, H. (2005). Global perspectives on e-waste. *Environmental Impact Assessment Review, 25*(5), 436–458. DOI: 10.1016/j.eiar.2005.04.001

Yuan, J., Chen, L., Chen, D., Guo, H., Bi, X., Ju, Y., Jiang, P., Shi, J., Yu, Z., Yang, J., Li, L., Jiang, Q., Sheng, G., Fu, J., Wu, T., & Chen, X. (2008). Elevated serum polybrominated diphenyl ethers and thyroid-stimulating hormone associated with lymphocytic micronuclei in Chinese workers from an E-waste dismantling site. *Environmental Science & Technology, 42*(6), 2195–2200. DOI: 10.1021/es702295f PMID: 18411489

Zheng, J., He, C.-T., Chen, S.-J., Yan, X., Guo, M.-N., Wang, M.-H., Yu, Y.-J., Yang, Z.-Y., & Mai, B.-X. (2017). Disruption of thyroid hormone (TH) levels and TH-regulated gene expression by polybrominated diphenyl ethers (PBDEs), polychlorinated biphenyls (PCBs), and hydroxylated PCBs in e-waste recycling workers. *Environment International, 102*, 138–144. DOI: 10.1016/j.envint.2017.02.009 PMID: 28245931

Zheng, X., Xu, F., Chen, K., Zeng, Y., Luo, X., Chen, S., Mai, B., & Covaci, A. (2015). Flame retardants and organochlorines in indoor dust from several e-waste recycling sites in South China: Composition variations and implications for human exposure. *Environment International, 78*, 1–7. DOI: 10.1016/j.envint.2015.02.006 PMID: 25677852

KEY TERMS AND DEFINITIONS

Bioleaching: A process that uses bacteria and fungi to extract valuable metals from electronic waste in an environmentally friendly manner.

Bioremediation: The use of microorganisms, such as bacteria and fungi, to detoxify pollutants in e-waste-contaminated environments.

Blockchain for E-Waste Tracking: A digital ledger system used to monitor e-waste from production to disposal, ensuring responsible recycling and preventing illegal dumping.

Carcinogenicity: The ability of certain e-waste chemicals, such as PAHs and heavy metals, to cause cancer through DNA mutations and cellular damage.

Circular Economy: An economic system focused on minimizing waste by designing products for reuse, repair, and recycling, reducing e-waste generation.

Electronic Waste (E-Waste): Discarded electrical or electronic devices, including computers, smartphones, and appliances, that pose environmental and health risks when improperly disposed of.

Extended Producer Responsibility (EPR): A policy approach that holds manufacturers accountable for the collection, recycling, and safe disposal of their electronic products.

Flame Retardants (PBDEs, PCBs): Chemicals used in electronics to reduce fire risk, but known to cause hormonal disruption and neurodevelopmental problems.

Heavy Metals: Toxic metals such as lead, mercury, cadmium, and chromium found in e-waste, which can cause severe health and environmental damage.

Hepatotoxicity: Liver damage caused by exposure to hazardous substances in e-waste, often indicated by elevated liver enzymes.

Informal E-Waste Recycling: Unregulated e-waste processing methods, such as open burning and acid leaching, are often carried out in developing countries, exposing workers to hazardous substances.

Nephrotoxicity: Kidney damage resulting from toxic heavy metals found in e-waste, affecting the body's ability to filter and excrete harmful substances.

Neurotoxicity: The harmful effect of toxic substances, such as lead and mercury in e-waste, on the nervous system, leading to cognitive impairments.

Persistent Organic Pollutants (POPs): Toxic chemicals that remain in the environment for long periods, accumulating in the food chain and causing serious health effects.

Phytoremediation: The use of plants to absorb, store, or break down pollutants from contaminated sites, including e-waste processing areas.

Right-to-Repair: Legislation requiring manufacturers to provide repair manuals, spare parts, and tools to consumers, extending the lifespan of electronics.

Supercritical Water Oxidation (SCWO): A high-temperature, high-pressure treatment process used to break down hazardous organic waste, including toxic compounds in e-waste.

Thermal Plasma Technology: A waste treatment technology using extremely high temperatures to convert e-waste into synthesis gas, which can be used for energy production.

Urban Mining: The process of recovering valuable metals like gold, silver, and copper from discarded electronic devices instead of traditional mining.

Chapter 11
The Growing Environmental Risks of E-Waste: A Review

Subhadra Rajpoot
https://orcid.org/0000-0003-3177-1871
Amity Institute of Environmental Toxicology Safety & Management, Amity University, Noida, India

ABSTRACT

One of the most current problems to emerge in the rich as well as the poor globally is electronic garbage, or e-waste. Trash is one of the most challenging and quickly growing problems in the industrial world. Electronic trash, sometimes referred to as e-waste, is a sort of garbage created by electronics. Old or end-of-life electronic equipment, including computers, laptops, TVs, generators, DVD players, cell phones, refrigerators, and other devices that are usually thrown away by their original owners because of their limited lifespan are referred to as e-waste. They have several potentially harmful ingredients that, if not adequately handled, might harm human health as well as the environment. It is difficult to deal with e-waste since it contains hazardous chemicals. According to previous studies, India produced 0.4 million tons of e-waste in 2010, and by 2013–2014, that amount might rise to 0.5–0.6 million tons.

DOI: 10.4018/979-8-3693-7383-5.ch011

INTRODUCTION

Trash is one of the most troublesome and growing problems in the developed world. Waste generated by electronics is known as electronic trash, or E-waste. E-waste refers to old or end-of-life electronic equipment, such as computers, laptops, TVs, generators, DVD players, cell phones, fridges and other devices that their original owners typically ditch after they reach the end of their lifespan. This contains many hazardous elements that may affect both humans and also the environment if they are not managed properly. The e-waste is a huge challenge as it includes harmful chemicals. The electronics industry is the largest and fastest growing industrial sector in the world. Just to maintain an expensive lifestyle, humans depend a great deal on modern-day technologies. Consequently, worldwide demand and consumption rate for these appliances have increased (Arya and Kumar, 2020). As billions of people use electronic gadgets, this forms the basis for the mass production of innovative and technology-based solutions that form the bedrock for the globalization of today international economy. Modern technology leads to tons of electrical waste around the globe. There are about seven billion people on the stone, according to a poll, but to five billion of the old Nori want services and access to be their basic needs. The survey shows approximately 7 billion people exist on the planet, but only 5 billion have about how to MEET their basic needs. Almost 6 billion people use mobile phones, more essential items. It is dangerous for the biosphere, due to its negative effects on human health. However, A Massive Production of E wasted by the Electronics industry is now suffering from the Negative effects. Electronic trash or E-waste is the name of discarded and obsolete devices left by consumers and includes computers, laptops, freezers, generators, televisions and other devices.

Three key factors which are making E-waste a major concern in health is mentioned below: a severe lack of awareness of how to manage E-waste (http://www.basel.int/), an increasing amount of electronic waste threatening the environment and humans. Direct or indirect, all forms of trash and garbage play a significant role in mental and physical health of people (Rautela et al., 2021). Thirdly automation has grown rapidly, leading to a decreasing volume of laws concerning e-waste recycling. One of the factors behind E-waste piling up as seen in Figure 1 illustrates. Various laws and regulations were put around the world regarding the importation and exportation of waste products and their related policies. Nobody has used serious measures on such a serious issue. Partly the reason for the indifference shown towards such wastes, which is increasing massively around the world, is the general notion that the electronic industry will define the future of communications, thus adding to daily economic development. Electronics is meant to be thrown away, which will allow them to maximize profits in the future and improve product performance. These are rarely novel items and for six months or a year of life, their written worth drops significantly, producing electronic waste (LeBel,2012). The present study attempts to analyze the scenario of generation of electronic trash in India and across the world to understand it better and tackle waste and adverse effects of these wastes on environment and human health.

There are different components of electronic items that can endanger human health, and the environment compared to regular municipal waste because they contain some hazardous materials [Santos, J 2021]. Nickel, beryllium and zinc are commonly found in circuit boards while lead, mercury and cadmium are commonly found in television and computer monitors. The Amazing Fleets work: The existence of such materials is leading e-waste recycling and disposal to be a worldwide concern. Some elements in electronic products contain toxic substances which are potentially posing risk to human beings and the environment as compared to regular municipal waste. Whereas nickel, beryllium and zinc are often found in circuit boards, lead, mercury and cadmium are commonly found in television and computer monitors. The presence of these materials makes the recycling and disposal of e-waste an important topic. This chapter is organized like this:

Section 1: Define and Introduction to e-waste
Section 2 — Issues/Regulations – Current State
Section 3: Provides a method of estimating how much e-waste is produced and recycled; Section
Section 4: Discusses the conclusions and limitations of this work.

DEFINITION OF E-WASTE

Published in Environmental Science and Pollution Research Introduction The use of electronic devices and waste generated by these electric and electronic equipments are increasing rapidly with the development of information technology and electronic industries [Adaaw, 2006]. With the high-speed revolution of communication and information technology, we are becoming a part of the new electronic appliances, which leads to fast depreciation of electronic gadgets. As such, e-waste generation is growing exponentially. According to Basel Action Network, "e-waste describes a broad and growing range of electronic devices, from computers to large household appliances, such as refrigerators and air conditioners, to cell phones and personal stereos, to consumer electronics."

Another definition is, an electrical-powered device no longer passes the needs of its current owner for the original use [Sinha, 2004]. World-wide, there is an annual production of 20–50 million tons of electronic equipment waste [Muyeed et al., 2011]. The right disposal and recycling of e-waste poses major issues in both rich and developing countries. The transfer of electronic waste from developed countries to underdeveloped countries has raised concerns regarding the environmental and health consequences of managing e-waste. This article summarizes e-waste, including the e-waste composition and the negative effects, local context, and possible disposal methods.

Categories of E-Waste

Electronic waste, also known as e-waste, is the waste of electronic instruments and devices. E-waste can be classified into different categories based on several parameters, which assists in identifying the types of waste generated and framing suitable recycling and disposal techniques. The main categories of e-waste are:

Big Home Appliances: This includes refrigerators, washers, dryers, and air conditioners. These products are leading sources of e-waste by volume and often contain toxic materials such as refrigerants and heavy metals.

Small Home Appliances: These include microwaves, toasters, coffee machines and electric kettles. Frequently a combination of metals, plastics, and electronic components.

IT and Telecommunication Devices: This category includes computers, laptops, servers, printers, telephones, and mobile phones. More carefully considered, the e-waste is rich in valuable metals (such as gold, silver, and palladium) and, thus, valuable targets for recycling activities [Balogun A.L.2022].

Consumer Electronics Consumer electronics include televisions, radios, cameras, and audio equipment. The evolution of technologies and consumer need for cutting-edge features lead to their regular replacement which accounts for a substantial part of e-waste.

Lighting equipment – this means fluorescent tubes and compact fluorescent lamps (CFLs), as well as light-emitting diode (LED) lamps Many of these items contain toxic materials, such as mercury, requiring proper disposal.

Power & Electrical Engineering Tools: Power tools such as power drills, power saws, sewing machines, or electrical garden tools fall under the category of electrical and electronic tools. While these items are often manufactured to last, they eventually become part of e-waste when their function no longer serves us.

Toys, Hobbies and Sports: Items include electronic toys, video game consoles and fitness machines. Electronics in these items means that they are part of the e-waste stream when disposed.

Medical Devices: Medical devices such as electronic thermometers, blood pressure monitors and imaging equipment contribute to e-waste. Medical data is sensitive and may pose biohazard risks, so proper disposal is critical.

Monitoring and Control Instruments: This category includes all instruments utilized in industrial processes, including thermostats, industrial robots, and laboratory apparatus. These devices have often specialized parts which need to be handled carefully during recycling.

Automatic Dispensers: ATMs, ticket machines, vending machines, etc. Such devices usually come with complex electronics and mechanisms that contribute to the e-waste challenge.

By categorizing e-waste, it can be easier to generate strategies to ensure e-waste is collected, recycled or disposed of properly — thereby reducing its environmental impact and allowing valuable resources to be recovered.

Figure 1. eWaste is growing in India

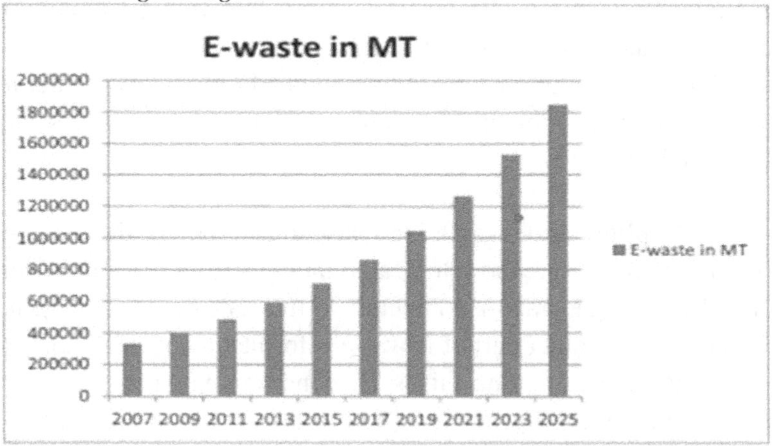

Source: Aniket Ingole (2018)

Sources of Waste

The nation's economy is functioning at a respectable level due to the increasing interest in electronic waste, this especially implies (laptops, PCs, solar cells, joint gadgets, cell phones). Hence, there is exponential growth in the IT sector. To promote new technology in the market, the public and commercial sector replaces obsolete electronic equipment with new ones, resulting in more e-waste rubbish. Products that generate Domestic E-waste include PCs, household appliances, small and large businesses, and institutions, as well as other industries. As homes contribute least to the domain of computers and mobile phones [Mundada et al., 2004]. Technological growth in recent years has occurred rapidly, and the ever-increasing consumer demand for faster, smarter, and more accessible gadgets has consequently caused a rise in e-waste generation. These electronics may be disposed of at the end of life, end of service or simply just surplus products that a manufacturer no longer needs. Moreover, the regular upgrading of electronic devices, along with the brief life cycles of many products, exacerbates the proliferation of e-waste. Consequently, the improper handling of electronic waste has emerged as a significant environmental and public health concern, emphasizing the need for proper recycling, disposal, and sustainable manufacturing methodologies. The largest contributors to e-waste are:

Consumer Electronics

- Mobile Phones and Smartphones: With rapid technological advancements and frequent upgrades, devices are replaced rapidly.
- Computers and Laptops: Home computers, laptops, and tablets that have gone out of date or are no longer working.
- Televisions and Monitors: CRT, LCD, and LED TV and computer monitors.
- Audio & Video Equipment — including DVD players, sound systems, and other such devices

Household Appliances: Considering households produce the least amount of e-waste. It's tricky to measure the actual amount of E-waste produced. Not domestic items like freezers, generators, computers or freezers, says a survey. It is about 20–21% alone, and the other part of E-waste is formed by other sectors. They can be Big Stuff: Fridges, washing machines, air-conditioning units and ovens.

Smaller Appliances: Microwaves, the toaster, coffee mills and the vacuum cleaner.

Office and Business Equipment: The business includes government agencies, the public and private sectors, multinational corporations, etc., providing the primary industry because the largest driver of E-waste development is this industry [Rautela et al., 2021]. According to the report, they are attacked on up to 79% of all installed PCs, which is the highest attack rate across all sectors. The number of old PCs being thrown away by companies and families is around 1.38 million.

- Printers and Scanners — Consumer and commercial grade.
- Copiers and Fax Machines: Usually found in office cubicles.
- Networking devices: Routers, switches and servers.

Medical Devices

- Diagnostic Equipment: MRI machines, X-ray machines, and similar imaging machines.

- Medical Imaging Devices: MRI, CT and ultrasound machines

Lighting Equipment

- Fluorescent Lamps: Contains toxic materials such as mercury.
- LED Bulbs: They are energy-efficient but still end up as e-waste after their usable life.
- Batteries & Accumulators: Rechargeable Batteries: From laptops, smart phone and other portable electronics Non-rechargeable Batteries: USBs most small devices go to.
- Cables and Circuit Boards: Power cords, USB cables, HDMI cables. Printed Circuit Boards (PCBs): Found in nearly all devices, they contain precious metals and toxic substances.

Some of the dangerous materials in e-waste include lead, mercury, cadmium, and brominated flame retardants, which can present major environmental and health hazards. These dangers are mitigated, and we can recover valuable materials from properly managed and recycled e-waste.

COMPOSITION OF E WASTE

Important, and possibly dangerous, materials are often found in e-waste. The e-waste composition is greatly influenced by age4, model and maker of the type of electronic equipment and its date of production. The precious metal content in scrap from IT and telecom systems is higher than that of home products [S, Rajpoot, 2021]. We also must consider responsibly processing e-waste in order not to waste limited and valuable materials. The changed composition of e-waste generated by technical advancements, especially for electronic parts are critical issues of strategy development, and parcels of this matter [S,Rajpoot 2024]. In this regard, e-waste recycling and disposal methods must evolve to meet the changing composition of e-waste [Köhler and Erdmann, 2004]. Several variables, such as the condition of the economy, a market that allows for reuse, the needs of the recycling industry, trash separation campaigns, and even the implementation of laws, influence the makeup of e-waste. Figure 2 shows the unique materials contained in waste electrical and electronic equipment.

Figure 2. The distinctive contents of a WEEE

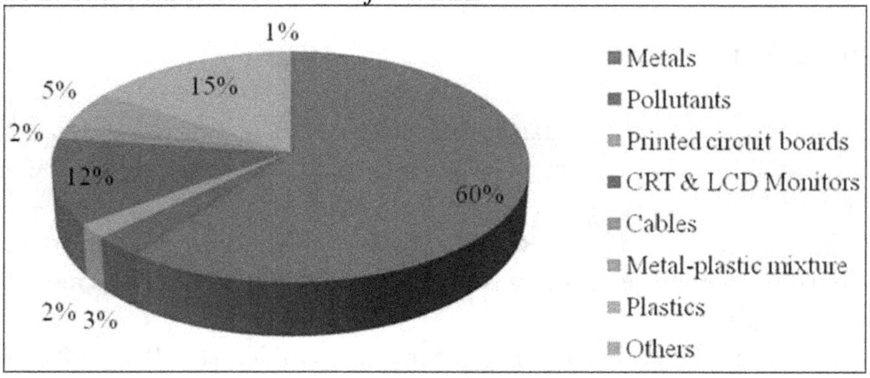

Source: Awasthi (2017)

Global E-Waste Production

The global production of e-waste is now outgrowing and outpacing recycling at an alarming rate of five-to-one, according to a new E-Waste Monitor from the United Nations (UN) published this week. According to a survey by the I.T.U. and UNITAR, the 62 million tons of e-waste produced between now and 2022 would fill 1.55 million 40-ton trucks or enough vehicles to create a bumper-to-bumper parade around the equator. Usually, as economies are growing and more people have access to technology, they expect the electronic waste to increase globally, as higher GDPs result in higher sales in electronic markets, thus producing more electronic waste. Around 57.4 million metric tons of e-waste were generated globally in 2021. That number is projected to rise by roughly 2 million metric tons each year to 74 million metric tons by 2030 if development trends hold. The largest producer of e-waste is Asia, followed by the Americas and Europe. Africa and Oceania produce the least e-waste. The global recycling rate is quite low in spite of the rising production of e-waste. Only 17.4% of e-waste was officially recorded as being collected and recycled in 2019. The other 82.6 percent is probably going to landfills, getting recycled informally or going unaccounted for." e-waste includes precious materials such as gold, silver, copper, and platinum. The total value of the raw materials from global e-waste was estimated to be about $57 billion in 2019. But the potential economic value is mostly squandered as a result of subpar recycling methods.

The world is changing rapidly in terms of information and technology, leading to the rise of technological products around the globe [Rautela et al., 2021; Gaidajis et al., 2010]. In reality, under-recognized in 2022, less than a quarter (22.3%) of the total e-waste mass was recognized as being effectively collected and recycled, which poses the threat of leaking toxins into communities around the world and leaving $62 billion worth of recoverable natural resources up for grabs. Globally, e-waste is growing by 2.6 million tons annually and is expected to reach 82 million tons by 2030, a further 33% increase95 on 2022 figures.) Examples include any unplugged or battery-powered device which can be regarded as e-waste and contribute to the health and pollution risk because of the hazardous compounds in their components (e.g. mercury) that can be harmful to the nervous system and brine— the amount of e-waste will be around 20–50 Mt each year, or 1–3 percent of the world's total garbage production (Perkins et al., 2014; Widmer et al., 2005). This is seen in Figure 2, showing the annual E-waste produced in metric tons. In 2010, the total amount contributed by pcs, mobile phones and televisions was found to be 5.5million tons according to an analysis. In high-income countries, e-waste represented 8% of all waste generated. The mass (kg) and number (N) of each of the electronic gadgets available in the market, its usage N, and the average life (L) of each item (years) measure its contribution to the annual production of E-waste, where MN/L = E.

E-Waste Production in India

Because electronic waste has become the fastest growing source of waste in the world, or E-waste. Rising living standards, technological developments and industrialization have been associated with the rapid growth of the E-waste stream. The world has produced about 53.6 million metric tons of electronic waste in 2020, but only 17–18% of waste is recycled, according to the global market. India is the third largest producer of E-waste in the world, contributing over 3.2 million metric tons of waste every year, preceded by USA and China. E-waste generation soared from 700,000 tons in 2017-18 to 1.6 million tons in 2021-22. This trend is only set to continue, with an estimated 14 million tons of e-waste by 2030. This projection has been made by the International Association of Electronic Recyclers (IAER). In 2010, there were 3 billion electrical and electronic appliances labeled WEEE. Around 20 to 50 million tons of e-waste are disposed of globally each year. If all waste that is generated (municipally) is considered, it accounts for 5% of the total waste. According to the Comptroller and Auditor-General's (CAG) Report, the country adds 7.2 MT industrial hazardous waste, 4 lakh tons electronic waste, 1.5 MT plastic waste, 1.7 MT medical waste and 48 MT municipal waste every year. Seventy percent of all e-waste generated in the country comes from ten states. 65 cities in the country to produce more than 60% of India's e-waste. Thereby gives Mumbai the first place as the top 10 cities that produce e-waste, after Mumbai, Delhi, Bengaluru, Chennai, Kolkata, Ahmedabad, Hyderabad, Pune, Surat, Nagpur. In

India, 70% of electronic or e-trash comes from the government, public and private (industrial) sectors. Household unit: 15%

E-waste management is a matter of serious concern which was addressed by PM Narendra Modi in December 2020 when he noted how important it is to cycle through electronic equipment as effectively as possible, keeping the old trash free from our conductive lands. His 'Garbage to Wealth' idea focuses on recycling garbage — a concept he launched in August 2021. According to the data from the Central Pollution Control Board of India (https://science.thewire.in), a total of about 1,014,961 tons of E-waste was generated in 2019 and 2020. Even though major production and consumption are present at high rates in the nation, official recycling and disposal protocols are only minimally practiced in a handful of cities and towns in the country. Millions of people in various sectors of activism have been harmed and therefore most governments have created recycling policies based on this fact.

However, since members of the informal sector are not informed and educated about the harmful health effect [Awasthi and Li, 2017], it is not intercessive. Figure 4 shows the many origins of the E-waste being produced in India. 15% from household activities, a further 15% through public, commercial and government sectors, and most of the remaining 70% also from electrical and electronic waste.

Figure 3. e waste production in India

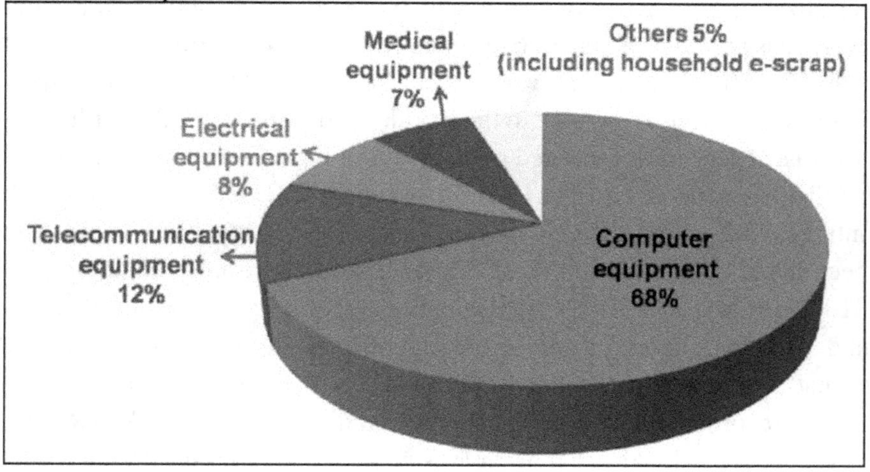

Source: Mudali (2021)

Impact Of E-Waste: Electronic or E-waste is one of the fastest-growing waste streams across the world and presents a significant challenge to the environment and human health. When e-waste is not properly disposed of, harmful substances, including lead, mercury, and cadmium, can leach into the environment, polluting soil, water and air. These pollutants are known to cause serious health problems, including respiratory problems, neurological damage and cancers. Furthermore, e-waste contains precious materials such as gold, silver, and copper, which, if not correctly recycled, represent a lost economic opportunity [Gupta, A 2022]. In many developing countries, the informal recycling sector compounds these problems, because workers pick apart electronic devices without adequate safety precautions, exposing them to toxic substances. Consequently, tackling the issue of e-waste necessitates a comprehensive solution that includes stringent regulations, sustainable recycling practices, and heightened public understanding of the need for responsible e-waste disposal and recycling-waste or electronic waste, which contains harmful substances indelibly contained in many electronic devices, can lead to serious damage to human health, animals, environment, soil etc. Here are some of the major effects.

Impact on Human Health

- Exposure to Toxic Substances: E-waste contains hazardous materials such as lead, mercury, cadmium, arsenic, and flame retardants. Exposure to these substances can occur through direct contact with e-waste components, inhalation of toxic fumes during improper recycling processes, or contamination of soil and water [Parvez, Sarker M et al., 2021].
- Respiratory Problems: Burning e-waste to recover valuable materials releases toxic fumes and fine particulate matter, leading to respiratory problems. Workers in informal e-waste recycling operations often suffer from chronic respiratory diseases due to prolonged exposure to these pollutants [Althaf et al., 2021].
- Neurological Effects: Heavy metals like lead and mercury can cause severe neurological damage, especially in children. Exposure to lead, for instance, can result in cognitive impairments, developmental delays, and behavioral issues
- Reproductive and Developmental Issues: Chemicals such as polybrominated diphenyl ethers (PBDEs) used as flame retardants in electronics can disrupt endocrine function, leading to reproductive health problems and developmental issues in children [Forti et al., 2020; Ghimire and Ariya, 2020]
- Cancer and Other Chronic Diseases: Long-term exposure to certain e-waste components, such as cadmium and PCBs (polychlorinated biphenyls), has been linked to an increased risk of cancer, kidney damage, and other chronic health conditions.
- Bioaccumulation: Many of the toxic substances in e-waste can bioaccumulate in the human body over time, leading to chronic health problems. For exam-

ple, mercury can accumulate in the brain and kidneys, causing severe damage over time.

- Occupational Hazards: Workers in informal e-waste recycling sectors often lack protective equipment and are regularly exposed to hazardous substances, resulting in a higher incidence of health issues compared to the general population.

Addressing the health impacts of e-waste requires proper handling and recycling practices, the use of protective equipment, and policies aimed at reducing the hazardous substances used in electronic devices. Additionally, public awareness and education about the dangers of e-waste can help mitigate its adverse health effects.

Figure 4. Impact of e waste on human health

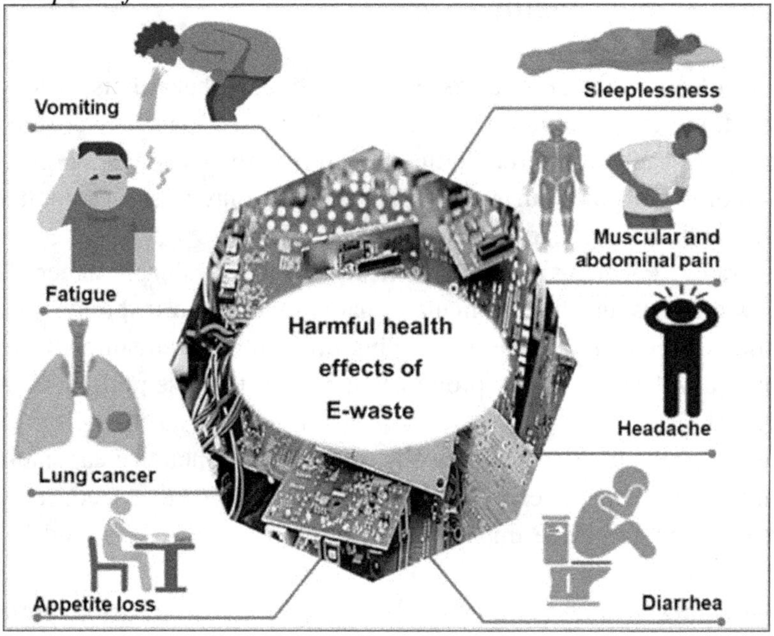

Source: Gratein (2023)

The Negative Effects on Air

Dismantling, splitting former or melting the e-waste emissions of air dust particles or its poisonous comes out, dioxins, pollution and manages damage systems in people by air. [Mundada, M. N., Kumar, S., & Shekdar, A. V. (2004]. Burning is a common method to dispose of low-value e-waste, but it can also be used to recover valuable metals such as copper from appliances. Burning e-waste also releases tiny particles that can travel hundreds of miles and threaten the health of both people and animals, raising the risk of cancer and chronic diseases. Waste electronics, highly integrated units, are often dismantled using chemical means, including acids and desoldering, and emit fumes where recycling is uncontrolled, all high-value components like gold and silver are removed. The most at risk from the toxic air effects of informal e-waste recycling are people who work with this stuff, but pollution from it can spread thousands of kilometers from recycling sites. species is more vulnerable to the air pollution caused by e-waste than to other species, threatening both these species and the biodiversity of areas in the grip of chronic pollution. Air pollution could also damage soil, plant species, and water quality over the long term, leading to irreversible damage to ecosystems. Depending on market needs, some e-waste recycling methods are in the explosive development stage, like e-waste collection points arranged by groups developing precious metals from e-waste in Guiyu, China How to develop e-waste in an environmentally friendly way is only in the exploration stage. Consequently, the surrounding air is heavily polluted with lead, which people inhale and then consume when exposed to water and soil [Wang K, Qian J, Liu L., 2020]. This could do greater neurological damage to nearby humans, wildlife and larger animals.

The Negative Effects on Soil

When e-waste is not appropriately disposed of—by dismantling, checking, or melting the materials—dust particles or toxins—such as dioxins—positioned into the air, bringing about pollution and destruction to respiratory frameworks. Burning is a common way of getting rid of low-value electronic waste, but it can also be used to extract high-value metals such as copper from devices. Burning e-waste produces tiny particles, too, that can travel hundreds of miles and cause deadly health threats to people and animals, raising the likelihood of cancer and chronic disease. Higher value components such as gold and silver from electronics heavily integrated into these components, when removed, become dead weight that can be processed in the similar manner as that of precious metals through chemicals like acids and desoldering that can also release fumes in areas in which recycling can be uncontrolled. E-waste, or electronic waste, is a danger to our soil health because it contains toxic materials. Most importantly, here are some major adverse impacts on the soil.

- Toxic Leachate: When e-waste is inappropriately disposed of hazardous substances like lead, mercury, cadmium and brominated flame retardants can seep into the soil. This leachate is harmful to plants and microorganisms, and can contaminate the soil and groundwater, rendering the land unusable for agriculture. [Twagirayezu, Gratien, et al]

- Heavy Metal Contamination: Metals such as lead and cadmium from e-waste can accumulate in the soil, causing nutrient imbalance and harming plant health. These metals can get absorbed by plants, making their way into the food chain and presenting health risks to animals and humans.
- Soil Acidification: This can occur when certain components in e-waste, including batteries, leach acids to the soil, thereby decreasing soil pH. This may lead to soil infertility and obliterate plant growth as well as microbial activity.
- Soil Microbial Community Disruption: Toxic materials such as heavy metals and solvents seep from e-waste to disrupt the microbial communities in the soil. Beneficial microbes are essential for nutrient cycling and soil structure maintenance, and this can affect soil health and its ability to sustain plant life. [Ankit, Lala Saha, Virendra Kumar et al., 2021]
- Decreased Soil Fertility: Contaminants leached from e-waste can bind with essential soil nutrients, preventing their uptake by plants. This results in a decline in soil fertility and agricultural output. [Ankit, Lala Saha, Virendra Kumar et al., 2021]
- Bioaccumulation: Soil organisms and plants absorb toxic substances from e-waste, causing bioaccumulation. Over time, such toxins can bioaccumulate, meaning they can reach higher concentrations in organisms at the top of the food chain, including humans, leading to severe health problems. [Ankit et al., 2021]

Effective management of e-waste, including recycling, recycling, responsible disposal, and policies to reduce e-waste creation, are necessary for mitigating these impacts.

The Negative Effects on Water

Heavy metals in e-waste, such as mercury, lithium, lead and barium, leach deeper into the soil and contaminate groundwater after polluting soil. This prevents the heavy metals from seeping into groundwater, preventing the heavy metals from entering lakes, rivers, ponds, and streams. Those pathways can make water toxic and acidified and threaten communities, plants and animals even when they're far from recycling sites. Finding potable water becomes a challenge. Acidification can kill both freshwater and marine life, destroying ecosystems and disrupting biodiversity. Even if acidification occurs in water sources, ecosystems can be affected to the extent that recovery is unlikely, if not impossible. [Gazzeh K.2022].

STRATEGY TO MANAGE E-WASTE

Managing e-waste in India requires a multi-faceted approach involving policy enforcement, infrastructure development, public awareness, and stakeholder collaboration. Here's a comprehensive strategy:

1. **Policy and Regulation:** Strengthening Legislation: Implement and enforce stringent e-waste management rules. The E-Waste (Management) Rules, 2016, should be updated regularly to incorporate best practices and new technologies. We must Ensure Extended Producer Responsibility (EPR) where manufacturers are responsible for the end-of-life management of their products. Provide incentives for companies and consumers who follow proper e-waste disposal methods and impose penalties for non-compliance.
2. **Infrastructure Development:** Establish a network of easily accessible e-waste collection centers across urban and rural areas. For proper recycling we must Develop advanced recycling facilities that can handle the diverse types of e-waste and ensure environmentally sound recycling practices and to Promote the establishment of certified e-waste handling and recycling enterprises.
3. **Public Awareness and Education** Awareness Campaigns: Conduct nationwide awareness campaigns to educate the public about the hazards of improper e-waste disposal and the benefits of proper recycling. School and College Integrate e-waste management topics into school and college curricula to educate the younger generation.
4. **Technological Innovations:** Invest in R&D for developing efficient e-waste recycling technologies and methods. Encourage the design of electronic products that are easier to recycle and have a longer lifespan [Leung, R.2020].
5. **Collaboration and Partnerships:** Foster partnerships between the government, private sector, and non-governmental organizations (NGOs) for effective e-waste management. Collaborate with international bodies to adopt the best global practices and technologies in e-waste management.
6. **Economic Models:** Promote a circular economy model where electronic products are designed for durability, reuse, and recycling. Develop financial models that make recycling economically viable, such as subsidizing recycling costs or offering tax benefits.
7. **Monitoring and Evaluation:** Establish robust systems for data collection on e-waste generation, collection, and recycling. Conduct regular audits and assessments to evaluate the effectiveness of e-waste management strategies and make necessary adjustments.

8. **Role of Informal Sector:** Integrate informal e-waste workers into the formal sector by providing training and certification programs. Offer support and training to informal sector workers to improve their skills and safety standards. Implementing these strategies can significantly improve e-waste management in India, leading to environmental sustainability and public health protection.

Countries with developed economies have implemented several strategies, including LCA, MFA, MCA, and EPR, to reduce the environmental effect of e-waste.

LIFE CYCLE ASSESSMENT (LCA)

Life Cycle Assessment (LCA) is a comprehensive method for evaluating the environmental impacts of a product, process, or service throughout its entire life cycle, from raw material extraction to disposal. For managing e-waste, the LCA strategy can help identify and mitigate environmental impacts associated with electronic products. A Life Cycle Assessment (LCA) strategy for managing e-waste involves a comprehensive evaluation of the environmental impacts associated with electronic products throughout their entire life cycle—from raw material extraction to production, use, and end-of-life disposal or recycling [Zeng et al.2013], the life cycle viewpoint is a simple method for examining a product's overall performance and figuring out what is ecologically sound at various life cycle stages. Several studies on the effects of LCA on the environment, product development, and eco-design have been conducted by various scholars and organizations. Because end-of-life e-waste treatment has a negative environmental impact, it is quite concerning. Electric goods are either recycled or disposed of in landfills after their useful lives are over. The LCA process begins with inventory analysis, which identifies and quantifies the energy, water, and materials used as well as the emissions and waste generated at each stage. This is followed by an impact assessment that evaluates the potential environmental effects, such as resource depletion, pollution, and greenhouse gas emissions. By identifying critical impact areas, stakeholders can develop targeted strategies to minimize adverse effects. For instance, improving product design for easier disassembly and recycling, promoting the use of sustainable materials, and enhancing collection and recycling infrastructures. Ultimately, an LCA approach to e-waste management helps in making informed decisions that support sustainability, reduce environmental footprints, and encourage responsible consumption and disposal of electronic products.

While LCA is also used in America to evaluate the adverse effects of e-waste management, it has been utilized in Asia to estimate the impact of e-waste and e-waste management (Kiddee et al., 2013). Compared to land filling or incineration, recycling is the most appropriate approach for managing e-waste, as determined by most research that use life cycle assessment (LCA) as their instrument. However, other studies concluded that recycling was not an effective alternative.

EXTENDED PRODUCER RESPONSIBILITY (EPR):

Extended Producer Responsibility (EPR) for e-waste management is an environmental policy approach where producers are given a significant responsibility—financial and/or physical—for the treatment or disposal of post-consumer products. EPR aims to incentivize manufacturers to design environmentally friendly products by making them accountable for the entire lifecycle of their products, especially for their take-back, recycling, and final disposal. In the context of e-waste, which includes discarded electrical or electronic devices, EPR shifts the responsibility from municipalities and consumers to the producers, encouraging them to develop more sustainable and efficient recycling processes. This policy not only reduces the environmental impact of e-waste but also promotes the development of a circular economy, where resources are reused and recycled, reducing the need for virgin materials and minimizing waste. Effective implementation of EPR for e-waste can lead to significant reductions in hazardous waste, conservation of natural resources, and the mitigation of environmental pollution.

The EU implemented the 2002 WEEE rule, which required items to be delivered for processing and recycling, in 2004. This law binds manufacturers to downstream handling of electronic waste and promotes ecologically sound e-waste recovery, recycling, and reuse at the end of its useful life (EU, 2002). Yet, unless an EPR system with a distinct role is established, all the legislation will be ineffective. When designing an EPR system, five main factors must be considered: producer responsibility, legal regulation, system coverage, system financing, and ensuring compliance (Widmer et al., 2005).

Material Flow Analysis (MFA)

Material Flow Analysis (MFA) for e-waste is a powerful tool used to quantify the flow and stocks of electrical and electronic equipment (EEE) throughout its lifecycle, from production and consumption to disposal and recycling. MFA helps in identifying the pathways of e-waste, the quantities involved at each stage, and the potential environmental impacts. By analyzing these flows, policymakers and stakeholders can better understand the sources and destinations of e-waste, enabling the development of effective strategies for its management. MFA reveals critical points where interventions can be applied to reduce waste generation, enhance recycling rates, and minimize environmental harm. It also aids in assessing the efficiency of current waste management systems and identifying opportunities for improvement. The insights gained from MFA can support the creation of circular economy models, where materials are continuously reused and recycled, reducing the demand for virgin resources and mitigating the environmental footprint of e-waste.

Numerous studies have been conducted on this subject, which is useful in understanding how recycling and disposal occur in various nations. A variety of techniques, including the market supply approach and the survey method, are being utilized to assess potential e-waste volumes. When there is fast economic development and little data available, MFA and economic evaluation might be helpful tools.

E-Waste Administration in India

The significant quantity of E-waste generated and recycled is a significant issue. There were just 20 recycling facilities in two major cities in India as of 2012. For example, the informal sector is entirely dedicated to the removal, classification, and reuse of E-waste. To extract valuable metals, garbage is dismantled in a dangerous manner in India (Turaga et al., 2019; Ministry of Housing and Urban Affairs, 2022). Technology used in the organized sector to treat e-waste is safe and environmentally friendly. Copper, plastic, iron, and aluminium are among the recyclable materials that make up more than 95% of e-waste. The Recycling Unit in the NCR region claims to separate over a dozen components out of trash. Five percent of the hazardous E-waste is sent to treatment and disposal facilities that have been certified by the government for processing and recycling. Statistics from the CPCB show that only a small number of the about 65 recyclers in the formal sector have licenses. To get rid of their electronic trash, several of these businesses have partnered with other organizations (Corwin, 2018).

The different phases of recycling e-waste in a formal setting are outlined below: The processing for e-waste is split into two stages: manual and mechanical. Nonferrous minerals are transformed into copper, aluminium, and other metals at metallurgical treatment plants. Efficiency hence beats recycling and recovery rates. There are both registered and unregistered facilities for the disposal of electronic waste in India, even though there is a lot of information about the larger, legal, informal recyclers of electronic rubbish that are not well documented [Heacock,2018]. Conversely, almost 90% of India's e-waste is managed by individuals in the unorganized sector. To focus on lucrative components like gold, iron, copper, and silver, informal units use riskier approaches. These detrimental behaviors are now directly resulting in several people suffering from a variety of health issues. Environmental harm results from the disposal of unwanted trash because of these actions (Heacock et al., 2018).

Conversely, most workers in the unorganized sector lack formal education and are not aware of the long-term risks associated with discharging hazardous chemicals into the environment and water supplies. Cathode ray tubes require hand removal of copper, metal, and glass. Circuit boards are embellished with microchips, condensers, and gold-plated brass pins. Heat treatment of these common components produces toxic vapors. Acid is used to separate gold and brass from gold-plated brass pins. Large amounts of acid are required to dissolve metallic parts found in microchips and condensers (Singh et al., 2020).

E-Waste Collection

Beyond what is already the case, further safety measures are needed to combat the growing threat that e-waste poses to public health. Worker protection, company strategy development, waste management method collection and control, and monitoring mechanism design are other actions that need to be implemented. Garbage management strategies may be substantially improved by establishing objectives, imposing penalties, or offering municipalities financial incentives to collect the garbage. All parties who participate in the collection and treatment of such garbage should be required to provide regular reports to a central organization to enhance the control and effectiveness of E-waste management. Electronic waste management transactions must be carried out via traceable bank accounts or credit cards to uphold accountability and transparency. It is necessary to create and maintain a central register for data collection and monitoring activities [Heacock et al., 2018].

Recycling electronic waste (e-waste) involves several methods to recover valuable materials and reduce environmental impact. Here are some common recycling methods for e-waste:

1. **Manual Disassembly:** Manual disassembly of e-waste involves the physical separation of electronic devices into their constituent parts by human labor. This process is crucial for the effective recycling and recovery of valuable materials such as metals, plastics, and electronic components. Unlike automated disassembly, manual disassembly allows for a higher degree of precision and care, ensuring that delicate components are not damaged and can be reused or sold as spare parts. It also enables the identification and removal of hazardous substances like batteries and capacitors, reducing environmental and health risks. However, manual disassembly is labor-intensive and can be costly, requiring skilled workers and proper safety measures [Lopes, A 2018]. Despite these challenges, it remains a vital step in the e-waste recycling chain, especially for complex devices and in regions where labor costs are lower. By maximizing material recovery and minimizing waste, manual disassembly contributes to

more sustainable e-waste management practices. Workers manually separate valuable components from e-waste, such as circuit boards, hard drives, and plastics.

Pros: Precise and can recover valuable components intact.

Cons: Labor-intensive and time-consuming.

2. **Mechanical Processing:** Mechanical processing of e-waste involves the physical dismantling and separation of electronic waste components to recover valuable materials and reduce environmental harm. This process is a crucial part of the recycling and management of e-waste, and it typically includes several stages [Ohajinwa et al.,2017]. Uses machinery to crush and shred e-waste into smaller pieces, followed by separation processes

Key Stages of Mechanical Processing

Collection and Transportation:

- Collection: E-waste is collected from various sources, including households, businesses, and recycling centers.
- Transportation: Collected e-waste is transported to processing facilities.

Dismantling:

- Manual Dismantling: Larger items, such as televisions and computers, are manually dismantled to remove hazardous components (like batteries and capacitors) and segregate different materials.
- Automated Dismantling: Machines may also be used to aid in the dismantling process, increasing efficiency and safety.

Shredding:

- Primary Shredding: E-waste is fed into shredders that break it down into smaller pieces, making it easier to handle and sort.
- Secondary Shredding: Further size reduction may be required to ensure that materials are small enough for effective separation.

Separation:

- Magnetic Separation: Magnets are used to separate ferrous metals (like steel) from non-ferrous metals and other materials.

- Eddy Current Separation: This process uses magnetic fields to separate non-ferrous metals (such as aluminium and copper) from non-metallic materials.
- Density Separation: Materials are separated based on their density using air or water, often in a float-sink tank.
- Electrostatic Separation: Differences in electrical conductivity are used to separate materials.

Recovery and Refining:

- Material Recovery: Separated materials, including metals, plastics, and glass, are collected for further processing.
- Refining: Some materials may require additional refining to remove impurities and achieve the desired purity levels.

Benefits of Mechanical Processing:

- Resource Recovery: Valuable materials like precious metals (gold, silver, platinum), base metals (copper, aluminium), and plastics are recovered for reuse.
- Environmental Protection: Proper processing reduces the release of hazardous substances (like lead, mercury, and cadmium) into the environment.
- Economic Value: Recovered materials can be sold, providing economic incentives for e-waste recycling.

Challenges and Considerations:

- Complexity: The diverse composition of e-waste makes processing complex and requires sophisticated technology.
- Hazardous Materials: Safe handling and disposal of hazardous components are essential to protect workers and the environment.
- Economic Viability: The cost of mechanical processing must be balanced against the value of recovered materials and environmental benefits.
- Regulation Compliance: Adherence to local and international regulations governing e-waste management is crucial.

In this chapter we are explaining the four main techniques typically considered for e waste management, and they are:

Recycling: Recycling of electronic waste (e-waste) is a critical component of sustainable waste management. E-waste encompasses discarded electronic devices such as smartphones, computers, and televisions, which often contain hazardous materials like lead, mercury, and cadmium. Proper recycling processes ensure that these toxic substances are safely managed, preventing environmental contamination and health risks. Additionally, e-waste recycling allows for the recovery of valuable materials like gold, silver, and copper, reducing the need for mining and conserving natural resources. Effective e-waste recycling not only mitigates the harmful impacts of electronic waste but also promotes a circular economy by reintroducing recycled materials into the manufacturing process, thereby reducing the overall environmental footprint of new electronic products (Barba-Gutierrez et al., 2008; Lu et al., 2006). This involves converting waste materials into new materials and objects. It reduces the consumption of fresh raw materials, decreases energy usage, lowers greenhouse gas emissions, and minimizes pollution.

Composting: This process involves the biological decomposition of organic waste, such as food scraps and yard waste, into a nutrient-rich material that can be used as a soil conditioner. It helps divert waste from landfills and provides a valuable product for agriculture and gardening. Composting e-waste is not a viable method for managing electronic waste due to the unique composition of e-waste, which includes metals, plastics, and hazardous materials [Heeks et al., 2015]. Unlike organic waste, which can decompose naturally through biological processes, e-waste contains toxic substances such as lead, mercury, and cadmium that pose significant environmental and health risks. These materials do not break down through composting and can contaminate the soil and water if improperly handled. Effective e-waste management requires specialized recycling processes that can safely extract and reuse valuable materials while preventing the release of hazardous substances [Smith, A 2023]. This involves dismantling electronic devices, separating different components, and using advanced technologies to recover metals and other materials. Proper e-waste recycling helps to conserve resources, reduce pollution, and prevent environmental degradation.

Landfilling: This is the most common method of waste disposal, where waste is buried in the ground. Landfilling e-waste, or electronic waste, is a significant environmental and health concern. E-waste comprises discarded electronic devices, such as computers, smartphones, and televisions, which contain hazardous materials like lead, mercury, cadmium, and brominated flame retardants. When these items are disposed of in landfills, toxic substances can leach into the soil and groundwater, posing severe risks to both the environment and human health. Additionally, e-waste contains valuable metals such as gold, silver, and palladium, which are lost when the waste is not properly recycled. Landfilling e-waste also contributes to the growing problem of landfill space scarcity. Despite these issues, many countries still rely heavily on landfilling due to insufficient recycling infrastructure and public awareness. Addressing the challenges of e-waste management requires robust policies promoting recycling and proper disposal, improved collection systems, and public education to reduce the environmental footprint of discarded electronics. Modern landfills are designed to minimize environmental impact, often incorporating systems for managing leachate and capturing methane gas. However, landfilling can still pose significant environmental challenges, such as groundwater contamination and methane emissions.

Incineration: This method involves burning waste at high temperatures to reduce its volume and mass, often recovering energy in the process. While it can significantly decrease the amount of waste that ends up in landfills, it can also produce harmful emissions if not properly managed. The incineration of electronic waste (e-waste) is a method of waste disposal that involves burning e-waste at high temperatures to reduce its volume and potentially recover energy. [Hoornweg D.,2012] This process can significantly decrease the physical bulk of e-waste, making it easier to manage. However, incineration of e-waste poses serious environmental and health risks. Burning e-waste releases toxic substances, including heavy metals like lead and mercury, as well as harmful chemicals such as dioxins and furans, into the atmosphere. These pollutants can have severe health implications for humans, causing respiratory issues, neurological damage, and even cancer. Furthermore, the environmental impact is substantial, as these toxins can contaminate air, soil, and water, affecting wildlife and ecosystems. Despite these risks, advancements in incineration technology aim to mitigate some of these negative effects through better filtration and gas cleaning systems. However, the consensus remains that recycling, and more sustainable waste management practices are preferred over incineration to handle the growing e-waste problem responsibly.

Biological procedure for E-waste: Utilizing live organisms to decompose and retrieve valuable elements from electronic trash is the biological process for managing e-waste. This technique, called phytoremediation or bioleaching, makes use of the skills of certain fungus, bacteria, and plants. By using microorganisms that create acids to solubilize metals, precious elements like gold, copper, and palladium may be extracted from e-waste through a process called bioleaching (Awasthi et al., 2019). Metals may be liberated from complicated materials and materials broken down by fungi, especially some species of Penicillium and Aspergillus. In a process known as phytoremediation, plants are used to extract heavy metals from the soil in areas where e-waste has been dumped. The metals are then recovered by harvesting the plants. This biological method is a promising one since it is less harmful to the environment, requires less harsh chemicals, and produces less pollutants. Bioleaching, Biosorption, Bioaccumulation, Biotransformation, Biomineralization, and Microbially-Enhanced Chemisorption of Metals are the six main processes involved in the micro remediation of hazardous metals.

It is still in the early stages of development, though, and has issues with scalability, efficiency, and the requirement for ideal circumstances for the organisms involved.

REFERENCES

A Closer Look: Lithium-Ion Batteries in E-waste. (n.d.). https://www.simslifecycle .com/ 2019/05/21/a-closer-look-lithium-ion-batteries-in-E-waste/

Abalansa, S., El Mahad, B., Icely, J., & Newton, A. (2021). Electronic waste, an environmental problem exported to developing countries: The GOOD, the BAD and the UGLY. *Sustainability (Basel)*, *13*(9), 5302. DOI: 10.3390/su13095302

Akenji, L., Hotta, Y., Bengtsson, M., & Hayashi, S. (2011). EPR policies for electronics in developing Asia: An adapted phase-in approach. *Waste Management & Research*, *29*(9), 919–930. DOI: 10.1177/0734242X11414458 PMID: 21730041

Ankit, L. S., Kumar, V., Tiwari, J., Sweta, S. R., & Jiwan, S. K. B. (2021). Electronic waste and their leachates impact on human health and environment: Global ecological threat and management". *Environmental Technology & Innovation*, *24*(November), 102049. DOI: 10.1016/j.eti.2021.102049

Ari, V. (2016). A review of technology of metal recovery from electronic waste. E-Waste in transition—From pollution to resource. Academic Press.

Aston, J.E., Peyton, B.M., Lee, B.D., Apel, W.A., 2010. Effects of ferrous sulphate, inoculum history, and anionic form on lead, zinc, and copper toxicity to Acid thiobacillus caldus strain BC13. Academic Press.

Awasthi, A. K., Hasan, M., Mishra, Y. K., Pandey, A. K., Tiwary, B. N., Kuhad, R. C., Gupta, V. K., & Thakur, V. K. (2019). Environmentally sound system for E-waste: Biotechnological perspectives. *Current Research in Biotechnology*, *1*, 58–64. DOI: 10.1016/j.crbiot.2019.10.002

Awasthi, A. K., & Li, J. (2017). Management of electrical and electronic waste: A comparative evaluation of China and India. *Renewable & Sustainable Energy Reviews*, *76*, 434–447. DOI: 10.1016/j.rser.2017.02.067

Awasthi, A. K., Zeng, X., & Li, J. (2016). Comparative examining and analysis of E-waste recycling in typical developing and developed countries. *Procedia Environmental Sciences*, *35*, 676–680. DOI: 10.1016/j.proenv.2016.07.065

Baldé, C. P., Forti, V., Gray, V., Kuehr, R., & Stegmann, P. (2017). The global E-waste monitor 2017: Quantities, flows and resources. United Nations University, International Telecommunication Union, and International Solid Waste Association.

Bandyopadhyay, A. (2008). Indian initiatives on E-waste management—A critical review. *Environmental Engineering Science*, *25*(10), 1507–1526. DOI: 10.1089/ees.2007.0342

Bhutta, M. K. S., Omar, A., & Yang, X. (2011). Electronic Waste: A Growing Concern in Today's Environment. *Economic Research International*, *2011*(1), 1–8. DOI: 10.1155/2011/474230

Ceballos, D. M., & Dong, Z. (2016). The formal electronic recycling industry: Challenges and opportunities in occupational and environmental health research. *Environment International*, *95*, 157–166. DOI: 10.1016/j.envint.2016.07.010 PMID: 27568575

Chatterjee, S. (2012). India's readiness on ROHS directives: A strategic analysis. *Global Journal of Science Frontier Research*, *10*(1), 14–26.

Cleaning Up Electronic Waste (E-Waste). (n.d.). US EPA. https://www.epa.gov/international-cooperation/cleaning-electronic-waste-E-waste

Corwin, J. E. (2018). "Nothing is useless in nature": Delhi's repair economies and valuecreation in an electronics "waste" sector. *Environment and Planning A.Economy and Space*, *50*(1), 14–30.

Damgaard, A., & Christensen, T. H. (2011). Recycling of metals. In *Solid waste technology and management* (pp. 234–242). Wiley.

Debnath, B., Saha, I., Mukherjee, T., Mitra, S., Das, A., & Das, A. (2021). Sorbents from waste materials: a circular economic approach. In *Sorbents Materials for Controlling Environmental Pollution* (pp. 285–322). Elsevier. DOI: 10.1016/B978-0-12-820042-1.00028-6

Decharat, S. (2018). Urinary mercury levels among workers in E-waste shops in Nakhon Si Thammarat Province, Thailand. *Journal of Preventive Medicine and Public Health = Yebang Uihakhoe Chi*, *51*(4), 196–204. DOI: 10.3961/jpmph.18.049 PMID: 30071707

Disposal of Asbestos, Chemicals, Oil, Trade and e-Waste. (n.d.). Parkes Shire Council. https://www.parkes.nsw.gov.au/environment/waste-recycling/waste-facilities/disposal-of-chemicals-oil-trade-and-E-waste/

El-Ghiaty, M. A., & El-Kadi, A. O. (2021). Arsenic: Various species with different effects on cytochrome P450 regulation in humans. EXCLI J., 20, 1184. https://eur-lex. europa.eu/

Forti, V., Baldé, C. P., Kuehr, R., & Bel, G. (2020). *The Global E-waste Monitor 2020. United Nations University (UNU), International Telecommunication Union (ITU) & International Solid Waste Association.* ISWA.

Gazzeh, K., Abubakar, I. R., & Hammad, E. (2022). Impacts of COVID-19 Pandemic on the Global Flows of People and Goods: Implications on the Dynamics of Urban Systems. *Land (Basel)*, *11*(3), 429. DOI: 10.3390/land11030429

Goyal, N., & Goyal, D. (2020). Exploring E-waste Management: Strategies and Implications. Handbook of Solid Waste Management: Sustainability through Circular Economy, 1-14.

Grant, K., Goldizen, F. C., Sly, P. D., Brune, M. N., Neira, M., Van den Berg, M., & Norman, R. E. (2013). Health consequences of exposure to E-waste: A systematic review. *The Lancet. Global Health*, *1*(6), 350–36. DOI: 10.1016/S2214-109X(13)70101-3 PMID: 25104600

Hoornweg, D., & Bhada-Tata, P. (2012). What a Waste: A Global Review of Solid Waste Management. Urban Development Series. World Bank.

Laurentiis, V., Corrado, S., & Sala, S. (2018). Quantifying household waste of fresh fruit and vegetables in the EU. *Waste Management (New York, N.Y.)*, *77*, 238–251. DOI: 10.1016/j.wasman.2018.04.001 PMID: 29655924

Leung, R., & Loo, P. (2020). Co-creating interactive dining experiences via interconnected and interoperable smart technology. *Asian Journal of Technology Innovation*, *30*(1), 45–67. DOI: 10.1080/19761597.2020.1822748

Leung, R., & Loo, P. (2020). Co-creating interactive dining experiences via interconnected and interoperable smart technology. *Asian Journal of Technology Innovation*, *30*(1), 45–67. DOI: 10.1080/19761597.2020.1822748

Mundada, M. N., Kumar, S., & Shekdar, A. V. (2004). E-waste: A new challenge for waste management in India. *The International Journal of Environmental Studies*, *61*(3), 265–279. DOI: 10.1080/0020723042000176060

Parvez. (2021). Health consequences of exposure to e-waste: an updated systematic review. *The Lancet. Planetary Health*, *5*(12), e905-e920. DOI: 10.1016/S2542-5196(21)00263-1

Rajpoot. (2020). A Review on ways to Manage Biomedical Waste at Different Locations in Faizabad. International Journal for Modern Trends in Science and Technology, 6(1), 33-36.

Rajpoot, S. (2021). Impact of Nanotechnology on Environment – A Review. *International Journal for Modern Trends in Science and Technology*, *7*(10), 159–164. DOI: 10.46501//IJMTST0710026

Rajpoot, S. (2024). *Eco-friendly solutions for sustainable waste management: An approach towards a cleaner and greener future*. Waste Management and Life Cycle Assessment for Sustainable Business Practice. DOI: 10.4018/979-8-3693-2595-7.ch003

Smith, A., Johnson, M., & Davis, R. (2023). Robotic Kitchen Systems for Food Waste Reduction in Restaurants. *International Journal of Robotics and Automation*, *38*(2), 87–103.

Tokareva, T. (2015). Food Wasting on Consumption Level: Analysis of the Situation in Latvia Compared to Global Tendencies. Economics and Business, 26, 113-118. DOI: 10.7250/eb.2014.028

Chapter 12
E-Waste Recycling Intention Among Generation Z:
An Integration of the Theory of Planned Behaviour

Tengku Nurshazwina Tengku Sahrum
https://orcid.org/0009-0005-7087-5157
Universiti Teknologi MARA, Malaysia

Safwah Malyanah Sahariddun
https://orcid.org/0009-0002-4722-1161
Universiti Teknologi MARA, Malaysia

Nurul Hidayana Mohd Noor
https://orcid.org/0000-0003-2262-2524
Universiti Teknologi MARA, Malaysia

ABSTRACT

The practice of recycling reduces the amount of waste produced, reduces the use of natural resources, and provides opportunities to produce new products using recycled materials. This study identified the determinants of e-waste recycling intention among Gen Z using the theory of planned behavior (TPB). This preliminary study used a quantitative method involving a total of 251 respondents. The analysis consists of descriptive and correlation analysis and multiple regression analysis. The result shows that Gen Z's intention to recycle e-waste is high. Next, significant relationships exist between attitude, social norms, moral norms, information, incentives, and e-waste recycling intention. Finally, the regression results indicate that

DOI: 10.4018/979-8-3693-7383-5.ch012

information strongly influences e-waste recycling intention. This study is essential to expand the understanding of e-waste recycling among the younger generation.

INTRODUCTION

The Sustainable Development Goals (SDGs) are an action plan to achieve competent and sustainable expectations for all. All 17 goals are interlinked, and to ensure no one is left behind, we must achieve them all by 2030. These goals are clearly explained to us, and as a shared responsibility to the environment, this journey requires effort and cooperation from all. E-waste management is a key and important concern in reaching the global SDGs. The SDGs will not be realized if the country's e-waste management is not made a national priority. Responsible consumption and production (SDG 12) target the environmental management of all waste, incorporating e-waste, by 2030. This goal aligns with Thrust 7, Malaysian National Security Policy 2021-2025.

In achieving SDG 12, Malaysia faces several challenges in managing e-waste. These comprise a shortage of knowledge among the public, insufficient infrastructure, and the non-existence of an inclusive e-waste management system. Malaysia is a rapidly developing country in terms of electrical and electronic industries. The electrical and electronics manufacturing industry is growing increasingly rapidly and faster than other industries (Rautela et al., 2021). The rapid development in the electrical and electronic industry causes the price of this equipment to be sold at a cheap price (Forti et al., 2020). As a result, the demand for electrical and electronic goods is high. Because the demand is too high, e-waste has been created. There are several reasons for the increase in the amount of e-waste. Device enhancements are among the leading causes of the increase in e-waste (Ismail & Hanafiah, 2020). Technological innovation creates new electrical and electronic equipment, such as smartphones and smartwatches. Second, there is a surge in the number of users in most countries with an expansion in population. Next is the faster usage life of equipment due to the development of technology (Ismail & Hanafiah, 2020). Almost every day, manufacturers produce new devices.

E-waste is the fastest-growing waste compared to other solid waste (Forti et al., 2020). The United Nations Environment Program (2020) defines e-waste as the waste material of electronic devices, equipment, and electrical components that do not work, such as computers, laptops, smartphones, televisions, washing machines, printers, and others (as cited in Liu et al., 2023). E-waste can be categorized into several types such as 1) electrical equipment, 2) IT equipment (computers, printers, scanners, and others), 3) entertainment equipment (microphones, speakers, and others), 4) telecommunication equipment (telephone, cellular phone, battery, and others), 5) media, audio, and video (CD, DVD, cassette, and others), 6) use of printers (cartridges, ink, and laser printers), 7) circuit boards (main components found in electrical equipment), and 8) battery (Thakur & Kumar, 2022). Society depends on using electrical equipment and electronics because they can help facilitate daily routine affairs. In the First Schedule (Environmental Quality Act 1974 [Act 127] 2005), Malaysia has listed e-waste as scheduled waste. This regulation has established that no individual can throw away e-waste in a landfill before recycling. Scheduled materials such as electrical and electronic waste can corrode and react to toxic substances.

Electrical and electronic waste is growing as a global issue because the aggregate of e-waste generated is increasing promptly every year. Corresponding to the Global E-Waste Monitor report (2020), Malaysian residents are estimated to produce 364 kilo metric tons, an average of 7.3kg per capita, producing electrical and electronic waste (as cited in Forti et al., 2020). The increase in the volume of e-waste hurts human health and the environment. Efficient e-waste management is essential to curb this waste from harming society's and the environment's well-being (Silva et al., 2023). Youth have an essential role in fostering sustainability practices and proper waste management. Therefore, the level of youth involvement in e-waste management practices is essential to study. The youth nowadays use a lot of electrical and electronic appliances like air conditioners, laptops, smartphones, and televisions to facilitate all their daily affairs. In Malaysia, recycling awareness still needs to be improved (Umar et al., 2021). Initiatives to implement the e-waste management program in Malaysia, such as collecting electrical and electronic items that are not used, have been implemented but are still at an early stage (Noor et al., 2023a).

The main problem related to e-waste in Malaysia is the need for more awareness among Malaysians about e-waste recycling (Pariatamby & Bhatti, 2020). In addition, there are still many areas for improvement in this equipment's management and disposal control mechanisms despite the increase in electrical and electronic equipment (Yu et al., 2022). Although recycling education has been conducted at the school level, only a few people in the community, especially the youth, practice e-waste recycling daily. Most Malaysians have thrown away electrical and electronic equipment at will in the garbage disposal area. Believe it or not, everyone can make a big difference in decreasing the quantity of e-waste they produce. By practicing waste reduction, we can lower the volume of e-waste in landfills.

Therefore, this study aims to find the level of Gen Z's intentions toward e-waste recycling. Second is the factors that affect Gen Z's e-waste recycling intention. The third is to suggest approaches to increase e-waste recycling behavior. Gen Z are those born between 1997 and 2012, the demographic group that replaced the Millennial Generation and preceded the Alpha Generation (Twenge, 2023). Gen Z, who grew up in the age of digital technology, has greater access to information about environmental issues. GenZ tends to be highly aware of environmental issues, with various information accessible via the Internet (Chaturvedi et al., 2020). This helps them understand the impact of human activities on nature and motivates them to act (Mohr & Mohr, 2017; Szabo & Webster, 2021). A better insight into intentions will assist in funneling the younger generation from informal discarding to ecological recycling methods in the future by addressing the following questions:

- What is Gen Z's intention toward e-waste recycling?
- What drives Gen Z's intentions toward e-waste recycling?
- What is the most vital variable to predict Gen Z's intention toward e-waste?

This study focuses on Gen Z's intention to recycle e-waste in Malaysia. The study results can give information and consequently help the government and local council plan better policies, regulations, or legislation. The findings also can be used as a reference for non-governmental organizations (NGOs) and communities to develop a more sustainable community life. Moreover, the study also aims to direct several research gaps. First is a gap related to the theory used, the Theory of Planned Behaviour (TPB), that requires further expansion or interpretation. The theory proclaims that human behavior is influenced by three constructs, namely a) attitudes that lead to desired or unwanted behavior, b) subjective norms, which are the effects or perceptions of social pressure or people around them, and c) perceptions of behavioral control, which contribute to growing self-confidence to control the behavior. These three constructs contribute to the advancement of intentions and subsequently executing the behavior (Ajzen, 1991). However, the theory did not mention specific elements of each element in predicting the intention and behavior. Some past findings oppose the existing theoretical framework (e.g., Muthukumari et al., 2024). This study continues two original TPB elements, which include attitude and social norms and includes three specific predictors, namely, moral norms, information, and incentives, which could significantly persuade Gen Z e-waste recycling intention. Moreover, there is inadequate empirical evidence in e-waste management since most studies have heavily examined household domestic, plastic, and solid waste (e.g., Fadhullah et al., 2022). Thus, this empirical gap requires further data collection to support prior understanding.

LITERATURE REVIEW AND THEORETICAL FRAMEWORK

Theory of Planned Behavior (TPB)

The theory of planned behavior (TPB) by Ajzen (1991) is a development from the Theory of Reason Action (TRA); the conceptual basis emerges as the main framework for understanding and predicting human social behavior. Ajzen (1991) states that the intention to perform various types of behavior can be predicted by three determinants: attitudes toward behavior, subjective norms, and perceived behavioral control. Attitude toward behavior is the intensity to which a person has positive or negative reactions about the behavior of concern. Individual attitude responses can be measured positively or negatively when performing e-waste recycling. Ajzen (1991) defined attitude as a characteristic related to beliefs, feelings, and behavioral tendencies toward critical social objects, events, or group symbols. One must consider one's attitude to evaluate individual behavior because the compatibility principle states that some attitudes can positively affect behavior. The principle of compatibility explains the difference between attitude and behavior. To ensure that an individual is dedicated to engaging in pro-environmental behavior such as waste segregation, their attitude should be positive and support the efforts (Noor et al., 2023b).

As for subjective norms, Piscitelli and D'Uggento (2022) found that friends are decisive in e-waste recycling among teenagers. For adults, it has also been found that spouses and peers are encouragers in implementing a desired behavior. The TPB is analogous to TRA, except it improves a new construct: perceived behavioral control, which implies the assessment of control over the execution of a given behavior. It is also influenced by the effects of dualistic beliefs: control beliefs and perceived facilitation. Control beliefs include abilities, possessions, and prospects, while perceived achievement from a set of results is given. The intention also goes hand in hand with the perception of behavioral control, which accounts for significant behavioral differences. The TPB guides researchers to comprehend how its components can change individual behavior. For this study, we will discuss how the younger generation is changing their behavior to recycle e-waste. According to Ajzen (1991), intention plays a vital role in conducting behavior. This intention indicates how willing the user is to try to act.

E-Waste

Today's all-digital lifestyle makes us very dependent on various electronic items. We often buy the latest gadgets to be more up-to-date. As a result, the consumption of electronic goods continues to increase, and old electronic goods become garbage, commonly known as "e-waste." E-waste is the term for electronic items that are no longer used and thrown away, either because they are damaged or outdated (Ismail & Hanafiah, 2021). For example, mobile phones, laptops, computers, and other gadgets. In Malaysia, e-waste is categorized as scheduled waste with code SW110 under the First Schedule, Environmental Quality Regulations (Scheduled Waste) 2005, Environmental Quality Act 1974 enforced by the Environmental Department (Kang et al., 2020).

E-waste production is separated into two traces: from the industrial sector and houses. Code SW110 is interpreted as waste from electrical and electronic installations containing components such as accumulators, mercury switches, glass from cathode ray tubes and activated glass or other polychlorinated biphenyl capacitors, or which are contaminated with cadmium, mercury, lead, nickel, chromium, copper, lithium, silver, manganese, or polychlorinated biphenyls (Kang et al., 2020). E-waste has become a wide-reaching problem. The more electrical and electronic goods are generated, the more e-waste must be placed. Much e-waste is disposed of short of using environmentally friendly processes such as dropping it into rivers, getting into landfills, burning it at will, or awarding it to the wrong party. In that case, this e-waste can endanger life, affect our health, and cause the deterioration of nature's quality. The best compliment to this problem is correctly assembling and including used and unwanted electronics. Only through recycling can these chemicals be securely contained. Many materials can be reused for other industrial purposes. At the same time, unusable parts are effectively stored or disposed of in a way that has the most minor consequences for human health and the environment (Shaharudin et al., 2023). It takes an organized attempt from municipalities, private industry, and individuals to maintain e-waste from landfills and incinerators. Municipalities should leap at the occasion to increase the surrounding environment, decrease negative impacts, and help establish a better world. Awareness of e-waste disposal among Malaysians still needs to be improved.

Determinants of E-Waste Recycling

Corresponding to the UN's Global E-Waste Monitor (2020), a record 53.6 million metric tons of e-waste were produced in 2019 and is anticipated to catch 74 million metric tons in 2030 (as cited in Forti et al., 2020). Only 17.4% (53.6 million metric tons) of the collected are recycled, leaving $57 billion (USD) worth of elevated-value materials (to end up in landfills, even though they can be recaptured in the recycling procedure. The environmental outlays of e-waste are clear and accompanied by substantial economic costs. Generation Z is said to be a technologically literate group that is motivated, and they bring cultural diversity, habits, and behaviors that hinder how they act, work, communicate, and adapt to their environment.

According to Chaturvedi et al. (2020), Gen Z shows green aspirations through their commute, location, transport, working methods, and policies. Gen Z is also said to have a higher green lifestyle. The aspiration of attitude and responsibility towards the environment from Gen Z is essential because they play a direct role in the knowledge-based solutions to environmental problems in the future. Recycling is essential for society and nature. Society needs to act quickly to practice e-waste recycling because the amount of waste occasionally increases. Many circumstances contribute to recycling behavior. In general, it is separated into (i) behavioral factors and (ii) non-behavioral factors.

Attitude and E-Waste Recycling Intention

Attitudes greatly influence behavior, and it is essential to assess attitudes in any consumer behavior research (Azjen, 1991). Ajzen (1991) defines attitude as the level at which a person makes a good or bad judgment about their behavior. Several studies have shown that attitudes are essential in individual decisions to influence pro-environmental behavior (e.g., Li & Wu, 2020; Tamar et al., 2021) and e-waste recycling (e.g., Islam et al., 2025a; Islam et al., 2025b). People believe that positive behavior toward the environment will produce positive consequences while having negative behavior will also have negative consequences for the environment (Ajzen, 1991). The findings of a study by Ahmed et al. (2021), who used TPB to understand the factors underlying the actual consumption of organic food, confirmed that product-specific attitudes and willingness to pay, together with perceived availability, are essential factors that influence behavior. Candrianto et al. (2023) define environmental attitudes as the tendency to learn to respond consistently by favoring or not favoring the environment. Environmental attitudes are fundamental in prompting and promoting human behavior towards the ecosystem. Individuals with a positive attitude toward the environment will apply it daily (Tian et al., 2020).

Yaacob et al.'s (2025) study of 213 higher education students from Malaysia proves that attitude substantially predicts recycling practices. Dhir et al. (2021) argued that there is a significant relationship between attitude and e-waste recycling. If the person's initial observation of the e-waste recycling program is adverse, recycling declines. Studies by Yadav et al. (2022) showed that residents more interested in recycling activities will produce more recyclable e-waste than residents less interested in recycling activities. Individual participation in environmental conservation, such as recycling, is a significant stride in environmental protection attempts. The failure to inform the public about stabilizing progress in monetary, collective, and environmental sustainability has become a primary apprehension.

Social Norms and E-Waste Recycling Intention

Subjective norms are acceptance and focus on influences such as friends and peers that significantly impact individual behavior over aspects of biology, personality, family, religion, culture, and other influences (Azjen, 1991). Subjective norms are also denoted as social distribution and the completion of selected attitudes from fundamental people to the individual. Religion is essential in social formation. The social effects can appear from interpersonal stimuli such as family, neighbors, or friends and external inducements such as the mass media and press. If that social expectancy advances a person to implement the behavior in question, they are inclined to implement it. Research visuals show that subjective norms affect respondents' ability to recycle e-waste (Aboelmaged, 2021).

The results of previous research prove that subjective norms positively influence the individual's intention to perform e-waste recycling (Theerthaana & Raj, 2025). Alam et al. (2025) found that friends, family members, and employees are significant components of subjective norms that can influence a young consumer's behavior. The study showed that subjective norms influenced respondents to implement e-waste segregation at home due to someone close to them encouraging them to carry out e-waste recycling activities. To ensure the environment is safe, the public must care for and consider cleanliness inside and outside their homes. Keeping the environment clean needs to be rooted in the home. Every community member needs to be highly aware of maintaining cleanliness; for example, they should not throw e-waste away. The community's involvement in environmental protection should be applied (Swanson & Ferrari, 2023). NGOs can play an important role not only in welfare work but also in environmental care. Therefore, NGOs can also play a role in encouraging community involvement in e-waste recycling activities. The community should obtain information from parties such as the government, private sector, or NGOs to further increase awareness of the importance of recycling activities. At the same time, facilities such as material collection points, transportation, or recycling bins are among the elements that should be provided to encourage consumers to participate in recycling programs (Mohamed & Rasnan, 2021).

Moral Norms and E-Waste Recycling Intention

Morals are a set of values and principles that govern human behavior in everyday life. Morals are views about right and wrong, good and bad, and just and unjust. Past studies have discovered that morals are very relevant in taking care of the environment, including e-waste recycling (Islam et al., 2025a; Islam et al., 2025b). Values and beliefs held by individuals drive changes in attitude and behavior. According to Vuong (2021), value is the conceptualization of desire that guides a person to choose something, assess action, and guide actions and assessments. In other words, value publishing and asserting attitudes and being a driving force for people to do something or be an obstacle not to do something. Values are learned and not inherited.

Such virtues should be cultivated from the beginning when a person is still young and refined through school education. The educational process that students go through affects environmental morals and values. Appreciating good values is critical to strengthening positive attitudes and behaviors, such as pro-environmental behavior and e-waste recycling (Noor et al., 2024; Noor, 2024). According to Tembhare et al. (2021), concern for ecology is an essential predictor of youth intention to recycle e-waste. Ecocentrism emphasizes the importance of maintaining the balance of the ecosystem without favoring one particular entity (Taylor et al., 2020). So, in this theory, humans, animals, plants, and other ecosystems are considered to have the same rights and interests.

Environmental ethics can help maintain the sustainability of nature by encouraging awareness and responsibility in interacting with the environment (Fang et al., 2022). By applying the principles of environmental ethics, such as respecting the rights of all living things, using natural resources wisely, and engaging in environmental conservation efforts, we can contribute to maintaining the sustainability of nature and the balance of the ecosystem (Robinson, 2011). Paying attention to environmental ethics can have a beneficial influence on the environment and human life. Excessive use of natural resources without considering the balance of the ecosystem can result in environmental damage and species extinction (Fang et al., 2022). In addition, air, water, and soil pollution due to irresponsible human actions can threaten the health of humans and other living things. Therefore, we all need to pay attention to environmental ethics and be responsible in every action we take towards the natural environment. To maintain the sustainability of nature and the balance of the ecosystem, every individual needs to play an active and responsible role in protecting the environment (Carolan, 2020).

Information and E-Waste Recycling Intention

Information is the result of processing, collecting, and analyzing data that can add knowledge to the recipient of the information. A lack of knowledge and awareness about the recycling program has affected e-waste recycling behavior (Nguyen et al., 2025). Knowledge, skills, and pure values, as well as the right actions in dealing with the issue of waste disposal, need to be implemented through various programs or activities (Thakur & Kumar, 2022). Gen Z tends to have a high level of awareness of environmental issues, with a variety of information that can be accessed through the Internet (Mohr & Mohr, 2017; Szabo & Webster, 2021). This helps them comprehend the influence of human behavior on nature and persuades them to act. Past studies have discovered that the intensity of recycling knowledge helps develop e-waste recycling practices (Güngördü et al., 2025; Nguyen et al., 2025). Sabbir et al. (2023) show that respondents need more exposure or explanation to improve their recycling knowledge. Most respondents from the study felt they still needed clarification on the three types of trash. Therefore, it can be explained that those who know about recycling but need to use their knowledge have low responsibility and attitude toward e-waste management.

According to a study conducted by Delcea et al. (2020), women have a higher awareness and understanding of the importance of the environment related to the recycling system than men. This is because women are used to doing housework, which makes them more aware of materials that can and cannot be used. Efforts should be made to raise awareness about recycling practices, especially among the younger generation. Gen Z must be aware of and sensitive to environmental issues and adopt e-waste recycling as a lifestyle. Education and information on the use of e-waste recycling facilities and efforts can affect the quality and civilization of a nation in this world (Mohd Noor et al., 2022). With proper information, people can have intelligence, noble character, personality, spiritual strength, and skills that benefit themselves and society (Liu et al., 2023). Therefore, education is fundamental to raising knowledge and consciousness of environmental problems, including recycling. Educational institutions need to play a more effective role and convince their students that a change of mindset, practical efforts, and environmental legislation are needed if we want to have a chance of success in reducing the adverse effects of the threatening climate crisis (Fang et al., 2022). More organizations and offices need to take up recycling efforts in their places so that this recycling attitude can continue and be reinforced in the workplace, where people spend a large amount of their time (Mohamed & Rasnan, 2021).

Incentives and E-Waste Recycling Intention

An incentive spurs someone to do something or perform in a specific condition. Two types of incentives stimulate human choices. These are intrinsic and extrinsic incentives. Intrinsic incentives urge someone to do something out of their pursuits or desires, deprived of external pressure or promised rewards. Conversely, extrinsic incentives are guided by rewards such as salary raises to succeed in specific solutions or avoid punishment (Gao & Chen, 2024). Incentives are given to encourage people to recycle more, usually with vouchers or payments to individuals (Venturi et al., 2025). The government could incentivize organizations and contractors to apply e-waste recycling (Sabbaghi, 2025). For example, introducing a financial subsidy scheme can persuade manufacturers to embrace good e-waste recycling practices.

For example, in Hong Kong, to enable and persuade the public to use community recycling services, the GREEN$ (Greeny Coins) smart card is introduced. Public members could bring at least 2 kg of recyclables and receive gift items by displaying the GREEN$ smart card (Gibovic & Bikfalvi, 2021). In Malaysia, electronic waste administration was established in 2005 under the authority of the Department of Environment. DRB-Hicom Environmental Services Sdn. Bhd. has worked together with Alam Flora Sdn. Bhd. by running the 3R On Wheels program. The patrons can experience cash inducements and Petronas Mesra reward points by vending e-waste and other recyclable items. Several buy-back centers have been set up in Putrajaya, Cyberjaya, and Kuala Lumpur to make it easier for the community to participate in recycling activities. Nonetheless, there are still people who are less bothered about the environment by dropping e-waste.

In Malaysia, there are no legal requirements for Malaysians to recycle e-waste. Denmark is one of the countries that have successfully implemented recycling practices. The waste collection system implemented in this country is very efficient (Christensen et al., 2022). Manufacturers are required to pay a deposit to the government when producing products that can be recycled. The deposit will be refunded when the manufacturer recycles the product after use. Moreover, those who want to dispose of any waste will be taxed. They also have a waste subsidy program where the authority will give financial aid to those who want to develop recycling technology. Moreover, those who wish to dispose of non-degradable solid waste such as plastic packaging, batteries, and plastic bags will be taxed (Christensen et al., 2022). Based on the above argument, the following assumptions are developed:

H1: Attitude significantly influences Gen Z e-waste recycling intention.
H2: Social norms significantly influence Gen Z e-waste recycling intention.
H3: Moral norms significantly influence Gen Z e-waste recycling intention.
H4: Information significantly influences Gen Z e-waste recycling intention.
H5: Incentives significantly influence Gen Z e-waste recycling intention.

Figure 1 portrays the research model for the study.

Figure 1. Conceptual model

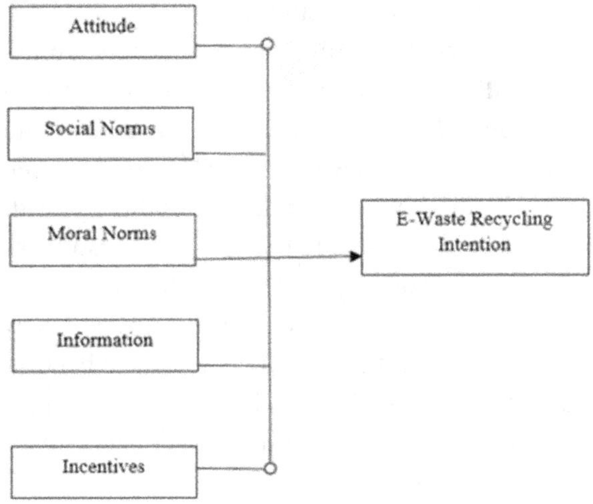

METHODOLOGY

The researchers used a survey and convenience sampling techniques in this study. The study used a questionnaire method distributed via Google Forms. The research's study population is Gen Z in Malaysia. Malaysian Gen Z comprises people born between 1997 and 2012. Nonetheless, only people born between 1997 and 2006, which means people at least 18 years old, are counted in this study. Next, this sample selection is supported by Roscoe (1975), who states that a sample of 30–500 people is appropriate for a study. Thus, 300 respondents were selected for this study. To reach the sample, questionnaires were distributed, and it is an accurate method to collect data if the sample is spread out (scattered). The survey questions were modified and referenced from former research (i.e., Rakhmawati et al., 2023; Shen et al., 2019). Table 1 summarizes the items used in the study.

Table 1. Measurement of variables

Variables	Items
Attitude	
	E-waste recycling makes me feel very satisfied.
	E-waste recycling makes a contribution to society.

continued on following page

Table 1. Continued

Variables	Items
	E-waste recycling is everyone's responsibility.
	E-waste recycling is beneficial.
Social Norms	
	Most people who are important to me would support my proper waste disposal/recycling behavior.
	Most people who are important to me would want me to dispose of my e-waste properly.
	Most people who matter to me would approve of proper disposal of e-waste disposal/recycling behavior.
Moral Norms	
	I am willing to put extra effort into properly disposing of my e-waste.
	I would feel guilty if I did not recycle my e-waste correctly.
	I feel a personal solid obligation to dispose of my e-waste correctly.
Information	
	I learn about e-waste management through formal education at school/university.
	I know about e-waste management through social media, advertisements, and pamphlets.
	I know where to recycle the e-waste in my living area.
	I know the benefits we can gain from proper e-waste management.
	I love to listen to environmental issues sharing.
	I love to search for environmental information online to enhance my understanding of e-waste.
Incentive	
	Awarding shopping vouchers would stimulate the behavior of e-waste management.
	Non-financial incentives such as giving appreciation letters for e-waste management will help foster e-waste management behavior among the community.
	Encouragement in e-waste management among the public can be done if the government allocates a budget to build the centers in housing areas.
	Payment out of the spoilt electronic products submission could interest the individuals to participate in e-waste management.
	Government incentives such as tax holidays would attract many corporates to engage with e-waste management initiatives.
E-Waste Recycling Intention	
	I am willing to contact a professional recyclers for proper waste disposal in the future.
	I would prefer to purchase an electronic products that promise recycling / proper disposal in the future.
	I am willing to tell my friends about my experiences of disposing of e-waste for recycling.

The survey data will be recorded and evaluated using Statistical Package for Social Science (SPSS) version 29.0. The notion of normality of the data influences the model being corresponded. If the data is not standard, it will disturb the value of the statistic or F, which then influences the confidence interval and prediction interval. The Skewness value should fall over within the range of -3 to +3. Moreover, for kurtosis, the bounds of -10 to +10 need to be presumed (Kline, 2005). Cronbach's alpha coefficient determines consistency internally as an indicator of reliability scale (Hair, 2010). The coefficient variation varies from 0 to 1, and a value of less than 0.6 shows poor or unacceptable internal reliability. Cronbach's alpha surpasses or is equal to 0.7, which is deemed reliable, demonstrating the reliability strength. In other words, the closer Cronbach's alpha is to one, the greater the internal consistency. Researchers have constructed five (5) hypotheses, which are assessed using regression analysis to prophesy the value of the dependent variable from one or more independent variables. Through testing the hypothesis, the researchers also employed Pearson's correlation investigation to weigh the strength of the correlation between the variables. The Pearson correlation coefficient is applied to assess the intensity of the relationship between the variables. The coefficient assists the researchers in finding if pairs of variables are strongly correlated.

Usable questionnaires were received from 251 respondents. Based on Table 2, most of the respondents are female (72.9%), and the rest are male respondents (27.1%). Next, most respondents are from the 20 to 24-year-old group (74.1%). Meanwhile, the highest academic level is a bachelor's degree (80.9%). Finally, most respondents live in urban areas (55.8%).

Table 2. Demographic profile

No.	Profile	Frequency (n)	Percentage (%)
1	**Gender**		
	Male	68	27.1
	Female	183	72.9
2	**Age**		
	18-20 years old	20	8.0
	20-24 years old	186	74.1
	25-30 years old	45	17.9
3	**Highest Academic Qualification**		
	Diploma	28	11.2
	Bachelor's Degree	203	80.9
	Other	20	8.0

continued on following page

Table 2. Continued

No.	Profile	Frequency (n)	Percentage (%)
4	**Living Area**		
	Urban	140	55.8
	Semi-Urban	93	37.1
	Rural	18	7.2

FINDINGS AND DISCUSSION

As shown in Table 3, the variables used are acceptable and reliable and can be used in the following study (Hair et al., 2010). For Skewness and Kurtosis, the variables used are normally distributed (Kline, 2005). To answer the first research question, "What is Gen Z's intention toward e-waste recycling?" the researchers have interpreted the mean level. In explaining the items' level, the mean scale between 1.00 – 2.50 is reflected as low, 2.51 – 3.50 is medium, and 3.51 – 5.00 is considered high. Table 3 shows the respondents' intention to recycle e-waste at a high level, with a mean value of 3. A high level implies that respondents are persuaded to participate in e-waste recycling. Moreover, mean values for attitude, social norms, moral norms, information, and incentive demonstrate that the reception of variables is high-level.

Table 3. Normality & reliability results

Variable	Mean	SD	Skewness	Kurtosis	α
Attitude	4.209	0.714	-1.179	2.888	0.899
Social Norms	4.069	0.789	-1.004	1.721	0.912
Moral Norms	4.015	0.811	-0.729	0.649	0.924
Information	3.758	0.846	-0.656	0.430	0.881
Incentive	4.089	0.707	-0.679	0.725	0.884
E-Waste Recycling Intention	3.447	0.920	-0.348	-0.280	0.819

Table 4. Pearson correlation results

		E-Waste Recycling Intention
Attitude	Pearson Correlation Sig. (1-tailed) N	0.521*** 0.000 251
Social Norms	Pearson Correlation Sig. (1-tailed) N	0.449*** 0.000 251
Moral Norms	Pearson Correlation Sig. (1-tailed) N	0.447** 0.000 251
Information	Pearson Correlation Sig. (1-tailed) N	0.711** 0.000 251
Incentive	Pearson Correlation Sig. (1-tailed) N	0.381** 0.000 251

To answer the second research question, "What drives Gen Z's intentions toward e-waste recycling?" correlation analyses were used. Based on Table 4, there is a significant positive relationship between attitude ($r = 0.521; p < 0.050$), social norms ($r = 0.449; p < 0.050$), moral norms ($r = 0.447; p < 0.050$), information ($r = 0.711; p < 0.050$), and incentive ($r = 0.381; p < 0.050$) towards Gen Z e-waste recycling intention.

Table 5. Regression results

Variables	Beta (β)	Sig. (p)	Tolerance	VIF
Attitude	0.335	0.000	0.511	1.956
Social Norms	-0.132	0.067	0.349	2.867
Moral Norms	-0.033	0.637	0.365	2.739
Information	0.689	0.000	0.549	1.821
Incentive	-0.071	0.234	0.497	2.012
R^2 Adjusted R^2 F Change Sig.	0.564 0.555 63.279 0.000			

Finally, multiple regression was used to answer the third research question, "What is the most vital variable to predict Gen Z's intention toward e-waste?" The study has observed multicollinearity, which refers to the circumstance of a linear relationship between independent variables. A linear relationship between independent variables can occur in a perfect linear relationship (perfect) and an imperfect linear relationship (imperfect). Likewise, the value of variance inflation factor (VIF) and tolerance (TOL) can be used to detect the existence of multicollinearity in the regression model. If the VIF value goes above four or the tolerance is fewer than 0.2, multicollinearity has appeared (Hair et al., 2010). The inferences revealed that there is no multicollinearity. Then, the R-Square value of 0.564 displays that the % of the independent variable impact from attitude, social norms, moral norms, information, and incentive on e-waste recycling intention is 56.4%. That means all independent variables have a ratio of effect on e-waste recycling intention, amounting to 56.4%. The remaining 43.6% is affected by another variable. Subsequent, the regression findings can be outlined as follows: 1) attitude (β =0.335, p=0.000); 2) social norms (β =-0.132, p=0.067); 3) moral norms (β =-0.033, p=0.637); 4) information (β =0.689, p=0.000); and 5) incentive (β =-0.071, p=0.234). These results show that only attitude and information substantially affect Gen Z e-waste recycling intention. Therefore, the study agrees on H1 and H4 and rejects H2, H3, and H5.

RESEARCH AMD PRACTICAL IMPLICATIONS

Hypothetically, this study has supported new insights into the streaming literature on e-waste recycling by expanding the TPB elements into specific elements such as attitude, social norms, moral norms, information, and incentives. The first result was that the level of e-waste recycling intention among Malaysian Gen Z was high. Gen Z, who grew up amid awareness of climate change and the need to protect the sustainability of nature, has played a significant role in fighting for environmental preservation (Mohr & Mohr, 2017; Szabo & Webster, 2021). The result shows significant relationships between attitude, social norms, moral norms, information, incentive, and Gen Z e-waste recycling intention. Next, the regression results indicate that attitude and information significantly influence Gen Z e-waste recycling intention. Information has become the most significant predictor with the highest beta value obtained. The availability of information and entertainment at Gen Z's fingertips via smartphones has caused many of them to find it difficult to escape technology (Szymkowiak et al., 2021). This is proven by the high time spent using social media and consuming digital content. In this study, this dependence positively impacts Gen Z's awareness of e-waste recycling. Growing up in a dynamic digital era, Gen Z has broad access to technology as a very effective learning tool (Mohr & Mohr, 2017). Technology is a handy tool for expanding their knowledge, but they need more than digital skills to become independent (Szabo & Webster, 2021). This includes a willingness to seek knowledge actively.

Gen Z also needs to have a good attitude toward recycling. The formation and change of attitudes will be determined by two factors, namely, internal factors and external factors. Information is one of the external factors that could significantly influence attitude (Fang et al., 2022; Liu et al., 2023). Thus, attitude and information simultaneously affect Gen Z e-waste recycling behavior. As research implications, the study's results have proved that there is a generational difference in e-waste recycling. For example, Bhutto et al. (2023) have discovered that personality factors and social influence theory influence e-waste recycling intentions among Gen Y. Severo et al. (2017) also highlight that generations have different perceptions of environmental issues. Baby Boomers presented greater awareness of environmental sustainability, and Gen Y put environmental practices into action. These differences could explain the failure of educational awareness programs or the appropriate strategies that the governing bodies and authorities should apply.

As practical suggestions, more comprehensive education and awareness of the significance of e-waste recycling systems should be expanded. This can be done through educational campaigns connecting schools, universities, and society. The government has executed the MyEwaste application, which can make it effortless for users or companies to find information and e-waste collection sites. E-waste will be placed on premises authorized by the Department of Environment. It also proposes to upsurge the public's awareness and understanding of the e-waste management system in Malaysia and to guarantee that the management and assemblage of e-waste can be carried out in a more environmentally friendly way and diminish the influence on human health and environmental pressures.

However, public recycling centers must be upgraded and expanded to support growing demand. More recycling centers nationwide need to be built to make it easier for people to recycle more quickly and efficiently. The location of the recycling center and color bins must be more strategic to facilitate the public bringing recycled materials. The place must be close to residential areas, schools, and government agencies. Research and development in e-waste recycling technology must be increased to make the recycling process more efficient and effective. Innovations in recycling technology can help improve the quality of recycled products and expand the range of materials that can be recycled (Gebremariam et al., 2020). Creating a program in schools or universities where students are encouraged to create creativity and innovation toward recycled goods could be a significant strategy. Moreover, compliance with the law and strict enforcement are essential to ensure that e-waste recycling practices in Malaysia run smoothly. Companies and individuals who violate the law on recycling should be subject to appropriate legal action.

The media plays a role as community educators. In newspapers, a section dedicated to news discusses an environmental issue and how to perform e-waste recycling that the public can read to increase knowledge. The media is responsible for disseminating information related to recycling. Through this field, people who know the environment or people who do not have an environmental background but have an interest in the environment are trained to become environmental warriors. On a broad scale, technology has democratized access to education, opening opportunities for people from all backgrounds to improve their skills and understanding of various topics, such as e-waste management, even when local schools and courses are unavailable. Then, more incentives need to be introduced. The Ministry of Housing and Local Government is drafting a solid waste management transition policy towards a circular economy emphasizing sustainable use and production practices. The "Trash to Cash" initiative is a method of buying trash at a specific price.

Therefore, the Solid Waste Management and Public Cleansing Corporation (SWCorp) has set a target for the national recycling rate to be increased to 40 percent by 2025 compared to the current 32 percent rate. The Kuala Lumpur City Hall also carried out an initiative to exchange waste for house rent deductions and obtain necessities through the 1Community 1Recycling (1C1R) program. Through this initiative using the barter system concept, the residents who bring recycled goods to the 1C1R kiosk will get reward points that can be collected and then exchanged for house rent payments and daily necessities. For example, every kilogram (kg) of aluminum that can be recycled represents 12 reward points that can be exchanged for a packet of sugar, while every kg of used cooking oil is given reward points worth RM2. Such strategies should be expanded in other states. Moreover, using environmentally friendly packaging and quickly recyclable materials can simplify recycling and reduce pollution. In addition, the transition to the economic cycle, where most products are designed with the initiative to be easily recycled, is an opportunity to maximize this practice.

FUTURE RESEARCH SUGGESTIONS

There are restraints to this study that should be contemplated when assessing the results. The first limitation of this study is that the results only concentrate on the responses from the respondents in Peninsular Malaysia. Hence, the outcomes of this study cannot be generalized. Future researchers should expand the study to several states and areas. Second, this study only examines the influence of attitude, social norms, moral norms, information, and incentives on Gen Z e-waste recycling intention. It is hoped that future researchers will emphasize other affecting factors, such as cultural attitudes or government policies, which could provide a more comprehensive model. Finally, future researchers should handle research by mixing quantitative and qualitative methods.

CONCLUSION

The findings show that Gen Z's intention to recycle e-waste is high. Next, significant relationships exist between attitude, social norms, moral norms, information, incentives, and e-waste recycling intention. Finally, the regression outcomes indicate that information strongly influences e-waste recycling intention. E-waste is an expected part of our modern existence. As earnings and living measures increase in the emerging world, the expanding volume of e-waste will become more demanding to control. There are quick fixes that can stop the problem at its source. A simple yet effective solution that everyone can help with is reducing and reusing electronic items. Some recommendations include building and improving public recycling centers throughout the country, imposing e-waste collection by declaring the "polluter pays" principle, establishing a recycling fund to help invest in better recycling infrastructure, and increasing community awareness of the importance of e-waste recycling practices through educational and promotional campaigns. The public should also reconsider the need to dispose of items such as TVs or computers that can be repaired. If society ignores all the red flags concerning the environmental waves of e-waste, our natural resources will immediately run out, and the environment and people's health will hurt.

REFERENCES

Aboelmaged, M. (2021). E-waste recycling behavior: An integration of recycling habits into the theory of planned behavior. *Journal of Cleaner Production, 278*, 124182. DOI: 10.1016/j.jclepro.2020.124182

Ahmed, N., Li, C., Khan, A., Qalati, S. A., Naz, S., & Rana, F. (2021). Purchase intention toward organic food among young consumers using theory of planned behavior: Role of environmental concerns and environmental awareness. *Journal of Environmental Planning and Management*, *64*(5), 796–822. DOI: 10.1080/09640568.2020.1785404

Ajzen, I. (1991). The theory of planned behavior. *Organizational Behavior and Human Decision Processes*, *50*(2), 179–211. DOI: 10.1016/0749-5978(91)90020-T

Alam, S. S., Haque, I. S., Kokash, H. A., Ahmed, S., & Ahsan, M. N. (2025). Drivers of waste separation behavior in urban Bangladesh: Leveraging social norms and environmental awareness for circular economy success. *Circular Economy and Sustainability*, 1–33. DOI: 10.1007/s43615-025-00526-2

Bhutto, M. Y., Rūtelionė, A., Šeinauskienė, B., & Ertz, M. (2023). Exploring factors of e-waste recycling intention: The case of generation Y. *PLoS One*, *18*(10), e0287435. DOI: 10.1371/journal.pone.0287435 PMID: 37856490

Candrianto, C., Aimon, H., & Sentosa, S. U. (2023). The role of knowledge, awareness, and environmental attitudes in green product management. *Global Journal of Environmental Science and Management*, *9*(1), 101–112. DOI: 10.22034/gjesm.2023.01.08

Carolan, M. S. (2020). *Society and the environment: Pragmatic solutions to ecological issues*. Routledge. DOI: 10.4324/9781003004912

Chaturvedi, P., Kulshreshtha, K., & Tripathi, V. (2020). Investigating the determinants of behavioral intentions of Generation Z for recycled clothing: An evidence from a developing economy. *Young Consumers*, *21*(4), 403–417. DOI: 10.1108/YC-03-2020-1110

Christensen, T. B., Johansen, M. R., Buchard, M. V., & Glarborg, C. N. (2022). Closing the material loops for construction and demolition waste: The circular economy on the island Bornholm, Denmark. *Resources, Conservation & Recycling Advances*, *15*, 200104. DOI: 10.1016/j.rcradv.2022.200104

Delcea, C., Crăciun, L., Ioană , C., Ferruzzi, G., & Cotfas, L. A. (2020). Determinants of individuals' E-waste recycling decision: A case study from Romania. *Sustainability (Basel), 12*(7), 2753. DOI: 10.3390/su12072753

Dhir, A., Koshta, N., Goyal, R. K., Sakashita, M., & Almotairi, M. (2021). Behavioral reasoning theory (BRT) perspectives on e-waste recycling and management. *Journal of Cleaner Production, 280*, 124269. DOI: 10.1016/j.jclepro.2020.124269

Fadhullah, W., Imran, N. I. N., Ismail, S. N. S., Jaafar, M. H., & Abdullah, H. (2022). Household solid waste management practices and perceptions among residents in the East Coast of Malaysia. *BMC Public Health, 22*(1), 1–20. DOI: 10.1186/s12889-021-12274-7 PMID: 34983455

Fang, W. T., Hassan, A. A., & LePage, B. A. (2022). Environmental ethics: Modelling for values and choices. In *The Living Environmental Education: Sound Science Toward a Cleaner, Safer, and Healthier Future* (pp. 151–174). Singapore: Springer Nature Singapore. DOI: 10.1007/978-981-19-4234-1_6

Forti, V., Baldé, C. P., Kuehr, R., & Bel, G. (2020). The global e-waste monitor 2020. United Nations University (UNU), International Telecommunication Union (ITU) & International Solid Waste Association (ISWA). https://www.greene.gov.in/wpcontent/uploads/2020/12/2020120929.pdf

Gao, M., & Chen, Y. (2024). Multi-agent collaborative model of e-waste recycling. *Journal of Environmental Management, 349*, 119457. DOI: 10.1016/j.jenvman.2023.119457 PMID: 37918229

Gebremariam, A. T., Di Maio, F., Vahidi, A., & Rem, P. (2020). Innovative technologies for recycling End-of-Life concrete waste in the built environment. *Resources, Conservation and Recycling, 163*, 104911. DOI: 10.1016/j.resconrec.2020.104911

Gibovic, D., & Bikfalvi, A. (2021). Incentives for plastic recycling: How to engage citizens in active collection. Empirical evidence from Spain. *Recycling, 6*(2), 29. DOI: 10.3390/recycling6020029

Güngördü Belbağ, A., & Belbağ, S. (2025). Consumer intentions to purchase remanufactured products: A behavioral reasoning theory approach. *Journal of Social Marketing, 15*(1), 1–19. DOI: 10.1108/JSOCM-02-2024-0033

Hair, J. F., Anderson, R. E., Babin, B. J., & Black, W. C. (2010). *Multivariate data analysis: A global perspective* (Vol. 7). Academic Press.

Islam, M. H., Islam, M. T., Hoque, M. R., Haque, S. Z., Barua, B., Islam, M. M., & Hossain, M. F. (2025a). Understanding young consumers' e-waste recycling behaviour in Bangladesh: A developing country perspective. *Recycling*, *10*(1), 24. DOI: 10.3390/recycling10010024

Islam, M. H., Sabbir, M. M., Mahbub, M., Jhilik, T. T., & Shirmin, S. (2025b). Sustaining e-waste recycling behavior among young consumers to implement circular economy: An extended model of Theory of Reasoned Goal Pursuit. *Cleaner Waste Systems*, *10*, 100212. DOI: 10.1016/j.clwas.2025.100212

Ismail, H., & Hanafiah, M. M. (2020). A review of sustainable e-waste generation and management: Present and future perspectives. *Journal of Environmental Management*, *264*, 110495. DOI: 10.1016/j.jenvman.2020.110495 PMID: 32250915

Ismail, H., & Hanafiah, M. M. (2021). Evaluation of e-waste management systems in Malaysia using life cycle assessment and material flow analysis. *Journal of Cleaner Production*, *308*, 127358. DOI: 10.1016/j.jclepro.2021.127358

Kang, K. D., Kang, H., Ilankoon, I. M. S. K., & Chong, C. Y. (2020). Electronic waste collection systems using Internet of Things (IoT): Household electronic waste management in Malaysia. *Journal of Cleaner Production*, *252*, 119801. DOI: 10.1016/j.jclepro.2019.119801

Kline, T. (2005). Psychological testing: A practical approach to design and evaluation. *Sage (Atlanta, Ga.)*. Advance online publication. DOI: 10.4135/9781483385693

Li, Q., & Wu, M. (2020). Tourists' pro-environmental behavior in travel destinations: Benchmarking the power of social interaction and individual attitude. *Journal of Sustainable Tourism*, *28*(9), 1371–1389. DOI: 10.1080/09669582.2020.1737091

Liu, K., Tan, Q., Yu, J., & Wang, M. (2023). A global perspective on e-waste recycling. *Circular Economy*, *2*(1), 100028. DOI: 10.1016/j.cec.2023.100028

Mohamed, A. F., & Rasnan, M. I. (2021). Enhancing role and participation of industry and community for sustainable e-waste recovery for sustainable consumption and production (SCP): Case study Kuala Lumpur Malaysia. *EcoDesign and Sustainability II: Social Perspectives and Sustainability Assessment*, 135-145. DOI: 10.1007/978-981-15-6775-9_9

Mohd Noor, N. H., Mustafa, M. A. S., Saharom, M., & Shamsol Kamal, N. S. S. (2022). Understanding Malaysian household waste separation: An extended theory of planned behavior. *Malaysian Journal of Sustainable Environment*, *9*(1), 19–35. DOI: 10.24191/myse.v9i1.17285

Mohr, K. A., & Mohr, E. S. (2017). Understanding Generation Z students to promote a contemporary learning environment. *Journal on Empowering Teaching Excellence*, *1*(1), 9. DOI: 10.15142/T3M05T

Muthukumari, W. A. C. S., Ahn, J., & Kim, M. (2024). Impact of information publicity on Korean residents' E-waste recycling intentions: Applying the norm activation model and theory of planned behavior. *Heliyon*, *10*(14), e34319. Advance online publication. DOI: 10.1016/j.heliyon.2024.e34319 PMID: 39816366

Nguyen, H. T. T., Lee, C. H., Nguyen, H. T. T., Phan, T. T., Nguyen, V. V., & Hung, R. J. (2025). Factors affecting community participation in an e-waste recycling program. *International Journal of Environment and Waste Management*, *36*(1), 95–110. DOI: 10.1504/IJEWM.2025.143501

Noor, N. H. M. (2024). Defeating Global Glut of Clothing: An Examination of Sustainable Fashion Consumption of Young Consumers. In *Driving Green Marketing in Fashion and Retail* (pp. 26-49). IGI Global. DOI: 10.4018/979-8-3693-3049-4.ch002

Noor, N. H. M., Hassan, N. H., & Ismadi, R. (2023a). Towards sustainable consumption: An examination of the usage of recyclable and biodegradable products among Malaysians. *Asian People Journal*, *6*(1), 190–206. DOI: 10.37231/apj.2023.6.1.491

Noor, N. H. M., Soleman, N. A. F., & Azuan, A. S. K. (2023b). To recycle or not to recycle? Factors affecting Malaysian residents' intention for recycling e-waste. *Malaysian Journal of Social Sciences and Humanities*, *8*(2), e002102–e002102. DOI: 10.47405/mjssh.v8i2.2102

Noor, N. H. M., Zaini, S. M., Pudin, S., & Sidek, S. J. (2024). Encouraging pro-environmental behavior: The function of environmental connectedness, environmental ethics, & environmental knowledge. *Muallim Journal of Social Sciences and Humanities*, *8*(2), 17–32. DOI: 10.33306/mjssh/272

Pariatamby, A., & Bhatti, M. S. (2020). Waste management challenges in Malaysia. In *Sustainable waste management challenges in Developing Countries* (pp. 186–222). IGI Global. DOI: 10.4018/978-1-7998-0198-6.ch008

Piscitelli, A., & D'Uggento, A. M. (2022). Do young people really engage in sustainable behaviors in their lifestyles? *Social Indicators Research*, *163*(3), 1467–1485. DOI: 10.1007/s11205-022-02955-0

Rakhmawati, T., Damayanti, S., Jati, R. K., & Astrini, N. J. (2023). An extended TPB model of waste-sorting intention: A case study of Indonesia. *Management of Environmental Quality*, *34*(5), 1248–1268. DOI: 10.1108/MEQ-11-2022-0309

Rautela, R., Arya, S., Vishwakarma, S., Lee, J., Kim, K. H., & Kumar, S. (2021). E-waste management and its effects on the environment and human health. *The Science of the Total Environment, 773*, 145623. DOI: 10.1016/j.scitotenv.2021.145623 PMID: 33592459

Robinson, J. G. (2011). Ethical pluralism, pragmatism, and sustainability in conservation practice. *Biological Conservation, 144*(3), 958–965. DOI: 10.1016/j.biocon.2010.04.017

Roscoe, J. T. (1975). *Fundamental Research Statistics for the Behavioral Sciences* (2nd ed.). Holt, Rinehart, and Winston.

Sabbaghi, M. (2025). Perceptions of value in retired smartphones: The role of monetary incentives in influencing end-of-hibernation decisions. *Waste Management (New York, N.Y.), 193*, 199–206. DOI: 10.1016/j.wasman.2024.12.015 PMID: 39671746

Sabbir, M. M., Khan, T. T., Das, A., Akter, S., & Hossain, M. A. (2023). Understanding the determinants of consumers' reverse exchange intention as an approach to e-waste recycling: A developing country perspective. *Asia-Pacific Journal of Business Administration, 15*(3), 411–439. DOI: 10.1108/APJBA-11-2021-0565

Severo, E. A., de Guimarães, J. C. F., Brito, L. M. P., & Dellarmelin, M. L. (2017). Environmental sustainability and sustainable consumption: The perception of baby boomers, generation x and y in Brazil. *Revista de Gestão Social e Ambiental, 11*(3), 92–110. DOI: 10.24857/rgsa.v11i3.1266

Shaharudin, M. R., Said, R., Hotrawaisaya, C., Nik Abdul Rashid, N. R., & Azman Perwira, N. F. S. (2023). Linking determinants of the youth's intentions to dispose of portable e-waste with the proper disposal behavior in Malaysia. *The Social Science Journal, 60*(4), 680–694. DOI: 10.1080/03623319.2020.1753157

Shen, L., Si, H., Yu, L., & Si, H. (2019). Factors influencing young people's intention toward municipal solid waste sorting. *International Journal of Environmental Research and Public Health, 16*(10), 1708. DOI: 10.3390/ijerph16101708 PMID: 31096698

Silva, J. A. P., Lima, G. G., Camilo-Cotrim, C. F., Bailão, E. F. L. C., Caramori, S. S., Nabout, J. C., & Almeida, L. M. (2023). Impact of e-waste toxicity on health and nature: Trends, biases, and future directions. *Water, Air, and Soil Pollution, 234*(5), 320. DOI: 10.1007/s11270-023-06328-2

Swanson, H. L., & Ferrari, J. R. (2023). Predictors of e-waste: Considerations for community psychology prevention and intervention. *Journal of Community Psychology, 51*(5), 2001–2009. DOI: 10.1002/jcop.22986 PMID: 36586134

Szabo, S., & Webster, J. (2021). Perceived greenwashing: The effects of green marketing on environmental and product perceptions. *Journal of Business Ethics*, *171*(4), 719–739. DOI: 10.1007/s10551-020-04461-0

Szymkowiak, A., Melović, B., Dabić, M., Jeganathan, K., & Kundi, G. S. (2021). Information technology and Gen Z: The role of teachers, the internet, and technology in the education of young people. *Technology in Society*, *65*, 101565. DOI: 10.1016/j.techsoc.2021.101565

Tamar, M., Wirawan, H., Arfah, T., & Putri, R. P. S. (2021). Predicting pro-environmental behaviors: The role of environmental values, attitudes, and knowledge. *Management of Environmental Quality*, *32*(2), 328–343. DOI: 10.1108/MEQ-12-2019-0264

Taylor, B., Chapron, G., Kopnina, H., Orlikowska, E., Gray, J., & Piccolo, J. J. (2020). The need for ecocentrism in biodiversity conservation. *Conservation Biology*, *34*(5), 1089–1096. DOI: 10.1111/cobi.13541 PMID: 32424955

Tembhare, S. P., Bhanvase, B. A., Barai, D. P., & Dhoble, S. J. (2021). E-waste recycling practices: A review on environmental concerns, remediation, and technological developments with a focus on printed circuit boards. *Environment, Development and Sustainability*, 1–83. DOI: 10.1007/s10668-021-01819-w

Thakur, P., & Kumar, S. (2022). Evaluation of e-waste status, management strategies, and legislations. *International Journal of Environmental Science and Technology*, *19*(7), 6957–6966. DOI: 10.1007/s13762-021-03383-2

Theerthaana, P., & Raj, S. V. (2025). An investigation on adoption of AI-enabled e-waste recycling kiosks in the post-pandemic era. *Electronic Government. International Journal (Toronto, Ont.)*, *21*(1), 68–80. DOI: 10.1504/EG.2025.143120

Tian, H., Zhang, J., & Li, J. (2020). The relationship between pro-environmental attitude and employee green behavior: The role of motivational states and green work climate perceptions. *Environmental Science and Pollution Research International*, *27*(7), 7341–7352. DOI: 10.1007/s11356-019-07393-z PMID: 31884553

Twenge, J. M. (2023). Generations: The Real Differences Between Gen Z, Millennials, *Gen X,Boomers, and Silents—and What They Mean for America's Future*. Simon and Schuster.

Umar, U. A., Shafiq, N., & Ahmad, F. A. (2021). A case study on the effective implementation of the reuse and recycling of construction & demolition waste management practices in Malaysia. *Ain Shams Engineering Journal*, *12*(1), 283–291. DOI: 10.1016/j.asej.2020.07.005

Venturi, S., Zulauf, K., Cuel, R., & Wagner, R. (2025). Trash to treasure: Gamification and informed recycling behavior. *Resources, Conservation and Recycling, 215*, 108108. DOI: 10.1016/j.resconrec.2024.108108

Vuong, Q. H. (2021). The semiconducting principle of monetary and environmental values exchange. *Economics and Business Letters, 10*(3), 284–290. DOI: 10.17811/ebl.10.3.2021.284-290

Yaacob, N. S., Azman, N. A. N. N., Salleh, S. M., & Alrazi, B. (2025). Assessing the relationship of knowledge and attitudes on e-waste recycling practices among students of higher learning institutions. *Journal of Nusantara Studies, 10*(1), 353–372. DOI: 10.24200/jonus.vol10iss1pp353-372

Yadav, R., Panda, D. K., & Kumar, S. (2022). Understanding the individuals' motivators and barriers of e-waste recycling: A mixed-method approach. *Journal of Environmental Management, 324*, 116303. DOI: 10.1016/j.jenvman.2022.116303 PMID: 36191502

Yu, P. L., Ab Ghafar, N., Adam, M., & Goh, H. C. (2022). Understanding the human dimensions of recycling and source separation practices at the household level: An evidence in Perak, Malaysia. *Sustainability (Basel), 14*(13), 8023. DOI: 10.3390/su14138023

KEY TERMS AND DEFINITIONS

Attitude: A complex mental state involving beliefs, feelings val,ues, and dispositions to act in specific ways.

E-Waste: The term for discarded electrical and electronic devices that can be hazardous to the environment and human health.

Gen Z: Those born between 1997 and 2012, the demographic group that replaced the Millennial Generation and preceded the Alpha Generation.

Incentive: Something that encourages a person or organization to do something.

Information: Knowledge communicated or received concerning a particular fact or circumstance.

Moral Norms: The rules of morality that people ought to follow.

Recycling: The recovery and reprocessing of waste materials for use in new products.

Social Norms: The rules and regulations that determine the behavior and actions of people within a particular group, culture, or structure.

Theory of Planned Behavior: A framework for studying human action and behavior. Specifically, the intention is influenced by three determinants, namely a) attitude that leads to a desired or unwanted behavior, b) subjective norms, which are affected by perceptions of social pressure or people around, and c) notions of perceived behavioral control to increase self-confidence to control the behavior.

Chapter 13
The Link Between E-Waste and Linear Economy Through a Bootstrap Rolling Window Analysis

Mikail Pehlivan
https://orcid.org/0000-0002-7273-333X
Trakya University, Turkey

Şahin Akkaya
https://orcid.org/0000-0003-2394-7232
Istanbul University, Turkey

ABSTRACT

It is known that the dominant economic doctrine in the current economic system is neoclassical economics. The fact that waste electrical and electronic equipment (e-waste) is the fastest growing solid waste in the world may be an important indicator of this linear production and consumption economy model. Therefore, this study empirically tests the relationship between e-waste and (linear) economic growth. According to this analysis, the relationship between total e-waste and economic growth in the countries included in the study (USA, Germany, Australia, Brazil, China, Ghana, South Korea, India, Switzerland, Japan, Türkiye) was tested with panel bootstrap causality analysis using annual data for the period 1960-2020. According to the findings of the analysis, a bidirectional causality is found between e-waste and economic growth in all sub-samples for all countries. In this respect, it can be stated that within the framework of the sustainable production and consumption

DOI: 10.4018/979-8-3693-7383-5.ch013

paradigm, a transition from a linear economy to a circular economy seems essential.

INTRODUCTION

Technological development is an important factor in increasing economic welfare for all countries (Romer 1990). Today, electrical and electronic equipments (e-products) have become indispensable in all aspects of modern life. E-products, including information and communication technology* (ICT), represent an important factor that will shape further development. Similar to capital, new technology plays a key role in economic growth in all countries, and this is particularly evident in the last few decades (Erumban and Das 2016; Jorgenson et al. 2016).

In 2010, internet use contributed 1.9% to Gross Domestic Product (GDP) in a group of 30 developing/emerging countries (EC) and 3.4% in developed countries (DC) (Nottebohm et al. 2012). However, this technological progress not only improves the living standards of the society, but also brings along the problem of waste generation (Panambunan-Ferse and Breiter 2013; Umair et al. 2015; Chatterjee and Abraham 2017). Waste electrical and electronic equipments (e-waste) from technological development represents the fastest growing waste group with an annual increase of 4.5% (Yang et al. 2017; Qu et al. 2019). However, while e-product use is increasing, its impact on countries' economic growth varies (Prasad 2012; Hofman et al. 2016; Donou-Adonsou 2019). Similarly, the increase in e-waste also differs across countries, such that the rate of increase in EC is faster than that of the DC (Sthiannopkao and Wong 2013).

Recycling, a central concept in the circular economy, aims to minimize environmental impact. The recycling rate often indicates the percentage of waste directed toward material recovery (Oughton et al. 2022), reflecting progress in environmental conservation and efficient resource management. Countries with higher recycling rates generally produce less waste (Islam et al. 2019), emit fewer greenhouse gases (Magazzino et al. 2020), and experience increased economic growth and job opportunities (Liu et al. 2020).

Recycling provides three key advantages: i) economic, ii) environmental, and iii) public health and safety. To begin with, e-waste holds substantial economic value due to the precious metals it contains. However, fewer than 22% of global e-waste is processed for recycling, and only 1% of the demand for rare earth elements is fulfilled through e-waste recycling (Baldé et al. 2024). Furthermore, inadequate e-waste management results in the loss of valuable materials and leads to an increased need for mining to obtain primary resources. The concentration of metals in e-waste far exceeds that found in traditional mining operations. Research indicates that the quality of global ore is declining, forcing mining operations to extract more complex and finer-grained ores to meet the global metal demand (Lèbre and Corder 2015). Secondly, the rising amount and volume of e-waste pose serious risks to the environment and human health, mainly due to the presence of various toxic heavy metals and organic contaminants (Kiddee et al. 2013). In addition to the direct emissions from the harmful substances within e-waste, secondary and tertiary pollutants can be released as a result of improper disposal and unregulated recycling methods (Schluep et al. 2009). Third, as highlighted earlier, e-waste contains dangerous metals and chemicals that present risks not only to the environment but also to human health and safety. E-waste affects human health in two key ways. One way is through the pollutants emitted by e-waste, which contaminate the soil, water, and air, allowing harmful substances to enter the food chain and eventually impacting human health. The second way is through direct exposure to e-waste during disposal, which can lead to both short-term and long-term health problems (Garlapati 2016; Scruggs et al. 2016).

Conversely, the linear "take-make-dispose" model is based on the assumption that resources and energy are easily accessible and abundant. However, this model is becoming increasingly misaligned with the reality of today's world, where resources are finite and environmental concerns are rising. Even with more efficient use of resources, this approach does not change the fact that these resources are limited; it only delays the inevitable outcomes of resource depletion. Therefore, there is a growing recognition that a comprehensive transformation of the entire operating system is necessary to adopt a more sustainable and circular approach to resource management (Rashid et al. 2013).

Despite the significant increase in e-waste worldwide, the main drivers are still being studied (Boubellouta and Kusch-Brandt 2021a). As seen in the studies, some factors stand out in the increase in e-waste. As a matter of fact, these factors (urbanization, imports of ICT products, renewable energy consumption, industrialization, population growth), which are thought to be related to e-waste, have been subjected to empirical tests (Boubellouta and Kusch-Brandt 2021a; Boubellouta and Kusch-Brandt 2022; Kalia et al. 2022). The analysis presented in this study differs from previous studies discussed in the Literature Review section in several key ways, offering significant contributions to the existing literature. These differences include: (i) the methodology employed, (ii) the dataset used, (iii) the results of the variables analyzed on a country-by-country basis, (iv) the specific years in which the causality of the variables is observed, and (v) the direction of the effect of the variables on one another over time.

While differences in economic growth are due to different levels of development across countries, it is not possible that the different volumes of e-waste generated are due to factors unrelated to GDP. The aim here is to better understand the nature of the relationship between e-waste and economic growth within the framework of countries at different levels of economic development. The insights gained from this relationship are expected to be useful in creating a sustainable vision for the future of e-waste management.

In light of the analysis, the paper is structured as follows: The "Background Information and Literature Review" section first examines the relationship between the linear and circular economy paradigms in the context of e-waste, followed by a review of empirical studies on e-waste. The "Data" and "Econometric Methodology" sections describe the data used in the study, the test conducted, and the rationale behind the selection of the test. The following section presents the "Empirical Findings and Discussion". Finally, the paper concludes with the study's findings and their implications.

BACKROUND INFORMATION AND LITERATURE REVIEW

Economic growth, urbanization, and industrialization have significantly contributed to the increase in both the volume and variety of solid and hazardous waste. Moreover, the expansion of international waste trade, driven by globalization—where developing nations often act as the importers—has intensified waste management challenges. These developments not only have detrimental effects on public health and the environment but also lead to the loss of valuable materials and resources due to inefficient waste management practices. In this regard, an integrated waste management approach is essential in both international and national strategies for sustainable development. In the context of product life cycles, the focus is typically on waste prevention and reduction. When solid and hazardous waste remains, effective and efficient management strategies can lead to their improvement, reuse, and recycling.

Within this framework, or in a broader context, two main approaches are employed to manage e-waste. The first is the "closed-loop" (circular economy) system, which facilitates material recovery, while the second is the "open-loop" (linear economy) system, which, though resulting in energy loss, remains a viable alternative (Ayres 1994; Flygansvaer 1999). For a closed-loop system, as it pertains to sustainability, it is crucial for designers and manufacturers to consider the entire product life cycle—starting from raw material extraction and production to usage and disposal. This approach ensures that products at the end of their life cycle can either be repurposed to recover raw materials or reintroduced into the production process (Villalba et al. 2002; White et al. 2003; King et al. 2006).

Linear Economy, Circular Economy, and E-Waste

In the linear economic system that started with the Industrial Revolution, a significant amount of industrial waste has been released into the environment as a result of the increase in industrial production with the development of technology. In parallel with the rapidly developing technology from the Industrial Revolution to the present day, there have been changes in people's needs. Today, e-products have become the basic needs of people (Crocker 2013). Therefore, end-of-life (EoL) e-products have become the fastest growing type of waste in the world, while the disposal of e-waste is a global environmental problem that needs to be addressed (Schluep et al. 2009; Lundgren 2012). E-products are composed of complex materials and compositions, some of which contain poisons. Therefore, if they are disposed of in primitive ways, such as incineration or creating landfills (dumpsites), they can cause serious damage to the environment and human health (Wena et al. 2009; Rajput 2013).

In the late 19th century, as the economy shifted from manpower to machine-driven industry, manufacturers realized that their factories could now produce more goods than could be easily distributed and consumed. But this overproduction made possible by the industrial age brings with it a dilemma: (i) the problem of demand and (ii) distribution. That is, while demand must be constantly created and sustained, the goods produced must be easily and profitably delivered from the factory to consumers. During this period, technological advances in communication and transportation solved the distribution problem. Repeated consumption was the solution to the problem of sustainable demand. Otherwise, the fall in demand in the face of overproduction leads the economy to stagnation (Slade 2007).

As the 20th century began, it was a well-known technique to produce with low-quality materials in order to reduce costs. But some manufacturers realized that this process also stimulated demand. In other words, repeated consumption can be achieved with this technique. When consumer demand fell radically during the Depression (1929), manufacturers deliberately used low-quality materials to shorten the life of products and force consumers to buy new ones. This technique used by manufacturers to maintain repeated consumption is called planned/technological/functional obsolescence, and the maintenance of repeated consumption by influencing consumer behavior through developments in marketing is called psychological obsolescence (Packard 1963; Slade 2007). Today, as a result of companies' desire to overcome the demand problem brought about by overproduction with various complex marketing strategies and planned and psychological obsolescence techniques, shortening the lifespan of e-products leads to repeated consumption and also causes an increase in e-waste.

A traditional (linear) supply chain can be defined as a process that covers all stages from producer to consumer. When it comes to recycling, there is a reverse flow of materials from the consumer to the producer site. This circular system, referred to as reverse logistics (RL) or reverse manufacturing, is the process that includes all logistics activities involving the transformation of waste materials used by the consumer back into marketable products. In terms of distribution planning, RL involves the physical transportation of the product that is no longer used by the consumer to the producer. The next stage is the conversion of the product into a reusable state by the producer (Figure 1) (Fleischmann et al. 1997).

Figure 1. Product buy-back options

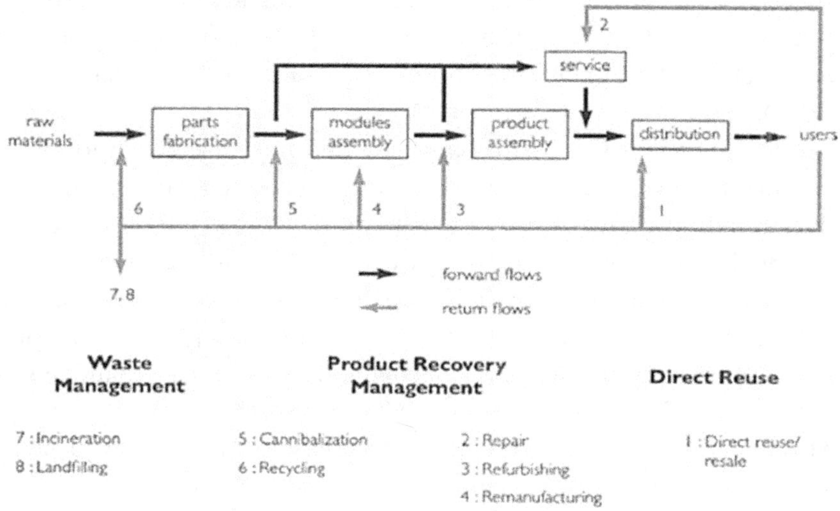

Source: Thierry et al. (1995: p. 118)

While forward (linear) logistics (FL) or forward manufacturing are often referred to as pull systems, RL is a combination of pull and push systems, since the customer on both sides of the chain is both the supplier/disposer and the reuser. In other words, while in FL the focus is on serving the demand site, in RL the focus is on the entire recovery process from the supply (push) to the reuse (demand=pull) of the waste quantity (Figure 2) (Lourenço and Soto 2002; White et al. 2003).

Figure 2. Forward (linear) and reverse (circular) manufacturing cycle

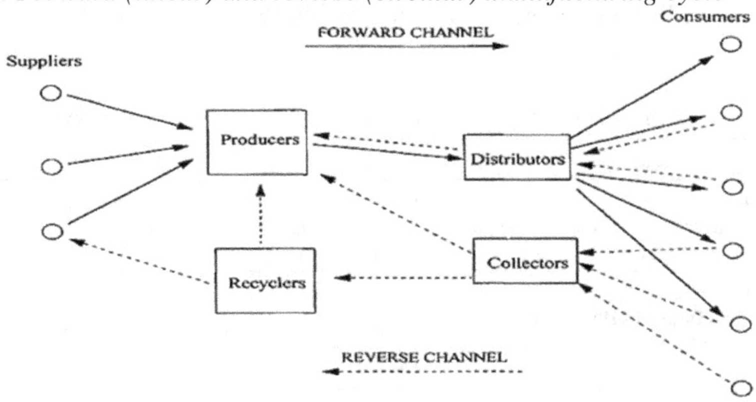

Source: Fleischmann et al. (1997: p. 5)

Supply chain management (SCM) research has traditionally focused on the transformation of raw materials and the movement of goods towards final consumers. RL studies date back to the 1960s and 70s (Seitz and Wells 2006). Over time, growing consumer concerns about the environment, increasing government regulation of the volume of waste from manufacturing operations, and the significant economic potential in this area have increased interest in RL management and safe product disposal at EoL (Neto et al. 2008; Guide and Wassenhove 2009; O'Reilly and Kumar 2016). In terms of how environmental and economic perspectives should be incorporated to improve the performance of RL operations, the concept of RL is gaining attention both in academia and in practice (John et al. 2017). Moreover, firms have started to realize that the implementation of RL operations has the potential to generate large amounts of revenue or profit while minimizing costs to remain sustainable and competitive in the market (Pandian and Abdul-Kader 2017; Morgan et al. 2018). To reduce the increasing e-waste, it is crucial to follow proper recovery processes. For this purpose, RL can be applied for the recycling of used returned e-product.

To design a sustainable RL network, attention is drawn to the triple bottom line approach introduced by Elkington (1997), which focuses on ensuring environmental safety, economic development and social welfare. It should be noted that in addition to ensuring economic and environmental sustainability, the role of social indicators should also be considered in designing and managing a RL network (Sarkis et al. 2010). In this respect, it can be stated that eight factors are effective within the framework of the above-mentioned triple bottom line when establishing a RL system for the efficient and effective processing of e-waste. These are (i) legislation, (ii) customer demand, (iii) strategic benefit/cost, (iv) environmental concern, (v) volume and quality, (vi) incentives, (vii) resources (viii) integration and coordination (Tibben-Lembke and Rogers 2002; Rahman and Subramanian 2012). Table 1 presents RL studies, including e-products, within the framework of these factors.

Table 1. Some studies on the factors affecting the reverse logistics system

Factors	Impact
(i) Legislation (Carter and Ellram 1998; Knemeyer et al. 2002; Walker et al. 2008)	Legislation is one of the drivers of successful RL and also serves as a motivation for cost-effective innovative sustainable practices.
(ii) Customer Demand (Lamming and Hampson 1996; Walton et al. 1998; Vachon and Klassen 2006; Alvarez-Gil et al. 2007)	Customer demand forces companies to be socially responsible and incorporate environmental management practices into their supply chains.
(iii) Strategic Benefit/Cost (Ravi et al. 2005; Ferguson and Toktay 2006; Guide et al. 2006; Guide and Wassenhove 2009; Dat et al. 2012)	The strategic costs of implementing RL provide significant direct and indirect benefits to the firm.
(iv) Environmental Concern (Beamon 1999; Pokharel and Mutha 2009)	Stimulating environmental awareness through RL leads to a competitive advantage for the firm.
(v) Volume and Quality (Dowlatshahi 2005)	The quality of the returned product benefits from the RL transportation cost. Quality-based pricing increases the volume of returns.
(vi) Incentives (Savaskan et al. 2004)	Properly setting incentives among actors in RL can influence return rates and customer behavior.
(vii) Resources (Daugherty et al. 2005; Richey et al. 2005)	Effective allocation of available resources for RL operations, combined with appropriate product strategy, can provide a firm with wealth.
(viii) Integration and Coordination (Carter and Carter 1998; New et al. 2002; Ayvaz et al. 2015)	The integration of RL coordination and information support system can increase the speed of adaptation and profitability of the firm.

Source: Adapted from Rahman and Subramanian 2012.

The ultimate goal of a RL system is to reduce the resources used. This resource reduction includes both minimizing the materials used in the product and minimizing the waste and energy generated through the design of more environmentally sound products. Subsequently, the next objective is to reuse materials and then recycle as much waste as possible. Disposal may be the last option and even then, incineration may be preferable as some form of energy recovery is likely (Carter and Ellram 1998). While e-products are suitable for the establishment of a RL system, it is inherently more difficult than a forward supply chain due to uncertain factors such as the quantity, cost, duration and quality of e-waste (Table 2).

Table 2. Forward and reverse logistics comparison

Forward Logistics	Reverse Logistics
The forecasts are relatively clear.	Predictions are more difficult.
Transportation is "from one to many".	Transportation is "many to one".
Product quality is standard.	Product quality is not standardized.
Product packaging is single sample.	The product package is mostly damaged.
The route is clear.	The route is not clear.
The price is usually standard.	Pricing depends on many factors.
Forward distribution costs are closely monitored by the accounting system.	RL costs are less obvious.
Inventory management is consistent.	Inventory management is not consistent.
Agreements between the parties are clear and understandable.	Agreements with parties are more complicated.
Marketing methods are defined.	Marketing is more complex due to the influence of many factors.
Real-time information is available to monitor the product.	Processes are more difficult to trace.

Source: Tibben-Lembke and Rogers 2002.

Since collection is an important step for transportation and recovery activities, the estimation of the number of product returns is vital for the optimal performance of the system. The design of the RL network, the layout of recovery facilities and the planning of processes are significantly influenced by the estimation of product returns. Therefore, the implementation of RL network design requires the determination of an appropriate network structure for the flows of returned products. The EoL product in this chain comes from several places and is combined in several places or in one place. It is then necessary to determine the appropriate physical locations, the number of layers required, the number of hubs needed, their maximum capacity and the transportation between stations in the network to transfer the returned products from consumers to recovery facilities and potential markets.

Figure 3. Simplified demonstration of the e-waste reverse logistics network

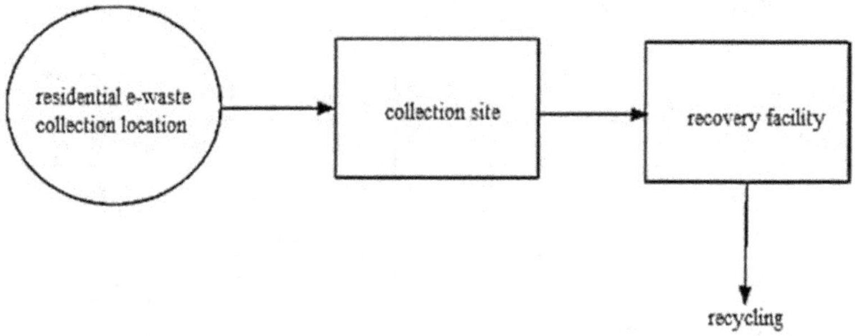

Source: Made by the authors

As shown in Figure 3, the first layer is the collection facilities where customers can return their used products, the second is the recovery centers, which include recycling and remanufacturing centers for processing these used products. The last layer is the markets where the recovery materials can be sold. Depending on the different types of returned products and the various network designs, there are other add-ons to RL networks, such as dismantling and disposal centers. Considering this structure, the cost minimization/profit maximization functions (Dat et al. 2012; Kilic et al. 2015; Demirel et al. 2016; John et al. 2018) and environmental health aimed by the RL setup can be combined with industrial symbiosis.

Eco-industrial park (EIP), which leverage the benefits of geographic proximity and collaboration, exemplify practical instances of industrial symbiosis. The overall framework of these parks can be characterized by the exchange and dissemination of information through an established network, creating mutual opportunities for securing necessary inputs, generating value-added objectives and optimizing technical processes for non-product outputs, and acting as a tool for competitive advantage and eco-innovation (Chertow 2000; Lombardi and Laybourn 2012).

The persistence of linear production and consumption models in the current economic system remains a significant obstacle to the development of EIP practices, particularly regarding financial aspects. In particular, the need to align regulations, investment mechanisms, and practices with traditional linear models including the assessment of financial risks creates challenges for the adoption of emerging circular economic approaches like EIP projects. Therefore, government participation in financing EIP projects, especially those involving large-scale infrastructure, seems crucial in facilitating the realization of circular systems (Korhonen et al. 2004; Vermeulen 2006).

E-waste is highly compatible with EIP systems in terms of its composition and economic value. Within the realm of e-waste recycling, EIP initiatives provide the essential labor, technology, and economic cycles required to transform e-products into raw materials that can be reused in manufacturing processes. While these raw materials might not necessarily be reincorporated into the same product type, the aim is to identify efficient uses for recycled materials. This process includes building networks across different production chains to identify optimal matches. For example, copper extracted from dismantling a computer within an EIP may not be used in new computers, but could instead be repurposed for the construction or telecommunications sectors. This model offers a potential win-win situation for participating companies (Marconi et al. 2018).

Material resources are an important source of income and employment, forming the physical basis of the economy. Their value varies across countries in terms of their physical and chemical properties, abundance and rarity. Economic growth means increased demand for raw materials, energy and other natural resources, and if the product life cycle (extraction/acquisition, transportation, production, consumption, recovery and disposal) is not managed appropriately, there is an increase in the quantities that end up as waste. Thus, the use of raw materials and related production and consumption processes have environmental, economic and social consequences within countries and beyond national borders (OECD 2021). Today, countries are faced with a growing problem that closely concerns both the economy and the environment. This problem is EoL e-products. In the current economic system, e-waste is the fastest growing category of hazardous solid waste in the world. Addressing this problem requires international cooperation, economic incentives that protect labor, and management approaches that minimize negative impacts on the environment and human health.

Literature Review

When the relationship between e-waste and GDP is investigated, it is seen that the studies on the relationship between e-waste and GDP do not go back very far and therefore this field is developing. In this respect, the studies examining the relationship between e-waste and GDP are as follows:

According to the study conducted by Lu et al. (2015) on China for the period 2000-2012, there is a positive relationship between e-waste generation per capita and GDP per capita.

Kusch and Hills (2017) analyzed the relationship between e-waste and GDP with 2014 data of 50 countries in the Pan-European region. According to the results of this analysis, there is a statistically significant economic relationship between e-waste and GDP. Accordingly, it is stated that an increase of 1000 international dollars in GDP according to purchasing power parity (PPP) leads to an increase of 0.5 kg in e-waste, but the linear relationship is more significant when GDP according to PPP is used than nominal GDP.

Kumar et al. (2017) found a linear relationship between GDP and e-waste in a study conducted in 2014 using 50 countries with the highest GDP and population. The relationship between e-waste and population is also confirmed, but GDP is a much stronger factor than population in explaining the amount of e-waste generated. In addition, it is also stated that the amount of e-waste production increases as the wealth of individuals and thus their purchasing power increases.

Awasthi et al. (2018) examined the relationship between GDP and e-waste using data from 2009 to 2014 for 28 European countries. Accordingly, it was found that an increase of 1000 international dollars in GDP according to PPP causes an additional 0.27 kg of e-waste production and an increase of 7.7 kg of e-waste for each additional citizen.

Namlis and Komilis (2019) found a strong positive linear correlation (equal to r=0.92 and statistically significant at the 5% level) between GDP per capita and e-waste in 10 European countries for the period 2008-2015. On the other hand, it is stated that the phenomenon of unemployment, which is related to consumption, is negatively related to the amount of e-waste.

Xavier et al. (2021) conducted an analysis of e-waste management with a focus on the Americas (Brazil and Canada in particular) to determine how legal, economic and environmental criteria influence e-waste management options. The results showed a linear relationship between GDP and e-waste generation.

Boubellouta and Kusch-Brandt (2020) investigated the relationship between e-waste and the Environmental Kuznets Curve (EKC) using data from 2000 to 2016 for 30 European countries (EU28+2). According to the results of the analysis, it is stated that the amount of e-waste increases with the increase in GDP, but the amount of e-waste decreases after a certain level of GDP growth. A similar result was also found in the study conducted by the same authors for 27 European Union (EU) countries for the period 2008-2016. It is also stated that the export of ICT products has a negative effect on the amount of e-waste (Boubellouta and Kusch-Brandt 2021). In another study, the relationship between e-waste and EKC was analyzed using 2016 data from 174 countries. As a result of this analysis, it was observed that there is an inverted-U-shaped relationship between economic growth and environmental degradation (e-waste). That is, e-waste generation initially increases with increasing GDP, but when GDP reaches a certain level (turning point), the amount of e-waste generated does not increase despite the increase in economic growth. However, the calculated inflection points occur at very high GDP levels and therefore most countries around the world are expected to witness a drastic increase in e-waste if current models are not radically transformed to prevent e-waste growth. In addition, population, urbanization, industrialization and access to electricity were used as control variables in the analysis and it was stated that urbanization and industrialization have a strong and positive correlation with respect to total e-waste generation, access to electricity is a particularly strong factor with respect to per capita e-waste generation, while population has a positive effect on total e-waste generation and a negative effect on per capita e-waste generation (Boubellouta and Kusch-Brandt 2021a). The same authors investigated the relationship between e-waste and EKC using data from 2000 to 2015 for 30 European countries (EU28+2). According to the analysis on the determinants of e-waste generation, it was found that in addition to population and urbanization, renewable energy consumption and trade openness also had a positive effect on the amount of e-waste, while the relationship between GDP was found to have an inverted-U shape (Boubellouta and Kusch-Brandt 2022). Again, Boubellouta and Kusch-Brandt (2023) investigated the relationship between e-waste recycling rate and GDP using data from 30 European countries from 2008 to 2018. Here, it is stated that the relationship in question is in the form of an N-shaped curve. In other words, the e-waste recycling rate first increases with economic growth, then decreases in EC, and then starts to increase again as EC continue to grow.

According to the results of Kalia et al. (2022) using 172 countries, unlike other studies, GDP per capita does not directly affect e-waste generation. While this result is also stated for literacy rate, internet access and urbanization variables, it is stated that increased internet use in EC leads to an increase in e-waste generation, while high literacy rates in DC suppress e-waste generation. For urbanization, it is stated that e-waste generation is higher in countries that do not have a regulatory legal framework and have a high urban population.

Azwardi et al. (2023) investigated the relationship between e-waste and GDP using data from 2000 to 2020 for 28 EU countries. The results of the analysis indicate that e-waste recycling has a positive and statistically significant effect on GDP per capita, while e-waste generation per capita has no statistically significant effect on economic growth.

Yilmaz and Koyuncu (2023) investigated the relationship between e-waste recycling and globalization in terms of its sub-dimensions (economic, political and social) using the data of 30 European countries (EU28 + 2) from 2008 to 2018. According to the results of the analysis, it is stated that GDP has a positive effect on e-waste recycling.

Studies show that there is a strong interdependence between e-waste and GDP and that other variables mentioned above also have an impact. However, more research is needed to reveal the nature of the interdependence. These studies have used similar regression models (OLS - pooled ordinary least squares; DOLS - dynamic ordinary least squares; FMOLS - fully modified ordinary least squares; GMM – generalized method of moments; PQR - panel quantile regression; 2SLS - two-stage least squares). The analysis conducted in this study differs from previous analyses in several ways, as mentioned in the Introduction section.

DATA

The main objective of this part of the study, which includes empirical analysis, is to reveal the relationship between total e-waste and economic growth within the framework of the countries† examined (USA, Germany, Australia, Brazil, China, Ghana, South Korea, India, Switzerland, Japan, Türkiye). In this analysis, annual data for the period 1960-2020 are used‡. The general outlook of the series of the variables used for each country is shown in Figure 4.

Figure 4. E-Waste and GDP in selected countries over the years

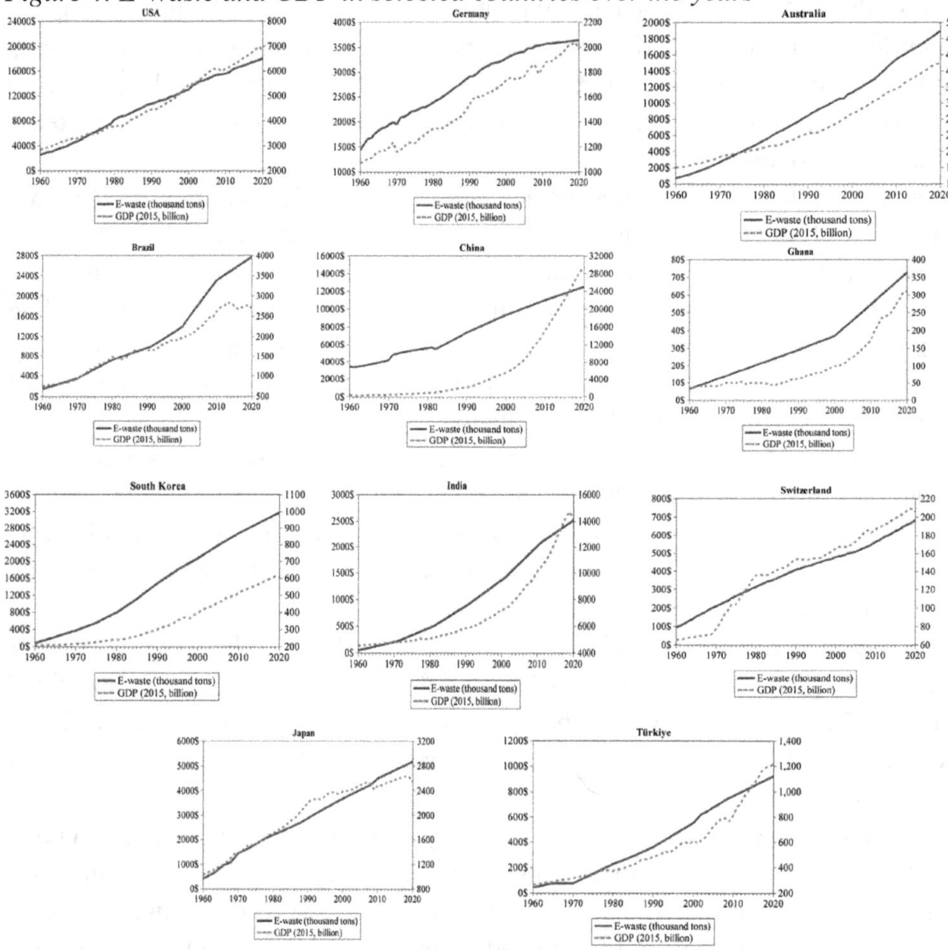

Source: E-waste data were calculated by the authors, based on the formula explained in the paper, [E = (M.N) / L]. Data for GDP were taken from the WB and OECD databases.

The total e-waste (kg) series was calculated by the author for each country using the formula [E=(M.N)/L][§]. Where E (kg/year) represents the amount of e-waste generated, M (kg) is the weight of the product, N is the number of products in use, and L (years) is the average lifetime of the product. The information considered for use in the formulation, taking into account the classification in the EU E-Waste Directive, is given in Table 3. For economic growth, data on real GDP (2015, $) are taken from the World Bank (WB) and Organisation for Economic Co-operation and Development (OECD) databases. Both variables are transformed into natural logarithmic form before being subjected to econometric tests.

Table 3. Information used to calculate the amount of e-waste

E-product	Average unit weight (kg)	Average lifespan (years)
Television	30	5
Telephone	1	5
Cell Phone	0.1	2
Computer	25	3
VCD/DVD Player	5	5
Refrigerator	35	10
Deep Freezer	35	10
Dishwasher	50	10
Washing Machine	65	8
Air-conditioning	55	12
Photocopy Machine	60	8
Radio	2	10
Microwave	15	7

Source: Adapted by Robinson 2009.

ECONOMETRIC METHODOLOGY

The bootstrap panel causality test, which is used in the empirical analysis of the causal relationship between e-waste and economic growth, has some advantages. First of all, it does not require unit root and cointegration preliminary tests that need to be performed before causality analysis. Therefore, in this analysis, the variables in the model do not need to be stationary. In this analysis, it is possible to obtain bootstrap critical values for each country separately. Second, this test takes into account cross-section dependence and takes into account the heterogeneity of the panel. Thus, it is possible to test the causality relationship separately in the context of panel data (Kónya 2006).

The hypothesis of this study is as follows: Bidirectional causality exists between total e-waste and economic growth.

Accordingly, the theoretical model used to test the hypothesis is as follows:

$$\ln tew_{it} = \alpha + \beta gdp_{it} + \varepsilon_{it} \tag{1}$$

$$\ln gdp_{it} = \alpha + \beta tew_{it} + \varepsilon_{it} \tag{2}$$

In the model, *tew* represents the total amount of e-waste, *gdp* represents economic growth, $i = 1,...,N$ countries, $t = 1,...,T$ time dimension and ε is the error term.

Whether or not cross-sectional dependence between the series is taken into account significantly affects the results of the analysis (Breusch-Pagan 1980). Therefore, cross-section dependence should be tested before proceeding with the analysis. The Lagrange Multiplier (LM) test developed by Breusch-Pagan (1980), which is a pioneering test used to test for cross-section dependence when the time dimension is large and the cross-section dimension is small, is as follows:

$$\text{LM} = \sum_{i=1}^{N-1} \sum_{j=i+1}^{N} T_{ij}\rho_{ij}^2 \sim X^2_{N(N-1)/2} \tag{3}$$

However, when both time and cross-sectional dimensions are large, the Cross-Section Dependence (CD_{LM}) test developed by Pesaran (2004) can be used to investigate whether there is cross-sectional dependence. The CD_{LM} is as follows:

$$\text{CD}_{\text{LM}} = \left(\frac{1}{N(N-1)})^{\frac{1}{2}} \sum_{i=1}^{N-1} \sum_{j=i+1}^{N-1} (T\rho_{ij}^2 - 1 \right) \sim N(0,1) \tag{4}$$

Considering that the CD_{LM} test may lead to size distortions in the case of small T and large N, Pesaran (2004) developed the CD test statistic as a more general test that can be applied in this case. The CD test is as follows:

$$CD = \sqrt{\left(\frac{2T}{N(N-1)}\right)} \sum_{i=1}^{N-1} \sum_{j=i+1}^{N} (\rho_{ij} - 1) \sim N(0,1) \tag{5}$$

Pesaran et al. (2008) developed the Adjusted LM (LM_{adj}) test, which can be applied in case of dimensional distortions in the CD test. The LM_{adj} test statistic is as follows:

$$LM_{adj} = \sqrt{(\frac{2}{N(N-1)})}\sum_{i=1}^{N-1}\sum_{j=i+1}^{N}[\rho_{ij}^2\left(\frac{(T-k)\rho_{ij}^2-\mu_{Tij}}{\upsilon_{Tij}}\right] \sim N(0,1) \quad (6)$$

Where k is the number of regressors and μ_{Tij} and υ_{Tij} are the mean and variance of $(T-k)\rho_{ij}^2$, respectively (Pesaran et al. 2008). The hypotheses of the LM_{adj} test are as follows:

H_0: There is no cross-section dependence.
H_1: There is cross-section dependence.

For panel causality analysis, it is important to decide whether the slope coefficients are homogeneous or heterogeneous. In order to test the homogeneity/heterogeneity of the slope coefficient in the analysis, the pioneering Swamy (1970) test was developed by Pesaran and Yamagata (2008). The statistics of Swamy (1970) test and Pesaran and Yamagata (2008) delta test are as follows:

$$\mathcal{S} = \sum_{i=1}^{N}(\beta_i - \beta_{WFE})\frac{'X_i' M_T X_i}{\sigma_i^2}(\beta_i - \beta_{WFE}) \quad (7)$$

$$\Delta = \sqrt{N}\left(\frac{N^{-1}\mathcal{S} - k}{\sqrt{2k}}\right) \quad (8)$$

$$\Delta_{adj} = \sqrt{N}(\frac{N^{-1}\mathcal{S} - E(z_{iT})}{\sqrt{Var(z_{iT})}} \quad (9)$$

Hypotheses of the delta test:

H_0: Slope coefficients are homogeneous. ($\beta_i=\beta$)
H_1: Slope coefficients are heterogeneous. ($\beta_i\neq\beta$)

After determining the presence of cross-section dependence and country-specific heterogeneity in the model, the panel causality test proposed by Kónya (2006) can be applied. The following set of equations is estimated in this causality test based on the Wald Test with country-specific bootsrap critical values and Zellner's (1962) Seemingly Unrelated Regressions (SUR) estimation:

$$tew_{1,t} = \alpha_{1,1} + \sum_{l=1}^{ltew_1}\beta_{1,1,l}tew_{1,t-1} + \sum_{l=1}^{lgdp_1}\gamma_{1,1,l}gdp_{1,t-1} + \varepsilon_{1,1,t} \quad (10)$$

$$tew_{2,t} = \alpha_{1,2} + \sum_{l=1}^{ltew_1}\beta_{1,2,l}tew_{2,t-1} + \sum_{l=1}^{lgdp_1}\gamma_{1,2,l}gdp_{2,t-1} + \varepsilon_{1,2,t}$$

$$tew_{N,t} = \alpha_{1,N} + \sum_{l=1}^{ltew_1} \beta_{1,N,l} tew_{N,t-1} + \sum_{l=1}^{lgdp_1} \gamma_{1,N,l} gdp_{N,t-1} + \varepsilon_{1,N,t}$$

and

$$gdp_{1,t} = \alpha_{2,1} + \sum_{l=1}^{ltew_1} \beta_{2,1,l} tew_{1,t-1} + \sum_{l=1}^{lgdp_1} \gamma_{2,1,l} gdp_{1,t-1} + \varepsilon_{2,1,t} \quad (11)$$

$$gdp_{2,t} = \alpha_{2,2} + \sum_{l=1}^{ltew_1} \beta_{1,2,l} tew_{2,t-1} + \sum_{l=1}^{lgdp_1} \gamma_{2,2,l} gdp_{2,t-1} + \varepsilon_{2,2,t}$$

$$gdp_{N,t} = \alpha_{2,N} + \sum_{l=1}^{ltew_1} \beta_{2,N,l} tew_{N,t-1} + \sum_{l=1}^{lgdp_1} \gamma_{2,N,l} gdp_{N,t-1} + \varepsilon_{2,N,t}$$

Where l is the optimal lag length determined using the Akaike Information Criterion. To test the causality from *gdp* to *tew* for the first unit of the panel, the significance of $\gamma_{1,1,l}$ is tested using the cross-sectional bootstrap critical values and the Wald Test in Equation 3. Similar causality relationships can be examined by testing related terms.

Global economic and political events can change the structure of causal relationships (Foon 2008; Yilanci and Ozgur 2019) and Kónya's (2006) bootstrap panel causality test only shows the results of the entire sample. Therefore, we investigate the stability of the causal relationship by applying a rolling window bootstrap panel causality test on the subsample for, $t = \tau - 1 + l, \tau - 1, \tau = l, l + 1, ..., T$, where the constant l denotes the size of the subsample. To control for small sample bias, bootstrap simulations are used to obtain the required critical values and p-values. An important element here is to determine the length of the sub-period (number of windows) as suggested by Brooks and Hinich (1998). In this direction, the number of windows is determined by taking Caspi (2017) study into account. The number of windows in the model is determined by the equation below.

$$ss = \left[T\left(0.01 + \frac{1.8}{\sqrt{T}}\right)\right] \quad (12)$$

EMPIRICAL FINDINGS AND DISCUSSION

Table 4 presents the findings of cross-section dependence and homogeneity tests for the model. Accordingly, it is observed that there is cross-section dependence in the model and the slope parameters of the variables used in the model are heterogeneous. Then, bootstrap panel causality test is applied to obtain the causality between *gdp* and *tew* considering the whole sample and the findings are presented in Table 5 for each country.

Table 4. Cross-section dependence and homogeneity tests

Tests	Direction of Causality	
	gdp-tew	tew-gdp
LM	1011.841*	223.619*
CD_{LM}	91.231*	16.077*
CD	21.974*	5.943*
LM_{adj}	15.529*	15.001*
Δ	91.811*	84.025*
Δ_{adj}	94.116*	86.135*

Note: *, indicates 1% significance level.

Table 5. Bootstrap panel causality test results

Countries	Wald Statistics	Bootstrap Probability Value	Critical Values		
			%1	%5	%10
Panel A					
H_0: gdp ≠> tew					
USA	0.026	0.902	9.413	5.274	3.762
Germany	5.377	0.137	14.123	8.574	6.454
Australia	0.210	0.744	9.445	5.491	3.982
Brazil	1.533	0.903	29.124	20.725	16.884
China	11.486*	0.003	8.447	4.872	3.610
Ghana	0.324	0.759	14.117	8.489	6.418
South Korea	13.778*	0.000	7.245	4.115	2.901
India	0.288	0.736	14.648	8.298	5.948
Switzerland	0.107	0.994	17.407	12.166	10.199
Japan	2.322	0.448	14.903	9.345	7.044
Türkiye	0.055	0.875	11.057	6.953	5.047
Panel B					
H_0: tew ≠> gdp					
USA	3.089	0.875	13.726	10.671	9.400
Germany	8.344	0.956	44.601	34.481	29.726

continued on following page

Table 5. Continued

Countries	Wald Statistics	Bootstrap Probability Value	Critical Values		
			%1	%5	%10
Australia	0.273	0.984	8.883	6.670	5.621
Brazil	0.001	1.000	11.554	9.281	8.116
China	10.103	0.305	17.887	14.534	12.976
Ghana	0.533	0.783	4.313	3.176	2.642
South Korea	2.718	0.628	12.956	9.675	7.971
India	11.195***	0.060	14.084	11.502	10.276
Switzerland	0.368	0.908	7.056	5.088	4.277
Japan	2.386*	0.002	1.479	0.833	0.576
Türkiye	2.821	0.812	9.846	8.009	7.034

Note: *, **, *** indicate 1%, 5% and 10% significance level, respectively. Critical values are obtained with 10000 bootstrap cycles.

Panel A in Table 5 shows the test results of the null hypothesis that "*gdp* variable is not the cause of *tew* variable" for each country, while Panel B shows the test results of the null hypothesis that "*tew* variable is not the cause of *gdp* variable". Table 5 shows that the causality relationship from *gdp* to *tew* is statistically significant only for China and South Korea. On the other hand, the causality from *tew* to *gdp* is statistically significant only for India and Japan.

In addition to the unit root properties and cointegration relationship in the models, economic, structural and technological changes also affect the causality tests. In other words, the nature and/or existence of the causal link may change over time (Yilanci and Ozgur 2019). In this context, an analysis was conducted on the sub-sample determined based on Equation (12) to determine the relationship between the variables. Accordingly, the sub-sample size was set as 15 years and the optimal lag length in the sub-sample was determined using the Akaike Information Criterion.

First of all, according to the general bootstrap panel causality test in Table 5, causality from *gdp* to *tew* is observed only in China and South Korea. However, as seen in Figure 5, according to the bootstrap panel causality test performed on sub-samples, it is found that there is a hidden causality from *gdp* to *tew* for all countries. According to Figure 5, this causality is observed for the USA in 1983-5, 1993, 2001, 2004-6, 2009, 2016, 2020; Germany in 1990, 2004, 2015; Australia in 1993; Brazil in 1985, 1995-6, 2003, 2005-6, 2008-11, 2017-18; China in 1976, 1979, 1983-6, 1993, 1996-2002, 2007, 2010, 2016, 2019; Ghana in 1976, 1979-80, 1982-3, 1991-5, 1997, 2014, 2016-8, 2020; South Korea in 1989-93, 2013, 2016-7; India in 1985, 1996, 2006, 2008, 2013, 2015-6, 2020; Switzerland in 1976, 1979, 1983, 1988-90,

2004, 2007, 2016, 2019-20; Japan in 1989-90, 1995-2001, 2007, 2016, 2019; and Türkiye in 1979, 1981, 1985-6, 1989-91, 2007, 2013.

Figure 5. Causality from gdp to tew by countries

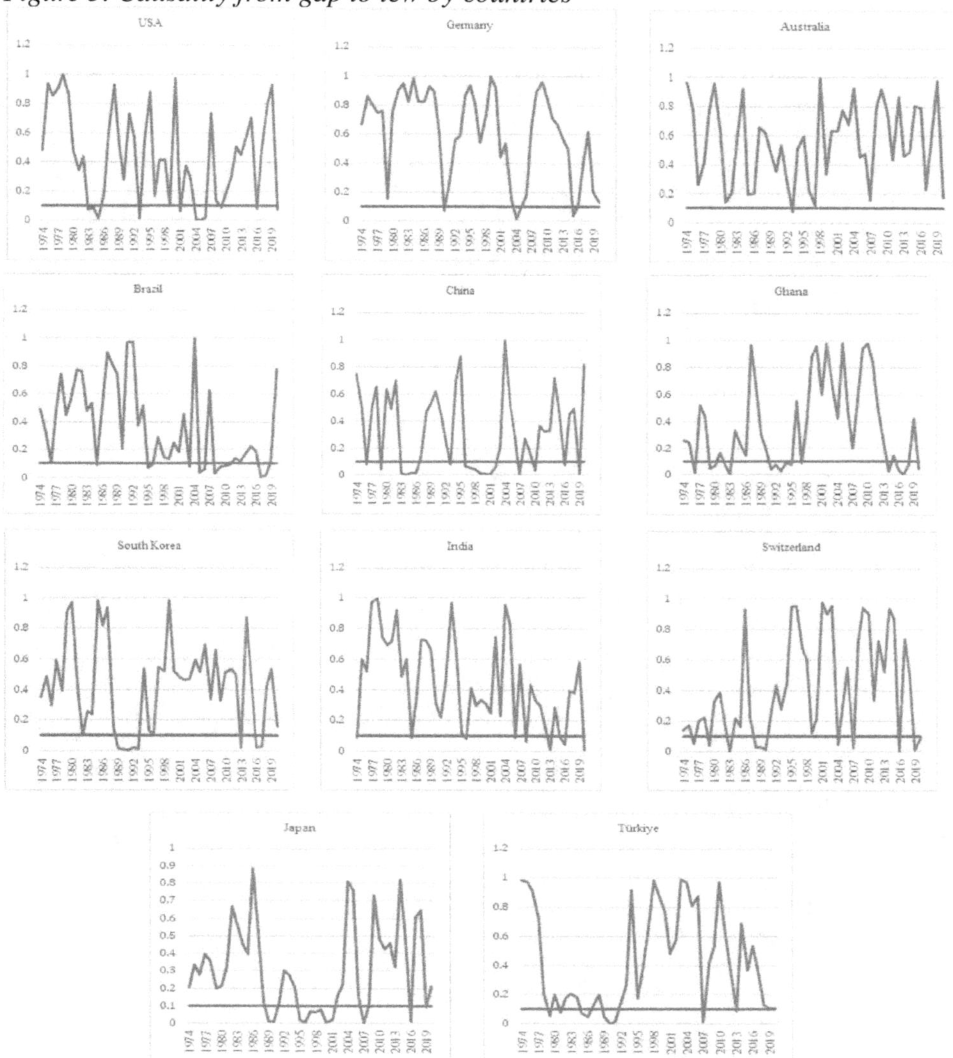

On the other hand, according to the general bootstrap panel causality test in Table 5, the causality from *tew* to *gdp* is observed only in India and Japan. However, as seen in Figure 6, according to the bootstrap panel causality test performed on the sub-sample, it is found that there is a hidden causality from *tew* to *gdp* for all countries. According to Figure 6, this causality is observed for the USA in 2010; Germany in 1983, 1989; Australia in 1982; Brazil in 2006; China in 1983, 1990,

2002, 2007, 2010-2; Ghana in 1983-4, 1988, 1992-6, 2005; South Korea in 2019; India in 1992, 2008, 2010-11; Switzerland in 1975, 1992, 2004; Japan in 2001, 2003, 2010; and Türkiye in 1976, 2001. The analysis results for all countries show that there is a bidirectional causality from *gdp* to *tew* and from *tew* to *gdp*.

Figure 6. Causality from tew to gdp by countries

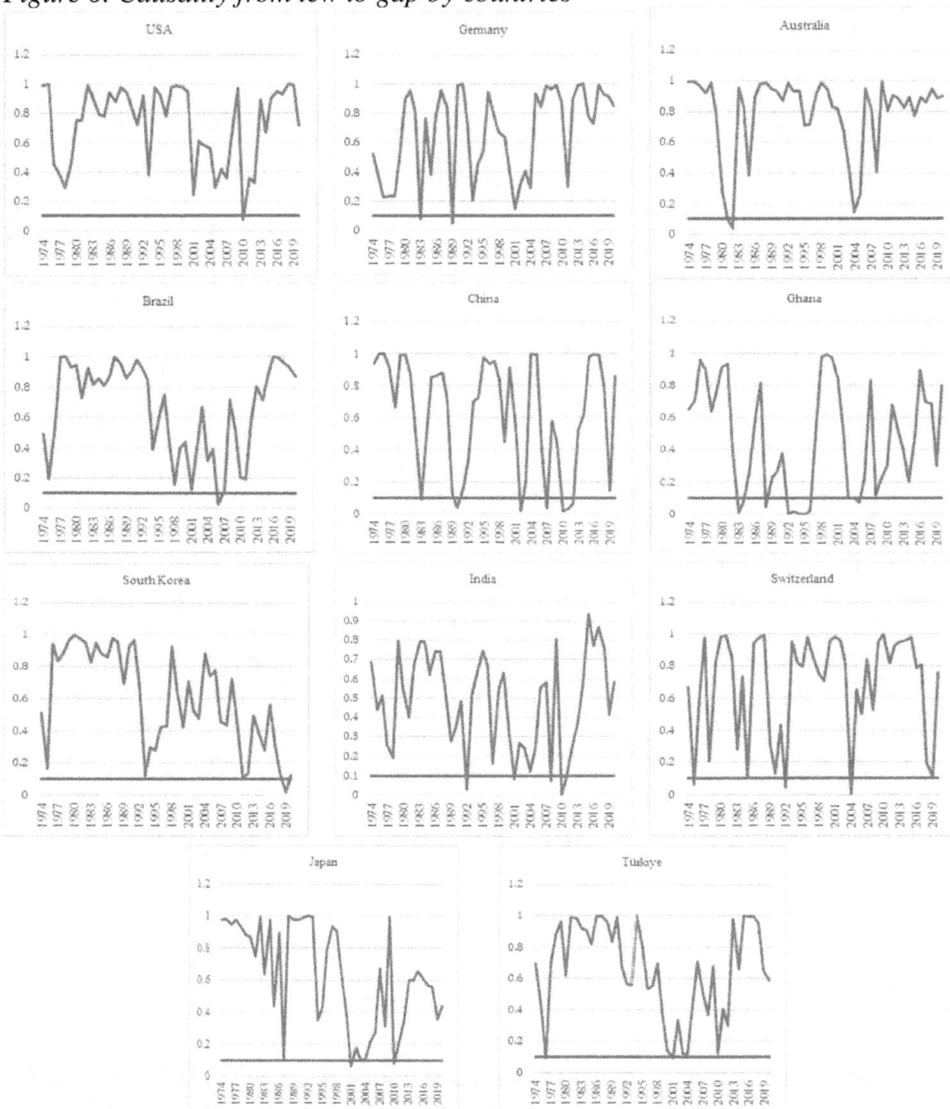

As a result of the analysis, a bidirectional causality relationship was found between the variables for all countries. In Figure 7, bootstrap coefficient estimates (Balcilar et al. 2010; Aslan et al. 2018; Ozcan et al. 2018; Yilanci and Ozgur 2019) are calculated and the effect from *gdp* to *tew* in Panel [1] and from *tew* to *gdp* in Panel [2] for all countries.

Figure 7a. The impact of gdp on tew and tew on gdp by country

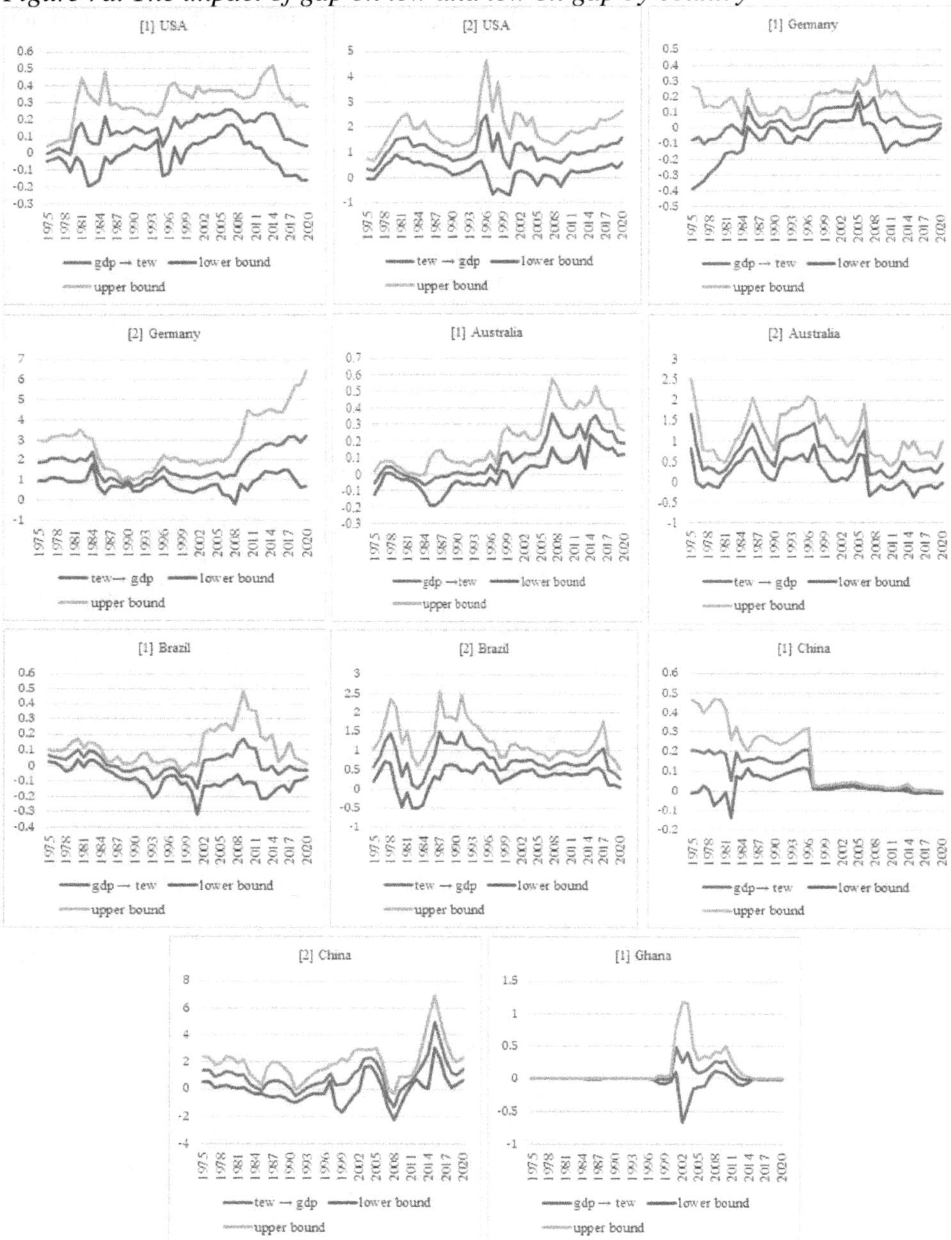

Figure 7b. The impact of gdp on tew and tew on gdp by country

According to the analysis outputs in Panel [1], countries can be evaluated as follows: As in the studies conducted by Lu et al. (2015), Kusch and Hills (2017), Kumar et al. (2017), Awasthi et al. (2018), Namlis and Komilis (2019) and Xavier et al. (2021), a causality relationship from *gdp* to *tew* is found for all countries on a variable basis. In addition, it is clearly observed that the effect of *gdp* on *tew* is positive for Germany and Brazil since 1985 and 2003, respectively, for the USA, Australia, China, Ghana, South Korea, India, Japan and Türkiye throughout the entire sub-sample, and negative only for Switzerland since 1991. In this framework, it can be stated that economic growth increases the volume of e-waste in all countries except for Switzerland. Considering Switzerland's e-waste awareness, management success and population, it seems logical that economic growth is not a triggering factor in e-waste increase.

According to the analysis outputs in Panel [2], countries can be evaluated as follows: In addition to the existence of a causality relationship from *tew* to *gdp* for all countries, it is observed that the effect of *tew* on *gdp* is positive throughout the entire sub-sample in all countries analyzed. Considering that e-waste is a valuable resource and an economy emerges through its recovery, it is expected to contribute to economic growth. As a matter of fact, the analysis outputs support this expectation.

When Panel [1] for all countries is analyzed from the perspective of the EKC hypothesis, that is, when the effect of *gdp* on *tew* is analyzed in terms of trend direction, the *inverted-U* form is not supported. This implies that the EKC hypothesis is not supported when *tew* is taken as an indicator for environmental pollution. As a matter of fact, this is consistent with the fact that the effect of *tew* on *gdp* is positive in all countries in Panel [2]. Otherwise, the effect of *tew* on *gdp* representing environmental pollution would be expected to be negative. According to the analysis output, it can be stated that *tew* is related to resource allocation rather than environmental pollution and provides information in this respect. The results obtained for the EKC hypothesis are not in line with the results of Boubellouta and Kusch-Brandt (2020; 2021; 2021a; 2022). However, while the results in the aforementioned studies are obtained over variables, in this study, it is analyzed country-specific. In addition, Ozcan et al. (2018) state that the validity of the EKC hypothesis is significant in the case of unidirectional causality. However, here, a bidirectional causality is found between *gdp* and *tew*.

On the other hand, it can be stated that e-waste has an increasing importance in terms of both resource allocation and the environment. By adopting a circular economy approach, e-waste can be brought under control and contribute to economic growth, while also reducing the use of primary resources. Otherwise, in the projection given in Figure 8 for the years 2020-2060, the amount of e-waste is estimated to increase in the future, and this situation holds the possibility of a significant financial loss for countries, as well as the possibility that the pressure on the environment will become more prominent.

Figure 8. Estimation of e-waste amount in selected countries for the period 2010-2060

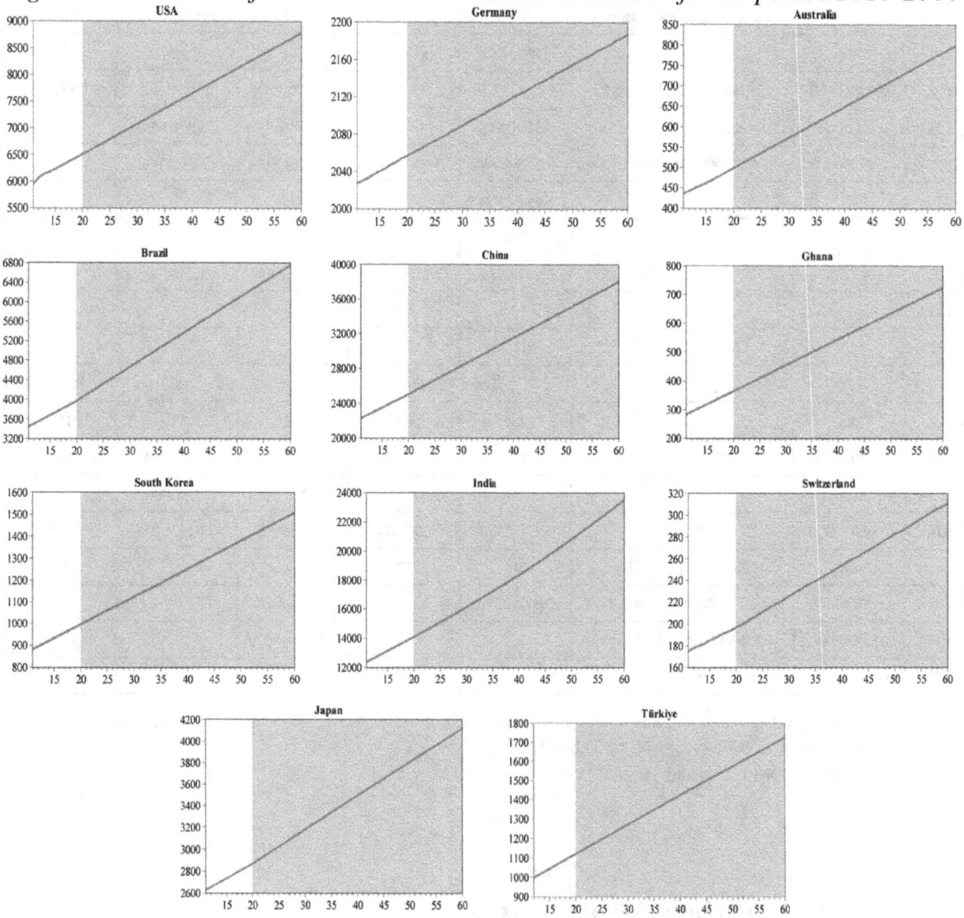

Source: Made by the authors based on e-waste data estimated for the period 1960-2020.
Note: Left scale is thousand tons; gray area is estimated area.

In order to illustrate the above-mentioned financial loss, the potential revenue that countries can earn from e-waste products was calculated using the potential increase in the volume of e-waste shown in Figure 8 and the data in Table 6 and Table 7, and these calculated data are shown in Figure 9. It should be noted that the change in potential revenue is directly related to the number of products included in the calculation and the prices of all minerals.

Table 6. Material composition in e-waste products (%)

	Basic Metals and Plastics				Precious Metals			
	Fe	**Cu**	**Al**	**Plastics**	**Ag**	**Au**	**Pd**	**Pt**
CE	0.65580800	0.35258900	0.14068700	0.76629700	0.00670600	8.66052548	18.71531571	3.16000000
LHA	0.51000000	0.04000000	0.02000000	0.38000000	0.00016000	3.78000000	1.99400000	
SHA	0.29000000	0.08000000	0.05000000	0.53000000	0.00013200	1.68000000	2.03100000	
	Rare Metals							
	Co	**Ga**	**Ta**	**Nd**	**Pr**	**Bi**	**Ba**	
CE	0.04575900	1.77000000	0.00186300	0.00445000	7.91000000	0.00037300	0.00468269	
LHA								
SHA								
	Heavy Metals							
	As	**Be**	**Bi**	**Cd**	**Cr**	**Ni**	**Pb**	**Hg**
CE	0.00002700	2.37008800	0.00052400	1.40000000	0.00332900	0.02037800	0.02208500	0.00790700
LHA					0.00010000	0.00050000	0.01500000	
SHA	0.08013900				0.00024800	0.00098900	0.01976200	
	Sb	**Sn**	**Zn**	**Br**	**Cl**	**Si**		
CE	0.00254600	0.01936700	0.04022000	0.02365300	0.00361504	0.03953600		
LHA	0.00090000	0.02400000	0.03200000					
SHA	0.00059400	0.01619800	0.01028100	0.00163700	0.00405656			

Note: CE, consumer electronics; LHA, large household appliances; SHA, small household appliances. (Classification is based on the products in Table 3).

Source: Adapted from Islam and Huda 2019.

Table 7. Average market price of metals ($/tonne)

Basic Metals and Plastics				Precious Metals			
Fe	**Cu**	**Al**	**Plastics**	**Ag**	**Au**	**Pd**	**Pt**
120	8134	2348	890	990000	48000000	32500000	40133000
Rare Metals							
Co	**Ga**	**Ta**	**Nd**	**Pr**			
58283	1137000	178000	176	107908			
Heavy Metals and others							
Cd	**Ni**	**Pb**	**Sb**	**Sn**	**Zn**		
1220	27200	2164	9922	23	2539		

Source: Adapted from Islam and Huda 2019.

Figure 9. Potential revenue from e-waste for countries (million, $)

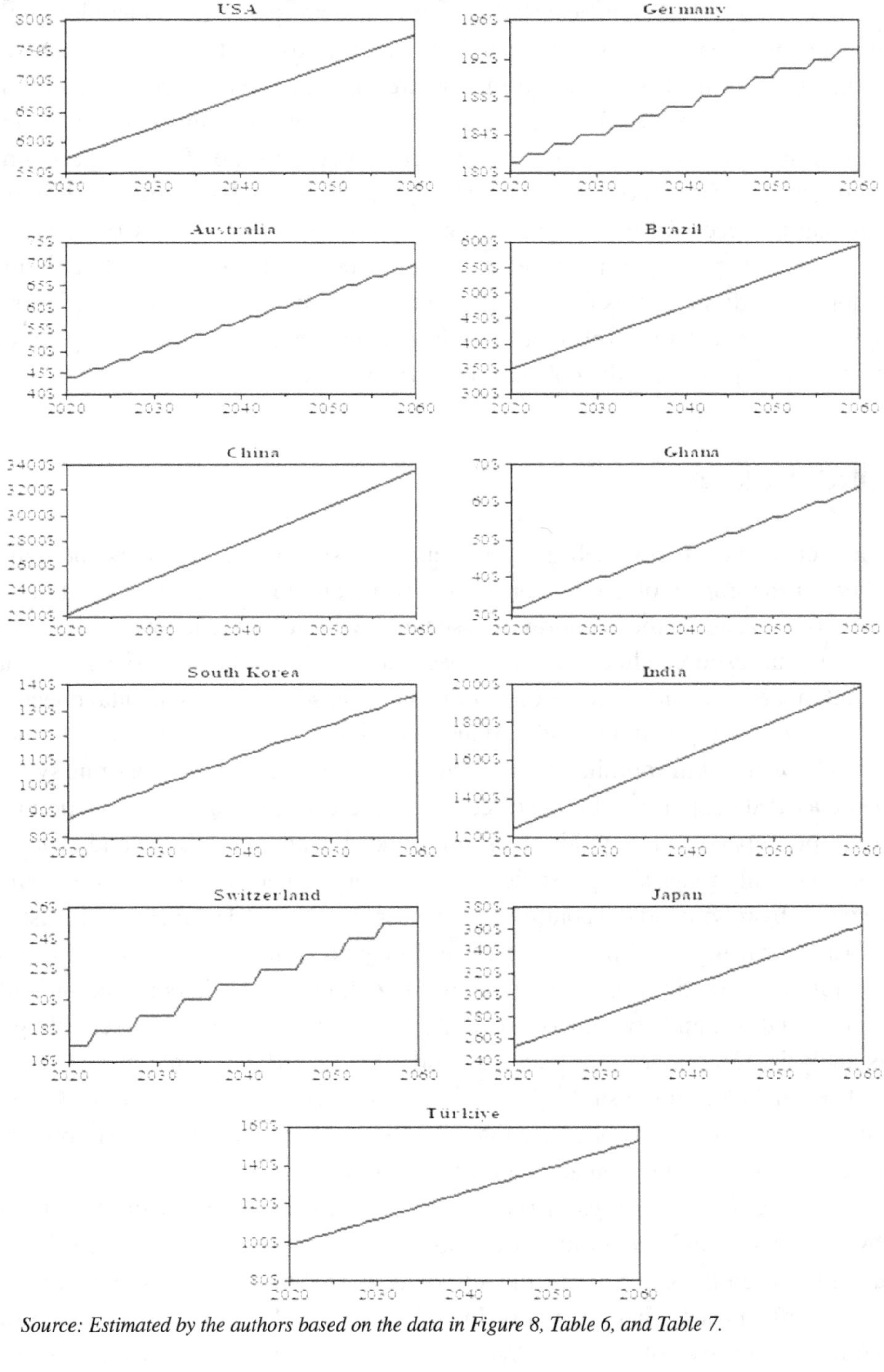

Source: Estimated by the authors based on the data in Figure 8, Table 6, and Table 7.

In the light of this information, it can be said that integrating the circular economy into e-waste has significant advantages for countries. These include; (i) the inclusion of a material resource that is currently mostly idle into the economy, (ii) the establishment of a good infrastructure for resource recovery from e-waste provides a positive external benefit for the environment, (iii) supporting the double dividend hypothesis** with the revenues generated by states from the economy generated by e-waste, the scope of which is well defined on legal grounds. It is also clear that a successful e-waste management practice in the context of economy, environment, technology and governance will support the development of circular economy paradigm. In this framework, countries should strengthen their efforts to implement an effective and efficient e-waste management plan and support practices that proactively improve the collection and recovery of e-waste.

CONCLUSION

In recent years, there has been a growing interest among researchers and policy makers in the impact of e-waste on the economy and the environment. While the impact of e-waste on the economy is associated with resource recovery and hence efficiency in resource allocation, its impact on the environment is associated with the fact that e-waste has become one of the main causes of environmental concerns due to its increasing volume, high production rate and toxic environmental burden.

It is known that the dominant economic doctrine in the current economic system is neoclassical economics. In this direction, while determining the production and consumption perspective within the framework of economic growth and development targets, high waste/output societies have emerged that tend to increase resource utilization by relying on the continuity of natural resources. The fact that e-waste is the fastest growing solid waste in the world may be an important indicator of this linear production and consumption economy model. Therefore, this study empirically tests the relationship between e-waste and (linear) economic growth. According to this analysis, the relationship between total e-waste and economic growth in the countries included in the study (USA, Germany, Australia, Brazil, China, Ghana, South Korea, India, Switzerland, Japan, Türkiye) was tested with panel bootstrap causality analysis using annual data for the period 1960-2020.

According to the findings of the analysis, a bidirectional causality was found between e-waste and economic growth in all sub-samples for all countries. Subsequently, when the direction of this causality is analyzed, it is observed that the variables affect each other positively. In addition, Panel [1] in Figure 7 is analyzed from the perspective of the EKC hypothesis and the *inverted-U* form is not supported. However, the expected turning points may occur at very high GDP levels.

Therefore, in most countries around the world, a significant increase in e-waste is expected if the current linear economic model is not transformed to prevent the increase in e-waste. The potential loss of revenue due to ineffective management of this volumetric increase, which will have a negative impact on the environment, is shown in Figure 8 and Figure 9 respectively.

In this framework, countries can overcome their disadvantageous position in the face of the increase in e-waste by abandoning the linear economy (neoclassical economics in the supply chain) model and adopting and supporting an effective management style that adopts the circular economy (ecological economics) model, thereby eliminating the negative externalities generated by e-waste and using it as a tool to achieve efficiency in resource allocation. The advantages of countries in the current situation can be listed as follows: (i) The inclusion of a material resource that is currently mostly idle into the economy, (ii) the establishment of a good infrastructure for resource recovery from e-waste provides a positive external benefit for the environment, (iii) supporting the double dividend hypothesis with the revenues generated by states from the economy generated by e-waste, the scope of which is well defined on legal grounds. It is also clear that a successful e-waste management practice in the context of economy, environment, technology and governance will support the development of circular economy paradigm. In this framework, countries should strengthen their efforts to implement an effective and efficient e-waste management plan and support practices that proactively improve the collection and recovery of e-waste. The solution to the e-waste problem, which can be seen as a consequence of the linear production and consumption economy model, can only be possible by changing the system of thought that creates it. In this direction, it can be said that within the framework of sustainable production and consumption paradigm, it is essential to transition from linear economy to circular economy.

In conclusion, to provide guidance for future research, it can be suggested that investigations into the recovery of raw materials and key components from e-waste will promote recycling efforts in the sector. This will also help avoid the accumulation of e-waste in open spaces or its disposal through burning. Future studies could broaden the scope by including a larger number of countries (especially given the limited research on e-waste in certain developed and developing nations, such as France, Canada, and South Africa) and adopting a mixed-methods approach (empirical, descriptive, semi-structured interviews, etc.) to examine e-waste generation and management practices in those countries. Such research would offer valuable insights into e-waste production, recovery, regulation, and the environmental and health consequences in these regions. Incorporating interdisciplinary approaches in the development of e-waste policies is essential, as it would enable policymakers to better understand e-waste management challenges and address the environmental and health risks associated with it. Finally, as the recovery of batteries and electronic components from electric vehicles (which can be categorized as smart devices) becomes more relevant, it can be anticipated that e-waste will increasingly become the focus of interdisciplinary research within the framework of the circular economy.

ACKNOWLEDGMENT

This study is based on the thesis titled "Public Policies Towards Waste Electrical and Electronic Equipments: Implications for Türkiye" prepared at Istanbul University, Institute of Social Sciences, PhD program in Public Finance.

REFERENCES

Adeleye, N., & Eboagu, C. (2019). Evaluation of ICT development and economic growth in Africa. *NETNOMICS: Economic Research and Electronic Networking*, *20*(1), 31–53. DOI: 10.1007/s11066-019-09131-6

Aghaei, M., & Rezagholizadeh, M. (2017). The impact of information and communication technology (ICT) on economic growth in the OIC countries. *Economic and Environmental Studies*, *17*(2), 257–278. DOI: 10.25167/ees.2017.42.7

Akkaya, Ş., & Bakkal, U. (2005). *Çevre Vergileri ve Çifte Yarar*. İstanbul Üniversitesi İktisat Fakültesi Mecmuası, 55(2), 1-22.

Albiman, M., & Sulong, Z. (2017). The linear and non-linear impacts of ICT on economic growth, of disaggregate income groups within SSA region. *Telecommunications Policy*, *41*(7-8), 555–572. DOI: 10.1016/j.telpol.2017.07.007

Alvarez-Gil, M., Berrone, P., Husillos, F., & Lado, N. (2007). Reverse logistics, stakeholders' influence, organizational slack, and managers' posture. *Journal of Business Research*, *60*(5), 463–473. DOI: 10.1016/j.jbusres.2006.12.004

Appiah-Otoo, I., & Song, N. (2021). The impact of ICT on economic growth-Comparing rich and poor countries. *Telecommunications Policy*, *45*(2), 1–15. DOI: 10.1016/j.telpol.2020.102082

Aslan, A., Destek, M., & Okumus, I. (2018). Bootstrap rolling window estimation approach to analysis of the Environment Kuznets Curve hypothesis: Evidence from the USA. *Environmental Science and Pollution Research International*, *25*(3), 2402–2408. DOI: 10.1007/s11356-017-0548-3 PMID: 29124642

Awasthi, A., Cucchiella, F., D'Adamo, I., Li, J., Rosa, P., Terzi, S., Wei, G., & Zeng, X. (2018). Modelling the correlations of e-waste quantity with economic increase. *The Science of the Total Environment*, *613-614*, 46–53. DOI: 10.1016/j.scitotenv.2017.08.288 PMID: 28898811

Ayres, R. (1994). *Industrial metabolism: theory and policy. In the greening of industrial ecosystems*. National Academy Press.

Ayvaz, B., Bolat, B., & Aydın, N. (2015). Stochastic reverse logistics network design for waste of electrical and electronic equipment. *Resources, Conservation and Recycling*, *104*, 391–404. DOI: 10.1016/j.resconrec.2015.07.006

Azwardi, A., Igamo, A., & Wijaya, W. (2023). *The Concept of Waste Management on Economic Development in the European Union.* International Journal of Energy Economics and Policy, 13(1), 1-6. DOI: 10.32479/ijeep.13667

Balcilar, M., Ozdemir, Z., & Arslanturk, Y. (2010). Economic growth and energy consumption causal nexus viewed through a bootstrap rolling window. *Energy Economics*, *32*(6), 1398–1410. DOI: 10.1016/j.eneco.2010.05.015

Baldé, C. P., Kuehr, R., Yamamoto, T., McDonald, R., D'Angelo, E., Althaf, S., . . . Wagner, M. (2024). Global e-waste monitor 2024. International Telecommunication Union (ITU) and United Nations Institute for Training and Research (UNITAR).

Beamon, B. (1999). Designing the green supply chain. *Logistics Information Management*, *12*(4), 332–342. DOI: 10.1108/09576059910284159

Boubellouta, B., & Kusch-Brandt, S. (2020). Testing the environmental Kuznets Curve hypothesis for e-waste in the EU28+2 countries. *Journal of Cleaner Production*, *277*, 1–11. DOI: 10.1016/j.jclepro.2020.123371

Boubellouta, B., & Kusch-Brandt, S. (2021). *Relationship between economic growth and mismanaged e-waste: Panel data evidence from 27 EU countries analyzed under the Kuznets curve hypothesis.* Waste Management, 85-97.

Boubellouta, B., & Kusch-Brandt, S. (2021a). Cross-country evidence on environmental Kuznets Curve in waste electrical and electronic equipment for 174 countries. *Sustainable Production and Consumption*, *25*, 136–151. DOI: 10.1016/j.spc.2020.08.006

Boubellouta, B., & Kusch-Brandt, S. (2022). Determinants of e-waste composition in the EU28+ 2 countries: A panel quantile regression evidence of the STIRPAT model. *International Journal of Environmental Science and Technology*, *19*(11), 1–18. DOI: 10.1007/s13762-021-03892-0

Boubellouta, B., & Kusch-Brandt, S. (2023). Driving factors of e-waste recycling rate in 30 European countries: New evidence using a panel quantile regression of the EKC hypothesis coupled with the STIRPAT model. *Environment, Development and Sustainability*, *25*(8), 7533–7560. DOI: 10.1007/s10668-022-02356-w

Breusch, T., & Pagan, A. (1980). The lagrange multiplier test and its applications to model specification in econometrics. *The Review of Economic Studies*, *47*(1), 239–253. DOI: 10.2307/2297111

Brooks, C., & Hinich, M. (1998). Episodic nonstationarity in exchange rates. *Applied Economics Letters*, *5*(11), 719–722. DOI: 10.1080/135048598354203

Carter, C., & Carter, J. (1998). Interorganizational determinants of environmental purchasing: Initial evidence from the consumer products industries. *Decision Sciences*, *29*(3), 659–684. DOI: 10.1111/j.1540-5915.1998.tb01358.x

Carter, C., & Ellram, L. (1998). Reverse logistics: A review of the literature and framework for future investigation. *Journal of Business Logistics*, *19*(1), 85–102.

Caspi, I. (2017). Rtadf Testing For Bubbles With Eviews. *Journal of Statistical Software*, *81*(1), 1–16.

Chatterjee, A., & Abraham, J. (2017). Efficient management of e-wastes. *International Journal of Environmental Science and Technology*, *14*(1), 211–222. DOI: 10.1007/s13762-016-1072-6

Chertow, M. (2000). Industrial symbiosis: Literature and taxonomy. *Annual Review of Environment and Resources*, *25*, 313–337.

Crocker, R. (2013). *From access to excess: consumerism, 'compulsory' consumption and behaviour change. In motivating change: sustainable design and behaviour in the built environment.* Routledge Publishing.

Dat, L., Linh, D., Chou, S., & Yu, V. (2012). Optimizing reverse logistic costs for recycling end-of-life electrical and electronic products". *Expert Systems with Applications*, *39*(7), 6380–6387. DOI: 10.1016/j.eswa.2011.12.031

Daugherty, P., Richey, R., Genchev, S., & Chen, H. (2005). Reverse logistics: Superior performance through focused resource commitments to information technology. *Transportation Research Part E, Logistics and Transportation Review*, *41*(2), 77–92. DOI: 10.1016/j.tre.2004.04.002

Demirel, E., Demirel, N., & Gökçen, H. (2016). A mixed integer linear programming model to optimize reverse logistics activities of end-of-life vehicles in Turkey. *Journal of Cleaner Production*, *112*, 2101–2113. DOI: 10.1016/j.jclepro.2014.10.079

Donou-Adonsou, F. (2019). Technology, education, and economic growth in Sub-Saharan Africa. *Telecommunications Policy*, *43*(4), 353–360. DOI: 10.1016/j.telpol.2018.08.005

Dowlatshahi, S. (2005). A strategic framework for the design and implementation of remanufacturing operations in reverse logistics. *International Journal of Production Research*, *43*(16), 3455–3480. DOI: 10.1080/00207540500118118

Edquist, H., & Henrekson, M. (2017). Swedish lessons: How important are ICT and R&D to economic growth? *Structural Change and Economic Dynamics*, *42*, 1–12. DOI: 10.1016/j.strueco.2017.05.004

Elkington, J. (1997). *Cannibals with Forks: The Triple Bottom Line of 21st Century Business*. Capstone Publishing.

Endres, A., & Radke, V. (2018). *Economics for Environmental Studies: A Strategic Guide to Micro- and Macroeconomics* (2nd ed.). Springer. DOI: 10.1007/978-3-662-54828-8

Erumban, A., & Das, D. (2016). Information and communication technology and economic growth in India. *Telecommunications Policy*, *40*(5), 412–431. DOI: 10.1016/j.telpol.2015.08.006

Ferguson, M., & Toktay, L. (2006). The Effect of Competition on Recovery Strategies. *Production and Operations Management*, *15*(3), 351–368. DOI: 10.1111/j.1937-5956.2006.tb00250.x

Fernandez-Portillo, A., Almodovar-Gonzalez, M., & Hernandez-Mogollon, R. (2020). Impact of ICT development on economic growth. A study of OECD European union countries. *Technology in Society*, *63*, 1–9. DOI: 10.1016/j.techsoc.2020.101420

Fleischmann, M., Bloemhof-Ruwaard, J. M., Dekker, R., Laan, E., Nunen, J. A., & Wassenhove, L. (1997). Quantitative models for reverse logistics: A review. *European Journal of Operational Research*, *103*(1), 1–17. DOI: 10.1016/S0377-2217(97)00230-0

Flygansvaer, B. (1999). The need for cooperation and coordination in a reverse distribution channel - a model. In Proceedings of the second international working seminar on re-use (pp. 55-70). Eindhoven University of Technology Press.

Foon, T. (2008). A re-examination of the role of foreign direct investment and exports in Malaysia's economic growth: A time series analysis, 1970-2006. *International Journal of Management Studies*, *15*, 47–67.

Fullerton, D., & Metcalf, G. (1997). *Environmental Taxes and The Double-Dividend Hypothesis: Did You Really Expect Something for Nothing?* Cambridge: NBER Working Paper No 6199.

Garlapati, V. (2016). E-waste in India and developed countries: Management,recycling, business and biotechnological initiatives. *Renewable & Sustainable Energy Reviews*, *54*, 874–881. DOI: 10.1016/j.rser.2015.10.106

Ghosh, S. (2016). Does mobile telephony spur growth? Evidence from Indian states. *Telecommunications Policy*, *40*(10-11), 1020–1031. DOI: 10.1016/j.telpol.2016.05.009

Goulder, L. (1995). Environmental taxation and the "double dividend": A reader's guide. In Public economics and the environment in an imperfect world (pp. 277-314). New York: Springer Science+Business Media.

Guide, V., Souza, G., Wassenhove, L., & Blackburn, J. (2006). Time value of commercial product returns. *Management Science*, *8*(8), 1200–1214. DOI: 10.1287/mnsc.1060.0522

Guide, V., & Wassenhove, L. (2009). The evolution of closed-loop supply chain research. *Operations Research*, *57*(1), 10–18. DOI: 10.1287/opre.1080.0628

Haftu, G. (2019). Information communications technology and economic growth in Sub-Saharan Africa: A panel data approach. *Telecommunications Policy*, *43*(1), 88–99. DOI: 10.1016/j.telpol.2018.03.010

Hofman, A., Aravena, C., & Aliaga, V. (2016). Information and communication technologies and their impact in the economic growth of Latin America, 1990-2013. *Telecommunications Policy*, *40*(5), 485–501. DOI: 10.1016/j.telpol.2016.02.002

Hong, J. (2017). Causal relationship between ICT R&D investment and economic growth in Korea. *Technological Forecasting and Social Change*, *116*, 70–75. DOI: 10.1016/j.techfore.2016.11.005

Hussain, A., Batool, I., Akbar, M., & Nazir, M. (2021). Is ICT an enduring driver of economic growth? Evidence from South Asian economies. *Telecommunications Policy*, *45*(8), 1–12. DOI: 10.1016/j.telpol.2021.102202

Hwang, W., & Shin, J. (2017). ICT-specific technological change and economic growth in Korea. *Telecommunications Policy*, *41*(4), 282–294. DOI: 10.1016/j.telpol.2016.12.006

Ishida, H. (2015). The effect of ICT development on economic growth and energy consumption in Japan. *Telematics and Informatics*, *32*(1), 79–88. DOI: 10.1016/j.tele.2014.04.003

Islam, R., Nazifa, T., Yuniarto, A., Uddin, A., Salmiati, S., & Shahid, S. (2019). An empirical study of construction and demolition waste generation and implication of recycling. *Waste Management (New York, N.Y.)*, *95*, 10–21. DOI: 10.1016/j.wasman.2019.05.049 PMID: 31351595

Islam, T., & Huda, N. (2019). E-waste in Australia: Generation estimation and untapped material recovery and revenue potential. *Journal of Cleaner Production, 237*, 1–21. DOI: 10.1016/j.jclepro.2019.117787

John, S., Sridharan, R., & Kumar, P. (2017). Multi-period reverse logistics network design with emission cost. *International Journal of Logistics Management, 28*(1), 127–149. DOI: 10.1108/IJLM-08-2015-0143

John, S., Sridharan, R., Kumar, P., & Krishnamoorthy, M. (2018). Multi-period reverse logistics network design for used refrigerators. *Applied Mathematical Modelling, 54*, 311–331. DOI: 10.1016/j.apm.2017.09.053

Jorgenson, D., Ho, M., & Samuels, J. (2016). The impact of information technology on postwar US economic growth. *Telecommunications Policy, 40*(5), 398–411. DOI: 10.1016/j.telpol.2015.03.001

Kalia, P., Zia, A., & Mladenovic, D. (2022). Examining country development indicators and e-waste under the moderating effect of country development levels and e-waste policy. *International Journal of Quality & Reliability Management, 39*(7), 1601–1616. DOI: 10.1108/IJQRM-09-2021-0335

Kiddee, P., Naidu, R., & Wong, M. H. (2013). Electronic waste management approaches: An overview. *Waste Management (New York, N.Y.), 33*(5), 1237–1250. DOI: 10.1016/j.wasman.2013.01.006 PMID: 23402807

Kilic, H., Cebeci, U., & Ayhan, M. (2015). Reverse logistics system design for the waste of electrical andelectronic equipment (WEEE) in Turkey. *Resources, Conservation and Recycling, 95*, 120–132. DOI: 10.1016/j.resconrec.2014.12.010

King, A. M., Burgess, S. C., Ijomah, W., & McMahon, C. A. (2006). Reducing waste: Repair, recondition, remanufacture or recycle? *Sustainable Development (Bradford), 14*(4), 257–267. DOI: 10.1002/sd.271

Knemeyer, A., Ponzurick, T., & Logar, C. (2002). A qualitative examination of factors affecting reverse logistics systems for end-of-life computers. *International Journal of Physical Distribution & Logistics Management, 32*(6), 455–479. DOI: 10.1108/09600030210437979

Kolstad, C. (2011). *Environmental Economics* (2nd ed.). Oxford University Press.

Kónya, L. (2006). Exports and growth: Granger causality analysis on OECD countries with a panel data approach. *Economic Modelling, 23*(6), 978–992. DOI: 10.1016/j.econmod.2006.04.008

Korhonen, J., Malmborg, F., Strachan, P., & Ehrenfeld, J. (2004). Management and policy aspects of industrial ecology: An emerging research agenda. *Business Strategy and the Environment, 13*(5), 289–305. DOI: 10.1002/bse.415

Kumar, A., Holuszko, M., & Espinosa, D. (2017). E-waste: An overview on generation, collection, legislation and recycling practices. *Resources, Conservation and Recycling, 122*, 32–42. DOI: 10.1016/j.resconrec.2017.01.018

Kumar, R., Stauvermann, P., & Samitas, A. (2016). The effects of ICT on output per worker: A study of the Chinese economy. *Telecommunications Policy, 40*(2-3), 102–115. DOI: 10.1016/j.telpol.2015.06.004

Kurniawati, M. (2020). The role of ICT infrastructure, innovation and globalization on economic growth in OECD countries, 1996-2017. *Journal of Science and Technology Policy Management, 11*(2), 193–215. DOI: 10.1108/JSTPM-06-2019-0065

Kusch, S., & Hills, C. (2017). The Link between e-Waste and GDP—New Insights from Data from the Pan-European Region. *Resources, 6*(15), 1–10. DOI: 10.3390/resources6020015

Lamming, R., & Hampson, J. (1996). The Environment as a supply chain management issue". *British Journal of Management, 7*(Special Issue), 45–62. DOI: 10.1111/j.1467-8551.1996.tb00147.x

Latif, Z., Danish, Y. M., Latif, S., Ximei, L., Pathan, Z. H., Salam, S., & Jianqiu, Z. (2018). The dynamics of ICT, foreign direct investment, globalization and economic growth: Panel estimation robust to heterogeneity and cross-sectional dependence. *Telematics and Informatics, 35*(2), 318–328. DOI: 10.1016/j.tele.2017.12.006

Lèbre, É., & Corder, G. (2015). Integrating industrial ecology thinking into the management of mining waste. *Resources, 4*(4), 765–786. DOI: 10.3390/resources4040765

Liu, Y., Park, S., Yi, H., & Feiock, R. (2020). Evaluating the employment impact of recycling performance in Florida. *Waste Management (New York, N.Y.), 101*, 283–290. DOI: 10.1016/j.wasman.2019.10.025 PMID: 31648169

Lombardi, D., & Laybourn, P. (2012). Redefining industrial symbiosis: Crossing academic-practitioner boundaries. *Journal of Industrial Ecology, 16*(1), 28–37. DOI: 10.1111/j.1530-9290.2011.00444.x

Lourenço, H. R., & Soto, J. P. (2002). *Reverse Logistics Models and Applications: A Recoverable Production Planning Model.* Universitat Pompeu Fabra: Grup de Recerca en Logística Empresarial.

Lu, C., Zhang, L., Zhong, Y., Ren, W., Tobias, M., Mu, Z., Ma, Z., Geng, Y., & Xue, B. (2015). An overview of e-waste management in China. *Journal of Material Cycles and Waste Management*, *17*(1), 1–12. DOI: 10.1007/s10163-014-0256-8

Lundgren, K. (2012). *The Global Impact of E-Waste: Addressing the Challenge*. ILO.

Magazzino, C., Mele, M., & Schneider, N. (2020). The relationship between municipal solid waste and greenhouse gas emissions: Evidence from Switzerland. *Waste Management (New York, N.Y.)*, *113*, 508–520. DOI: 10.1016/j.wasman.2020.05.033 PMID: 32546447

Marconi, M., Gregori, F., Germani, M., Papetti, A., & Favi, C. (2018). An approach to favor industrial symbiosis: The case of waste electrical and electronic equipment. *Procedia Manufacturing*, *21*, 502–509. DOI: 10.1016/j.promfg.2018.02.150

Morgan, T., Tokman, M., Richey, R., & Defee, C. (2018). Resource commitment and sustainability: A reverse logistics performance process model. *International Journal of Physical Distribution & Logistics Management*, *48*(2), 164–182. DOI: 10.1108/IJPDLM-02-2017-0068

Myovella, G., Karacuka, M., & Haucap, J. (2020). Digitalization and economic growth: A comparative analysis of Sub-Saharan Africa and OECD economies. *Telecommunications Policy*, *44*(2), 1–12. DOI: 10.1016/j.telpol.2019.101856

Namlis, K., & Komilis, D. (2019). *Influence of four socioeconomic indices and the impact of economic crisis on solid waste generation in Europe*. Waste Management, 190-200.

Neto, J., Bloemhof-Ruwaard, J., Nunen, J., & Heck, E. (2008). Designing and evaluating sustainable logistics networks. *International Journal of Production Economics*, *111*(2), 195–208. DOI: 10.1016/j.ijpe.2006.10.014

New, S., Green, K., & Morton, B. (2002). An analysis of private versus public sector responses to the environmental challenges of the supply chain. *Journal of Public Procurement*, *2*(1), 93–105. DOI: 10.1108/JOPP-02-01-2002-B004

Nguyen, T., Pham, T., & Tram, H. (2020). Role of information and communication technologies and innovation in driving carbon emissions and economic growth in selected G-20 countries. *Journal of Environmental Management*, *261*, 1–10. DOI: 10.1016/j.jenvman.2020.110162 PMID: 32148259

Niebel, T. (2018). ICT and economic growth - Comparing developing, emerging and developed countries. *World Development*, *104*, 197–211. DOI: 10.1016/j.worlddev.2017.11.024

Nottebohm, O., Manyika, J., Bughin, J., Chui, M., & Syed, A. (2012). *Online and Upcoming: The Internet's Impact on Aspiring Countries*. McKinsey & Company.

O'Reilly, S., & Kumar, A. (2016). Closing the loop: An exploratory study of reverse ready-made garment supply chains in Delhi NCR. *International Journal of Logistics Management, 27*(2), 486–510. DOI: 10.1108/IJLM-03-2015-0050

OECD. (2021). *Environment at a Glance Indicators*. OECD Publishing.

Olamide, E., Ogujiuba, K., Maredza, A., & Semosa, P. (2022). Poverty, ICT and economic growth in SADC region: A panel cointegration evaluation. *Sustainability (Basel), 14*(15), 1–15. DOI: 10.3390/su14159091

Oughton, C., Kurup, B., Anda, M., & Ho, G. (2022). Industrial symbiosis to circular economy: What does the literature reveal for a successful complex industrial area? *Circular Economy and Sustainability, 2*(4), 1317–1344. DOI: 10.1007/s43615-022-00153-1

Ozcan, B., Apergis, N., & Shahbaz, M. (2018). A revisit of the environmental Kuznets curve hypothesis for Turkey: New evidence from bootstrap rolling window causality. *Environmental Science and Pollution Research International, 25*(32), 32381–32394. DOI: 10.1007/s11356-018-3165-x PMID: 30229495

Packard, V. (1963). *The Waste Makers*. Longmans.

Panambunan-Ferse, M., & Breiter, A. (2013). Assessing the side-effects of ICT development: E-waste production and management: A case study about cell phone end-of-life in Manado, Indonesia. *Technology in Society, 35*(3), 223–231. DOI: 10.1016/j.techsoc.2013.04.002

Pandian, G., & Abdul-Kader, W. (2017). Performance evaluation of reverse logistics enterprise - an agent-based simulation approach. *International Journal of Sustainable Engineering, 10*(6), 384–398. DOI: 10.1080/19397038.2017.1370032

Pesaran, M. (2004). *General Diagnostic Tests for Cross Section Dependence in Panels*. Munich: CESifo Working Paper, No. 1229.

Pesaran, M., Ullah, A., & Yamagata, T. (2008). A bias-adjusted LM test of error cross-section independence. *The Econometrics Journal, 11*(1), 105–127. DOI: 10.1111/j.1368-423X.2007.00227.x

Pesaran, M., & Yamagata, T. (2008). Testing slope homogeneity in large panels. *Journal of Econometrics, 142*(1), 50–93. DOI: 10.1016/j.jeconom.2007.05.010

Pokharel, S., & Mutha, A. (2009). Perspectives in reverse logistics: A review. *Resources, Conservation and Recycling*, *53*(4), 175–182. DOI: 10.1016/j.resconrec.2008.11.006

Pradhan, R., Mallik, G., & Bagchi, T. (2018). Information communication technology (ICT) infrastructure and economic growth: A causality evinced by cross-country panel data. *IIMB Management Review*, *30*(1), 91–103. DOI: 10.1016/j.iimb.2018.01.001

Prasad, P. (2012). *Information communication technology (ICT) - its waste and consequences.* Int. J. Environmental Technology and Management, 15(3/4/5/6), 363-376.

Qu, Y., Wang, W., Liu, Y., & Zhu, Q. (2019). Understanding residents' preferences for e-waste collection in China - A case study of waste mobile phones. *Journal of Cleaner Production*, *228*, 52–62. DOI: 10.1016/j.jclepro.2019.04.216

Raheem, I., Tiwari, A., & Balsalobre-Lorente, D. (2020). The role of ICT and financial development in CO2 emissions and economic growth. *Environmental Science and Pollution Research International*, *27*(2), 1912–1922. DOI: 10.1007/s11356-019-06590-0 PMID: 31760620

Rahman, S., & Subramanian, N. (2012). Factors for implementing end-of-life computer recycling operations in reverse supply chains. *International Journal of Production Economics*, *140*(1), 239–248. DOI: 10.1016/j.ijpe.2011.07.019

Rajput, J. E. (2013). E-Waste-Concept, Problems and Solution Policies in India. *International Journal of Management and Business*, *3*(1), 31–36.

Rashid, A., Asif, F., Krajnik, P., & Nicolescu, C. (2013). Resource conservative manufacturing: An essential change in business and technology paradigm for sustainable manufacturing. *Journal of Cleaner Production*, *57*, 166–177. DOI: 10.1016/j.jclepro.2013.06.012

Ravi, V., Shankar, R., & Tiwari, M. (2005). Analyzing alternatives in reverse logistics for end-of-life computers: ANP and balanced scorecard approach. *Computers & Industrial Engineering*, *48*(2), 327–356. DOI: 10.1016/j.cie.2005.01.017

Richey, R., Genchev, S., & Daugherty, P. (2005). The role of resource commitment and innovation in reverse logistics performance. *International Journal of Physical Distribution & Logistics Management*, *35*(4), 233–257. DOI: 10.1108/09600030510599913

Robinson, B. H. (2009). E-Waste: An Assessment of Global Production and Environmental Impacts. *The Science of the Total Environment*, *408*(2), 183–191. DOI: 10.1016/j.scitotenv.2009.09.044 PMID: 19846207

Romer, P. (1990). *Endogenous Technological Change*. NBER.

Salanie, B. (2003). *The Economics of Taxation*. The MIT Press.

Sarkis, J., Helms, M., & Hervani, A. (2010). Reverse logistics and social sustainability. *Corporate Social Responsibility and Environmental Management*, *17*(6), 337–354. DOI: 10.1002/csr.220

Savaskan, R., Bhattacharya, S., & Wassenhove, L. (2004). Closed-Loop Supply Chain Models with Product Remanufacturing. *Management Science*, *50*(2), 239–252. DOI: 10.1287/mnsc.1030.0186

Schluep, M., Hageluekenb, C., Kuehrc, R., Magalinic, F., Maurerc, C., Meskersb, C., & Wang, F. (2009). *Recycling: From E-Waste to Resources*. United Nations Environment Programme & United Nations University.

Scruggs, C., Nimpuno, N., & Moore, R. (2016). Improving information flow onchemicals in electronic products and E-waste to minimize negativeconsequences for health and the environment. *Resources, Conservation and Recycling*, *113*, 149–164. DOI: 10.1016/j.resconrec.2016.06.009

Seitz, M., & Wells, P. (2006). Challenging the implementation of corporate sustainability: The case of automotive engine remanufacturing. *Business Process Management Journal*, *12*(6), 822–836. DOI: 10.1108/14637150610710954

Sepehrdoust, H., & Ghorbanseresht, M. (2019). Impact of information and communication technology and financial development on economic growth of OPEC developing economies. *Kasetsart Journal of Social Sciences*, *40*, 546–551.

Slade, G. (2007). *Made to Break Technology and Obsolescence in America*. Harvard University Press. DOI: 10.4159/9780674043756

Stanley, T., Doucouliagos, H., & Steel, P. (2018). Does ICT generate economic growth? A meta-regression analysis. *Journal of Economic Surveys*, *32*(3), 705–726. DOI: 10.1111/joes.12211

Sthiannopkao, S., & Wong, M. (2013). Handling e-waste in developed and developing countries: Initiatives, practices, and consequences. *The Science of the Total Environment*, *463-464*, 1147–1153. DOI: 10.1016/j.scitotenv.2012.06.088 PMID: 22858354

Swamy, P. (1970). Efficient Inference in a Random Coefficient Regression Model. *Econometrica*, *38*(2), 311–323. DOI: 10.2307/1913012

Thierry, M., Salomon, M., Nunen, J., & Wassenhove, L. (1995). Strategic Issues in Product Recovery Management. *California Management Review*, *37*(2), 114–134. DOI: 10.2307/41165792

Tibben-Lembke, R., & Rogers, D. (2002). Differences between forward and reverse logistics in a retail environment. *Supply Chain Management, 7*(5), 271–282. DOI: 10.1108/13598540210447719

Toader, E., Firtescu, B., Roman, A., & Anton, S. (2018). Impact of Information and Communication Technology Infrastructure on Economic Growth: An Empirical Assessment for the EU Countries. *Sustainability (Basel), 10*(10), 1–22. DOI: 10.3390/su10103750

Tsachtsiris, G., Magoutas, A., & Papadogonas, T. (2022). ICT and Economic Growth in EU: A macro Level Comparison of Estimated ICT Output Elasticities. *Journal of Global Information Technology Management, 25*(3), 202–216. DOI: 10.1080/1097198X.2022.2094182

Umair, S., Björklund, A., & Petersen, E. (2015). Social impact assessment of informal recycling of electronic ICT waste in Pakistan using UNEP SETAC guidelines. *Resources, Conservation and Recycling, 95*, 46–57. DOI: 10.1016/j.resconrec.2014.11.008

Usman, A., Ozturk, I., Hassan, A., Zafar, S., & Ullah, S. (2021). The effect of ICT on energy consumption and economic growth in South Asian economies: An empirical analysis. *Telematics and Informatics, 58*, 1–9. DOI: 10.1016/j.tele.2020.101537

Vachon, S., & Klassen, R. (2006). Extending green practices across the supply chain: The impact of upstream and downstream integration. *International Journal of Operations & Production Management, 26*(7), 795–821. DOI: 10.1108/01443570610672248

Vermeulen, W. (2006). The social dimension of industrial ecology: On the implications of the inherent nature of social phenomena. *Progress in Industrial Ecology, 3*(6), 574–598. DOI: 10.1504/PIE.2006.012754

Villalba, G., Segarra, M., Fernández, A., Chimenos, J., & Espiell, F. (2002). A proposal for quantifying the recyclability of materials. *Resources, Conservation and Recycling, 37*(1), 39–53. DOI: 10.1016/S0921-3449(02)00056-3

Walker, H., Sisto, L., & McBain, D. (2008). Drivers and barriers to environmental supply chain management practices: Lessons from the public and private sectors. *Journal of Purchasing and Supply Management, 14*(1), 69–85. DOI: 10.1016/j.pursup.2008.01.007

Walton, S., Handfield, R., & Melnyk, S. (1998). The Green supply chain: Integrating suppliers into environmental management processes. *International Journal of Purchasing and Materials Management, 34*(1), 1–11. DOI: 10.1111/j.1745-493X.1998.tb00042.x

Ward, M., & Zheng, S. (2016). Mobile telecommunications service and economic growth: Evidence from China. *Telecommunications Policy*, *40*(2-3), 89–101. DOI: 10.1016/j.telpol.2015.06.005

Wena, L., Lin, C., & Lee, S. (2009). Review of recycling performance indicators: A study on collection rate in Taiwan. *Waste Management (New York, N.Y.)*, *29*, 2248–2256. PMID: 19398319

White, C. D., Masanet, E., Rosen, C. M., & Beckman, S. L. (2003). Product recovery with some byte: An overview of management challenges and environmental consequences in reverse manufacturing for the computer industry. *Journal of Cleaner Production*, *11*, 445–458.

Xavier, L., Ottoni, M., & Lepawsky, J. (2021). Circular economy and e-waste management in the Americas: Brazilian and Canadian frameworks. *Journal of Cleaner Production*, *297*, 1–13. DOI: 10.1016/j.jclepro.2021.126570

Yang, C., Tan, Q., Liu, L., Dong, Q., & Li, J. (2017). Recycling tin from electronic waste: A problem that needs more attention. *ACS Sustainable Chemistry & Engineering*, *5*(11), 9586–9598. DOI: 10.1021/acssuschemeng.7b02903

Yi, Z. (1986). Changes in family structure in China: A simulation study. *Population and Development Review*, *12*(4), 675–703. DOI: 10.2307/1973431

Yilanci, V., & Ozgur, O. (2019). Testing the environmental Kuznets curve for G7 countries: Evidence from a bootstrap panel causality test in rolling windows. *Environmental Science and Pollution Research International*, *26*(24), 24795–24805. DOI: 10.1007/s11356-019-05745-3 PMID: 31240642

Yilmaz, R., & Koyuncu, C. (2023). The Impact of Globalization on the Rate of E-waste Recycling: Evidence From European Countries. *Amfiteatru Economic*, *25*(62), 180–195. DOI: 10.24818/EA/2023/62/180

Zellner, A. (1962). An Efficient Method of Estimating Seemingly Unrelated Regressions and Tests for Aggregation Bias. *Journal of the American Statistical Association*, *57*(298), 348–368. DOI: 10.1080/01621459.1962.10480664

KEY TERMS AND DEFINITIONS

Circular Economy: The concept of prolonging the usable life of goods, materials, or resources is known as the "circular economy". The use of resources inside closed-loop systems, lowering pollution or preventing resource leakage while maintaining economic growth, is a key component of the circular economy idea. Furthermore, this strategy is the most recent attempt to conceive how to combine economic activity and environmental well-being.

E-Waste: E-waste covers a wide and growing range of consumer electronics such as refrigerators, air conditioners, cell phones, stereos, computers, etc. that are discarded by their users, from large household appliances to small appliances.

Externality: An externality is, in general, a phenomenon in which economic actors (individuals or firms) are positively or negatively affected by economic activities that take place outside of themselves, thereby pushing the perfectly competitive market away from its optimal equilibrium.

Industrial Symbiosis: As a subfield of industrial ecology, industrial symbiosis necessitates a steadfast focus on the movement of resources and energy across local and regional economies. In an effort to gain a competitive edge, industrial symbiosis brings together previously disparate businesses through the physical exchange of resources, energy, and/or by-products. Collaboration and the synergistic opportunities provided by geographic closeness are the keys to industrial symbiosis.

Linear Economy: Raw resources are taken out, processed, merged, sold, disseminated, sold again, utilized, and eventually disposed away in a linear economy.

Reverse Logistics: Reverse logistics plays a key role in how a circular economy operates. This is based on the notion that wastes from after use need to be moved back upstream in order to be reprocessed and regain their value. This relates to the idea of take-back programs.

Sustainable Production and Consumption: Sustainable production and consumption systems involve sustainable efforts and activities that policy makers refer to as "top-down" and companies refer to as "bottom-up". "Top-down" activities result in economic intervention by the government, while 'bottom-up' efforts are realized when firms incorporate sustainable production and consumption commitments into their business activities.

ENDNOTES

* For some recent studies examining the relationship between progress in ICT and economic growth, see Ishida 2015; Ghosh 2016; Kumar et al. 2016; Ward and Zheng 2016; Aghaei and Rezagholizadeh 2017; Albiman and Sulong 2017;

Edquist and Henrekson 2017; Hong 2017; Hwang and Shin 2017; Latif et al. 2018; Niebel 2018; Pradhan et al. 2018; Stanley et al. 2018; Toader et al. 2018; Adeleye and Eboagu 2019; Haftu 2019; Sepehrdoust and Ghorbanseresht 2019; Fernandez-Portillo et al. 2020; Kurniawati 2020; Myovella et al. 2020; Nguyen et al. 2020; Raheem et al. 2020; Appiah-Otoo and Song 2021; Hussain et al. 2021; Usman et al. 2021; Olamide et al. 2022; Tsachtsiris et al. 2022.

† The selection of these countries was based on (i) their representativeness to countries with similar socio-economic structures (ii) and the fact that they are the most studied countries in e-waste management research.

‡ Considering all countries, the widest time period was determined as from 1960 to 2020 due to data limitations.

§ While the data for the variables M (kg) and L (year) used in the formula for calculating the amount of e-waste were created using Robinson (2009), the variable N was based on the number of households and the number of households for countries was created using the Helgi Library database and Yi (1986).

** Environmental taxes are an important policy instrument in terms of both protecting environmental quality at low cost and generating public revenue. However, in addition to these benefits, additional benefits can also be achieved in terms of the use of the revenue generated. It is from this perspective that the double dividend hypothesis came into focus in the 1990s, as environmental and climate change issues came to the forefront. That is, this perspective implies that (i) pollution abatement generates welfare gains and public revenue, while (ii) this revenue is used to reduce distortions in the economy caused by the income and substitution effects of other taxes. Naturally, this hypothesis assumes that the current tax system is not optimal. Accordingly, for example, in order to reduce the cost of labor, which is a productive factor, using the revenue generated by environmental taxes to reduce employers' contributions to the social insurance system is expected to have an employment-incentive effect. (Goulder 1995; Fullerton and Metcalf 1997; Salanie 2003; Akkaya and Bakkal, 2005; Kolstad 2011; Endres and Radke 2018). As a matter of fact, it can be stated that e-waste is quite suitable for the double dividend hypothesis in terms of composition and value.

Chapter 14
Understanding E-Waste Disposal Behavior Across Diverse Groups: A Social Cognitive Perspective

Varanasi Rahul
https://orcid.org/0000-0003-2407-7654
Jain University, India

N. Sathyanarayana
https://orcid.org/0000-0002-4185-7751
Jain University, India

Usha Prabhu
https://orcid.org/0000-0001-6219-5358
Jain University, India

Y. Fathima
https://orcid.org/0000-0002-0045-1421
Jain University, India

V. Lava Kumar
https://orcid.org/0000-0002-3539-9280
GITAM University, India

ABSTRACT

This chapter seeks to examine the factors and effects of e-waste disposal behaviour among various demographic groups. Social cognitive theory (SCT) is utilised to formulate and evaluate a conceptual model that examines the impact of psychological

DOI: 10.4018/979-8-3693-7383-5.ch014

factors, observational learning, and self-regulation on individuals' e-waste disposal behaviour. This study employs a structured survey methodology via a questionnaire to investigate the e-waste disposal behaviours of various demographic groups across different regions of India. Multiple regression analysis is utilised as the statistical method for data examination. The results of the multiple regression analysis reveal that psychological factors, observational learning, and self-regulation are positively correlated with the e-waste disposal behaviours of various demographic groups. This study has constraints, including common method bias and social desirability bias, owing to its dependence on self-report measures.

INTRODUCTION

A growing quantity of WEEE is a direct result of the increased usage and modernisation of electronic devices and equipment brought about by economic development. Worldwide, e-waste production increased by 21% in only 5 years, reaching 53.6 million Mts in 2009, with recycling rates hovering around 17%. E-waste generally exerts more significant environmental repercussions than municipal garbage. Conversely, poisonous chemicals, along with substantial quantities of other heavy metals and chemical additives, constitute electronic waste. Severe environmental damage results from improperly disposed of electronic garbage, which leads to unintentional burying and burning. Arain et al. (2020) electronic trash is chemically different from regular household or municipal trash, and it contains several dangerous chemicals that, if not disposed of properly, can hurt people and the environment. These substances include lead, polybrominated biphenyls, and diphenyl. Reusing, reprocessing, recycling, and sometimes burning or putting it in a landfill are all ways that electrical waste is handled. Recycling is expected to be a primary method for reducing pollution, conserving natural resources, and improving energy efficiency. Awasthi et al. (2022) say that taking part in successful waste management, such as recycling, depends on more than just caring about the environment, supporting environmental goals, or being open to recycling technologies.

Formal treatment of e-waste is a key part of long-term solid waste control in developing nations for the very reason that it is directly linked to environmental issues due to the existence of informal sectors in large numbers (Bharali & Kumar 2018). The global challenges of sustainable development and the production of large amounts of waste started to raise calls for a circular economy. Policy and business experts have come up with a concept in which, through a sustainability development strategy (Islam et al. 2020), solutions to environmental degradation and resource scarcity issues are proposed. Borthakur & Govind (2016) the increased dependence of the masses on this device, which comes in all ranges and sizes and suits many needs, has attained the discovery of innovative approaches to develop new products with the latest features required by limited life spans, as well as its quick phasing out of the old and obsolete. Gollwitzer (1999) makes use of a circular economic model as one way to fix this issue. In contrast to the linear production and consumption that occurs in a traditional model, wherein manufactured products are consumed and then discarded, the concept of the CE follows the trajectory of a closed-loop system where reuse, repair, and recycling are given importance. The CE can contribute substantially to reducing the environmental impacts of discarded electronics by prolonging product life and extracting valuable minerals from electronic trash (Carver & Scheier 1998).

Transition to a CE also includes a re-design of products to be easily disassembled and recycled. Other take-back schemes include extending manufacturers' responsibility for product collection and recycling when at the end of their life. Finally, it is also important to realize that governments should have stronger policies that foster the CE by forcing businesses to adopt sustainable production methods. In India, the move to a circular economy will also bring the country's informal recycling industry into the formal economy. Thus, the absorption of these informal workers into a formal mechanism, proper training, and providing all protection gear, while introducing advanced recycling technologies, will help India improve the environmental and social performance of its e-waste management system. New technological advancement has provided new tools for dealing with the current challenges of e-waste recycling. For example, automated sorting systems can help in the segregation of different types of electronic components more effectively, thus improving the quality of recovered materials. Development is underway on chemical recycling technologies that can extract metals from e-waste without causing environmental destruction. Attractive technology constitutes urban mining, an area that addresses the extraction of discarded precious metals through electronic gadgets, as opposed to traditional mining operations. In addition to reducing demand for new raw materials, urban mining reduces the environmental impacts typical of mining processes, and environments that more often are subjected to deforestation, destruction of habitat, and release of toxic chemicals into the atmosphere.

Another novelty in the management of electronic scrap involves tracking life cycles with blockchain technology. By recording various data on the production, sale, and disposal of electronic items on a blockchain, manufacturers and governments ensure greater transparency and accountability at every stage of recycling. Regarding the basics, blockchain can also provide certification that these materials meet certain environmental standards, further encouraging sustainable practices within the electronics industry.

At the international level, Basel Convention-type treaties are looming in terms of preventing the illegitimate dumping of e-waste into developing nations. Most developed countries lack any capacity in terms of recycling and have traditionally shipped the generated e-waste to poor countries for unsafe handling. Most international agreements help in stopping this practice once their enforcement is enhanced and ensure that waste is disposed of or handled responsibly.

In turn, the future management of WEEE will depend on both technological developments and shifting consumer behaviour. Indeed, the expected proliferation of connected devices through ongoing IoT and 5G network growth will further accelerate rates of WEEE generation. Addressing this burgeoning waste stream will require governments, businesses, and consumers to come together to encourage sustainability. Consumers play a very significant role in overcoming this challenge of electronic waste. More responsible habits of consumption could include repairing, rather than replacing electronic gadgets, and proper disposal of e-waste through certified recycling programs that reduce environmental harm from electronic waste.

Research Framework

The research framework is designed to investigate the disposal behaviour of e-waste in the light of Social Cognitive Theory. This paper explores how these important psychological factors, such as self-efficacy, observation learning, outcome expectation, and self-regulation influence the practices of waste disposal among individuals. The research design uses a mixed methods approach; therefore, the researcher administered surveys and interviews to participants to collect data. Regression analysis (Figure 1) is applied to spot the relationship between variables. These findings will contribute to best practices and inform strategies for sustainable e-waste management (Chatterjee et al. 2016).

E-Waste in India

As the largest consumer of electronic goods globally, India encounters significant issues in managing e-waste.

E-waste generation: Based on the findings of the Global E-waste Monitor 2020 study, India produced around 3.2 million metric tonnes of electronic garbage in the year 2019. This made it the third biggest country in the world that made E-waste. The amount of electronic trash made in India is expected to rise by 21% each year.

E-waste Collecting and Recycling: The recycling and collecting of e-waste in India will continue to pose significant obstacles. According to various estimates, merely approximately 10% of the total electronic trash produced in the country is subjected to scientific recycling. India only has a few scientifically approved sites for recycling e-waste. Most of the time, the informal sector takes care of getting rid of e-waste, which is unsafe and bad for the environment (Choi et al. 2018).

Legislative Framework: The E-Waste (Management) Rules 2016 were made by the Indian government. They explain how to properly handle and control electronic waste.

Disposal Behaviour of E-Waste in India

The disposal behaviour of E-waste depends on many socio-economic, cultural, and environmental factors. Researchers in India have discovered numerous elements that influence people's decisions about how to dispose of electronic waste, including their level of awareness, attitudes, motivation, and potential hurdles. While studying the determinant of E-waste disposal behaviour among Indian households, (Patnaik 2008) presented some identified factors that affect disposal choices. Rati & Kameswari (2017) assessed the disposal behaviour of college students and discussed various attributes affecting the E-waste disposal decision. Kaur (2018) investigated the E-waste disposal behaviour of Indians, identifying awareness, Knowledge, and environmental attitudes that determine disposal practices.

Social Cognitive Theory (SCT)

It is a psychological theory explaining how people learn, develop, and change their behaviour through the interaction of cognitive processes, social influence, and environmental factors. This particular theory predominantly focuses on the belief that people cognitively learn through experience and by observing or modelling others. The four steps of observational learning involve attention, retaining, reproduction, and motivation (Deoghare & Shukla 2018).

People selectively attend to and remember certain behaviours and then recreate or imitate them based on perceived consequences and the presence of positive role models (Dougherty et al. 2001).

It influences the choices that people make, including the degree of effort and persistence to realize certain difficulties (Rani et al. 2016). A second central constituent in SCT is that of self-regulation. This includes ways in which people set their goals, monitor realization through self-observation and self-feedback, and regulate and control their behaviours (Echegaray 2016).

The goals are focused on the psychological factors that influence how people in India get rid of their electronic waste. The goals are focused on the psychological factors that influence how people in India get rid of their electronic waste. This study will, therefore, be trying to find out the different psychosocial determinants of e-waste disposal behaviour in the various groups within the country. To investigate the role of observational learning in shaping these practices, putting into perspective how people can learn from other people's behaviour. Also, the research will try to find the relationship between self-regulation with responsible decisions regarding the dumping of e-waste by individuals. In the end, the study will show how these factors can be used to make better plans for dealing with electronic waste in India.

RELATED STUDIES ON GETTING RID OF E-WASTE

E-waste disposal habits and consumer awareness: A fundamental impact is the awareness of consumers concerning the management practices entrusted with e-waste. An awareness relating to environmental problems associated with the inappropriate disposal of e-waste seems a necessary determinant for responsible behaviour about its disposal (Nixon et al., 2022). A social cognitive framework would provide a clear look at how personal and societal perceptions influence an individual's decision in the recycling or inappropriate disposal of electronics. Individuals who are more ecologically conscious and aware of the risks connected to e-waste hazards associated with e-waste are more likely to act responsibly by recycling the e-waste or participating in a take-back activity (Mmereki et al., 2016).

Socioeconomic Disparities in E-Waste Disposal: Socioeconomic factors are the most instrumental in dictating disposal behaviour. Because of low economic levels and the lack of infrastructure for the proper disposal of e-waste, one may see that the low-income categories either do not dispose of waste appropriately or do not dispose at all. On second thought, it can so happen that relatively affluent communities may become more engaged with recycling processes regarding e-waste but are generating e-waste on account of high consumption rates and the replacement of electronic equipment quite frequently. The social cognitive perspective supports understanding this difference in the way variations in environmental awareness and resource availability shape the respective behaviours of different groups (Jayaraman et al., 2020).

Regulatory Frameworks and Policy Implementation: Improvement in e-waste disposal rates surely calls for effective policy implementation. However, countries vary widely in terms of regulatory frameworks relating to e-waste management, studies indicate. These countries with developed economies have better policies and facilities for recycling, while their developing counterparts have mostly been plagued by informal avenues in the same direction, which in most cases are hazardous in both environmental and health aspects. In that respect, the social cognitive framework does help in identifying how regulations and policy interventions shape individual behaviours and collective norms (Castro & Bassin, 2022).

Behavioural Interventions and Technological Solutions: Interventions involving the proper disposal of e-waste must address the socio-psychological drivers of behaviour. Indeed, research support is adequate for educational and economic motivators of appropriate disposal practices. For instance, a financial incentive can be influential enough to make an individual participate in a recycling program or return used-up electronic products for proper waste disposal (Hage & Söderholm, 2008). In this respect, it is also important to state that technology further plays a critical role in the simplification of e-waste management, whereby, through the development of apps and other digital platforms, it is increasingly becoming easier for any consumer to find information on where and how to dispose of e-waste (Borthakur & Govind, 2017).

Cultural aspects of E-waste disposal: The cultural background, which reflects all parts of human behaviour, has a greater influence on the disposal behaviour of end-users. The norms and values of a culture may make the communities more responsive to efforts at e-waste management. For example, if the culture enforces environmental sustainability, people will adopt responsible methods of disposal. On the other hand, in countries where economic development imperatives override environmental concerns, people may be less inclined to manage e-waste properly due to other socio-economic pre-occupations. Social cognitive theory frames such differences by emphasizing the role of social norms, values and collective self-efficacy in shaping E-waste disposal approaches (Tansel, 2017).

Issues in E-Waste Management: Despite developing efforts in managing e-waste, there are still considerable challenges in the global E-waste disposal system. The complexity in the characteristics of WEEE, with contents of hazardous materials like mercury and lead, and valuable components such as precious metals, makes the recycling processes equally sophisticated and resource-intensive. As (Li et al. 2015) have noted, this is aside from the fact that e-waste management policies are not standardized across countries. For example, while Europe has the WEEE Directive, many developing countries still depend on an informal recycling sector that causes environmental degradation and health risks as (Heacock et al. 2016) have pointed out.

Gender and E-Waste Disposal Behaviour: Disposal habits regarding electronic waste are significantly impacted by gender. Research has shown that women are more concerned about environmental sustainability and participate in recycling and proper electronic waste disposal than men do (Domina & Koch, 2002). The basis for this gender difference is elaborated from a social cognitive perspective, where differences in socialization, environmental awareness, and perceived responsibility ultimately result in different behaviours between men and women.

Impact of Technological Development on E-waste Generation: Technological development, as related to consumer electronics, proportionately increases the rate of generation of e-waste. The continuous launching of new devices and the short span of life of products lead to a greater generation of electronic waste. Due to this fact, when manufacturers start introducing new technologies into the market at a rapid rate, consumers tend to dispose prematurely and hence increase volumes of e-waste generation. Similarly, the social cognitive theory will be able to explain how marketing strategies linked with consumer behaviour interact toward accelerating the production of e-waste (Kumar & Holuszko, 2016).

Public Involvement in e-waste Programs: Public participation is very key to ensuring that e-waste management programs are indeed effective. This can be facilitated by the community through awareness creation, educational programs, and incentive-based schemes that enhance the rate of recycling. However, mostly, public participation may be fairly low due to unawareness or perceived inconveniences. According to the social cognitive theory, increased community involvement along with collective self-efficacy will enhance participation in e-waste programs.

EPR policies have been enacted in several countries based on the theory of incentivizing producers to design more recyclable products and to take back used electronics to improve e-waste management. However, policy effectiveness has varied and, thus, greater success has been realized in some countries over others (Lindhqvist & Lifset, 2003).

MATERIALS AND METHODS

Field of Research

Major cities from the South of India were considered for this study. These include Bangalore of Karnataka, Hyderabad of Telangana, Vijayawada of Andhra Pradesh, and Chennai of Tamil Nadu. People in these cities are very famous for using major parts of electronic devices in the state. The population as per the census 2011 in Bangalore was 96,21,551, with 23,93,845 households, Hyderabad- 39,43,323, with 8,49,051 households, Vijayawada- 10,21,806 with 2,76,081 Households and Chennai- 46,46,732 with 11,54,982 households. The data has been collected through both online and offline questionnaires. Major study samples can range from 30 to 500. A questionnaire was sent online via a Google form to 200 persons and personally interviewed 138 persons. While 200 online questioners have been dispatched 126 persons responded online and out of them, after proper scrutiny 94 responses were retained. Online 94 and Offline 138 total samples of 232 were finalized for further analysis.

Questioner Design

The survey tool for this study was created, utilised, and validated following an extensive examination of initial research efforts. The overall score was determined by measuring questions related to Social Cognitive Theory and participants' perspectives on E-Waste disposal using a 5-point Likert scale. A rating of 1 signifies a distinctly negative perspective, while a rating of 5 denotes a high level of agreement. The initial eight questions are primarily focused on demographic information. The questions numbered 9 to 25 are designed to examine the application of SCT concerning the E-waste disposal behaviours of respondents. The final set of questions aims to assess the respondents' knowledge and habits related to E-waste disposal.

Hypothesis

Psychological Determinants of Behaviour:

Outcome expectations: Expectations regarding outcomes in SCT pertain to the beliefs an individual holds about the possible results or repercussions of their actions. Within the context of SCT, outcome expectations can be categorized as either positive or negative (Fu et al. 2017). Researchers have repeatedly analyzed how an individual's outcome expectations impact his motivational, goal-setting, self-efficacy beliefs and, finally, his behavioural choices. Individuals will more probably be disposed toward proper E-Waste disposal when they can perceive some positive outcomes or benefits associated with such behaviours (Gollwitzer 1999). Positive outcome expectations may include environmental conservation, preservation of resources, avoidance of legal implications, social approval, or personal satisfaction. Other studies explored the impact of outcome expectations on intentions to carry out E-Waste recycling, responsible disposable practices, or participation in E-Waste management programs. Likewise, understanding an individual's outcome expectations allows for better facilitation of interventions and strategies associated with responsible E-Waste disposal (Islam et al. 2020), (Nejati et al. 2017).

Self-Efficacy: It is "an individual's belief in his or her capability to perform a task or behaviour or to bring about specific desirable outcomes effectively". Self-efficacy plays a major role in motivation, decision-making, and behavioural change (Echegaray 2016). High self-efficacy beliefs are associated with lots of desirable effects: a person is more likely to take up a challenge, and shows greater persistence and better performance. Self-efficacy has also been applied to understand and enhance people's capabilities in adopting new behaviours, surmounting barriers, and creating desirable consequences (Kaur et al. 2018). Those people who perceive themselves as more self-efficacious in E-Waste disposal and also perform the behaviours that are suitable for E-Waste disposal are most likely to perform the actual disposal. They also tend to have enough confidence in their ability to make the right decisions about E-Waste management, thus applying measures needed to overcome the identified obstacles or challenges. The studies conducted on self-efficacy in e-waste disposal put much focus on perceived competence and the association it has with the behaviour of disposal (Rani et al., 2016).

Collective efficacy: Collective efficacy represents a group's belief in their capabilities in processing a certain task or attaining a common goal as a unit (Kumar et al. 2017). There are, however, four major influencing sources of information which dictate collective efficacy, diverse experiences, emotional arousal, verbal persuasion and shared experiences (Lai et al. 2009).

> **H_{11}:** - Psychological determinants of behavior have a positive impact on the recycling behavior of e-waste consumers.

Figure 1. Research model

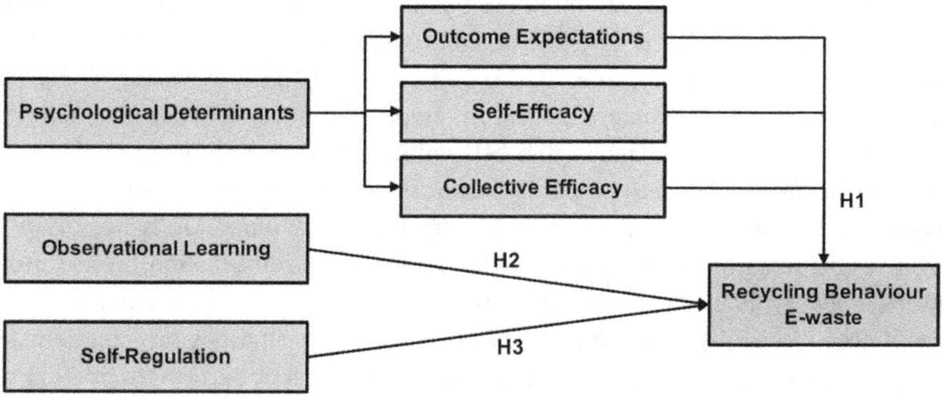

Observational Learning

Here, attention, retention and reproduction, and motivation are the four major processes involved. First of all, individuals pay attention to the model's behaviour and its consequences. They have to attend to the relevant components and understand the sequence of events that the model uses in acting (Lundgren & Yan .2017). Secondly, the individual must retain the observed information in memory if they are to reproduce it at a later stage of life. Observational Learning on E-Waste Disposal greatly influences the disposal behaviour among individuals. Individuals learn from observing the actions and behaviours of others about how they go about E-Waste disposal (Maddux & Rogers 1983). These may include observations made concerning family members, friends, colleagues, and social network influencers (Memon et al. 2017). Observationally, an individual learns appropriate or preferential methods of E-Waste disposal by following the correct procedures of recycling, making use of collection centres, or responsible consumption of electronic products (Miura et al. 2004). Under the context of e-waste disposal, observational learning studies have placed a focus on the elements of social modelling along with information diffusion. Interestingly, peers, family members, and neighbours are different models used to examine the influence on individuals in disposing of E-Waste.

H_{12}: - Observational learning has a positive impact on the recycling behaviour of E-Waste consumers.

Self - Regulation

Self-regulation is the process by which individuals set goals, monitor their progress relative to the attainment of those goals, and change their behaviours in pursuit of desired outcomes. One takes as an assumption that self-regulation underlines goal-directed behaviour and personal agency (Nema et al. 2016). This effective self-regulation will result in individuals being more likely to adopt appropriate E-Waste disposal behaviours, such as recycling their electronic devices and not dumping illegally by searching for appropriate channels to dispose of waste (Nahm et al. 2010). Self-regulation revolves around setting disposal-related goals, self-monitoring of disposal behaviours by an individual, evaluation of the performance of those behaviours, and reinforcement of positive disposal practices through self-rewards or self-punishments. In this respect, studies have documented the role of self-regulation in driving E-waste disposal behaviour. Results showed that persons having high self-regulatory abilities demonstrated more responsible E-Waste disposal behaviour, and persons with low self-regulation would act by proper disposal practices (Oreg et al. 2011).

H_{13}: - Self-regulation has a positive impact on the recycling behavior of E-Waste consumers.

RESULTS AND DISCUSSION

Demographic Profile of Respondents

Table 1. Demographic results

Variable		N	%	Variable		N	%
Gender	Men	123	53.2	**Employment**	Private Sector	85	36.79
	Women	108	46.98		Government Sector	22	9.52
Age	≤ 30 years	71	30.73		Professional	32	13.85
	31 to 40 years	122	52.81		Self-Employed	39	16.88
	41 to 50 years	18	7.8		Retired	4	1.73
	51 to 60 years	16	6.92	**Family Size**	1-2 individuals	28	12.12
	Over 60 years	4	1.73		3-4 individuals	157	67.96
Marital Status	Married	83	35		5-6 individuals	46	19.91
	Single	148	65	**Region**	Andhra Pradesh	53	22.94
Education	Graduate	96	41.55		Karnataka	58	25.1
	Postgraduate	89	38.53		Kerala	41	17.74
	Higher Secondary	18	7.79		Telangana	43	18.61
	Higher School	28	12.12		Tamil Nadu	36	15.58

Table 1 illustrates that 123 (53.2%) of the respondents were men, while 108 (46.98%) were women. 71 individuals (30.73%) were aged 30 years or younger, while 122 individuals (52.81%) fell within the age range of 31 to 40 years. A total of 18 individuals (7.80%) fell within the age range of 41 to 50 years, while 16 individuals (6.92%) were aged between 51 and 60 years. Additionally, 4 individuals (1.73%) were over the age of 60. In terms of marital status, 83 (35%) of the participants were married, whereas 148 (65%) were single. A significant portion of the survey respondents are graduates, totaling 96 individuals (41.55%). Additionally, 89 respondents (38.53%) hold postgraduate degrees, while 18 participants (7.79%) completed higher secondary school education, and 28 individuals (12.12%) have completed higher school education. The majority of the respondents are employed in the private sector. 85 (36.79%), 22 (9.52%) are from the Government sector, 32 (13.85%) were professionals, 39 (16.88%) were self-employed, and 4 (1.73%) were retired employees. In examining the family size, it was found that 157 homes (67.96%) were occupied by 3-4 individuals, 28 homes (12.12%) were occupied by 1-2 individuals, and 46 homes (19.91%) were occupied by 5-6 individuals. The study gathered responses from South India, with participation as follows: 58 (25.10%)

from Karnataka, 53 (22.94%) from Andhra Pradesh, 43 (18.61%) from Telangana, 36 (15.58%) from Tamil Nadu, and 41 (17.74%) from Kerala.

Impact of Social Cognitive Theory on Disposal Behavior

Three key parameters of Social Cognitive Theory, Psychological determinants, Observational learning, and self-regulation were considered for the analysis and multiple regression analysis method was performed (Sengupta et al. 2023). Findings obtained through multiple regression analysis revealed significant relationships between three variables and the respondents' intention to engage in recycling practices (Stacey et al. 2015).

Psychological Determinants of Behavior and Disposal Behavior

To determine the relationship of psychological determinants of behavior under SCT on disposal behavior which is a dependent variable, and independent variables of outcome expectations, self-efficacy and collective efficacy were put into equation (Table 2). The presence of H1 is significantly and positively linked with disposal behavior. The values (R=0.570, R^2 = 0.325, Adjusted R^2 = 0.278, and F= 6.841) are positively associated with the disposal behavior of respondents.

Table 2. Multiple regression analysis of psychological determinants of behaviour and disposal behaviour

Model	R	R Square	Adjusted R Square	Std. Error of the Estimate	Change Statistics				
					R Square Change	F Change	df1	df2	Sig. F Change
1	.570[a]	.325	.278	.56414	.325	6.841	15	213	.000

a. Predictors: (Constant), PD_CE5, PD_SE4, PD_OE3, PD_OE4, PD_SE2, PD_OE1, PD_SE1, PD_CE2, PD_SE5, PD_OE5, PD_OE2, PD_CE1, PD_CE3, PD_CE4, PD_SE3

The above-mentioned findings will be some extent supported by the studies of Bandura (Dougherty et al. 2001) on self-efficacy which states that individuals with higher self-efficacy are more likely to engage in proper disposal practices and overcome barriers or challenges associated with disposal (Sinha-Khetriwal et al. 2005). These psychological determinants operate within a dynamic interaction with the social and environmental context in which individuals live. They influence individuals' perceptions, motivations and abilities, ultimately shaping their disposal behavior (Uma et al. 2020). Interventions and strategies aimed at promoting responsible disposal behavior should consider these psychological determinants to efficiently change individuals' attitudes, beliefs and behaviors towards waste disposal (Oreg et al. 2011).

Observational Learning in Disposal Behavior:

To determine the impact of observational learning on disposal behavior, independent variables of observational learning and dependent variables of disposal behavior were put into the equation (Table 3). The values (R = 0.419, R^2 = 0.176, Adjusted R^2 = 0.157 and F = 9.562) are positively associated with the disposal behavior of respondents. Here H2 is also positively linked with disposal behavior.

Table 3. Multiple regression analysis of observational learning in disposal behaviour

Model	R	R Square	Adjusted R Square	Std. Error of the Estimate	Change Statistics				
					R Square Change	F Change	df1	df2	Sig. F Change
1	.419[a]	.176	.157	.60830	.176	9.562	5	224	.000

a. Predictors: (Constant), OL 5, OL 2, OL 1, OL 4, OL 3

Observational learning occurs when individuals observe others' behaviors and learn from their actions. Responsible ways of getting rid of E-Waste are more likely to be adopted by others if they see others doing the same and getting good results. Interventions aimed at promoting responsible E-Waste disposal can leverage observational learning by providing opportunities for individuals to observe and learn from positive disposal behaviors (Nema et al. 2016).

Self-Regulation in Disposal Behavior

To examine the influence of self-regulation on disposal behavior, independent variables of self-regulation and dependent variables of disposal behavior were put into equation (**Table 4**). The values (R = 0.758, R^2 = 0.575, Adjusted R^2 = 0.566 and F = 60.664) are positively associated with the disposal behavior of respondents. H3 is positively associated with disposal behaviors.

Table 4. Multiple regression analysis of Self-regulation in disposal behaviour

Model	R	R Square	Adjusted R Square	Std. Error of the Estimate	Change Statistics				
					R Square Change	F Change	df1	df2	Sig. F Change
1	.758[a]	.575	.566	.43673	.575	60.664	5	224	.000

a. Predictors: (Constant), SR 5, SR 2, SR 1, SR 3, SR 4

Research has shown that self-regulation significantly predicts disposal behavior in various domains, including E-Waste disposal. Studies have found a positive association between self-regulation and responsible disposal behaviors, highlighting the importance of self-regulatory processes in influencing individuals' actions and choices regarding E-Waste disposal (Patnaik et al. 2018). Also, self-regulation contributes to the development of individuals' attitudes and belief formation regarding E-Waste disposals. It may influence the shift in pro-environment values and personal responsibilities that eventually lead them to proper disposal behavior (Rahul & Rajan. 2023).

Practical Issues in e-Waste Disposal

Electronic waste disposal in India is beset with several practical issues that threaten seriously its effective and sustainable management. Various practical issues in the disposal of e-waste are highlighted below as well:

Informal Sector and Lack of Formal Recycling Infrastructure

In India, the paired informal industry does most of the e-waste recycling by unsafe unregulated ways of extraction (Ravi Varma et al. 2023). Most of these ways of extraction in informal ways are extremely unsafe, where even open burning and acid leaching release toxic fumes into the atmosphere (Wood & Bandura 1989). The inability of formal recycling infrastructure leads to inappropriate dismantling, hazardous materials exposure, and ineffective resource recovery (Lundgren & Yan 2017).

Ignorance and Inappropriate Disposal Methods

One aspect of this is that most consumers and businesses are not aware of the environmental and health hazards associated with E-Waste that pose a challenge to E-Waste disposal. Most consumers do not have any idea of appropriate methods of disposal and structures of recycling electronic devices. Much e-waste is still disposed of via inappropriate methods such as landfilling and informal recycling (Rathi & Kameswari 2017).

Limited Collection and Recycling Facilities

Inadequate collection and recycling facilities in the country are also a big challenge: a limited number of authorized collection centres and hence the recycling units, whereas the rural pocket has negligible coverage (Rizzolatti & Craighero 2004).

The reasons are: This leads to inadequate collection and recycling rates, leaving a significant portion of E-Waste untreated or improperly managed (Zimmerman 2000). The amount of e-waste that can be collected and recycled is limited by bad infrastructure and complicated logistics.

Shortcoming- Legislative Framework

Even though the Indian government has rules for managing e-waste since 2016, it is still a challenge how to effectively implement this legislative framework into practice. Compliance with this regulation is often inadequate, and proper mechanisms for monitoring and enforcement are usually lacking, leading to non-compliance by stakeholders and ineffective E-Waste disposal practices (Zhong 2022).

Lack of Integration and Collaboration

Effective E-waste management requires collaboration among different stakeholders, including manufacturers, consumers, recyclers, and Government agencies (Schunk 2003). However, because of the lack of coordination and integration by and among these particular stakeholders, comprehensive E-waste management strategies remain a challenge. More co-ordination and co-operation are expected to guarantee appropriate E -e-waste collection, recycling and disposal (Vijayan et al. 2023).

LIMITATIONS

Most of the limitations that always seem to occur with these studies do relate to generalizability. Sample bias is one such limitation that might be considered for this study. In the current study, it would appear that the characteristics and representativeness of the sample are going to limit the findings. As the data is self-reported, it may be subjected to social desirability bias, where participants respond with answers that they perceive as socially desirable and hence may incur biased responses. The cross-sectional design has been adopted for studying disposal behaviour, which constrains any establishment of the causal relationship between the social cognitive theory factors and disposal behaviours.

DISCUSSION

The findings from the present study are surely going to provide the necessary insight into the predicting factors of e-waste disposal behaviours, particularly for South India. By considering the role of psychological determinants, observational learning, and also self-regulation framed under Social Cognitive Theory or SCT, the present study depicts the main driving factors for responsible practices related to e-waste management. The analysis of findings regarding the worldwide context of e-waste disposal practices and their implications for e-waste management in various countries will be detailed in the discussion section.

Determinants of E-Waste Disposal-Related Psychological Factors

This study confirms that psychological determinants like outcome expectations, self-efficacy, and collective efficacy are important in determining disposal behaviour. Similarly, the same patterns could be depicted internationally. For instance, in Brazil, studies by Echegaray (2016) depicted that persons with high self-efficacy engage in proper recycling behaviour due to perceived benefits such as environmental conservation and legality. This psychological framework is supported by research that was done in China and demonstrating high self-efficacy translating to increased participation in the recycling program. As indicated, this trend can be appreciated within the context of developed countries such as the United States where e-waste disposal behaviour is usually driven by a mix of self-efficacy and regulatory frameworks (Fu et al. 2017). For instance, extended producer responsibility is a policy that places liability on manufacturers for electronic waste emanating from electrical goods they have produced; assurance of available and efficient channels for disposal instils confidence in recycling behaviour.

Observational Learning and Its Influence on E-Waste Disposal

Indeed, it was considered that observational learning could be an important variable responsible for behavioural change in e-waste disposal. Again, this shows agreement with the Social Cognitive Theory, which postulates that people learn behaviours through observation and imitation of others when it seems ventilative to them (Bandura, 1997). In the context of e-waste management, observing may take place through public campaigns, social networks, or visible role models who perform proper waste disposal. The role of observational learning in encouraging e-waste recycling is well documented internationally and across various settings. For example, university students in Australia whose peers had engaged in recycling were more likely to do so themselves (Islam et al., 2020). In Japan, where community-based initiatives have visibly demonstrated appropriate modes of e-waste disposal, there is a significant rise in awareness and participation in e-waste recycling (Nahm et al., 2010). Observational learning, however, depends on the extent to which such behaviours are visible and accessible. The effectiveness of observational learning is equally high in Switzerland, for instance, where recycling facilities are available and visible, and public campaigns are rampant. However, in regions where this setup is much less formalized, such as most parts of Africa and South Asia, less formal means of recycling are more practised, and a lot of toxic waste is disposed of in a rather observational manner where people also mimic these hazardous ways of disposal.

Self-Regulation in E-Waste Disposal Behaviour

The results of this study, about self-regulation, also offer insights into the role of personal goal-setting and behaviour monitoring in e-waste disposal practices. Bandura (1989) suggests that in self-regulation, individuals set goals, monitor their progress, and change their behaviours to conform with the set goals. With high self-regulatory capacity, individuals can easily engage in proper waste disposal practices in the context of e-waste since they can plan and act in manners commensurate with their environmental goals. Several self-regulation principles have been underscored the world over in many e-waste management programs. Self-regulation on the part of consumers, for instance, provides a holistic way of ensuring that even in the European Union, where a WEEE Directive is promoting the recycling of electronic waste, the consumers will follow proper guidelines in their disposal. In developing nations like India and Nigeria, these frameworks might be less effective, and an absence of available options for disposal and formal infrastructure has undermined self-regulation. According to (Sander & Zack, 2023). The study also reveals that self-regulation may be informed by environmental beliefs and accessibility to mechanisms of providing feedback. In locations with assured recycling programs, the feedback for such a disposal action comes immediately through financial benefits, community recognition, or environmental advantages. A person in regions that lack this may fail to sustain self-regulated behaviours and resort to poor disposal or hoarding of e-waste.

Practical Implications for International E-Waste Management

The following are some key practical implications of the present study for e-waste management in India as well as internationally: First, the influential role of psychological factors in shaping e-waste disposal behaviour implies a strategic focus of interventions in that direction. Public campaigns emphasizing the positive results derived from recycling and spelling out what to do can nurture these mental determinants, as witnessed in countries like Germany, which boasts of some of the highest rates of e-waste recycling in the world, windshield. Observational learning provides us with a sound tool to encourage the spread of e-waste disposal behaviour. It is an investment by governments and organizations in community-based recycling programs and public demonstrations of the right management of e-waste that allows individuals the chance to model responsible behaviour. Such initiatives have had tremendous success across regions like the Netherlands, where local community-based operations have contributed to higher rates of recycling. Self-regulation can be fostered by the presence of a formal structure for recycling and by feedback.

Limitations and Avenues for Future Research

Though this study has provided a very useful insight into the e-waste disposal behaviour, some limitations must be put into the forefront. The geographical focus of this study has confined it to South India, which may limit the generalization of findings across other regions with dissimilar cultural, economic, or environmental settings. Moreover, reliance on self-reported information is prone to social desirability bias-for instance, there is a likelihood that the respondents might overstate the positive disposing behaviour. In this regard, future studies should overcome such limitations by expanding the sample size in other regions, too and using more objective measures for disposal behaviour, such as the actual tracking of the disposal of trash through waste collection data. Future research could also investigate how various emerging technologies, such as blockchain and AI, can contribute to the introduction of improved sets of practices for e-waste management and enhance the rate of participation by individuals in recycling programs.

BEST METHODS FOR MANAGING E-WASTE AROUND THE WORLD:

E-waste, which stands for electronic waste, is one of the world's quickest-growing types of trash. E-waste is a particularly intractable management problem due to the toxic materials it contains and its generation rate, impelled by rapid technological changes. Various countries and regions have taken different approaches to dealing with electronic waste and formalized regulatory regimes for novel recycling technologies. This section is devoted to outstanding international best practices in e-waste management, including successful models and strategies of various countries and regions.

European Union: The WEEE Directive

The European Union is conventionally considered a leader in e-waste management due to its far-reaching legislative umbrella. The European Union adopted the WEEE Directive for the first time in 2003, and a more ambitious revised version was enacted in 2012 covering collection, recycling, and recovery targets. The Directive was conceived according to the Extended Producer Responsibility concept: producers were to bear responsibility for the whole life cycle of the product. The producers should pay for the collection, treatment, recycling, and environmentally sound disposal when these e-products have reached their end of life. In this respect, the WEEE Directive has been especially successful in devising an organized collection and treatment scheme in EU member states. Each country has to determine ways of ensuring that the waste electronic is returned by customers at no cost to either the retailers or other collecting points. The Directive stipulates specific recycling and recovery rates involving various electronic product groups, large household appliances, IT equipment, and lighting. On the other hand, one can also agree that the WEEE Directive has one strong point-it can be remodelled whenever new challenges, posed by emerging technologies, are to be confronted. The addition of smartphones, tablets, and other portable devices in the new Directive, for example, reflects the EU's commitment to moving the curve ahead of the growing problem of e-waste. The EU set binding targets and held producers responsible, thus it could achieve one of the highest e-waste collection and recycling rates globally (Gao et al., 2016).

Japan: An Approach to Innovation in E-Waste Recycling

Another country with a significant e-waste management system is Japan, where attention was focused on technological innovation and consumer participation. Japan in 2001 called for the manufacturer to take back and recycle four types of appliances: washing machines, televisions, air conditioners, and refrigerators. The law did not concern just the producers themselves but also the consumers who had to pay a fee when recycling their old appliances. Waste management in Japan is primarily inspired by the highly technological landscape in the country. Advanced recycling facilities mark the country right on target towards efficient separation and recovery of valuable materials from electronic waste. For example, both Hitachi and Panasonic have modern recycling plants that use automated, efficient dismantling and sorting systems for e-waste components, improving the rate of recycling (Zhong, 2022).

The United States: The Patchwork of State-Level Policies

The United States does not have any federal legislation on the handling of e-waste, unlike the EU and Japan. Instead, regulation of e-waste in the U.S. is through a patchwork of state-level policies. By 2023, twenty-five states have passed legislation requiring the recycling of electronic waste. Like the EU's WEEE Directive, producers are required to finance e-waste-collection and recycling programs under the EPR legislation introduced by several US states including California, New York, and Maine. California has led the way by proposing laws to regulate electronic waste through the enactment of the Electronic Waste Recycling Act in 2003. It imposes that for some covered electronic devices, such as box-type TVs and computer monitors, consumers should bear a forward collection rate upon purchase. This covers, to a certain extent, the State's e-waste recycling.

China: Tackling the World's Greatest E-Waste Challenge

Management of ever-increasing waste from abandoned electronic devices is a gigantic task for China, which is ranked as the world's biggest generator of e-waste. In the year 2020-21, it generated about 10.1 metric million tons of e-waste, which exceeded more than 20% of the total global volume. According to (Islam et al. 2020), several initiatives have been taken by the Government to improve this condition by formalizing such a vast informal recycling sector.

Different policies on e-waste management have been enacted in China, importantly all guided by the principle of EPR. The regulations place responsibility on manufacturers to establish systems for the recycling of products at their end-of-life. The "Circular Economy Promotion Law," which went into effect in 2011, advances the recycling and reuse practices of companies during the product life cycle of goods. This may reduce waste generation and enhance the resource efficiency of the key factors in addressing the increasing volumes of e-waste generated by the country. It has also invested much in the renovation of electronic waste infrastructure.

It has also partnered with private enterprises regarding the setting up of high-tech recycling facilities for electronic waste to recover valuable material. For instance, urban mining, which is very well described as taking metals out of discarded electronics, has emerged as a cottage industry in China. This has drastically reduced demand for primary mining and its attendant environmental burdens. While progress has been made, China's e-waste management network still has yet to face up to key challenges that involve huge problems in the so-called informal sector. Indeed, in general, informal recyclers use hazardous practices involving the open burning of materials and acid leaching, where very serious health and environmental risks are possible.

India

Confronting Unofficial Recycling Similar to China, India, the world's third-largest producer of e-waste, encounters the problem of a significant amount of its generated e-waste being handled through informal channels. For instance, according to Borthakur and Govind (2016), formal recycling facilities in India processed only 10-15% of the 3.2 million metric tonnes of e-waste that was created that year. The majority of the e-waste is handled by waste recyclers, who generally lack proper equipment and/or training to safely handle hazardous materials. The launching of the E-Waste Management Rules in 2016 undergirds India's electronic waste management framework. These rules imply that producers are required to put in place a take-back scheme and ensure proper electronic waste disposal. However, these rules are poorly implemented, with the informal sector continuing to dominate the landscape in the recycling of e-waste, as posited by (Sander & Zack, 2023).

Various initiatives have, therefore, been waged to formalize the informal sector, coupled with enhancing e-waste management. The integration of informal workers into the formal system of recycling, wherein training on the job, safety equipment, and modern technologies for recycling are provided, has been one of the better approaches in this regard. Public awareness campaigns are also being advocated, together with extended producer responsibility schemes, to encourage proper e-waste disposal among consumers.

CONCLUSION

One of the crucial challenges that nations all over the world, replete with mounting quantities of discarded electronic devices, are faced with is managing electronic waste. McCart's theory incorporates the following: Social Cognitive Theory introduces important factors that affect e-waste disposal behaviours, such as self-efficacy, observation learning, and self-regulation.

From an international perspective, there are a few stellar examples of success in the management of e-waste. The WEEE Directive of the European Union is one example, where comprehensive regulatory frameworks place the onus of responsibility on producers and encourage consumer involvement. This is where the concept of EPR, or Extended Producer Responsibility, comes in, whereby the responsibility for dealing with e-waste is left to those who produce it. The advantages of this system are that it not only maximizes rates of recycling but also-and for the first time-provides producers with a significant monetary incentive to design more environmentally friendly products. Similarly, Japan's attention to technological innovation in recycling has enabled it to achieve high rates of material recovery, setting an example for other countries. The fractured approach adopted by the United States beneath a missing federal policy serves to illustrate the challenges facing policymakers due to this lack of a unified federal approach. While states have taken steps in the right direction, like California's legislation on recycling, the dearth of policy at the national level has produced inconsistent results in other areas. Developing countries like India and China are experiencing much more serious challenges for e-waste management because of the preponderance of recycling by their informal sectors. Hazardous recycling practices here pose significant environmental and health risks; formalizing the informal sector is thus one critical step toward sustainable e-waste management.

The future of e-waste management will be defined by a combination of regulatory policies, technological innovation in design, and changes in consumer behaviour. As the growth of connected devices increases and technologies continue to become rapidly obsolete, so does the volume of electronic waste go upward; it is, therefore, prudent that governments, industries, and consumers work together on sustainable practices that will encourage this. All these will be supported by education campaigns that raise public awareness, strict enforcement of regulations concerning electronic waste, and the development of accessible recycling infrastructure.

REFERENCES

Arain, A. L., Pummill, R., Adu-Brimpong, J., Becker, S., Green, M., Ilardi, M., Van Dam, E., & Neitzel, R. L. (2020). Analysis of e-waste recycling behaviour based on a survey at a Midwestern US university. *Waste Management (New York, N.Y.)*, *105*, 119–127. DOI: 10.1016/j.wasman.2020.02.002 PMID: 32045839

Awasthi, A. K., Awasthi, M. K., Mishra, S., Sarsaiya, S., & Pandey, A. K. (2022). Evaluation of E-waste materials linked potential consequences to the environment in India. *Environmental Technology & Innovation*, *28*, 102477. DOI: 10.1016/j.eti.2022.102477

Bandura, A. (1986). *Social Foundations of Thought and Action*.

Bandura, A. (1998). Health promotion from the perspective of social cognitive theory. *Psychology & Health*, *13*(4), 623–649. DOI: 10.1080/08870449808407422

Bharali, A., & Kumar, S. (2018). E-waste management behaviour among urban households in India: An empirical investigation. *Journal of Materials*, C.

Borthakur, A., & Govind, M. (2016). Emerging trends in consumers' E-waste disposal behaviour and awareness: A worldwide overview with special focus on India. *Resources, Conservation and Recycling*. Advance online publication. DOI: 10.1016/j.resconrec.2016.11.011

Bouma, A. J., van Wilgen, P., Lemmink, K. A., Stewart, R., Dijkstra, A., & Diercks, R. L. (2018). Barrier-belief lifestyle counselling in primary care: A randomized controlled trial of efficacy. *Patient Education and Counseling*, *101*(12), 2134–2144. DOI: 10.1016/j.pec.2018.07.015 PMID: 30072044

Carver, C. S., & Scheier, M. F. (1998). *On the self-regulation of behaviour*. Cambridge University Press.

Chatterjee, S., Kumar, V., & Gupta, D. (2016). E-waste management in India: Issues and strategies. *Resources, Conservation and Recycling*, *107*, 28–36.

Choi, J.-H., Yoo, H.-E., Chung, H., Lee, H.-S., Lee, M.-J., & Chang, H.-J. (2018). Evaluation of Effectiveness of Food Safety Education Program Based on Social Cognitive Theory for the Elderly in the Middle Class. Academic Press.

Deoghare, P. R., & Shukla, S. (2018). Analysis of e-waste disposal behaviour among Indian IT professionals: An empirical study. *Procedia Computer Science*, *133*, 423–431.

Dougherty, C. M., Johnson-Crowley, N. R., Lewis, F. M., & Thompson, E. A. (2001). Theoretical development of nursing interventions for sudden cardiac arrest survivors using social cognitive theory. *Advances in Nursing Science*, *24*(1), 78–86. DOI: 10.1097/00012272-200109000-00009 PMID: 11554535

Echegaray, F. (2016). Assessing the intention-behaviour gap in electronic waste recycling: The case of Brazil. *Journal of Cleaner Production*, 1–11. DOI: 10.1016/j.jclepro.2016.05.064

Fu, F., Liu, Y., Wang, Y., & Chen, J. (2017). Analysis of the influencing factors on individuals' recycling intention for e-waste: A case study of China. *Journal of Cleaner Production*, *162*, 1566–1575.

Gao, C., Wang, X., & Zhang, M. (2016). Investigating the determinants of e-waste recycling behaviour: Evidence from a Chinese household survey. *Journal of Cleaner Production*, *136*, 141–150.

Gist, M. E., & Mitchell, T. R. (1992). Self-efficacy: A theoretical analysis of its determinants and malleability. *Academy of Management Review*, *17*(2), 183–211. DOI: 10.2307/258770

Gollwitzer, P. M. (1999). Implementation intentions: Strong effects of simple plans. *The American Psychologist*, *54*(7), 493–503. DOI: 10.1037/0003-066X.54.7.493

Islam, M. T., Dias, P., & Huda, N. (2020). Young consumers' e-waste awareness, consumption, disposal, and recycling behaviour: A case study of university students in Sydney, Australia. *Journal of Cleaner Production*. Advance online publication. DOI: 10.1016/j.jclepro.2020.124490

Kaur, H., Singh, R., & Khanna, S. (2018). E-waste disposal behaviour and factors influencing it: A study of Indian consumers. *Journal of Environmental Planning and Management*, *61*(2), 326–344.

Kumar, N., Sah, B., & Jain, V. (2017). E-waste management: A review of recent research. *Resources, Conservation and Recycling*, *122*, 55–69.

Lai, K. H., Sarathy, R., & Grover, V. (2009). Information diffusion and adoption of electric vehicles: The case of Hong Kong. *Journal of Management Information Systems*, *26*(2), 241–273.

Lent, R. W., Brown, S. D., & Hackett, G. (1994). Toward a unifying social cognitive theory of career and academic interest, choice, and performance. *Journal of Vocational Behavior*, *45*(1), 79–122. DOI: 10.1006/jvbe.1994.1027

Li, L., Li, B., Liu, Y., Li, X., & Ma, C. (2017). Determinants of individuals' electronic waste recycling behaviour: A comparative analysis of China and South Korea. *Resources, Conservation and Recycling*, *122*, 146–156.

Locke, E. A., & Latham, G. P. (2002). Building a practically useful theory of goal setting and task motivation: A 35-year odyssey. *The American Psychologist*, *57*(9), 705–717. DOI: 10.1037/0003-066X.57.9.705 PMID: 12237980

Lundgren, K., & Yan, M. (2017). The informal e-waste sector: Potential benefits and drawbacks of its integration into the formalized system. *Environmental Science & Policy*, *69*, 39–47.

Maddux, J. E., & Rogers, R. W. (1983). Protection motivation and self-efficacy: A revised theory of fear appeals and attitude change. *Journal of Experimental Social Psychology*, *19*(5), 469–479. DOI: 10.1016/0022-1031(83)90023-9

Memon, M. A., Rahman, Z., & Kumar, A. (2017). A conceptual model for e-waste recycling behaviour among consumers. *Journal of Cleaner Production*, *141*, 370–381.

Miura, S.-i., Yamaguchi, Y., Urata, H., Himeshima, Y., Otsuka, N., & Tomita, S.. (2004). Efficacy of a multicomponent program (PACE+ Japan) for lifestyle modification in patients with essential hypertension. *Hypertension Research*, *27*(11), 859–864. DOI: 10.1291/hypres.27.859 PMID: 15824468

Nahm, E.-S., Barker, B., Resnick, B., Covington, B., Magaziner, J., & Brennan, P. F. (2010). Effects of a social cognitive theory-based hip fracture prevention website for older adults. *Computers, Informatics, Nursing*, *28*(6), 371–379. DOI: 10.1097/NCN.0b013e3181f69d73 PMID: 20978408

Nejati, M., Saboury, A., Nejati, M., & Sabouri, M. S. (2017). Factors influencing the intention to use electronic waste recycling services: An integration of the theory of planned behaviour and protection motivation theory. *Resources, Conservation and Recycling*, *122*, 521–530.

Nema, A. K., Garg, A., & Yadav, R. K. (2016). E-waste management in India: Issues and challenges. *Environmental Science and Pollution Research International*, *23*(2), 1787–1800.

Oreg, S., Vakola, M., & Armenakis, A. (2011). Change recipients' reactions to organizational change: A 60-year review of quantitative studies. *The Journal of Applied Behavioral Science*, *47*(4), 461–524. DOI: 10.1177/0021886310396550

Patnaik, S., Chakraborty, S., & Dey, B. L. (2018). Determinants of e-waste disposal behaviour among Indian households: An empirical study. *Journal of Environmental Management*, *213*, 116–126.

Rahul, V., & Rajan, U. (2023). Sickle effect: A hybrid model of product life cycle and supply chain management for achieving sustainable development. *Materials Today: Proceedings*. Advance online publication. DOI: 10.1016/j.matpr.2023.06.075

Rani, M., Singh, R., & Kumar, A. (2016). Informal electronic waste recycling: A sector review with special focus on India. *Journal of Cleaner Production*, *129*, 431–448.

Rathi, N., & Kameswari, V. L. (2017). Determinants of e-waste disposal behaviour among college students in India: An empirical analysis. *Waste Management & Research*, *35*(6), 650–658.

Ravi Varma, V., Mathi Krishnan, M., Parayitam, S., Duraisami, S. P. A., & Saravanaselvan, N. R. (2023). Exploring e-waste recycling behaviour intention among households: Evidence from India. *Cleaner Materials*, *7*, 100174. DOI: 10.1016/j.clema.2023.100174

Rizzolatti, G., & Craighero, L. (2004). The mirror-neuron system. *Annual Review of Neuroscience*, *27*(1), 169–192. DOI: 10.1146/annurev.neuro.27.070203.144230 PMID: 15217330

Sander, W. E., & Zack, S. A. (2023). The veterinarian's role in and attitude to the disposal of unwanted pharmaceuticals. *Environmental Challenges*, *11*, 100718. DOI: 10.1016/j.envc.2023.100718

Schunk, D. H. (2003). Self-efficacy for reading and writing: Influence of modelling, goal setting, and self-evaluation. *Reading & Writing Quarterly*, *19*(2), 159–172. DOI: 10.1080/10573560308219

Sengupta, D., Ilankoon, I. M. S. K., Kang, K. D., & Chong, M. N. (2023). Circular economy and household e-waste management in India. Part II: A case study on informal e-waste collectors (Kabadiwalas) in India. *Minerals Engineering*, *200*, 108154. DOI: 10.1016/j.mineng.2023.108154

Sinha-Khetriwal, D., Kraeuchi, P., & Widmer, R. (2005). Export of used electronic goods from Switzerland to developing countries: Environmental impacts and their implications. *Environmental Impact Assessment Review*, *25*(5), 525–539.

Stacey, F. G., James, E. L., Chapman, K., Courneya, K. S., & Lubans, D. R. (2015). A systematic review and meta-analysis of social cognitive theory-based physical activity and nutrition behaviour change interventions for cancer survivors. *Journal of Cancer Survivorship: Research and Practice*, *9*(2), 305–338. DOI: 10.1007/s11764-014-0413-z PMID: 25432633

Uma, K. E., Nwaka, I. D., Nwogu, M. U., & Obidike, P. C. (2020). What are the triggers of household decision-making on waste disposal choices? A gender-differentiated analysis. *Heliyon*, *6*(12), e05588. DOI: 10.1016/j.heliyon.2020.e05588 PMID: 33319089

Vijayan, R. V., Krishnan, M. M., Parayitam, S., Duraisami, S. P. A., & Saravanaselvan, N. R. (2023). Exploring e-waste recycling behaviour intention among households: Evidence from India. *Cleaner Materials*, *7*, 100174. DOI: 10.1016/j.clema.2023.100174

Wood, R., & Bandura, A. (1989). Social cognitive theory of organizational management. *Academy of Management Review*, *14*(3), 361–384. DOI: 10.2307/258173

Zhong, H., Zhou, S., Zhao, Z., Zhang, H., Nie, J., & Simayi, P. (2022). An empirical study on the types of consumers and their preferences for E-waste recycling with a points system. *Cleaner and Responsible Consumption*, *7*, 100087. DOI: 10.1016/j.clrc.2022.100087

Zhong, H., Zhou, S., Zhao, Z., Zhang, H., Nie, J., & Simayi, P. (2022). An empirical study on the types of consumers and their preferences for E-waste recycling with a points system. *Cleaner and Responsible Consumption*, *7*, 100087. DOI: 10.1016/j.clrc.2022.100087

Zhou, J., & Fan, T. (2019). Understanding the factors influencing patient e-Health literacy in online health communities (OHCs): A social cognitive theory perspective. *International Journal of Environmental Research and Public Health*, *16*(14), 2455. DOI: 10.3390/ijerph16142455 PMID: 31295924

Zimmerman, B. J. (2000). Self-efficacy: An essential motive to learn. *Contemporary Educational Psychology*, *25*(1), 82–91. DOI: 10.1006/ceps.1999.1016 PMID: 10620383

Compilation of References

A Closer Look: Lithium-Ion Batteries in E-waste. (n.d.). https://www.simslifecycle.com/ 2019/05/21/a-closer-look-lithium-ion-batteries-in-E-waste/

Abalansa, S., El Mahad, B., Icely, J., & Newton, A. (2021). Electronic waste, an environmental problem exported to developing countries: The GOOD, the BAD and the UGLY. *Sustainability (Basel)*, *13*(9), 5302. DOI: 10.3390/su13095302

Abdelbasir, S. M., Hassan, S. S. M., Kamel, A. H., & El-Nasr, R. S. (2018). Status of electronic waste recycling techniques: A review. *Environmental Science and Pollution Research International*, *25*(17), 16533–16547. DOI: 10.1007/s11356-018-2136-6 PMID: 29737485

Abhilasha, A., Sreenivasulu, A., & Manimozhi, T. (2022). The Model of Smart Sensing Device For Sensitive Nanoclusters Modification in Sensing Properties. 2022 2nd International Conference on Advance Computing and Innovative Technologies in Engineering (ICACITE), 1043-1047. DOI: 10.1109/ICACITE53722.2022.9823846

Aboelazm, E. A., Mohamed, N., & Ali, G. A. (2021). *Makhlouf AS CK recycling of cobalt oxides electrodes from spent lithium-ion batteries by electrochemical method*. Waste Recycling Technologies for Nanomater Manufacturing. DOI: 10.1007/978-3-030-68031-2_4

Aboelmaged, M. (2021). E-waste recycling behavior: An integration of recycling habits into the theory of planned behavior. *Journal of Cleaner Production*, *278*, 124182. DOI: 10.1016/j.jclepro.2020.124182

Achillas, C., Aidonis, D., Vlachokostas, C., Karagiannidis, A., Moussiopoulos, N., & Loulos, V. (2013). Depth of manual dismantling analysis: A cost–benefit approach. *Waste Management, 33*(4), 948-956. https://doi.org/https://doi.org/10.1016/j.wasman.2012.12.024

Acquah, A. A., D'Souza, C., Martin, B., Arko-Mensah, J., Nti, A. A., Kwarteng, L., Takyi, S., Quakyi, I. A., Robins, T. G., & Fobil, J. N. (2019). Processes and Challenges associated with informal electronic waste recycling at Agbogbloshie, a suburb of Accra, Ghana. *Proceedings of the Human Factors and Ergonomics Society ... Annual Meeting. Human Factors and Ergonomics Society.Annual meeting, 63*(1), 938–942. DOI: 10.1177/1071181319631219

Acquah, A. A., D'Souza, C., Martin, B., Arko-Mensah, J., Quakyi, I. A., Basu, N., Robins, T. G., & Fobil, J. N. (2021). Work-related exposures and musculoskeletal disorder symptoms among informal E-waste recyclers at Agbogbloshie, Ghana. *Proceedings of the 21st Congress of the International Ergonomics Association (IEA 2021).* Volume IV, *Healthcare and Healthy Work. International Ergonomics Association. Congress, 222*, 677–681. DOI: 10.1007/978-3-030-74611-7_93

Adanu, S. K., Gbedemah, S. F., & Attah, M. K. (2020). Challenges of adopting sustainable technologies in e-waste management at Agbogbloshie, Ghana. *Heliyon, 6*(8), 04548. DOI: 10.1016/j.heliyon.2020.e04548 PMID: 32802976

Adeleye, N., & Eboagu, C. (2019). Evaluation of ICT development and economic growth in Africa. *NETNOMICS: Economic Research and Electronic Networking, 20*(1), 31–53. DOI: 10.1007/s11066-019-09131-6

Aghaei, M., & Rezagholizadeh, M. (2017). The impact of information and communication technology (ICT) on economic growth in the OIC countries. *Economic and Environmental Studies, 17*(2), 257–278. DOI: 10.25167/ees.2017.42.7

Agrawal, A., & Sahu, K. K. (2010). Problems, prospects and current trends of copper recycling in India: An overview. *Resources, Conservation and Recycling, 54*(7), 401–416. DOI: 10.1016/j.resconrec.2009.09.005

Agrawal, R. (2023). Industrial Automated Multipurpose Robot Using WIFI. *Proceedings of the 4th International Conference for Emerging Technology (INCET)*, 1-8.

Agyei-Mensah, S., & Oteng-Ababio, M. (2012). Perceptions of health and environmental impacts of e-waste management in Ghana. *International Journal of Environmental Health Research, 22*(6), 500–517. DOI: 10.1080/09603123.2012.667795 PMID: 22428915

Ahirwar, R., & Tripathi, A. K. (2021). E-waste management: A review of recycling process, environmental and occupational health hazards, and potential solutions. *Environmental Nanotechnology, Monitoring & Management, 15*, 100409. DOI: 10.1016/j.enmm.2020.100409

Ahmed, M., & Jones, C. (2020). E-Waste in Developing Countries: A Challenge for Sustainability. *The Journal of Development Studies*, *36*(4), 104–121.

Ahmed, N., Li, C., Khan, A., Qalati, S. A., Naz, S., & Rana, F. (2021). Purchase intention toward organic food among young consumers using theory of planned behavior: Role of environmental concerns and environmental awareness. *Journal of Environmental Planning and Management*, *64*(5), 796–822. DOI: 10.1080/09640568.2020.1785404

Ahmed, S., & Kumar, R. (2018). Community Involvement in E-Waste Management: Case Studies from Asia. *Journal of Environmental Sciences (China)*, *32*(5), 573–589.

Ahmed, S., & Lee, K. (2023). Carbon footprint assessment of smartphones: A comprehensive review. *Journal of Cleaner Production*, *379*, 134579.

Ajzen, I. (1991). The theory of planned behavior. *Organizational Behavior and Human Decision Processes*, *50*(2), 179–211. DOI: 10.1016/0749-5978(91)90020-T

Akenji, L., Hotta, Y., Bengtsson, M., & Hayashi, S. (2011). EPR policies for electronics in developing Asia: An adapted phase-in approach. *Waste Management & Research*, *29*(9), 919–930. DOI: 10.1177/0734242X11414458 PMID: 21730041

Akhter, M., Al Mansur, A., Islam, M. I., Lipu, M. S. H., Karim, T. F., Abdolrasol, M. G. M., & Alghamdi, T. A. H. (2024). Sustainable Strategies for Crystalline Solar Cell Recycling: A Review on Recycling Techniques, Companies, and Environmental Impact Analysis. *Sustainability*, *16*(13), 5785. https://www.mdpi.com/2071-1050/16/13/5785

Akimoto, Y., Iizuka, A., & Shibata, E. (2018). High-voltage pulse crushing and physical separation of polycrystalline silicon photovoltaic panels. *Minerals Engineering*, *125*, 1–9. https://doi.org/https://doi.org/10.1016/j.mineng.2018.05.015

Akkaya, Ş., & Bakkal, U. (2005). *Çevre Vergileri ve Çifte Yarar*. İstanbul Üniversitesi İktisat Fakültesi Mecmuası, 55(2), 1-22.

Akortia, E., Olukunle, O. I., Daso, A. P., & Okonkwo, J. O. (2017). Soil concentrations of polybrominated diphenyl ethers and trace metals from an electronic waste dump site in the Greater Accra Region, Ghana: Implications for human exposure. *Ecotoxicology and Environmental Safety*, *137*, 247–255. DOI: 10.1016/j.ecoenv.2016.12.008 PMID: 27978452

Akram, R., Natasha, F. S., Hashmi, M. Z., Wahid, A., Adnan, M., & Nasim, W. (2019). Trends of electronic waste pollution and its impact on the global environment and ecosystem. *Environmental Science and Pollution Research International*, *26*(17), 16923–16938. DOI: 10.1007/s11356-019-04998-2 PMID: 31025281

Akram, S. V., Alshamrani, S. S., Singh, R., Rashid, M., Gehlot, A., AlGhamdi, A. S., & Prashar, D. (2021). Blockchain enabled automatic reward system in solid waste management. *Security and Communication Networks*, *2021*(1), 6952121. DOI: 10.1155/2021/6952121

Al Mubarak, F., Rezaee, R., & Wood, D. A. (2024). Economic, Societal, and Environmental Impacts of Available Energy Sources: A Review. *Eng, 5*(3), 1232-1265. https://doi.org/https://doi.org/10.3390/eng5030067

Alam, S. S., Haque, I. S., Kokash, H. A., Ahmed, S., & Ahsan, M. N. (2025). Drivers of waste separation behavior in urban Bangladesh: Leveraging social norms and environmental awareness for circular economy success. *Circular Economy and Sustainability*, 1–33. DOI: 10.1007/s43615-025-00526-2

Albiman, M., & Sulong, Z. (2017). The linear and non-linear impacts of ICT on economic growth, of disaggregate income groups within SSA region. *Telecommunications Policy*, *41*(7-8), 555–572. DOI: 10.1016/j.telpol.2017.07.007

Al-Ezzi, A. S., & Ansari, M. N. M. (2022). Photovoltaic Solar Cells: A Review. *Applied System Innovation*, *5*(4), 67. https://doi.org/https://doi.org/10.3390/asi5040067

Ali, A., Malik, S. A., Shafiullah, M., Malik, M. Z., & Zahir, M. H. (2023). Policies and regulations for solar photovoltaic end-of-life waste management: Insights from China and the USA. *Chemosphere*, *340*, 139840. https://doi.org/https://doi.org/10.1016/j.chemosphere.2023.139840 PMID: 37597624

Ali, A., Shahid, M., Qadir, S. A., Islam, M. T., Khan, M. W., & Ahmed, S. (2024). Solar PV End-of-Life Waste Recycling: An Assessment of Mechanical Recycling Methods and Proposed Hybrid Laser and High Voltage Pulse Crushing Method. *Preprints*. https://doi.org/http://dx.doi.org/10.20944/preprints202411.0560.v1

Althaf, S., Babbitt, C. W., & Chen, R. (2021). The evolution of consumer electronic waste in the United States. *Journal of Industrial Ecology*, *25*(3), 693–706. DOI: 10.1111/jiec.13074

Alvarez-Gil, M., Berrone, P., Husillos, F., & Lado, N. (2007). Reverse logistics, stakeholders' influence, organizational slack, and managers' posture. *Journal of Business Research*, *60*(5), 463–473. DOI: 10.1016/j.jbusres.2006.12.004

Amoabeng Nti, A. A., Arko-Mensah, J., Botwe, P. K., Dwomoh, D., Kwarteng, L., Takyi, S. A., Acquah, A. A., Tettey, P., Basu, N., Batterman, S., Robins, T. G., & Fobil, J. N. (2020). Effect of Particulate Matter Exposure on Respiratory Health of e-Waste Workers at Agbogbloshie, Accra, Ghana. *International Journal of Environmental Research and Public Health*, *17*(9), 3042. DOI: 10.3390/ijerph17093042 PMID: 32349371

An, D., Yang, Y., Chai, X., Xi, B., Dong, L., & Ren, J. (2015). Mitigating pollution of hazardous materials from WEEE of China: Portfolio selection for a sustainable future based on multi-criteria decision making. *Resources, Conservation and Recycling*, *105*, 198–210. DOI: 10.1016/j.resconrec.2015.10.025

Andeobu, L., Wibowo, S., & Grandhi, S. (2023). Informal E-waste recycling practices and environmental pollution in Africa: What is the way forward? *International Journal of Hygiene and Environmental Health*, *252*, 114192. https://doi.org/https://doi.org/10.1016/j.ijheh.2023.114192 PMID: 37348165

Angaye, T. C. N., Godbless, N. O., & Ogidi, O. I. (2019). Environmental and health risk assessment of particulate matter associated with sporting activities in Yenagoa metropolis, Nigeria. *General Internal medicine and clinical. Innovations*, *4*(3), 1–3. DOI: 10.15761/GIMCI.1000180

Angaye, T. C. N., & Ogidi, O. I. (2018). The environmental impacts of electronic wastes in Bayelsa State, Nigeria. *Med Crave Online Journal of Toxicology*, *4*(5), 331–332.

Appiah-Otoo, I., & Song, N. (2021). The impact of ICT on economic growth-Comparing rich and poor countries. *Telecommunications Policy*, *45*(2), 1–15. DOI: 10.1016/j.telpol.2020.102082

Arain, A. L., Pummill, R., Adu-Brimpong, J., Becker, S., Green, M., Ilardi, M., Van Dam, E., & Neitzel, R. L. (2020). Analysis of e-waste recycling behaviour based on a survey at a Midwestern US university. *Waste Management (New York, N.Y.)*, *105*, 119–127. DOI: 10.1016/j.wasman.2020.02.002 PMID: 32045839

Ari, V. (2016). A review of technology of metal recovery from electronic waste. E-Waste in transition—From pollution to resource. Academic Press.

Arora, N., Bhattacharjya, S., Bakshi, S. K., Anand, M., Dasgupta, D., Gupta, A., Prasad, N. S., Nanda, N., Pandey, S., Garud, S., & Bineesan, M. K. (2018). *Greening the Solar Power PV value chain*. European Union's Resource Efficiency Initiative (EU-REI) Project, Deutsche Gesellschaft für Internationale Zusammenarbeit (GIZ) GmbH www.resourceefficiencyindia.com

Arul Prishya, A. S., Chopra, L., & Manikanika. (2023). Comprehensive review on uses of silicon dioxide in solar cell. *Materials Today: Proceedings, 72*, 1471-1478. https://doi.org/https://doi.org/10.1016/j.matpr.2022.09.348

Asampong, E., Dwuma-Badu, K., Stephens, J., Srigboh, R., Neitzel, R., Basu, N., & Fobil, J. N. (2015). Health seeking behaviours among electronic waste workers in Ghana. *BMC Public Health, 15*(1), 1–9. DOI: 10.1186/s12889-015-2376-z PMID: 26474859

Aslan, A., Destek, M., & Okumus, I. (2018). Bootstrap rolling window estimation approach to analysis of the Environment Kuznets Curve hypothesis: Evidence from the USA. *Environmental Science and Pollution Research International, 25*(3), 2402–2408. DOI: 10.1007/s11356-017-0548-3 PMID: 29124642

Aston, J.E., Peyton, B.M., Lee, B.D., Apel, W.A., 2010. Effects of ferrous sulphate, inoculum history, and anionic form on lead, zinc, and copper toxicity to Acid thiobacillus caldus strain BC13. Academic Press.

Atta, U. R. K., & Ahmad, R. W. (2022). A Blockchain-Based IoT-Enabled E-Waste Tracking and Tracing System for Smart Cities. *IEEE Access : Practical Innovations, Open Solutions, 10*, 86256–86269. DOI: 10.1109/ACCESS.2022.3198973

Awasthi, A. K., Awasthi, M. K., Mishra, S., Sarsaiya, S., & Pandey, A. K. (2022). Evaluation of E-waste materials linked potential consequences to the environment in India. *Environmental Technology & Innovation, 28*, 102477. DOI: 10.1016/j.eti.2022.102477

Awasthi, A. K., Hasan, M., Mishra, Y. K., Pandey, A. K., Tiwary, B. N., Kuhad, R. C., & Thakur, V. K. (2019b). Environmentally sound system for E-waste: Biotechnological perspectives. *Current Research in Biotechnology, 1*, 58–64. DOI: 10.1016/j.crbiot.2019.10.002

Awasthi, A. K., Li, J., Koh, L., & Ogunseitan, O. A. (2019a). Circular economy and electronic waste. *Nature Electronics, 2*(3), 86–89. DOI: 10.1038/s41928-019-0225-2

Awasthi, A. K., Wang, M., Awasthi, M. K., Wang, Z., & Li, J. (2018). Environmental pollution and human body burden from improper recycling of e-waste in China: A short-review. *Environmental Pollution, 243*, 1310–1316. DOI: 10.1016/j.envpol.2018.08.037 PMID: 30268981

Awasthi, A., Cucchiella, F., D'Adamo, I., Li, J., Rosa, P., Terzi, S., Wei, G., & Zeng, X. (2018). Modelling the correlations of e-waste quantity with economic increase. *The Science of the Total Environment, 613-614*, 46–53. DOI: 10.1016/j.scitotenv.2017.08.288 PMID: 28898811

Awasthi, K. A., & Li, J. (2017). Management of electrical and electronic waste: A comparative evaluation of China and India. *Renewable & Sustainable Energy Reviews*, *76*, 434–447. DOI: 10.1016/j.rser.2017.02.067

Awasthi, K. A., Zeng, X., & Li, J. (2016). Comparative examining and analysis of e-waste recycling in typical developing and developed countries. *Procedia Environmental Sciences*, *35*, 676–680. DOI: 10.1016/j.proenv.2016.07.065

Ayres, R. (1994). *Industrial metabolism: theory and policy. In the greening of industrial ecosystems*. National Academy Press.

Ayvaz, B., Bolat, B., & Aydın, N. (2015). Stochastic reverse logistics network design for waste of electrical and electronic equipment. *Resources, Conservation and Recycling*, *104*, 391–404. DOI: 10.1016/j.resconrec.2015.07.006

Azwardi, A., Igamo, A., & Wijaya, W. (2023). *The Concept of Waste Management on Economic Development in the European Union.* International Journal of Energy Economics and Policy, 13(1), 1-6. DOI: 10.32479/ijeep.13667

Babaei, A., & Nasr Esfahani, A. (2024). A Review of Photovoltaic Waste Management from a Sustainable Perspective. *Electricity, 5*(4), 734-750. https://www.mdpi.com/2673-4826/5/4/36

Baghel, D. S., Sharma, S., Singh, S., Kumar, R., Singh, G., & Kohli, R. (2025). Advancing Hazardous Waste Management: Integrating Innovative Technologies in Transport and Disposal. *Solid Waste Management for Rural Regions*, 55–68. DOI: 10.4018/979-8-3693-8527-2.ch004

Bai, X., & Tang, Y. (2023). Life cycle assessment of e-waste recycling: A case study on carbon footprint reduction. *Journal of Environmental Management*, *324*, 116444.

Bajagain, R., Panthi, G., An, Y.-J., & Jeong, S.-W. (2020). Current Practices on Solar Photovoltaic Waste Management: An Overview of the Potential Risk and Regulatory Approaches of the Photovoltaic Waste. *Journal of Korean Society of Environmental Engineers*, *42*(12), 690–708. https://doi.org/https://doi.org/10.4491/KSEE.2020.42.12.690

Bakhiyi, B., Gravel, S., Ceballos, D., Flynn, M. A., & Zayed, J. (2018). Has the question of e-waste opened a Pandora's box? An overview of unpredictable issues and challenges. *Environment International*, *110*, 173–192. DOI: 10.1016/j.envint.2017.10.021 PMID: 29122313

Baklouti, A., Mifdal, L., Dellagi, S., & Chelbi, A. (2020). An Optimal Preventive Maintenance Policy for a Solar Photovoltaic System. *Sustainability*, *12*(10), 4266. https://doi.org/https://doi.org/10.3390/su12104266

Balcilar, M., Ozdemir, Z., & Arslanturk, Y. (2010). Economic growth and energy consumption causal nexus viewed through a bootstrap rolling window. *Energy Economics*, *32*(6), 1398–1410. DOI: 10.1016/j.eneco.2010.05.015

Baldé, C. P., Forti, V., Gray, V., Kuehr, R., & Stegmann, P. (2017). *The global e-waste monitor 2017: Quantities, flows and resources*. United Nations University (UNU), International Telecommunication Union (ITU), International Solid Waste Association (ISWA). collections.unu.edu/

Baldé, C. P., Forti, V., Gray, V., Kuehr, R., & Stegmann, P. (2017). The global E-waste monitor 2017: Quantities, flows and resources. United Nations University, International Telecommunication Union, and International Solid Waste Association.

Baldé, C. P., Forti, V., Gray, V., Kuehr, R., & Stegmann, P. (2017). *The global e-waste monitor 2017: Quantities, flows, and resources*. United Nations University. https://collections.unu.edu/eserv/UNU:6349/PDF

Baldé, C. P., Kuehr, R., Yamamoto, T., McDonald, R., D'Angelo, E., Althaf, S., . . . Wagner, M. (2024). Global e-waste monitor 2024. International Telecommunication Union (ITU) and United Nations Institute for Training and Research (UNITAR).

Bandura, A. (1986). *Social Foundations of Thought and Action*.

Bandura, A. (1998). Health promotion from the perspective of social cognitive theory. *Psychology & Health*, *13*(4), 623–649. DOI: 10.1080/08870449808407422

Bandyopadhyay, A. (2008). Indian initiatives on E-waste management—A critical review. *Environmental Engineering Science*, *25*(10), 1507–1526. DOI: 10.1089/ees.2007.0342

Bansal, A., Verghese, S., Massey, M. H., Massey, D. D., & Singh, R. (2023). A Review Of The Effects Of Environmental Heavy Metal Contamination From Electronic Waste On Human Health. *IJRAR-International Journal of Research and Analytical Reviews*, *10*(4), 40–82.

Bao, S., Pan, B., Wang, L., Cheng, Z., Liu, X., Zhou, Z., & Nie, X. (2020). Adverse effects in Daphnia magna exposed to e-waste leachate: Assessment based on life trait changes and responses of detoxification-related genes. *Environmental Research*, *188*, 109821. DOI: 10.1016/j.envres.2020.109821 PMID: 32798945

Baragde, D. B., & Jadhav, A. U. (2020). Circular economy model for the E-waste management sector. In *Handbook of Research on Entrepreneurship Development and Opportunities in Circular Economy* (pp. 216–230). IGI Global.

Baralla, G., Pinna, A., Tonelli, R., & Marchesi, M. (2023). Waste management: A comprehensive state of the art about the rise of blockchain technology. *Computers in Industry*, *145*(1), 103812. DOI: 10.1016/j.compind.2022.103812

Barasa, M., Bogdanov, D., Oyewo, A. S., & Breyer, C. (2018). A cost optimal resolution for Sub-Saharan Africa powered by 100% renewables in 2030. *Renewable & Sustainable Energy Reviews*, *92*, 440–457. DOI: 10.1016/j.rser.2018.04.110

Beamon, B. (1999). Designing the green supply chain. *Logistics Information Management*, *12*(4), 332–342. DOI: 10.1108/09576059910284159

Berkhout, F., & Hertin, J. (2004). De-materialising and re-materialising: Digital technologies and the environment. *Futures*, *36*(8), 903–920. DOI: 10.1016/j.futures.2004.01.003

Bertrand, S. (2021). *Fact Sheet | Climate, Environmental, and Health Impacts of Fossil Fuels*. Environmental and Energy Study Institute (EESI). Retrieved 04/03/2025 from https://www.eesi.org/papers/view/fact-sheet-climate-environmental-and-health-impacts-of-fossil-fuels-2021

Bharali, A., & Kumar, S. (2018). E-waste management behaviour among urban households in India: An empirical investigation. *Journal of Materials*, C.

Bhatia, M. S., & Srivastava, R. K. (2018). Analysis of external barriers to remanufacturing using grey-DEMATEL approach: An Indian perspective. *Resources, Conservation and Recycling*, *136*, 79–87. DOI: 10.1016/j.resconrec.2018.03.021

Bhutta, M. K. S., Omar, A., & Yang, X. (2011). Electronic Waste: A Growing Concern in Today's Environment. *Economic Research International*, *2011*(1), 1–8. DOI: 10.1155/2011/474230

Bhutto, M. Y., Rūtelionė, A., Šeinauskienė, B., & Ertz, M. (2023). Exploring factors of e-waste recycling intention: The case of generation Y. *PLoS One*, *18*(10), e0287435. DOI: 10.1371/journal.pone.0287435 PMID: 37856490

Bigum, M., Damgaard, A., Scheutz, C., & Christensen, T. H. (2017). Environmental impacts and resource losses of incinerating misplaced household special wastes (WEEE, batteries, ink cartridges and cables). *Resources, Conservation and Recycling*, *122*, 251–260. DOI: 10.1016/j.resconrec.2017.02.013

Birtürk, A., Keskin, T., & Çeliktaş, M. S. (2023). The effect of thermal treatments on photovoltaic module recycling. In Aloui, F., Varuvel, E. G., & Sonthalia, A. (Eds.), *Handbook of Thermal Management Systems* (pp. 659–671). Elsevier. https://doi.org/https://doi.org/10.1016/B978-0-443-19017-9.00015-5

Biyouki, Z. A., Zaman, A., Marinova, D., Minunno, R., & Askari Shayegan, M. (2024). Solar Photovoltaics Value Chain and End-of-Life Management Practices: A Systematic Literature Review. *Sustainability*, *16*(16), 7038. https://www.mdpi.com/2071-1050/16/16/7038

Bogar, E. (2023). Chaos Game Optimization-Least Squares Algorithm for Photovoltaic Parameter Estimation. *Arabian Journal for Science and Engineering*, *48*(5), 6321–6340. https://doi.org/https://doi.org/10.1007/s13369-022-07364-6

Bogar, Z. O., Capraz, O., & Güngör, A. (2019). An Overview of Methods Used for Estimating E-waste Amount. In Prasad, M. N. V., & Vithanage, M. (Eds.), *Electronic Waste Management and Treatment Technology* (pp. 53–75). Butterworth-Heinemann. https://doi.org/https://doi.org/10.1016/B978-0-12-816190-6.00003-0

Borthakur, A., & Govind, M. (2016). Emerging trends in consumers' E-waste disposal behaviour and awareness: A worldwide overview with special focus on India. *Resources, Conservation and Recycling*. Advance online publication. DOI: 10.1016/j.resconrec.2016.11.011

Bošnjaković, M., Stojkov, M., Katinić, M., & Lacković, I. (2023). Effects of Extreme Weather Conditions on PV Systems. *Sustainability*, *15*(22), 16044. https://doi.org/https://doi.org/10.3390/su152216044

Boubellouta, B., & Kusch-Brandt, S. (2021). *Relationship between economic growth and mismanaged e-waste: Panel data evidence from 27 EU countries analyzed under the Kuznets curve hypothesis.* Waste Management, 85-97.

Boubellouta, B., & Kusch-Brandt, S. (2020). Testing the environmental Kuznets Curve hypothesis for e-waste in the EU28+2 countries. *Journal of Cleaner Production*, *277*, 1–11. DOI: 10.1016/j.jclepro.2020.123371

Boubellouta, B., & Kusch-Brandt, S. (2021a). Cross-country evidence on environmental Kuznets Curve in waste electrical and electronic equipment for 174 countries. *Sustainable Production and Consumption*, *25*, 136–151. DOI: 10.1016/j.spc.2020.08.006

Boubellouta, B., & Kusch-Brandt, S. (2022). Determinants of e-waste composition in the EU28+ 2 countries: A panel quantile regression evidence of the STIRPAT model. *International Journal of Environmental Science and Technology*, *19*(11), 1–18. DOI: 10.1007/s13762-021-03892-0

Boubellouta, B., & Kusch-Brandt, S. (2023). Driving factors of e-waste recycling rate in 30 European countries: New evidence using a panel quantile regression of the EKC hypothesis coupled with the STIRPAT model. *Environment, Development and Sustainability*, *25*(8), 7533–7560. DOI: 10.1007/s10668-022-02356-w

Bouma, A. J., van Wilgen, P., Lemmink, K. A., Stewart, R., Dijkstra, A., & Diercks, R. L. (2018). Barrier-belief lifestyle counselling in primary care: A randomized controlled trial of efficacy. *Patient Education and Counseling*, *101*(12), 2134–2144. DOI: 10.1016/j.pec.2018.07.015 PMID: 30072044

Bradu, P., Biswas, A., Nair, C., Sreevalsakumar, S., Patil, M., Kannampuzha, S., Mukherjee, A. G., Wanjari, U. R., Renu, K., Vellingiri, B., & Gopalakrishnan, A. V. (2023). Recent advances in green technology and Industrial Revolution 4.0 for a sustainable future. *Environmental Science and Pollution Research International*, *30*(60), 124488–124519. DOI: 10.1007/s11356-022-20024-4 PMID: 35397034

Breivik, K., Armitage, J. M., Wania, F., & Jones, K. C. (2019). Tracking the global generation and exports of e-waste. *Environmental Science & Technology*, *53*(13), 7425–7435. DOI: 10.1021/acs.est.9b00244 PMID: 25007134

Breusch, T., & Pagan, A. (1980). The lagrange multiplier test and its applications to model specification in econometrics. *The Review of Economic Studies*, *47*(1), 239–253. DOI: 10.2307/2297111

Brooks, C., & Hinich, M. (1998). Episodic nonstationarity in exchange rates. *Applied Economics Letters*, *5*(11), 719–722. DOI: 10.1080/135048598354203

Brown, M., & Smith, T. (2017). Sustainable Design and E-Waste Reduction. *Journal of Cleaner Production*, *142*(1), 327–339.

Brown, T., & Kumar, A. (2023). The environmental impact of wearable electronics: Carbon footprint and mitigation strategies. *Environmental Science & Technology*, *57*(3), 1892–1903.

Bułkowska, K., Zielińska, M., & Bułkowski, M. (2023). Implementation of blockchain technology in waste management. *Energies*, *16*(23), 7742. DOI: 10.3390/en16237742

Bułkowska, K., Zielińska, M., & Bułkowski, M. (2024). Blockchain-Based Management of Recyclable Plastic Waste. *Energies*, *17*(12), 2937. DOI: 10.3390/en17122937

Burns, K. N., Sayler, S. K., & Neitzel, R. L. (2019). Stress, health, noise exposures, and injuries among electronic waste recycling workers in Ghana. *Journal of Occupational Medicine and Toxicology (London, England)*, *14*(1), 1. DOI: 10.1186/s12995-018-0222-9 PMID: 30647766

Burns, K. N., Sun, K., Fobil, J. N., & Neitzel, R. L. (2016). Heart rate, stress, and occupational noise exposure among electronic waste recycling workers. *International Journal of Environmental Research and Public Health, 13*(1), 140. DOI: 10.3390/ijerph13010140 PMID: 26797626

California Department of Toxic Substances Control. (2021). *Electronic waste recycling act of 2003*. Author.

Campbell, F. W., Belding, S. R., Baron, R., Xiao, L., & Compton, R. G. (2009). The hydrogen evolution reaction at a silver nanoparticle array and a silver macroelectrode compared : Changed electrode kinetics between the macro- and nanoscales. *The Journal of Physical Chemistry. C, Nanomaterials and Interfaces, 33*(33), 14852–14857. DOI: 10.1021/jp904723b

Candrianto, C., Aimon, H., & Sentosa, S. U. (2023). The role of knowledge, awareness, and environmental attitudes in green product management. *Global Journal of Environmental Science and Management, 9*(1), 101–112. DOI: 10.22034/gjesm.2023.01.08

Cao, J., Lu, B., Chen, Y., Zhang, X., Zhai, G., Zhou, G., Jiang, B., & Schnoor, J. L. (2016). Extended producer responsibility system in China improves e-waste recycling: Government policies, enterprise, and public awareness. Renewable and Sustainable Energy Reviews, 62, 882–894. ttps://DOI: 10.1016/j.rser.2016.04.078

Capraz, O., Gungor, A., Mutlu, O., & Sagbas, A. (2024). Optimal sizing of grid-connected hybrid renewable energy systems without storage: A generalized optimization model. *Energy Sources. Part A, Recovery, Utilization, and Environmental Effects, 46*(1), 13317–13350. https://doi.org/https://doi.org/10.1080/15567036.2020.1803454

Carolan, M. S. (2020). *Society and the environment: Pragmatic solutions to ecological issues*. Routledge. DOI: 10.4324/9781003004912

Carter, C., & Carter, J. (1998). Interorganizational determinants of environmental purchasing: Initial evidence from the consumer products industries. *Decision Sciences, 29*(3), 659–684. DOI: 10.1111/j.1540-5915.1998.tb01358.x

Carter, C., & Ellram, L. (1998). Reverse logistics: A review of the literature and framework for future investigation. *Journal of Business Logistics, 19*(1), 85–102.

Carver, C. S., & Scheier, M. F. (1998). *On the self-regulation of behaviour*. Cambridge University Press.

Caspi, I. (2017). Rtadf Testing For Bubbles With Eviews. *Journal of Statistical Software, 81*(1), 1–16.

Castiglione, A., Cimmino, L., Nardo, M. D., & Murino, T. (2023). A framework for achieving a circular economy using the blockchain technology in a sustainable waste management system. *Computers & Industrial Engineering*, *180*, 109263. Advance online publication. DOI: 10.1016/j.cie.2023.109263

Castro, F. D., & Bassin, J. P. (2022). *"Electronic waste: Environmental risks and opportunities," in Hazard. Waste Manag*. Elsevier.

Ceballos, D. M., & Dong, Z. (2016). The formal electronic recycling industry: Challenges and opportunities in occupational and environmental health research. *Environment International*, *95*, 157–166. DOI: 10.1016/j.envint.2016.07.010 PMID: 27568575

Chancerel, P., Meskers, C. E. M., Hagelüken, C., & Rotter, V. S. (2009). Assessment of precious metal flows during preprocessing of waste electrical and electronic equipment. *Journal of Industrial Ecology*, *13*(5), 791–810. DOI: 10.1111/j.1530-9290.2009.00171.x

Chang, S., & Park, H. (2023). Carbon footprint analysis of 5G network infrastructure. *The International Journal of Life Cycle Assessment*, *28*(1), 55–69.

Chang, Y.-H., Yeh, C.-H., & Chang, Y.-W. (2013). A new method selection approach for fuzzy group multicriteria decision making. *Applied Soft Computing*, *13*(4), 2179–2187. DOI: 10.1016/j.asoc.2012.12.009

Chang, Y.-H., Yeh, C.-H., & Wang, S.-Y. (2007). A survey and optimization-based evaluation of development strategies for the air cargo industry. *International Journal of Production Economics*, *106*(2), 550–562. DOI: 10.1016/j.ijpe.2006.06.016

Chatterjee and Kumar. (2023). S. Murthy and S. Ramakrishna, "Global E-Waste Management: Urban Mining and Circular Economy,". *Springer Environmental Science and Pollution Research*, *29*(8), 11367–11385.

Chatterjee, A., & Abraham, J. (2017). Efficient management of e-wastes. *International Journal of Environmental Science and Technology*, *14*(1), 211–222. DOI: 10.1007/s13762-016-1072-6

Chatterjee, S. (2012). India's readiness on ROHS directives: A strategic analysis. *Global Journal of Science Frontier Research*, *10*(1), 14–26.

Chatterjee, S., Kumar, V., & Gupta, D. (2016). E-waste management in India: Issues and strategies. *Resources, Conservation and Recycling*, *107*, 28–36.

Chaturvedi, P., Kulshreshtha, K., & Tripathi, V. (2020). Investigating the determinants of behavioral intentions of Generation Z for recycled clothing: An evidence from a developing economy. *Young Consumers, 21*(4), 403–417. DOI: 10.1108/YC-03-2020-1110

Chawla, R., Singhal, P., & Garg, A. K. (2020). Photovoltaic Review of all Generations: Environmental Impact and Its Market Potential. *Transactions on Electrical and Electronic Materials, 21*(5), 456-476. https://doi.org/https://doi.org/10.1007/s42341-020-00217-9

Cheema, H. A., Ilyas, S., Kang, H., & Kim, H. (2024). Comprehensive review of the global trends and future perspectives for recycling of decommissioned photovoltaic panels. *Waste Management, 174*, 187-202. https://doi.org/https://doi.org/10.1016/j.wasman.2023.11.025

Chen, A., Dietrich, K. N., Huo, X., & Ho, S. M. (2011). Developmental neurotoxicants in e waste: An emerging health concern. *Environmental Health Perspectives, 119*(4), 431–438. DOI: 10.1289/ehp.1002452 PMID: 21081302

Chen, I.-S. (2012). A revised Inno-Qual performance system for higher education: The integrated applications of DEMATEL and ANP. *The Journal of the Operational Research Society, 63*(4), 478–488. DOI: 10.1057/jors.2011.57

Chen, J., Huang, Y., Li, G., An, T., Hu, Y., & Li, Y. (2016). VOCs elimination and health risk reduction in e-waste dismantling workshop using integrated techniques of electrostatic precipitation with advanced oxidation technologies. *Journal of Hazardous Materials, 302*, 395–403. DOI: 10.1016/j.jhazmat.2015.10.006 PMID: 26489914

Chen, L., & Zhang, Y. (2021). AI and Robotics in E-Waste Management: Enhancing Recovery Efficiency. *Robotics and Automation Journal, 29*(2), 189–204.

Chen, P.-H., Chen, W.-S., Lee, C.-H., & Wu, J.-Y. (2024). Comprehensive Review of Crystalline Silicon Solar Panel Recycling: From Historical Context to Advanced Techniques. *Sustainability, 16*(1), 60. https://www.mdpi.com/2071-1050/16/1/60

Chen, Y., & Lin, Z. (2023). Assessing the carbon footprint of smart home devices: A comparative study. *Journal of Sustainable Technology, 21*(4), 567–582.

Chen, Y., & Wang, X. (2020). Recycling Technologies for E-Waste: Challenges and Opportunities. *Waste Management & Research, 38*(4), 445–460.

Chen, Z., Ming, X., Zhang, X., Yin, D., & Sun, Z. (2019). A rough-fuzzy DEMATEL-ANP method for evaluating sustainable value requirement of product service system. *Journal of Cleaner Production, 228*, 485–508. DOI: 10.1016/j.jclepro.2019.04.145

Chertow, M. (2000). Industrial symbiosis: Literature and taxonomy. *Annual Review of Environment and Resources, 25*, 313–337.

Chiang, P. F., Han, S., Claire, M. J., Maurice, N. J., Vakili, M., & Giwa, A. S. (2024). Sustainable Treatment of Spent Photovoltaic Solar Panels Using Plasma Pyrolysis Technology and Its Economic Significance. *Clean Technologies, 6*(2), 432-452. https://doi.org/https://doi.org/10.3390/cleantechnol6020022

Chi, X., Streicher-Porte, M., Wang, M. Y. L., & Reuter, M. A. (2011). Informal electronic waste recycling: A sector review with special focus on China. *Waste Management (New York, N.Y.), 31*(4), 731–742. DOI: 10.1016/j.wasman.2010.11.006 PMID: 21147524

Chi, X., Wang, W. Y. L., & Reuter, M. A. (2014). E-waste collection channels and household recycling behaviors in Taizhou of China. *Journal of Cleaner Production, 80*, 87–95. DOI: 10.1016/j.jclepro.2014.05.056

Choi, J.-H., Yoo, H.-E., Chung, H., Lee, H.-S., Lee, M.-J., & Chang, H.-J. (2018). Evaluation of Effectiveness of Food Safety Education Program Based on Social Cognitive Theory for the Elderly in the Middle Class. Academic Press.

Chou, C. C. (2003). The canonical representation of multiplication operation on triangular fuzzy numbers. *Computers & Mathematics with Applications (Oxford, England), 45*(10–11), 1601–1610. DOI: 10.1016/S0898-1221(03)00139-1

Chowdhury, M. S., Rahman, K. S., Chowdhury, T., Nuthammachot, N., Techato, K., Akhtaruzzaman, M., Tiong, S. K., Sopian, K., & Amin, N. (2020). An overview of solar photovoltaic panels' end-of-life material recycling. *Energy Strategy Reviews, 27*, 100431. https://doi.org/https://doi.org/10.1016/j.esr.2019.100431

Christensen, T. B., Johansen, M. R., Buchard, M. V., & Glarborg, C. N. (2022). Closing the material loops for construction and demolition waste: The circular economy on the island Bornholm, Denmark. *Resources, Conservation & Recycling Advances, 15*, 200104. DOI: 10.1016/j.rcradv.2022.200104

Chung, S.-s., Lau, K.-y., & Zhang, C. (2011). Generation of and control measures for, e-waste in Hong Kong. *Waste Management, 31*(3), 544-554. https://doi.org/https://doi.org/10.1016/j.wasman.2010.10.003

Clarkson, T. W., & Magos, L. (2006). The toxicology of mercury and its chemical compounds. *Critical Reviews in Toxicology, 36*(8), 609–662. DOI: 10.1080/10408440600845619 PMID: 16973445

Cleaning Up Electronic Waste (E-Waste). (n.d.). US EPA. https://www.epa.gov/international-cooperation/cleaning-electronic-waste-E-waste

CMU (Carnegie Mellon University). (2011). *EIO-LCA: Free*. Fast, Easy Life Cycle Assessment.

Cobbing, M. (2008). *Toxic Tech: Not in Our Backyard Uncovering the Hidden Flows of e-Waste*. Greenpeace International. https://www.greenpeace.org/global/

Cole, C., Gnanapragasam, A., Cooper, T., & Singh, J. (2019). An assessment of achievements of the WEEE Directive in promoting movement up the waste hierarchy: Experiences in the UK. *Waste Management (New York, N.Y.)*, *87*, 417–427. DOI: 10.1016/j.wasman.2019.01.046 PMID: 31109542

Cole, C., Gnanapragasam, A., Cooper, T., & Singh, J. (2019). Assessing barriers to reuse of electrical and electronic equipment: A UK perspective. *Resources, Conservation and Recycling, X. 1*, 100004. DOI: 10.1016/j.rcrx.2019.100004

Corwin, J. E. (2018). "Nothing is useless in nature": Delhi's repair economies and valuecreation in an electronics "waste" sector. *Environment and Planning A.Economy and Space*, *50*(1), 14–30.

Costa, M. M., Neto, J. F. B., Alberte, E. P. V., & Carneiro, A. P. (2022, November). Blockchain-based framework for improving waste management and circular economy in construction. *IOP Conference Series. Earth and Environmental Science*, *1101*(6), 062009. DOI: 10.1088/1755-1315/1101/6/062009

Crocker, R. (2013). *From access to excess: consumerism, 'compulsory' consumption and behaviour change. In motivating change: sustainable design and behaviour in the built environment*. Routledge Publishing.

Cudjoe, D., & Acquah, P. M. (2021). Environmental impact analysis of municipal solid waste incineration in African countries. *Chemosphere*, *265*, 129186. DOI: 10.1016/j.chemosphere.2020.129186 PMID: 33307505

Dagan, R., Dubey, B., Bitton, G., & Townsend, T. (2007). Aquatic toxicity of leachates generated from electronic devices. *Archives of Environmental Contamination and Toxicology*, *53*(2), 168–173. DOI: 10.1007/s00244-006-0205-1 PMID: 17549543

Dambhare, M. V., Butey, B., & Moharil, S. V. (2021). Solar photovoltaic technology: A review of different types of solar cells and its future trends. *Journal of Physics: Conference Series*, *1913*(1), 012053. https://doi.org/https://doi.org/10.1088/1742-6596/1913/1/012053

Damgaard, A., & Christensen, T. H. (2011). Recycling of metals. In *Solid waste technology and management* (pp. 234–242). Wiley.

Das, S., Lee, S. H., Kumar, P., Kim, K.-H., Lee, S. S., & Bhattacharya, S. S. (2019). Solid waste management: Scope and the challenge of sustainability. *Journal of Cleaner Production*, *228*, 658–678. DOI: 10.1016/j.jclepro.2019.04.323

Dat, L., Linh, D., Chou, S., & Yu, V. (2012). Optimizing reverse logistic costs for recycling end-of-life electrical and electronic products". *Expert Systems with Applications*, *39*(7), 6380–6387. DOI: 10.1016/j.eswa.2011.12.031

Daugherty, P., Richey, R., Genchev, S., & Chen, H. (2005). Reverse logistics: Superior performance through focused resource commitments to information technology. *Transportation Research Part E, Logistics and Transportation Review*, *41*(2), 77–92. DOI: 10.1016/j.tre.2004.04.002

Davis, J. M., Akese, G., & Garb, Y. (2019). Beyond the pollution haven hypothesis: Where and why do e-waste hubs emerge and what does this mean for policies and interventions? *Geoforum*, *98*, 36–45. DOI: 10.1016/j.geoforum.2018.09.020

Davis, J., & Wong, A. (2023). Carbon emissions from data centers: Trends and reduction strategies. *Journal of Environmental Impact Assessment*, *15*(2), 243–259.

De Sousa, N. M., Oliveira, C. B., & Cunha, D. (2023). Photovoltaic electronic waste in Brazil: Circular economy challenges, potential and obstacles. *Social Sciences & Humanities Open*, *7*(1), 100456. https://doi.org/https://doi.org/10.1016/j.ssaho.2023.100456

de Souza, R. G., Clímaco, J. C., Sant'Anna, A. P., Rocha, T. B., do Valle, R. A., & Quelhas, O. L. (2016). Sustainability assessment and prioritisation of e-waste management options in Brazil. *Waste Management (New York, N.Y.)*, *57*, 46–56. DOI: 10.1016/j.wasman.2016.01.034 PMID: 26852754

Debnath, B., Saha, I., Mukherjee, T., Mitra, S., Das, A., & Das, A. (2021). Sorbents from waste materials: a circular economic approach. In *Sorbents Materials for Controlling Environmental Pollution* (pp. 285–322). Elsevier. DOI: 10.1016/B978-0-12-820042-1.00028-6

Decharat, S. (2018). Urinary mercury levels among workers in E-waste shops in Nakhon Si Thammarat Province, Thailand. *Journal of Preventive Medicine and Public Health = Yebang Uihakhoe Chi*, *51*(4), 196–204. DOI: 10.3961/jpmph.18.049 PMID: 30071707

Decharat, S., & Kiddee, P. (2020). Health problems among workers who recycle electronic waste in southern Thailand. *Osong Public Health and Research Perspectives*, *11*(1), 34–43. DOI: 10.24171/j.phrp.2020.11.1.06 PMID: 32149040

Delcea, C., Crăciun, L., Ioană , C., Ferruzzi, G., & Cotfas, L. A. (2020). Determinants of individuals' E-waste recycling decision: A case study from Romania. *Sustainability (Basel), 12*(7), 2753. DOI: 10.3390/su12072753

Demircan, B. G., & Yetilmezsoy, K. (2023). A hybrid fuzzy AHP-TOPSIS approach for implementation of smart sustainable waste management strategies. *Sustainability (Basel), 15*(8), 6526. DOI: 10.3390/su15086526

Demirel, E., Demirel, N., & Gökçen, H. (2016). A mixed integer linear programming model to optimize reverse logistics activities of end-of-life vehicles in Turkey. *Journal of Cleaner Production, 112*, 2101–2113. DOI: 10.1016/j.jclepro.2014.10.079

Demir, M. A., & Yakışık, H. (2024). Historical Development of Energy. *Journal of the Human and Social Science Researches, 13*(3), 1333–1353. https://doi.org/https://doi.org/10.15869/itobiad.1500506

Deng, R., Zhuo, Y., & Shen, Y. (2022). Recent progress in silicon photovoltaic module recycling processes. *Resources, Conservation and Recycling, 187*, 106612. https://doi.org/https://doi.org/10.1016/j.resconrec.2022.106612

Deng, W. J., Zheng, J. S., Bi, X. H., Fu, J. M., & Wong, M. H. (2007). Distribution of PBDEs in air particles from an electronic waste recycling site compared with Guangzhou and Hong Kong, South China. *Environment International, 33*(8), 1063–1069. DOI: 10.1016/j.envint.2007.06.007 PMID: 17631964

Deoghare, P. R., & Shukla, S. (2018). Analysis of e-waste disposal behaviour among Indian IT professionals: An empirical study. *Procedia Computer Science, 133*, 423–431.

Desai, N., & Pathak, M. (2024). End of Life Solar Panels: Assessment for Gujarat, India. https://doi.org/https://doi.org/10.2139/ssrn.4693717

Dhir, A., Koshta, N., Goyal, R. K., Sakashita, M., & Almotairi, M. (2021). Behavioral reasoning theory (BRT) perspectives on e-waste recycling and management. *Journal of Cleaner Production, 280*, 124269. DOI: 10.1016/j.jclepro.2020.124269

Dias, P., Javimczik, S., Benevit, M., & Veit, H. (2017). Recycling WEEE: Polymer characterization and pyrolysis study for waste of crystalline silicon photovoltaic modules. *Waste Management, 60*, 716-722. https://doi.org/https://doi.org/10.1016/j.wasman.2016.08.036

Dias, P., Javimczik, S., Benevit, M., Veit, H., & Bernardes, A. M. (2016). Recycling WEEE: Extraction and concentration of silver from waste crystalline silicon photovoltaic modules. *Waste Management, 57*, 220-225. https://doi.org/https://doi.org/10.1016/j.wasman.2016.03.016

Dias, P., Schmidt, L., Gomes, L. B., Bettanin, A., Veit, H., & Bernardes, A. M. (2018). Recycling Waste Crystalline Silicon Photovoltaic Modules by Electrostatic Separation. *Journal of Sustainable Metallurgy*, *4*(2), 176–186. https://doi.org/https://doi.org/10.1007/s40831-018-0173-5

Ding, Y., Zhang, S., Liu, B., Zheng, H., Chang, C., & Ekberg, C. (2019). Recovery of precious metals from electronic waste and spent catalysts: A review. *Resources, Conservation and Recycling*, *141*, 284–298. DOI: 10.1016/j.resconrec.2018.10.041

Disposal of Asbestos, Chemicals, Oil, Trade and e-Waste. (n.d.). Parkes Shire Council. https://www.parkes.nsw.gov.au/environment/waste-recycling/waste-facilities/disposal-of-chemicals-oil-trade-and-E-waste/

Domínguez, A., & Geyer, R. (2017). Photovoltaic waste assessment in Mexico. *Resources, Conservation and Recycling*, *127*, 29–41. https://doi.org/https://doi.org/10.1016/j.resconrec.2017.08.013

Domínguez, A., & Geyer, R. (2019). Photovoltaic waste assessment of major photovoltaic installations in the United States of America. *Renewable Energy*, *133*, 1188–1200. https://doi.org/https://doi.org/10.1016/j.renene.2018.08.063

Dong, Y., & Xu, L. (2023). Carbon footprint analysis of laptop manufacturing: A cradle-to-gate approach. *Journal of Industrial Ecology*, *27*(1), 101–115.

Donou-Adonsou, F. (2019). Technology, education, and economic growth in Sub-Saharan Africa. *Telecommunications Policy*, *43*(4), 353–360. DOI: 10.1016/j.telpol.2018.08.005

Dou, Y., & Sarkis, J. (2013). A multiple stakeholder perspective on barriers to implementing China RoHS regulations. Resources Conservation and Recycling, 81, 92–104. ttps://DOI: 10.1016/j.resconrec.2013.10.004

Dougherty, C. M., Johnson-Crowley, N. R., Lewis, F. M., & Thompson, E. A. (2001). Theoretical development of nursing interventions for sudden cardiac arrest survivors using social cognitive theory. *Advances in Nursing Science*, *24*(1), 78–86. DOI: 10.1097/00012272-200109000-00009 PMID: 11554535

Dowlatshahi, S. (2005). A strategic framework for the design and implementation of remanufacturing operations in reverse logistics. *International Journal of Production Research*, *43*(16), 3455–3480. DOI: 10.1080/00207540500118118

Duman, G. M., Kongar, E., & Gupta, S. M. (2019). Estimation of electronic waste using optimized multivariate grey models. *Waste Management, 95*, 241-249. https://doi.org/https://doi.org/10.1016/j.wasman.2019.06.023

Duman, G. M., Kongar, E., & Gupta, S. M. (2020). Predictive analysis of electronic waste for reverse logistics operations: A comparison of improved univariate grey models. *Soft Computing*, *24*(20), 15747–15762. https://doi.org/https://doi.org/10.1007/s00500-020-04904-w

Duran, A. S., Atasu, A., & Van Wassenhove, L. N. (2022). Cleaning after solar panels: Applying a circular outlook to clean energy research. *International Journal of Production Research*, *60*(1), 211–230. https://doi.org/https://doi.org/10.1080/00207543.2021.1990434

Dwivedy, M., & Mittal, R. K. (2010). Future trends in computer waste generation in India. *Waste Management, 30*(11), 2265-2277. https://doi.org/https://doi.org/10.1016/j.wasman.2010.06.025

Echegaray, F. (2016). Assessing the intention-behaviour gap in electronic waste recycling: The case of Brazil. *Journal of Cleaner Production*, 1–11. DOI: 10.1016/j.jclepro.2016.05.064

Eckhardt, B., & Kaifie, A. (2024). Bridging the knowledge gap! Health outcomes in informal e-waste workers. *Journal of Occupational Medicine and Toxicology (London, England)*, *19*(1), 11. DOI: 10.1186/s12995-024-00410-z PMID: 38622584

Edquist, H., & Henrekson, M. (2017). Swedish lessons: How important are ICT and R&D to economic growth? *Structural Change and Economic Dynamics*, *42*, 1–12. DOI: 10.1016/j.strueco.2017.05.004

Eguchi, A., Isobe, T., Ramu, K., Tue, N. M., Sudaryanto, A., Devanathan, G., Viet, P. H., Tana, R. S., Takahashi, S., Subramanian, A., & Tanabe, S. (2013). Soil contamination by brominated flame retardants in open waste dumping sites in Asian developing countries. *Chemosphere*, *90*(9), 2365–2371. DOI: 10.1016/j.chemosphere.2012.10.027 PMID: 23149186

Electronic Products Recycling Association (EPRA). (2022). *Annual report on electronic waste recycling in Canada*. Author.

El-Ghiaty, M. A., & El-Kadi, A. O. (2021). Arsenic: Various species with different effects on cytochrome P450 regulation in humans. EXCLI J., 20, 1184. https://eur-lex. europa.eu/

El-Khawad, L., Bartkowiak, D., & Kümmerer, K. (2022). Improving the end-of-life management of solar panels in Germany. *Renewable & Sustainable Energy Reviews*, *168*, 112678. https://doi.org/https://doi.org/10.1016/j.rser.2022.112678

Elkington, J. (1997). *Cannibals with Forks: The Triple Bottom Line of 21st Century Business*. Capstone Publishing.

Endres, A., & Radke, V. (2018). *Economics for Environmental Studies: A Strategic Guide to Micro- and Macroeconomics* (2nd ed.). Springer. DOI: 10.1007/978-3-662-54828-8

EPA. (2024). *End-of-Life Solar Panels: Regulations and Management*. United States Environmental Protection Agency. Retrieved 04/03/2024 from https://www.epa.gov/hw/end-life-solar-panels-regulations-and-management

Erumban, A., & Das, D. (2016). Information and communication technology and economic growth in India. *Telecommunications Policy*, *40*(5), 412–431. DOI: 10.1016/j.telpol.2015.08.006

European Commission. (2020). *Waste electrical and electronic equipment (WEEE) directive*. https://ec.europa.eu/environment/topics/waste-and-recycling/weee_en

European Environment Agency. (2020). *E-waste management in Europe: A circular economy perspective*. https://www.eea.europa.eu/publications

Evans, K., & Walker, G. (2019). Blockchain for E-Waste Tracking and Transparency. *Journal of Technology Management*, *56*(4), 288–302.

Fadhullah, W., Imran, N. I. N., Ismail, S. N. S., Jaafar, M. H., & Abdullah, H. (2022). Household solid waste management practices and perceptions among residents in the East Coast of Malaysia. *BMC Public Health*, *22*(1), 1–20. DOI: 10.1186/s12889-021-12274-7 PMID: 34983455

Fang, W. T., Hassan, A. A., & LePage, B. A. (2022). Environmental ethics: Modelling for values and choices. In *The Living Environmental Education: Sound Science Toward a Cleaner, Safer, and Healthier Future* (pp. 151–174). Singapore: Springer Nature Singapore. DOI: 10.1007/978-981-19-4234-1_6

Fang, H., & Li, J. (2023). Carbon footprint of electric vehicles: The role of electronic components. *Renewable & Sustainable Energy Reviews*, *169*, 112878.

Farji, M. (2021). Development of Photovoltaic Cells: A Materials Prospect and Next-Generation Futuristic Overview. *Brazilian Journal of Physics*, *51*(6), 1916–1928. https://doi.org/https://doi.org/10.1007/s13538-021-00981-w

Fatimah, Y. A., Govindan, K., Murniningsih, R., & Setiawan, A. (2020). Industry 4.0 based sustainable circular economy approach for smart waste management system to achieve sustainable development goals: A case study of Indonesia. *Journal of Cleaner Production*, *269*, 122263. DOI: 10.1016/j.jclepro.2020.122263

Fent, K. (2004). Ecotoxicological effects at contaminated sites. *Toxicology*, *205*(3), 223–240. DOI: 10.1016/j.tox.2004.06.060 PMID: 15464632

Ferguson, M., & Toktay, L. (2006). The Effect of Competition on Recovery Strategies. *Production and Operations Management, 15*(3), 351–368. DOI: 10.1111/j.1937-5956.2006.tb00250.x

Fernandez-Portillo, A., Almodovar-Gonzalez, M., & Hernandez-Mogollon, R. (2020). Impact of ICT development on economic growth. A study of OECD European union countries. *Technology in Society, 63*, 1–9. DOI: 10.1016/j.techsoc.2020.101420

Fiandra, V., Sannino, L., Andreozzi, C., & Graditi, G. (2019). End-of-life of silicon PV panels: A sustainable materials recovery process. *Waste Management, 84*, 91-101. https://doi.org/https://doi.org/10.1016/j.wasman.2018.11.035

Fleischmann, M., Bloemhof-Ruwaard, J. M., Dekker, R., Laan, E., Nunen, J. A., & Wassenhove, L. (1997). Quantitative models for reverse logistics: A review. *European Journal of Operational Research, 103*(1), 1–17. DOI: 10.1016/S0377-2217(97)00230-0

Flygansvaer, B. (1999). The need for cooperation and coordination in a reverse distribution channel - a model. In Proceedings of the second international working seminar on re-use (pp. 55-70). Eindhoven University of Technology Press.

Fogarasi, S., Imre-Lucaci, F., Egedy, A., Imre-Lucaci, Á., & Ilea, P. (2015). Eco-friendly copper recovery process from waste printed circuit boards using Fe^{3+}/Fe^{2+} redox system. *Waste Management (New York, N.Y.), 40*, 136–143. DOI: 10.1016/j.wasman.2015.02.030 PMID: 25816768

Foon, T. (2008). A re-examination of the role of foreign direct investment and exports in Malaysia's economic growth: A time series analysis, 1970-2006. *International Journal of Management Studies, 15*, 47–67.

Forti, V., Baldé, C. P., Kuehr, R., & Bel, G. (2020). The global e-waste monitor 2020. United Nations University (UNU), International Telecommunication Union (ITU) & International Solid Waste Association (ISWA). https://www.greene.gov.in/wpcontent/uploads/2020/12/2020120929.pdf

Forti, V., Balde, C. P., Kuehr, R., & Bel, G. (2020). *The Global E-waste Monitor 2020: Quantities, flows and the circular economy potential.* United Nations University (UNU)/United Nations Institute for Training and Research (UNITAR), International Telecommunication Union (ITU), International Solid Waste Association (ISWA). https://collections.unu.edu/

Forti, V., Baldé, C. P., Kuehr, R., & Bel, G. (2020). *The global e-waste monitor 2020: Quantities, flows and the circular economy potential.* United Nations University. https://collections.unu.edu/eserv/UNU:7737/PDF

Forti, V., Baldé, C. P., Kuehr, R., & Bel, G. (2020). The global e-waste monitor. Quantities, flows and the circular economy potential. https://www.itu.int/en/ITU-D/Environment/Documents/Toolbox/GEM_2020_def.pdf

Forti, V., Baldé, C., Kuehr, R. and Bel, G. "The Global E-waste Monitor 2020: Quantities, Flows, and the Circular Economy Potential", United Nations University/ United Nations Institute for Training and Research, International Telecommunication Union, and International Solid Waste Association.

Forti, V., Baldé, C. P., Kuehr, R., & Bel, G. (2020). *The Global E-waste Monitor 2020. United Nations University (UNU), International Telecommunication Union (ITU) & International Solid Waste Association*. ISWA.

Forti, V., Baldé, K., & Kuehr, R. (2018). *E-waste Statistics: Guidelines on Classifications* (2nd ed.). Reporting and Indicators.

Franzoni, A., Leggerini, C., & Bannò, M. (2024). A Predictive Framework for Photovoltaic Waste Quantities and Recovery Values: Insights and Application to the Italian Context. *Environmental and Climate Technologies, 28*(1), 243-257. https://doi.org/https://doi.org/10.2478/rtuect-2024-0020

Fraunhofer, I. S. E. (2024). *Photovoltaics Report*. Fraunhofer Institute for Solar Energy Systems, ISE with the support of PSE Projects GmbH. https://www.ise.fraunhofer.de/

Freitas, M. B. J. G., & Garcia, E. M. (2007). Electrochemical recycling of cobalt from cathodes of spent lithium–ion batteries. *Journal of Power Sources*, *171*(2), 953–959. DOI: 10.1016/j.jpowsour.2007.07.002

Frioui, S., Oumeddour, R., & Lacour, S. (2017). Highly selective extraction of metal ions from dilute solutions by hybrid electrodialysis technology. *Separation and Purification Technology*, *174*, 264–274. DOI: 10.1016/j.seppur.2016.10.028

Fu, F., Liu, Y., Wang, Y., & Chen, J. (2017). Analysis of the influencing factors on individuals' recycling intention for e-waste: A case study of China. *Journal of Cleaner Production*, *162*, 1566–1575.

Fujimori, T., Itai Goto, A., Asante, K. A., Otsuka, M., Takahashi, S., & Tanabe, S. (2016). Interplay of metals and bromine with dioxin-related compounds concentrated in e-waste open burning soil from Agbogbloshie in Accra, Ghana. *Environmental Pollution*, *201*, 155–163. DOI: 10.1016/j.envpol.2015.11.031 PMID: 26686056

Fullerton, D., & Metcalf, G. (1997). *Environmental Taxes and The Double-Dividend Hypothesis: Did You Really Expect Something for Nothing?* Cambridge: NBER Working Paper No 6199.

Gabus, A., & Fontela, E. (1973). Perceptions of the world problematique: Communication procedure, communicating with those bearing collective responsibility (DEMATEL Report No. 1). Battelle Geneva Research Centre.

Gabus, A., & Fontela, E. (1972). *World problems, an invitation to further thought within the framework of DEMATEL.* Battelle Geneva Research Centre.

Ganguly, R. (2016). E-Waste management in India-An overview. *International Journal of Earth Sciences and Engineering*, *9*(2), 574–588.

Gangwar, C., Choudhari, R., Chauhan, A., Kumar, A., Singh, A., & Tripathi, A. (2019). Assessment of air pollution caused by illegal e-waste burning to evaluate the human health risk. *Environment International*, *125*, 191–199. DOI: 10.1016/j.envint.2018.11.051 PMID: 30721825

Gao, C., Wang, X., & Zhang, M. (2016). Investigating the determinants of e-waste recycling behaviour: Evidence from a Chinese household survey. *Journal of Cleaner Production*, *136*, 141–150.

Gao, M., & Chen, Y. (2024). Multi-agent collaborative model of e-waste recycling. *Journal of Environmental Management*, *349*, 119457. DOI: 10.1016/j.jenvman.2023.119457 PMID: 37918229

Gao, S., & Zhang, M. (2023). Carbon footprint reduction in electronics manufacturing: A green supply chain approach. *Journal of Cleaner Production*, *380*, 134596.

Gardas, B. B., Raut, R. D., & Narkhede, B. (2018). Modelling the challenges to sustainability in the textile and apparel (T&A) sector: A Delphi-DEMATEL approach. *Sustainable Production and Consumption*, *15*, 96–108. DOI: 10.1016/j.spc.2018.05.001

Garlapati, V. K. (2016). E-waste in India and developed countries: Management, recycling, business and biotechnological initiatives. *Renewable & Sustainable Energy Reviews*, *54*(C), 874–881. DOI: 10.1016/j.rser.2015.10.106

Gautam, A., Shankar, R., & Vrat, P. (2021). End-of-life solar photovoltaic e-waste assessment in India: A step towards a circular economy. *Sustainable Production and Consumption*, *26*, 65–77. https://doi.org/https://doi.org/10.1016/j.spc.2020.09.011

Gazzeh, K., Abubakar, I. R., & Hammad, E. (2022). Impacts of COVID-19 Pandemic on the Global Flows of People and Goods: Implications on the Dynamics of Urban Systems. *Land (Basel)*, *11*(3), 429. DOI: 10.3390/land11030429

Gebremariam, A. T., Di Maio, F., Vahidi, A., & Rem, P. (2020). Innovative technologies for recycling End-of-Life concrete waste in the built environment. *Resources, Conservation and Recycling, 163*, 104911. DOI: 10.1016/j.resconrec.2020.104911

Geng, Y., & Doberstein, B. (2008). Developing the circular economy in China: Challenges and opportunities for achieving 'leapfrog development. *International Journal of Sustainable Development and World Ecology, 15*(3), 231–239. DOI: 10.3843/SusDev.15.3:6 PMID: 37309529

Gerold, E., & Antrekowitsch, H. (2024). Advancements and Challenges in Photovoltaic Cell Recycling: A Comprehensive Review. *Sustainability, 16*(6), 2542. https://doi.org/https://doi.org/10.3390/su16062542

Ghimire, H., & Ariya, P. A. (2020). E-wastes: Bridging the knowledge gaps in global production budgets, composition, recycling and sustainability implications. *Sustainable Chemistry, 1*(2), 154–182. DOI: 10.3390/suschem1020012

Ghosh, S. (2016). Does mobile telephony spur growth? Evidence from Indian states. *Telecommunications Policy, 40*(10-11), 1020–1031. DOI: 10.1016/j.telpol.2016.05.009

Gibovic, D., & Bikfalvi, A. (2021). Incentives for plastic recycling: How to engage citizens in active collection. Empirical evidence from Spain. *Recycling, 6*(2), 29. DOI: 10.3390/recycling6020029

Gist, M. E., & Mitchell, T. R. (1992). Self-efficacy: A theoretical analysis of its determinants and malleability. *Academy of Management Review, 17*(2), 183–211. DOI: 10.2307/258770

Golev, A., Schmeda-Lopez, D. R., Smart, S. K., Corder, G. D., & McFarland, E. W. (2016). Where next on e-waste in Australia? *Waste Management, 58*, 348-358. https://doi.org/https://doi.org/10.1016/j.wasman.2016.09.025

Golev, A., Schmeda-Lopez, D. R., Smart, S. K., Corder, G. D., & McFarland, E. W. (2016). Where next on e-waste in Australia? *Waste Management (New York, N.Y.), 58*, 348–358. DOI: 10.1016/j.wasman.2016.09.025 PMID: 27687078

Gollakota, A. R., Gautam, S., & Shu, C. M. (2020). Inconsistencies of e-waste management in developing nations-Facts and plausible solutions. *Journal of Environmental Management, 261*, 261. DOI: 10.1016/j.jenvman.2020.110234 PMID: 32148304

Gollwitzer, P. M. (1999). Implementation intentions: Strong effects of simple plans. *The American Psychologist, 54*(7), 493–503. DOI: 10.1037/0003-066X.54.7.493

Golsteijn, L., & Valencia Martinez, E. (2017). The Circular Economy of E-Waste in the Netherlands: Optimizing Material Recycling and Energy Recovery. *Journal of Engineering*, *2017*(1), 8984013. https://doi.org/https://doi.org/10.1155/2017/8984013

Gonda, L., D'Ans, P., & Degrez, M. (2019). A comparative assessment of WEEE collection in an urban and rural context: Case study on desktop computers in Belgium. *Resources, Conservation and Recycling*, *142*, 131–142. https://doi.org/https://doi.org/10.1016/j.resconrec.2018.11.008

Gopalakrishnan, P. K., Hall, J., & Behdad, S. 2020, August. A blockchain-based traceability system for waste management in smart cities. In International design engineering technical conferences and computers and information in engineering conference (Vol. 83952, p. V006T06A015). American Society of Mechanical Engineers. DOI: 10.1115/DETC2020-22553

Gopalakrishnan, P. K., Hall, J., & Behdad, S. (2021). Cost analysis and optimization of Blockchain-based solid waste management traceability system. *Waste Management (New York, N.Y.)*, *120*, 594–607. DOI: 10.1016/j.wasman.2020.10.027 PMID: 33288397

Gopalakrishnan, P., & Ramaguru, R. (2019). Blockchain based waste management. *International Journal of Engineering and Advanced Technology*, *8*(5), 2632–2635.

Goulder, L. (1995). Environmental taxation and the "double dividend": A reader's guide. In Public economics and the environment in an imperfect world (pp. 277-314). New York: Springer Science+Business Media.

Govindan, K., Shankar, K. M., & Kannan, D. (2016). Sustainable material selection for construction industry – a hybrid multi criteria decision making approach. *Renewable & Sustainable Energy Reviews*, *55*, 1274–1288. DOI: 10.1016/j.rser.2015.07.100

Goyal, N., & Goyal, D. (2020). Exploring E-waste Management: Strategies and Implications. Handbook of Solid Waste Management: Sustainability through Circular Economy, 1-14.

Grant, K., Goldizen, F. C., Sly, P. D., Brune, M. N., Neira, M., van den Berg, M., & Norman, R. E. (2013). Health consequences of exposure to e-waste: A systematic review. *The Lancet. Global Health*, *1*(6), e350–e361. DOI: 10.1016/S2214-109X(13)70101-3 PMID: 25104600

Guide, V., Souza, G., Wassenhove, L., & Blackburn, J. (2006). Time value of commercial product returns. *Management Science*, *8*(8), 1200–1214. DOI: 10.1287/mnsc.1060.0522

Guide, V., & Wassenhove, L. (2009). The evolution of closed-loop supply chain research. *Operations Research*, *57*(1), 10–18. DOI: 10.1287/opre.1080.0628

Gulyamov, S. (2024). Intelligent waste management using IoT, blockchain technology and data analytics. In *E3S Web of Conferences* (Vol. 501, p. 01010). EDP Sciences. DOI: 10.1051/e3sconf/202450101010

Gummadi, A. K., & Damacharla, P. (2024). An Intelligent Framework for Energy Management in IoT Networks Using Machine Learning. *Proceedings of the 5th International Conference on Smart Electronics and Communication (ICOSEC)*, 765-771.

Gumus, S., Egilmez, G., Kucukvar, M., & Park, Y. S. (2016). Integrating expert weighting and multi-criteria decision making into eco-efficiency analysis: The case of US manufacturing. *The Journal of the Operational Research Society*, *67*(4), 616–628. DOI: 10.1057/jors.2015.88

Güngördü Belbağ, A., & Belbağ, S. (2025). Consumer intentions to purchase remanufactured products: A behavioral reasoning theory approach. *Journal of Social Marketing*, *15*(1), 1–19. DOI: 10.1108/JSOCM-02-2024-0033

Guo, X., & Yan, K. (2017). Estimation of obsolete cellular phones generation: A case study of China. *The Science of the Total Environment*, *575*, 321–329. https://doi.org/https://doi.org/10.1016/j.scitotenv.2016.10.054 PMID: 27744159

Gupta, M., Sinha, S., & Chaturvedi, A. (2019). Economic feasibility of e-waste recycling in India: Policy implications. *Waste Management & Research*, *37*(4), 394–402. DOI: 10.1177/0734242X19831556 PMID: 30736727

Gupta, R., & Sharma, P. (2023). The environmental impact of electronic waste: Carbon footprint and recycling methods. *Waste Management (New York, N.Y.)*, *150*, 127–139.

Gupta, S. (2011). E-waste management: Teaching how to reduce, reuse and recycle for sustainable development-need of some educational strategies. *Journal of Education and Practice*, *2*(3), 2222–1735.

Gupta, S., & Singh, R. (2020). E-waste management: A review on technologies, challenges and future prospects. *Journal of Cleaner Production*, *265*, 121955.

Haftu, G. (2019). Information communications technology and economic growth in Sub-Saharan Africa: A panel data approach. *Telecommunications Policy*, *43*(1), 88–99. DOI: 10.1016/j.telpol.2018.03.010

Hai, H. T., Gwan, A. J., Jin, K. D., Sohn, J. S., Oh, K. C., & Saeng, C. H. (2007). Growth process of copper particles in the cuprous oxide-water system. *Materials Transactions*, *48*(5), 1137–1142. DOI: 10.2320/matertrans.48.1137

Hair, J. F., Anderson, R. E., Babin, B. J., & Black, W. C. (2010). *Multivariate data analysis: A global perspective* (Vol. 7). Academic Press.

Hakak, S., Khan, W. Z., Gilkar, G. A., Haider, N., Imran, M., & Alkatheiri, M. S. (2020). Industrial wastewater management using blockchain technology: Architecture, requirements, and future directions. *IEEE Internet of Things Magazine*, *3*(2), 38–43. DOI: 10.1109/IOTM.0001.1900092

Hakami, B. A. (2018). Electronic waste management approaches in Saudi Arabia. *American Journal of Environmental Engineering*, *8*, 4–10.

Han, X., & Chen, L. (2023). Assessing the carbon footprint of wearable health devices. *Journal of Environmental Health Science & Engineering*, *21*, 1023–1035.

Harikaran, M., Boopathi, S., Gokulakannan, S., & Poonguzhali, M. (2023). Study on the source of e-waste management and disposal methods. In *Sustainable Approaches and Strategies for E-Waste Management and Utilization* (pp. 39–60). IGI Global.

Hassanien, A. E., Bhattacharyya, S., Chakrabati, S., Bhattacharya, A., & Dutta, S. 2020. Emerging Technologies in Data Mining and Information Security. *Proceedings of IEMIS*, 1.

Hayakawa, K., Takatsuki, H., Watanabe, I., & Sakai, S.-I. (2004). Polybrominated diphenyl ethers (PBDEs), polybrominated dibenzo-p-dioxins/dibenzofurans (PBDD/Fs) and monobromo-polychlorinated dibenzo-p-dioxins/dibenzofurans (MoBPXDD/Fs) in the atmosphere and bulk deposition in Kyoto, Japan. *Chemosphere*, *57*(5), 343–356. DOI: 10.1016/j.chemosphere.2004.06.038 PMID: 15331261

Heacock, M., Kelly, C. B., Asante, K. A., Birnbaum, L. S., Bergman, Å. L., Bruné, M. N., Buka, I., Carpenter, D. O., Chen, A., Huo, X., Kamel, M., Landrigan, P. J., Magalini, F., Diaz-Barriga, F., Neira, M., Omar, M., Pascale, A., Ruchirawat, M., Sly, L., & Suk, W. A. (2016). E-Waste and Harm to Vulnerable Populations: A Growing Global Problem. *Environmental Health Perspectives*, *124*(5), 550–555. DOI: 10.1289/ehp.1509699 PMID: 26418733

Heacock, M., Kelly, C. B., & Suk, W. A. (2016). E-waste: The growing global problem and next steps. *Reviews on Environmental Health*, *31*(1), 131–135. DOI: 10.1515/reveh-2015-0045 PMID: 26820178

Hegge, K. K., & Agrawal, R. (2023). Wireless and Solar-Powered Multipurpose Robot for Industrial Automation: A Sustainable Solution. *Proceedings of the 8th International Conference on Communication and Electronics Systems (ICCES)*, 146-151.

Herceg, S., Pinto Bautista, S., & Weiß, K.-A. (2020). Influence of Waste Management on the Environmental Footprint of Electricity Produced by Photovoltaic Systems. *Energies*, *13*(9), 2146. https://doi.org/https://doi.org/10.3390/en13092146

He, W., Li, G., Ma, X., Wang, H., Huang, J., Xu, M., & Huang, C. (2006). WEEE recovery strategies and the WEEE treatment status in China. *Journal of Hazardous Materials*, *136*(3), 502–512. DOI: 10.1016/j.jhazmat.2006.04.060 PMID: 16820262

He, Y., & Zhao, X. (2023). Carbon footprint analysis of LED lighting: A life cycle assessment. *Journal of Environmental Management*, *326*, 116477.

Hicks, C., Dietmar, R., & Eugster, M. (2005). The recycling and disposal of electrical and electronic waste in China-legislative and market responses. *Environmental Impact Assessment Review*, *25*(5), 459–471. DOI: 10.1016/j.eiar.2005.04.007

Hofman, A., Aravena, C., & Aliaga, V. (2016). Information and communication technologies and their impact in the economic growth of Latin America, 1990-2013. *Telecommunications Policy*, *40*(5), 485–501. DOI: 10.1016/j.telpol.2016.02.002

Hong, J. (2017). Causal relationship between ICT R&D investment and economic growth in Korea. *Technological Forecasting and Social Change*, *116*, 70–75. DOI: 10.1016/j.techfore.2016.11.005

Hoornweg, D., & Bhada-Tata, P. (2012). What a Waste: A Global Review of Solid Waste Management. Urban Development Series. World Bank.

Huang, J., & Wu, Q. (2023). The carbon footprint of blockchain technology: An environmental perspective. *Environmental Research Letters*, *18*(2), 025009.

Huo, X., Zheng, X. B., Liu, Q., Zhang, T., Wang, Q. H., & Xu, X. J. (2019). Impact of informal e-waste recycling on human health. *Zhonghua Yu Fang Yi Xue Za Zhi [Chinese Journal of Preventive Medicine]*, *53*, 426–432. PMID: 30982282

Hussain, A., Batool, I., Akbar, M., & Nazir, M. (2021). Is ICT an enduring driver of economic growth? Evidence from South Asian economies. *Telecommunications Policy*, *45*(8), 1–12. DOI: 10.1016/j.telpol.2021.102202

Hwang, W., & Shin, J. (2017). ICT-specific technological change and economic growth in Korea. *Telecommunications Policy*, *41*(4), 282–294. DOI: 10.1016/j.telpol.2016.12.006

IEA. (2021). *World Energy Outlook 2021*. International Energy Agency. https://www.iea.org/

IEA. (2023). *World Energy Outlook 2023*. International Energy Agency. https://www.iea.org/

Ignatuschtschenko, E. (2017). E-waste management in China: Bridging the formal and informal sectors. *Journal of Chinese Governance*, *2*(4), 385–410. DOI: 10.1080/23812346.2017.1379629

Ikhlayel, M. (2016). Differences of methods to estimate generation of waste electrical and electronic equipment for developing countries: Jordan as a case study. *Resources, Conservation and Recycling*, *108*, 134–139. https://doi.org/https://doi.org/10.1016/j.resconrec.2016.01.015

Ikhlayel, M. (2017). Environmental impacts and benefits of state-of-the-art technologies for E-waste management. *Waste Management (New York, N.Y.)*, *68*, 458–474. DOI: 10.1016/j.wasman.2017.06.038 PMID: 28662843

Ilankoon, I. M. S. K., Ghorbani, Y., Chong, M. N., Herath, G., Moyo, T., & Petersen, J. (2018). E-waste in the international context–A review of trade flows, regulations, hazards, waste management strategies and technologies for value recovery. *Waste Management (New York, N.Y.)*, *82*, 258–275. DOI: 10.1016/j.wasman.2018.10.018 PMID: 30509588

IPCC. (2023). *Climate Change 2023: Synthesis Report* (Team, C. W., Lee, H., & Romero, J., Eds.)., DOI: 10.59327/IPCC/AR6-9789291691647

Iqbal, M., & Ahmed, R. (2023). Carbon emissions from electronic product lifecycle: A critical review. *Journal of Environmental Sciences (China)*, *137*, 57–73.

IRENA & CPI. (2023). *Global landscape of renewable energy finance 2023*. International Renewable Energy Agency and Climate Policy Initiative. https://www.climatepolicyinitiative.org/

IRENA. (2020). *Global Renewables Outlook: Energy Transformation 2050 (Edition: 2020)*. International Renewable Energy Agency. https://www.irena.org/

IRENA. (2023). *Renewable power generation costs in 2022*. International Renewable Energy Agency. https://www.irena.org/

IRENA. (2024). *Renewable energy statistics 2024*. International Renewable Energy Agency. https://www.irena.org/

Ishida, H. (2015). The effect of ICT development on economic growth and energy consumption in Japan. *Telematics and Informatics*, *32*(1), 79–88. DOI: 10.1016/j.tele.2014.04.003

Islam, M. H., Islam, M. T., Hoque, M. R., Haque, S. Z., Barua, B., Islam, M. M., & Hossain, M. F. (2025a). Understanding young consumers' e-waste recycling behaviour in Bangladesh: A developing country perspective. *Recycling*, *10*(1), 24. DOI: 10.3390/recycling10010024

Islam, M. H., Sabbir, M. M., Mahbub, M., Jhilik, T. T., & Shirmin, S. (2025b). Sustaining e-waste recycling behavior among young consumers to implement circular economy: An extended model of Theory of Reasoned Goal Pursuit. *Cleaner Waste Systems*, *10*, 100212. DOI: 10.1016/j.clwas.2025.100212

Islam, M. T., Dias, P., & Huda, N. (2020). Young consumers' e-waste awareness, consumption, disposal, and recycling behaviour: A case study of university students in Sydney, Australia. *Journal of Cleaner Production*. Advance online publication. DOI: 10.1016/j.jclepro.2020.124490

Islam, M. T., & Huda, N. (2019). E-waste in Australia: Generation estimation and untapped material recovery and revenue potential. *Journal of Cleaner Production*, *237*, 117787. https://doi.org/https://doi.org/10.1016/j.jclepro.2019.117787

Islam, R., Nazifa, T., Yuniarto, A., Uddin, A., Salmiati, S., & Shahid, S. (2019). An empirical study of construction and demolition waste generation and implication of recycling. *Waste Management (New York, N.Y.)*, *95*, 10–21. DOI: 10.1016/j.wasman.2019.05.049 PMID: 31351595

Ismail, H., & Hanafiah, M. M. (2020). A review of sustainable e-waste generation and management: Present and future perspectives. *Journal of Environmental Management*, *264*, 110495. DOI: 10.1016/j.jenvman.2020.110495 PMID: 32250915

Ismail, H., & Hanafiah, M. M. (2021). Evaluation of e-waste management systems in Malaysia using life cycle assessment and material flow analysis. *Journal of Cleaner Production*, *308*, 127358. DOI: 10.1016/j.jclepro.2021.127358

Jain, A., & Sareen, R. (2006). E-waste assessment methodology and validation in India. *Journal of Material Cycles and Waste Management*, *8*, 40–45. https://doi.org/https://doi.org/10.1007/s10163-005-0145-2

Jaiswal, S. K., Mukti, S. K., & Rath, K. C. (2023). E-waste control and its recycling to build sustainable society in the global context. In Handbook of Research on Applications of AI, Digital Twin, and Internet of Things for Sustainable Development (pp. 200-222). IGI Global.

Jayaraman, K., Vejayon, S., Raman, S., & Mostafiz, I. (2019). The proposed ewaste management model from the conviction of individual laptop disposal practices-An empirical study in Malaysia. *Journal of Cleaner Production, 208*, 688–696. DOI: 10.1016/j.jclepro.2018.10.125

Jiang, B., Adebayo, A., Jia, J., Xing, Y., Deng, S., Guo, L., & Zhang, D. (2019). Impacts of heavy metals and soil properties at a Nigerian e-waste site on soil microbial community. *Journal of Hazardous Materials, 362*, 187–195. DOI: 10.1016/j.jhazmat.2018.08.060 PMID: 30240992

Jiang, L., Cheng, Z., Zhang, D., Song, M., Wang, Y., Luo, C., & Zhang, G. (2017). The influence of e-waste recycling on the molecular ecological network of soil microbial communities in Pakistan and China. *Environmental Pollution, 231*, 173–181. DOI: 10.1016/j.envpol.2017.08.003 PMID: 28800486

Jiang, P., Zhang, L., You, S., Fan, Y. V., Tan, R. R., Klemes, J. J., & You, F. (2023). Blockchain technology applications in waste management: Overview, challenge and Opportunities. *Journal of Cleaner Production, 421*, 138466. Advance online publication. DOI: 10.1016/j.jclepro.2023.138466

Jiang, Y., & Liu, S. (2023). Carbon footprint of gaming consoles: A lifecycle assessment. *Journal of Cleaner Production, 381*, 134613.

Jibiri, J. D. (2015). Reduce Reuse Recycle Behavioural Intention Model in Higher Education Institution Accommodation (Doctoral dissertation, Universiti Teknologi Malaysia).

Jinhui, L., Huabo, D., & Pixing, S. (2011). Heavy metal contamination of surface soil in electronic waste dismantling area: Site investigation and source-apportionment analysis. *Waste Management & Research, 29*(7), 727–738. DOI: 10.1177/0734242X10397580 PMID: 21406506

John, S., Sridharan, R., & Kumar, P. (2017). Multi-period reverse logistics network design with emission cost. *International Journal of Logistics Management, 28*(1), 127–149. DOI: 10.1108/IJLM-08-2015-0143

John, S., Sridharan, R., Kumar, P., & Krishnamoorthy, M. (2018). Multi-period reverse logistics network design for used refrigerators. *Applied Mathematical Modelling, 54*, 311–331. DOI: 10.1016/j.apm.2017.09.053

Johnson, L., & Roberts, P. (2019). The Role of Extended Producer Responsibility in E-Waste Management. *Journal of Environmental Management, 25*(3), 256–267.

Jones, K., & Smith, J. (2023). Environmental impact of e-commerce: Carbon footprint of electronic products. *Journal of Retailing and Consumer Services, 70*, 102090.

Jorgenson, D., Ho, M., & Samuels, J. (2016). The impact of information technology on postwar US economic growth. *Telecommunications Policy, 40*(5), 398–411. DOI: 10.1016/j.telpol.2015.03.001

Kaaya, I., Ascencio-Vásquez, J., Weiss, K.-A., & Topič, M. (2021). Assessment of uncertainties and variations in PV modules degradation rates and lifetime predictions using physical models. *Solar Energy, 218*, 354–367. https://doi.org/https://doi.org/10.1016/j.solener.2021.01.071

Kaipannan, S., Govindarajan, K., Sundaramoorthy, S., & Marappan, S. (2019). Waste toner-derived carbon/Fe3O4 nanocomposite for high-performance supercapacitor. *ACS Omega, 4*(14), 15798–15805. DOI: 10.1021/acsomega.9b01337 PMID: 31592452

Kalia, P., Zia, A., & Mladenovic, D. (2022). Examining country development indicators and e-waste under the moderating effect of country development levels and e-waste policy. *International Journal of Quality & Reliability Management, 39*(7), 1601–1616. DOI: 10.1108/IJQRM-09-2021-0335

Kalmykova, Y., Patrício, J., Rosado, L., & Berg, P. E. (2015). Out with the old, out with the new – The effect of transitions in TVs and monitors technology on consumption and WEEE generation in Sweden 1996–2014. *Waste Management, 46*, 511-522. https://doi.org/https://doi.org/10.1016/j.wasman.2015.08.034

Kang, K. D., Kang, H., Ilankoon, I. M. S. K., & Chong, C. Y. (2020). Electronic waste collection systems using Internet of Things (IoT): Household electronic waste management in Malaysia. *Journal of Cleaner Production, 252*, 119801. DOI: 10.1016/j.jclepro.2019.119801

Karlilar Pata, S., & Balcilar, M. (2024). Decarbonizing energy: Evaluating fossil fuel displacement by renewables in OECD countries. *Environmental Science and Pollution Research International, 31*(21), 31304–31313. https://doi.org/https://doi.org/10.1007/s11356-024-33324-8 PMID: 38630397

Karunambiga, K., Nusrat Jahan, S., Keerthana, S., Pavithra, D., & Umakarthikeyan, C. (2023). Smart Waste Management System Using IOT And Blockchain. *IJRAR-International Journal of Research and Analytical Reviews, 10*(3), 63–67.

Kastanaki, E., & Giannis, A. (2022). Forecasting quantities of critical raw materials in obsolete feature and smart phones in Greece: A path to circular economy. *Journal of Environmental Management, 307*, 114566. https://doi.org/https://doi.org/10.1016/j.jenvman.2022.114566 PMID: 35091243

Kaur, H., Singh, R., & Khanna, S. (2018). E-waste disposal behaviour and factors influencing it: A study of Indian consumers. *Journal of Environmental Planning and Management*, *61*(2), 326–344.

Kaya, M. (2019). *Electronic Waste and Printed Circuit Board Recycling Technologies*. Springer. DOI: 10.1007/978-3-030-26593-9

Khan, J., Kumar, A., Giri, A., Pal, D. B., Tripathi, A., & Giri, D. D. (2021). Impact of electronic waste pollutants on underground water. In *Groundwater Geochemistry* (pp. 265–281). Pollution and Remediation Methods. DOI: 10.1002/9781119709732.ch13

Khanna, R., Mukherjee, P. S., & Park, M. (2020a). A critical assessment on resource recovery from electronic waste: Impact of mechanical pre-treatment. *Journal of Cleaner Production*, *53*, 122319. DOI: 10.1016/j.jclepro.2020.122319

Khanna, R., Saini, R., Park, M., Ellamparuthy, G., Biswal, S. K., & Mukherjee, P. S. (2020b). Factors influencing the release of potentially toxic elements (PTEs) during thermal processing of electronic waste. *Waste Management (New York, N.Y.)*, *105*, 414–424. DOI: 10.1016/j.wasman.2020.02.026 PMID: 32126369

Khatun, A., & Dhara, N. (2022). E-Waste Management: A New Concern for Environmental Sustainability. In Smart Cities, Citizen Welfare, and the Implementation of Sustainable Development Goals (pp. 222-238). IGI Global.

Khoshnazm, M. J., Marzban, A., & Azimi, N. (2023). Performance enhancement of photovoltaic panels integrated with thermoelectric generators and phase change materials: Optimization and analysis of thermoelectric arrangement. *Energy*, *267*, 126556. https://doi.org/https://doi.org/10.1016/j.energy.2022.126556

Kiddee, P., Naidu, R., & Wong, M. H. (2013). Electronic waste management approaches: An overview. *Waste Management (New York, N.Y.)*, *33*(5), 1237–1250. DOI: 10.1016/j.wasman.2013.01.006 PMID: 23402807

Kiddee, P., Naidu, R., Wong, M. H., Hearn, L., & Muller, J. F. (2014). Field investigation of the quality of fresh and aged leachates from selected landfills receiving e-waste in an arid climate. *Waste Management (New York, N.Y.)*, *34*(11), 2292–2304. DOI: 10.1016/j.wasman.2014.06.018 PMID: 25070222

Kilic, H., Cebeci, U., & Ayhan, M. (2015). Reverse logistics system design for the waste of electrical andelectronic equipment (WEEE) in Turkey. *Resources, Conservation and Recycling*, *95*, 120–132. DOI: 10.1016/j.resconrec.2014.12.010

Kim, H., & Lee, S. (2023). Carbon footprint assessment of virtual reality devices. *Journal of Environmental Impact*, *45*(1), 89–103.

Kim, S., An, J., Choi, H., & Hong, T. (2023). Assessment the technical and economic performance of a window-integrated PV system using third-generation PV panels. *Energy and Building*, *286*, 112978. https://doi.org/https://doi.org/10.1016/j.enbuild.2023.112978

Kim, S., & Park, J. (2019). Extended Producer Responsibility and Its Impact on E-Waste Collection Rates. *Environmental Policy Review*, *35*(2), 98–112.

Kim, Y., & Lee, J. (2021). Public Awareness and Participation in E-Waste Recycling. *The Journal of Environmental Education*, *42*(1), 77–90.

King, A. M., Burgess, S. C., Ijomah, W., & McMahon, C. A. (2006). Reducing waste: Repair, recondition, remanufacture or recycle? *Sustainable Development (Bradford)*, *14*(4), 257–267. DOI: 10.1002/sd.271

Kline, T. (2005). Psychological testing: A practical approach to design and evaluation. *Sage (Atlanta, Ga.)*. Advance online publication. DOI: 10.4135/9781483385693

Knemeyer, A., Ponzurick, T., & Logar, C. (2002). A qualitative examination of factors affecting reverse logistics systems for end-of-life computers. *International Journal of Physical Distribution & Logistics Management*, *32*(6), 455–479. DOI: 10.1108/09600030210437979

Kojima, M., Yoshida, A., & Sasaki, S.Y. XU ET AL. (2009). Difficulties in applying extended producer responsibility policies in developing countries: Case studies in e-waste recycling in China and Thailand. *Journal of Material Cycles and Waste Management*, *11*(3), 263–269. DOI: 10.1007/s10163-009-0240-x

Kolamroudi, M. K., Ilkan, M., Egelioglu, F., & Safaei, B. (2023). Feature selection by ant colony optimization and experimental assessment analysis of PV panel by reflection of mirrors perpendicularly. *Renewable Energy*, *218*, 119238. https://doi.org/https://doi.org/10.1016/j.renene.2023.119238

Kolstad, C. (2011). *Environmental Economics* (2nd ed.). Oxford University Press.

Komoto, K., Lee, J.-S., Zhang, J., Ravikumar, D., Sinha, P., Wade, A., & Heath, G. (2018). *End-of-Life Management of Photovoltaic Panels: Trends in PV Module Recycling Technologies*. International Energy Agency Photovoltaic Power Systems Program (IEA-PVPS). https://iea-pvps.org/

Kong, S., Liu, H., Zeng, H., & Liu, Y. (2012). The status and progress of resource utilization technology of ewaste pollution in China. *Procedia Environmental Sciences*, *16*, 515–521. DOI: 10.1016/j.proenv.2012.10.071

Kónya, L. (2006). Exports and growth: Granger causality analysis on OECD countries with a panel data approach. *Economic Modelling*, *23*(6), 978–992. DOI: 10.1016/j.econmod.2006.04.008

Korhonen, J., Malmborg, F., Strachan, P., & Ehrenfeld, J. (2004). Management and policy aspects of industrial ecology: An emerging research agenda. *Business Strategy and the Environment*, *13*(5), 289–305. DOI: 10.1002/bse.415

Kosai, S., Kishita, Y., & Yamasue, E. (2020). Estimation of the metal flow of WEEE in Vietnam considering lifespan transition. *Resources, Conservation and Recycling*, *154*, 104621. https://doi.org/https://doi.org/10.1016/j.resconrec.2019.104621

Koshariya, A. K., Khatoon, S., Marathe, A. M., Suba, G. M., Baral, D., & Boopathi, S. 2023. Agricultural Waste Management Systems Using Artificial Intelligence Techniques. In AI-Enabled Social Robotics in Human Care Services (pp. 236-258). IGI Global. DOI: 10.4018/978-1-6684-8171-4.ch009

Koshta, N., Patra, S., & Singh, S. P. (2021). Estimation of E-waste at micro level for reverse logistics: A case of Delhi. *Journal of Cleaner Production*, *314*, 128063. https://doi.org/https://doi.org/10.1016/j.jclepro.2021.128063

Krebs-Moberg, M., Pitz, M., Dorsette, T. L., & Gheewala, S. H. (2021). Third generation of photovoltaic panels: A life cycle assessment. *Renewable Energy*, *164*, 556–565. https://doi.org/https://doi.org/10.1016/j.renene.2020.09.054

Kudari, J. M., & Samal, A. (2021). Smart Agriculture and Smart Farming Using IoT Technology. *Journal of Physics: Conference Series*, *2089*(1), 012038. DOI: 10.1088/1742-6596/2089/1/012038

Kumar, A., & Dixit, G. (2018a). Evaluating critical barriers to implementation of WEEE management using DEMATEL approach. *Resources, Conservation and Recycling*, *131*, 101–121. DOI: 10.1016/j.resconrec.2017.12.024

Kumar, A., & Dixit, G. (2018b). An analysis of barriers affecting the implementation of e-waste management practices in India: A novel ISM-DEMATEL approach. *Sustainable Production and Consumption*, *14*, 36–52. DOI: 10.1016/j.spc.2018.01.002

Kumar, A., Holuszko, M., & Espinosa, D. (2017). E-waste: An overview on generation, collection, legislation and recycling practices. *Resources, Conservation and Recycling*, *122*, 32–42. DOI: 10.1016/j.resconrec.2017.01.018

Kumar, N., Sah, B., & Jain, V. (2017). E-waste management: A review of recent research. *Resources, Conservation and Recycling*, *122*, 55–69.

Kumar, R., Stauvermann, P., & Samitas, A. (2016). The effects of ICT on output per worker: A study of the Chinese economy. *Telecommunications Policy*, *40*(2-3), 102–115. DOI: 10.1016/j.telpol.2015.06.004

Kuntawee, C., Tantrakarnapa, K., Limpanont, Y., Lawpoolsri, S., Phetrak, A., Mingkhwan, R., & Worakhunpiset, S. (2020). Exposure to heavy metals in electronic waste recycling in Thailand. *International Journal of Environmental Research and Public Health*, *17*(9), 2996. DOI: 10.3390/ijerph17092996 PMID: 32357423

Kurniawati, M. (2020). The role of ICT infrastructure, innovation and globalization on economic growth in OECD countries, 1996-2017. *Journal of Science and Technology Policy Management*, *11*(2), 193–215. DOI: 10.1108/JSTPM-06-2019-0065

Kusch, S., & Hills, C. (2017). The Link between e-Waste and GDP—New Insights from Data from the Pan-European Region. *Resources*, *6*(15), 1–10. DOI: 10.3390/resources6020015

Kusch, S., & Hills, C. D. (2017). The link between e-waste and GDP—New insights from data analysis. *Resources, Conservation and Recycling*, *123*, 115–123. DOI: 10.1016/j.resconrec.2016.12.005

Kwon, Y., & Park, J. (2023). The carbon footprint of electronic payment systems. *Journal of Cleaner Production*, *382*, 134630.

L. Fernandez, H. Kim, and Park J. (2022). "The Role of AI and Machine Learning in E-waste Sorting,". *Springer Journal of Cleaner Production*, *402*(2), 129874.

Ladou, J., & Lovegrove, S. (2008). Export of electronics equipment waste. *International Journal of Occupational and Environmental Health*, *14*(1), 1–10. DOI: 10.1179/oeh.2008.14.1.1 PMID: 18320726

Lai, K. H., Sarathy, R., & Grover, V. (2009). Information diffusion and adoption of electric vehicles: The case of Hong Kong. *Journal of Management Information Systems*, *26*(2), 241–273.

Lamichhane, M. (2017). A smart waste management system using IoT and blockchain technology [Master's Thesis]. ITMO University.

Lamming, R., & Hampson, J. (1996). The Environment as a supply chain management issue". *British Journal of Management*, *7*(Special Issue), 45–62. DOI: 10.1111/j.1467-8551.1996.tb00147.x

Landrigan, P. J., Fuller, R., Fisher, S., Suk, W. A., Sly, P., Chiles, T. C., & Bose-O'Reilly, S. (2019). Pollution and children's health. *The Science of the Total Environment*, *650*(Pt 2), 2389–2394. DOI: 10.1016/j.scitotenv.2018.09.375 PMID: 30292994

Landsiedel, R., Ma-Hock, L., Kroll, A., Hahn, D., Schnekenburger, J., Wiench, K., & Wohlleben, W. (2010). Testing metal-oxide nanomaterials for human safety. *Advanced Materials*, *22*(24), 2601–2627. DOI: 10.1002/adma.200902658 PMID: 20512811

Latif, Z., Danish, Y. M., Latif, S., Ximei, L., Pathan, Z. H., Salam, S., & Jianqiu, Z. (2018). The dynamics of ICT, foreign direct investment, globalization and economic growth: Panel estimation robust to heterogeneity and cross-sectional dependence. *Telematics and Informatics*, *35*(2), 318–328. DOI: 10.1016/j.tele.2017.12.006

Laurentiis, V., Corrado, S., & Sala, S. (2018). Quantifying household waste of fresh fruit and vegetables in the EU. *Waste Management (New York, N.Y.)*, *77*, 238–251. DOI: 10.1016/j.wasman.2018.04.001 PMID: 29655924

Lèbre, É., & Corder, G. (2015). Integrating industrial ecology thinking into the management of mining waste. *Resources*, *4*(4), 765–786. DOI: 10.3390/resources4040765

Lee, J.-K., Lee, J.-S., Ahn, Y.-S., Kang, G.-H., Song, H.-E., Kang, M.-G., Kim, Y.-H., & Cho, C.-H. (2018). Simple pretreatment processes for successful reclamation and remanufacturing of crystalline silicon solar cells. *Progress in Photovoltaics: Research and Applications*, *26*(3), 179–187. https://doi.org/https://doi.org/10.1002/pip.2963

Lent, R. W., Brown, S. D., & Hackett, G. (1994). Toward a unifying social cognitive theory of career and academic interest, choice, and performance. *Journal of Vocational Behavior*, *45*(1), 79–122. DOI: 10.1006/jvbe.1994.1027

Lepawsky, J. (2018). *Reassembling rubbish: Worlding electronic waste*. MIT Press. DOI: 10.7551/mitpress/11111.001.0001

Leung, R., & Loo, P. (2020). Co-creating interactive dining experiences via interconnected and interoperable smart technology. *Asian Journal of Technology Innovation*, *30*(1), 45–67. DOI: 10.1080/19761597.2020.1822748

Li, A., Li, B., Lu, B., Yang, D., Hou, S., & Song, X. (2022). Generation estimation and material flow analysis of retired mobile phones in China. *Environmental Science and Pollution Research International*, *29*(50), 75626–75635. https://doi.org/https://doi.org/10.1007/s11356-022-21153-6 PMID: 35657548

Li, L., Li, B., Liu, Y., Li, X., & Ma, C. (2017). Determinants of individuals' electronic waste recycling behaviour: A comparative analysis of China and South Korea. *Resources, Conservation and Recycling*, *122*, 146–156.

Ling, T. C., & Poon, C. S. (2011). Utilization of recycled glass derived from cathode ray tube glass as fine aggregate in cement mortar. *Journal of Hazardous Materials*, *192*(2), 451–456. DOI: 10.1016/j.jhazmat.2011.05.019 PMID: 21705136

Lin, K. P., Tseng, M.-L., & Pai, P.-F. (2018). Sustainable supply chain management using approximate fuzzy DEMATEL method. *Resources, Conservation and Recycling*, *128*, 134–142. DOI: 10.1016/j.resconrec.2016.11.017

Lin, M., Tang, J., Ma, S., Yu, Y., Li, G., Fan, R., Mai, B., & An, T. (2020). Insights into biomonitoring of human exposure to polycyclic aromatic hydrocarbons with hair analysis: A case study in e-waste recycling area. *Environment International*, *136*, 105432. DOI: 10.1016/j.envint.2019.105432 PMID: 31884415

Lin, Y. H., Wang, J., Niu, D., & Tao, X. (2024). Blockchain-driven framework for construction waste recycling and reuse. *Journal of Building Engineering*, *89*, 109355. DOI: 10.1016/j.jobe.2024.109355

Lin, Y., & Wang, J. (2023). Carbon footprint analysis of home entertainment systems. *Journal of Cleaner Production*, *383*, 134657.

Li, Q., & Wu, M. (2020). Tourists' pro-environmental behavior in travel destinations: Benchmarking the power of social interaction and individual attitude. *Journal of Sustainable Tourism*, *28*(9), 1371–1389. DOI: 10.1080/09669582.2020.1737091

Liu, F., & Chen, Z. (2020). The Influence of AI on E-Waste Sorting and Processing. *AI and Environmental Science*, *44*(1), 23–37.

Liu, H., & Ma, Q. (2023). The environmental impact of cloud computing: Carbon footprint and mitigation strategies. *Environmental Science & Technology*, *57*(5), 2468–2481.

Liu, K., Tan, Q., Yu, J., & Wang, M. (2023). A global perspective on e-waste recycling. *Circular Economy*, *2*(1), 100028. DOI: 10.1016/j.cec.2023.100028

Liu, Q., Li, K. Q., Zhao, H., Li, G., & Fan, F. Y. (2009). The global challenge of electronic waste management. *Environmental Science and Pollution Research International*, *16*(3), 248–249. DOI: 10.1007/s11356-009-0131-7 PMID: 19308473

Liu, X., Tanaka, M., & Matsui, Y. (2006). Electrical and electronic waste management in China: Progress and the barriers to overcome. Waste Management & Research. *Waste Management & Research*, *24*(1), 92–101. DOI: 10.1177/0734242X06062499 PMID: 16496875

Liu, Y., Luo, X., Zeng, Y., Tu, W., Deng, M., Wu, Y., & Mai, B. (2020). Species-specific biomagnification and habitat-dependent trophic transfer of halogenated organic pollutants in insect-dominated food webs from an e-waste recycling site. *Environment International*, *138*, 105674. DOI: 10.1016/j.envint.2020.105674 PMID: 32234680

Liu, Y., Park, S., Yi, H., & Feiock, R. (2020). Evaluating the employment impact of recycling performance in Florida. *Waste Management (New York, N.Y.), 101*, 283–290. DOI: 10.1016/j.wasman.2019.10.025 PMID: 31648169

Liu, Y., Zhang, J., Yang, X., Yang, W., Chen, Y., & Wang, C. (2020). Efficient recovery of valuable metals from waste printed circuit boards by microwave pyrolysis. *Chinese Journal of Chemical Engineering, 13*, 24–36. DOI: 10.1016/j.cjche.2020.11.008

Liu, Z., Wu, T., Wang, F., Osmani, M., & Demian, P. (2022). Blockchain enhanced construction waste information management: A conceptual framework. *Sustainability (Basel), 14*(19), 12145. DOI: 10.3390/su141912145

Li, W., & Achal, V. (2020). Environmental and health impacts due to e-waste disposal in China - A review. *The Science of the Total Environment, 737*, 139745. DOI: 10.1016/j.scitotenv.2020.139745 PMID: 32516663

Li, X., & Zhang, Y. (2023). Carbon footprint reduction in the production of electronic components. *Journal of Industrial Ecology, 27*(2), 221–237.

Locke, E. A., & Latham, G. P. (2002). Building a practically useful theory of goal setting and task motivation: A 35-year odyssey. *The American Psychologist, 57*(9), 705–717. DOI: 10.1037/0003-066X.57.9.705 PMID: 12237980

Lohmann, R., & Jones, K. C. (1998). Dioxins and furans in air and deposition: A review of levels, behaviour and processes. *The Science of the Total Environment, 219*(1), 53–81. DOI: 10.1016/S0048-9697(98)00237-X PMID: 9770324

Lohse, J., Winteler, S., & Wulf-Schnabel, J. (1998). *Collection targets for waste from electrical and electronic equipment (WEEE)*. Report compiled for the Directorate General (DG XI Evironment, Nuclear Sfety and Civil Potection of the Cmmission of the European Communities. https://oekopol.de/

Lombardi, D., & Laybourn, P. (2012). Redefining industrial symbiosis: Crossing academic-practitioner boundaries. *Journal of Industrial Ecology, 16*(1), 28–37. DOI: 10.1111/j.1530-9290.2011.00444.x

Long, E., Kokke, S., Lundie, D., Shaw, N., Ijomah, W., & Kao, C.-C. (2016). Technical solutions to improve global sustainable management of waste electrical and electronic equipment (WEEE) in the EU and China. *Journal of Remanufacturing, 6*(1), 1–27. DOI: 10.1186/s13243-015-0023-6

Lourenço, H. R., & Soto, J. P. (2002). *Reverse Logistics Models and Applications: A Recoverable Production Planning Model.* Universitat Pompeu Fabra: Grup de Recerca en Logística Empresarial.

Lu, C., Zhang, L., Zhong, Y., Ren, W., Tobias, M., Mu, Z., & Xue, B. (2015). An overview of e-waste management in China. *Journal of Material Cycles and Waste Management*, *17*(1), 1–12. DOI: 10.1007/s10163-014-0256-8

Lundgren, K. (2012). *The Global Impact of E-Waste: Addressing the Challenge*. ILO.

Lundgren, K., & Yan, M. (2017). The informal e-waste sector: Potential benefits and drawbacks of its integration into the formalized system. *Environmental Science & Policy*, *69*, 39–47.

Luo, Z., & Huang, Y. (2023). Carbon footprint of artificial intelligence: A life cycle assessment. *Journal of Environmental Management*, *328*, 116502.

Maani, T., Celik, I., Heben, M. J., Ellingson, R. J., & Apul, D. (2020). Environmental impacts of recycling crystalline silicon (c-SI) and cadmium telluride (CDTE) solar panels. *The Science of the Total Environment*, *735*, 138827. https://doi.org/https://doi.org/10.1016/j.scitotenv.2020.138827 PMID: 32464407

Maddux, J. E., & Rogers, R. W. (1983). Protection motivation and self-efficacy: A revised theory of fear appeals and attitude change. *Journal of Experimental Social Psychology*, *19*(5), 469–479. DOI: 10.1016/0022-1031(83)90023-9

Magazzino, C., Mele, M., & Schneider, N. (2020). The relationship between municipal solid waste and greenhouse gas emissions: Evidence from Switzerland. *Waste Management (New York, N.Y.)*, *113*, 508–520. DOI: 10.1016/j.wasman.2020.05.033 PMID: 32546447

Mahadev Madgule, N. (2023). Mechanical properties and microstructure of activated TIG welded similar joints of Inconel alloys by desirability approaches. *Materials Today: Proceedings*, *77*, 528–533. DOI: 10.1016/j.matpr.2022.12.250

Mahmoudi, S., Huda, N., & Behnia, M. (2019). Photovoltaic waste assessment: Forecasting and screening of emerging waste in Australia. *Resources, Conservation and Recycling*, *146*, 192–205. https://doi.org/https://doi.org/10.1016/j.resconrec.2019.03.039

Mahmoudi, S., Huda, N., & Behnia, M. (2021). Critical assessment of renewable energy waste generation in OECD countries: Decommissioned PV panels. *Resources, Conservation and Recycling*, *164*, 105145. https://doi.org/https://doi.org/10.1016/j.resconrec.2020.105145

Maphosa, V., & Maphosa, M. (2020). E-waste management in Sub-Saharan Africa: A systematic literature review. *Cogent Business & Management*, *7*(1), 1814503. DOI: 10.1080/23311975.2020.1814503

Marconi, M., Gregori, F., Germani, M., Papetti, A., & Favi, C. (2018). An approach to favor industrial symbiosis: The case of waste electrical and electronic equipment. *Procedia Manufacturing*, *21*, 502–509. DOI: 10.1016/j.promfg.2018.02.150

Martilla, J. A., & James, J. C. (1977). Importance-performance analysis. *Journal of Marketing*, *41*(1), 77–79. DOI: 10.1177/002224297704100112

Martin, C., & Davis, R. (2023). Carbon footprint of electronic toys: An environmental analysis. *Journal of Sustainable Technology*, *22*(1), 71–86.

Martinez, D., & Anderson, B. (2018). The Environmental Impact of E-Waste Disposal. *Journal of Environmental Protection*, *22*(3), 355–371.

Marwede, M., Berger, W., Schlummer, M., Mäurer, A., & Reller, A. (2013). Recycling paths for thin-film chalcogenide photovoltaic waste – Current feasible processes. *Renewable Energy*, *55*, 220–229. https://doi.org/https://doi.org/10.1016/j.renene.2012.12.038

Matthews, H. S., McMichael, F. C., Hendrickson, C. T., & Hart, D. J. (1997). *Disposition and End-of-Life Options for Personal Computers*. Carnegie Mellon University.

Ma, X., & Wu, Y. (2023). Carbon footprint assessment of smartwatches. *Journal of Environmental Impact Assessment*, *16*(2), 309–323.

Ma, X., Yuan, H., & Du, W. (2024). Blockchain-Enabled Construction and Demolition Waste Management: Advancing Information Management for Enhanced Sustainability and Efficiency. *Sustainability (Basel)*, *16*(2), 721. DOI: 10.3390/su16020721

Memon, M. A., Rahman, Z., & Kumar, A. (2017). A conceptual model for e-waste recycling behaviour among consumers. *Journal of Cleaner Production*, *141*, 370–381.

Mesbah, Y. I., Ahmed, N., Ali, B. A., & Allam, N. K. (2020). Recycling of Li−Ni−Mn−Co hydroxide from spent batteries to produce high-performance supercapacitors with exceptional stability. *ChemElectroChem*, *7*(4), 975–982. DOI: 10.1002/celc.202000081

Mihai, F. C. (2020). Electronic waste management in Romania: pathways for sustainable practices. Handbook of Electronic Waste Management Butterworth-Heinemann, 533–551. DOI: 10.1016/B978-0-12-817030-4.00024-3

Mingfei, X., Yaping, W., Jun, L., & Hua, X. (2016). Lead recovery and glass microspheres synthesis from waste CRT funnel glasses through carbon thermal reduction enhanced acid leaching process. *Journal of Hazardous Materials*, *305*, 51–58. DOI: 10.1016/j.jhazmat.2015.11.032 PMID: 26642446

Mishra, D., & Rhee, Y.H. (2010). Current research trends of microbiological leaching for metal recovery from industrial wastes. *Current research technology and education topics in applied microbiology and microbial biotechnology* 2, 1289-1292.

Mishra, S. (2019). Perceived and manifested health problems among informal E-waste handlers: A scoping review. *Indian Journal of Occupational and Environmental Medicine*, *23*(1), 7–14. DOI: 10.4103/ijoem.IJOEM_231_18 PMID: 31040583

Miura, S.-i., Yamaguchi, Y., Urata, H., Himeshima, Y., Otsuka, N., & Tomita, S.. (2004). Efficacy of a multicomponent program (PACE+ Japan) for lifestyle modification in patients with essential hypertension. *Hypertension Research*, *27*(11), 859–864. DOI: 10.1291/hypres.27.859 PMID: 15824468

Mohamed, A. F., & Rasnan, M. I. (2021). Enhancing role and participation of industry and community for sustainable e-waste recovery for sustainable consumption and production (SCP): Case study Kuala Lumpur Malaysia. *EcoDesign and Sustainability II: Social Perspectives and Sustainability Assessment*, 135-145. DOI: 10.1007/978-981-15-6775-9_9

Mohd Noor, N. H., Mustafa, M. A. S., Saharom, M., & Shamsol Kamal, N. S. S. (2022). Understanding Malaysian household waste separation: An extended theory of planned behavior. *Malaysian Journal of Sustainable Environment*, *9*(1), 19–35. DOI: 10.24191/myse.v9i1.17285

Mohr, K. A., & Mohr, E. S. (2017). Understanding Generation Z students to promote a contemporary learning environment. *Journal on Empowering Teaching Excellence*, *1*(1), 9. DOI: 10.15142/T3M05T

Moktadir, M. A., Ali, S. M., Rajesh, R., & Paul, S. K. (2018). Modeling the interrelationships among barriers to sustainable supply chain management in leather industry. *Journal of Cleaner Production*, *181*, 631–651. DOI: 10.1016/j.jclepro.2018.01.245

Monish, P., & Dhanabhakyam, M. (2021). Sustainability Strategies for Developing SMEs and Entrepreneurship. In *Handbook of Research on Sustaining SMEs and Entrepreneurial Innovation in the Post-COVID-19 Era* (pp. 527–547). IGI Global.

Morgan, T., Tokman, M., Richey, R., & Defee, C. (2018). Resource commitment and sustainability: A reverse logistics performance process model. *International Journal of Physical Distribution & Logistics Management*, *48*(2), 164–182. DOI: 10.1108/IJPDLM-02-2017-0068

Moyen Massa, G., & Archodoulaki, V.-M. (2023). Electrical and Electronic Waste Management Problems in Africa: Deficits and Solution Approach. *Environments*, *10*(3), 44. https://doi.org/https://doi.org/10.3390/environments10030044

Mundada, M. N., Kumar, S., & Shekdar, A. V. (2004). E-waste: A new challenge for waste management in India. *The International Journal of Environmental Studies*, *61*(3), 265–279. DOI: 10.1080/0020723042000176060

Muthukumari, W. A. C. S., Ahn, J., & Kim, M. (2024). Impact of information publicity on Korean residents' E-waste recycling intentions: Applying the norm activation model and theory of planned behavior. *Heliyon*, *10*(14), e34319. Advance online publication. DOI: 10.1016/j.heliyon.2024.e34319 PMID: 39816366

Muthusamy, P. D., Velusamy, G., Thandavan, S., Govindasamy, B. R., & Savarimuthu, N. (2022). Industrial Internet of things-based solar photo voltaic cell waste management in next generation industries. *Environmental Science and Pollution Research International*, *29*(24), 35542–35556. https://doi.org/https://doi.org/10.1007/s11356-022-19411-8 PMID: 35237911

Myovella, G., Karacuka, M., & Haucap, J. (2020). Digitalization and economic growth: A comparative analysis of Sub-Saharan Africa and OECD economies. *Telecommunications Policy*, *44*(2), 1–12. DOI: 10.1016/j.telpol.2019.101856

N. Veena, R. Kumar, and A. Singh, "E-Waste Management: A Circular Economy Perspective,". *Springer Journal of Waste Management & Research*, *41*(3), 321–340.

Nahm, E.-S., Barker, B., Resnick, B., Covington, B., Magaziner, J., & Brennan, P. F. (2010). Effects of a social cognitive theory-based hip fracture prevention website for older adults. *Computers, Informatics, Nursing*, *28*(6), 371–379. DOI: 10.1097/NCN.0b013e3181f69d73 PMID: 20978408

Namias, J. (2013). *The future of electronic waste recycling in the United States: obstacles and domestic solutions*. Columbia University.

Namlis, K., & Komilis, D. (2019). *Influence of four socioeconomic indices and the impact of economic crisis on solid waste generation in Europe*. Waste Management, 190-200.

Natarajan, S., Krishnamoorthy, K., & Kim, S.-J. (2022). Effective regeneration of mixed composition of spent lithium-ion batteries electrodes towards building supercapacitor. *Journal of Hazardous Materials*, *430*, 128496. DOI: 10.1016/j.jhazmat.2022.128496 PMID: 35739677

Nathan, T., Aziz, A., Noor, A. F., & Prabaharan, S. R. S. (2008). Nanostructured NiO for electrochemical capacitors : Synthesis and electrochemical properties. *Journal of Solid State Electrochemistry*, *12*(7-8), 1003–1009. Advance online publication. DOI: 10.1007/s10008-007-0465-3

Ndalloka, Z. N., Nair, H. V., Alpert, S., & Schmid, C. (2024). Solar photovoltaic recycling strategies. *Solar Energy*, *270*, 112379. https://doi.org/https://doi.org/10.1016/j.solener.2024.112379

Needhidasan, S., Samuel, M., & Chidambaram, R. (2014). Electronic waste –an emerging threat to the environment of urban India. *Journal of Environmental Health Science & Engineering*, *12*(1), 1–9. DOI: 10.1186/2052-336X-12-36 PMID: 24393372

Neitzel, R. L., Sayler, S. K., Arain, A. L., & Nambunmee, K. (2020). Metal levels, genetic instability, and renal markers in electronic waste workers in Thailand. *The International Journal of Occupational and Environmental Medicine*, *11*(2), 72–84. DOI: 10.34172/ijoem.2020.1826 PMID: 32218555

Nejati, M., Saboury, A., Nejati, M., & Sabouri, M. S. (2017). Factors influencing the intention to use electronic waste recycling services: An integration of the theory of planned behaviour and protection motivation theory. *Resources, Conservation and Recycling*, *122*, 521–530.

Nema, A. K., Garg, A., & Yadav, R. K. (2016). E-waste management in India: Issues and challenges. *Environmental Science and Pollution Research International*, *23*(2), 1787–1800.

Neto, J., Bloemhof-Ruwaard, J., Nunen, J., & Heck, E. (2008). Designing and evaluating sustainable logistics networks. *International Journal of Production Economics*, *111*(2), 195–208. DOI: 10.1016/j.ijpe.2006.10.014

New, S., Green, K., & Morton, B. (2002). An analysis of private versus public sector responses to the environmental challenges of the supply chain. *Journal of Public Procurement*, *2*(1), 93–105. DOI: 10.1108/JOPP-02-01-2002-B004

Nguyen, H. T. T., Lee, C. H., Nguyen, H. T. T., Phan, T. T., Nguyen, V. V., & Hung, R. J. (2025). Factors affecting community participation in an e-waste recycling program. *International Journal of Environment and Waste Management*, *36*(1), 95–110. DOI: 10.1504/IJEWM.2025.143501

Nguyen, R. T., Diaz, L. A., Imholte, D. D., & Lister, T. E. (2017). Economic assessment for recycling critical metals from hard disk drives using a comprehensive recovery process. *Journal of the Minerals Metals & Materials Society*, *69*(9), 1546–1552. DOI: 10.1007/s11837-017-2399-2

Nguyen, T., Pham, T., & Tram, H. (2020). Role of information and communication technologies and innovation in driving carbon emissions and economic growth in selected G-20 countries. *Journal of Environmental Management*, *261*, 1–10. DOI: 10.1016/j.jenvman.2020.110162 PMID: 32148259

Nguyen, T., & Tran, L. (2023). Carbon emissions from the production and disposal of electronic goods. *Journal of Environmental Sciences (China)*, *138*, 145–159.

Niebel, T. (2018). ICT and economic growth - Comparing developing, emerging and developed countries. *World Development*, *104*, 197–211. DOI: 10.1016/j.worlddev.2017.11.024

Nie, X., Fan, C., Wang, Z., Su, T., Liu, X., & An, T. (2015). Toxic assessment of the leachates of paddy soils and river sediments from e-waste dismantling sites to microalga, Pseudokirchneriella subcapitata. *Ecotoxicology and Environmental Safety*, *111*, 168–176. DOI: 10.1016/j.ecoenv.2014.10.012 PMID: 25450930

Nithya, R., Sivasankari, C., & Thirunavukkarasu, A. (2021). Electronic waste generation, regulation, and metal recovery: A review. *Environmental Chemistry Letters*, *19*(2), 1347–1368. DOI: 10.1007/s10311-020-01111-9

Nižetić, S., Jurčević, M., Čoko, D., & Arıcı, M. (2021). A novel and effective passive cooling strategy for photovoltaic panel. *Renewable & Sustainable Energy Reviews*, *145*, 111164. https://doi.org/https://doi.org/10.1016/j.rser.2021.111164

Nnorom, I. C., & Odeyingbo, O. A. (2020). Electronic waste management practices in Nigeria. In *Handbook of Electronic Waste Management* (pp. 323–354). Butterworth-Heinemann. DOI: 10.1016/B978-0-12-817030-4.00014-0

Noor, N. H. M. (2024). Defeating Global Glut of Clothing: An Examination of Sustainable Fashion Consumption of Young Consumers. In *Driving Green Marketing in Fashion and Retail* (pp. 26-49). IGI Global. DOI: 10.4018/979-8-3693-3049-4.ch002

Noor, N. H. M., Hassan, N. H., & Ismadi, R. (2023a). Towards sustainable consumption: An examination of the usage of recyclable and biodegradable products among Malaysians. *Asian People Journal*, *6*(1), 190–206. DOI: 10.37231/apj.2023.6.1.491

Noor, N. H. M., Soleman, N. A. F., & Azuan, A. S. K. (2023b). To recycle or not to recycle? Factors affecting Malaysian residents' intention for recycling e-waste. *Malaysian Journal of Social Sciences and Humanities*, *8*(2), e002102–e002102. DOI: 10.47405/mjssh.v8i2.2102

Noor, N. H. M., Zaini, S. M., Pudin, S., & Sidek, S. J. (2024). Encouraging pro-environmental behavior: The function of environmental connectedness, environmental ethics, & environmental knowledge. *Muallim Journal of Social Sciences and Humanities*, *8*(2), 17–32. DOI: 10.33306/mjssh/272

Nottebohm, O., Manyika, J., Bughin, J., Chui, M., & Syed, A. (2012). *Online and Upcoming: The Internet's Impact on Aspiring Countries*. McKinsey & Company.

NREL. (2021). *Documenting a Decade of Cost Declines for PV Systems*. The National Renewable Energy Laboratory. Retrieved 04/03/2025 from https://www.nrel.gov/news/program/2021/documenting-a-decade-of-cost-declines-for-pv-systems.html

Nti, A., Arko-Mensah, A. A., Botwe, J., Dwomoh, P. K., Kwarteng, D., & Takyi, L., & Fobil. (2020). Effect of particulate matter exposure on respiratory health of e-waste workers at Agbogbloshie, Accra, Ghana. *International Journal of Environmental Research and Public Health, 17*(9). PMID: 32349371

Nwagwu, W., & Okuneye, M. (2016). Awareness and attitudes of small-scale information technology business operators in Lagos, Nigeria toward E-waste hazards. *Journal of Global Information Technology Management*, *19*(4), 267–282. DOI: 10.1080/1097198X.2016.1246934

O'Reilly, S., & Kumar, A. (2016). Closing the loop: An exploratory study of reverse ready-made garment supply chains in Delhi NCR. *International Journal of Logistics Management*, *27*(2), 486–510. DOI: 10.1108/IJLM-03-2015-0050

Odili, J. B., Noraziah, A., Ambar, R., Wahab, M. H., & Fakheraldin, M. A. I. (2018). E-waste generation, awareness and management in third world countries: Prospects and challenges. The Eurasia Proceedings of Educational and Social Sciences, 9, 349–353.

OECD. (2021). *Environment at a Glance Indicators*. OECD Publishing.

Ogidi, O. I., Ajoko, I. T., & Tawariowei, A. M. (2024a). Sustainable Application of Genetic Ecotoxicological Techniques. In *Biomonitoring For Environmental Sustainability. In: Biomonitoring of Pollutants in the Global South*. Springer Nature Singapore Pte Ltd. DOI: 10.1007/978-981-97-1658-6_19

Ogidi, O. I., Angaye, T. C. N., & Carbom, H. E. (2024c). Microorganisms in the monitoring of terrestrial and aquatic ecosystems: applications, challenges and future prospects. In *Biomonitoring of Pollutants in the Global South*. Springer Nature Singapore Pte Ltd. DOI: 10.1007/978-981-97-1658-6_18

Ogidi, O. I., Enenebeaku, U. E., Okara, E., & Elumelu, S. A. (2021b). Toxic Metal Profiles, Carcinogenic and Non-Carcinogenic Human Health Risk Assessment of Some Locally Produced Beverages in Nigeria. *Journal of Toxicology and Risk Assessment*, *7*(1), 1–8. DOI: 10.23937/2572-4061.1510039

Ogidi, O. I., Njoku, C. O., Onimisi, A. M., & Onomedjeke, P. E. (2021a). Evaluation of Heavy Metal Contents and Potential Human Health Risk Assessment of Selected Canned Sardines Fish Sold in Yenagoa, Nigeria. *Archives of Ecotoxicology*, *3*(2), 39–43. DOI: 10.36547/ae.2021.3.2.39-43

Ogidi, O. I., & Tawariowei, A. M. (2023). Endocrine Disrupting Chemicals and their Role in Cancer-A review. *Molecular Sciences And Applications*, *3*, 5–14. DOI: 10.37394/232023.2023.3.2

Oguchi, M., Kameya, T., Yagi, S., & Urano, K. (2008). Product flow analysis of various consumer durables in Japan. *Resources, Conservation and Recycling*, *52*(3), 463–480. https://doi.org/https://doi.org/10.1016/j.resconrec.2007.06.001

Ohajinwa, C. M., van Bodegom, P. M., Vijver, M. G., Olumide, A. O., Osibanjo, O., & Peijnenburg, W. J. G. M. (2018). Prevalence and injury patterns among electronic waste workers in the informal sector in Nigeria. *Injury Prevention : Journal of the International Society for Child and Adolescent Injury Prevention*, *24*(3), 185–192. DOI: 10.1136/injuryprev-2016-042265 PMID: 28679520

Okereke, J. N., Ogidi, O. I., & Obasi, K. O. (2016). Environmental and Health Impact of Industrial Wastewater Effluents in Nigeria- A Review. *International Journal of Advanced Research in Biological Science*, *3*(6), 55–67.

Olamide, E., Ogujiuba, K., Maredza, A., & Semosa, P. (2022). Poverty, ICT and economic growth in SADC region: A panel cointegration evaluation. *Sustainability (Basel)*, *14*(15), 1–15. DOI: 10.3390/su14159091

Ongondo, F. O., Williams, I. D., & Cherrett, T. J. (2011). How are WEEE doing? A global review of the management of electrical and electronic wastes. *Waste Management (New York, N.Y.)*, *31*(4), 714–730. DOI: 10.1016/j.wasman.2010.10.023 PMID: 21146974

Oreg, S., Vakola, M., & Armenakis, A. (2011). Change recipients' reactions to organizational change: A 60-year review of quantitative studies. *The Journal of Applied Behavioral Science*, *47*(4), 461–524. DOI: 10.1177/0021886310396550

Orlins, S., & Guan, D. (2016). China's toxic informal e-waste recycling: Local approaches to a global environmental problem. *Journal of Cleaner Production*, *114*, 71–80. DOI: 10.1016/j.jclepro.2015.05.090

Orosz, T., Rassõlkin, A., Arsénio, P., Poór, P., Valme, D., & Sleisz, Á. (2024). Current Challenges in Operation, Performance, and Maintenance of Photovoltaic Panels. *Energies*, *17*(6), 1306. https://doi.org/https://doi.org/10.3390/en17061306

Oughton, C., Kurup, B., Anda, M., & Ho, G. (2022). Industrial symbiosis to circular economy: What does the literature reveal for a successful complex industrial area? *Circular Economy and Sustainability*, *2*(4), 1317–1344. DOI: 10.1007/s43615-022-00153-1

Owusu-Sekyere, K., Batteiger, A., Afoblikame, R., Hafner, G., & Kranert, M. (2022). Assessing data in the informal e-waste sector: The Agbogbloshie Scrapyard. *Waste Management, 139*, 158-167. https://doi.org/https://doi.org/10.1016/j.wasman.2021.12.026

Oyebode, O.J. (2013). "Solid waste management for sustainable development and public health: A case study of Lagos State in Nigeria," Univ. J. Public Health.

Oyebode, O. J. (2022). *"Sustainable waste management towards circular economy in Nigerian context: Challenges, prospects and way forward," in Effective Waste Management and Circular Economy*. CRC Press.

Ozcan, B., Apergis, N., & Shahbaz, M. (2018). A revisit of the environmental Kuznets curve hypothesis for Turkey: New evidence from bootstrap rolling window causality. *Environmental Science and Pollution Research International, 25*(32), 32381–32394. DOI: 10.1007/s11356-018-3165-x PMID: 30229495

Özgür Polat, L., & Gungor, A. (2021). WEEE closed-loop supply chain network management considering the damage levels of returned products. *Environmental Science and Pollution Research International, 28*(7), 7786–7804. https://doi.org/https://doi.org/10.1007/s11356-020-10249-6 PMID: 33040288

Ozsut Bogar, Z., & Gungor, A. (2023). Forecasting Waste Mobile Phone (WMP) Quantity and Evaluating the Potential Contribution to the Circular Economy: A Case Study of Turkey. *Sustainability, 15*(4), 3104. https://doi.org/https://doi.org/10.3390/su15043104

P. Singh, R. Gupta, and Mehta, S. (2020). "Digital Twin Technology for E-waste Tracking and Reduction,". *IEEE Transactions on Industrial Informatics, 19*(5), 7896–7908.

Packard, V. (1963). *The Waste Makers*. Longmans.

Pagnanelli, F., Moscardini, E., Granata, G., Abo Atia, T., Altimari, P., Havlik, T., & Toro, L. (2017). Physical and chemical treatment of end of life panels: An integrated automatic approach viable for different photovoltaic technologies. *Waste Management, 59*, 422-431. https://doi.org/https://doi.org/10.1016/j.wasman.2016.11.011

Paiano, A. (2015). Photovoltaic waste assessment in Italy. *Renewable & Sustainable Energy Reviews, 41*, 99–112. https://doi.org/https://doi.org/10.1016/j.rser.2014.07.208

Palitzsch, W., & Loser, U. (2011, 19-24 June 2011). A new and intelligent demetalization step of broken silicon cells and silicon cell production waste in the recycling procedure of crystalline si modules. *37th IEEE Photovoltaic Specialists Conference*, Seattle, WA, USA.

Panambunan-Ferse, M., & Breiter, A. (2013). Assessing the side-effects of ICT development: E-waste production and management: A case study about cell phone end-of-life in Manado, Indonesia. *Technology in Society*, *35*(3), 223–231. DOI: 10.1016/j.techsoc.2013.04.002

Panda, S. K., Behera, S., & Patnaik, S. S. (2021, September 20-22). 2021). Lifetime Estimation of DC-Link Capacitor in a Standalone PV System with an Integrated Backup Battery. 2021 International Symposium of Asian Control Association on Intelligent Robotics and Industrial Automation (IRIA). https://doi.org/https://doi.org/10.1016/j.rser.2010.11.037

Pandian, G., & Abdul-Kader, W. (2017). Performance evaluation of reverse logistics enterprise - an agent-based simulation approach. *International Journal of Sustainable Engineering*, *10*(6), 384–398. DOI: 10.1080/19397038.2017.1370032

Panwar, N. L., Kaushik, S. C., & Kothari, S. (2011). Role of renewable energy sources in environmental protection: A review. *Renewable & Sustainable Energy Reviews*, *15*(3), 1513–1524. https://doi.org/https://doi.org/10.1016/j.rser.2010.11.037

Pariatamby, A., & Bhatti, M. S. (2020). Waste management challenges in Malaysia. In *Sustainable waste management challenges in Developing Countries* (pp. 186–222). IGI Global. DOI: 10.4018/978-1-7998-0198-6.ch008

Pariatamby, A., Cheah, W. Y., Shrizal, R., Thamlarson, N., Lim, B. T., & Barasarathi, J. (2015). Enhancement of landfill methane oxidation using different types of organic wastes. *Environmental Earth Sciences*, *73*(5), 2489–2496. DOI: 10.1007/s12665-014-3600-3

Parulkar, A. (2021). Plastic Waste Management Using Blockchain. Academic Press.

Parvez, S. M., Jahan, F., Brune, M.-N., Gorman, J. F., Rahman, M. J., Carpenter, D., Islam, Z., Rahman, M., Aich, N., Knibbs, L. D., & Sly, P. D. (2021). Health consequences of exposure to e-waste: An updated systematic review. *The Lancet. Planetary Health*, *5*(12), e905–e920. DOI: 10.1016/S2542-5196(21)00263-1 PMID: 34895498

Pastuszak, J., & Węgierek, P. (2022). Photovoltaic Cell Generations and Current Research Directions for Their Development. *Materials (Basel)*, *15*(16), 5542. https://doi.org/https://doi.org/10.3390/ma15165542 PMID: 36013679

Patel, N., & Shah, R. (2023). Carbon footprint reduction strategies in the consumer electronics industry. *Renewable & Sustainable Energy Reviews*, *170*, 112913.

Patel, R., & Singh, L. (2018). Regulatory Frameworks in E-Waste Management: Comparative Analysis. *International Journal of Environmental Law*, *19*(3), 150–168.

Patnaik, S., Chakraborty, S., & Dey, B. L. (2018). Determinants of e-waste disposal behaviour among Indian households: An empirical study. *Journal of Environmental Management*, *213*, 116–126.

Pavankumar, P., & Darwante, N. K. (2022). Performance Monitoring and Dynamic Scaling Algorithm for Queue Based Internet of Things. *Proceedings of the International Conference on Innovative Computing, Intelligent Communication and Smart Electrical Systems (ICSES)*, 1-7.

Pender, J. P., Jha, G., Youn, D. H., Ziegler, J. M., Andoni, I., Choi, E. J., Heller, A., Dunn, B. S., Weiss, P. S., Penner, R. M., & Mullins, C. B. (2020). Electrode degradation in lithium-ion batteries. *ACS Nano*, *14*(2), 1243–1295. DOI: 10.1021/acsnano.9b04365 PMID: 31895532

Peng, H., Michael, S., Reid, X., & Le, C. (2015). Consumption of rice and fish in an electronic waste recycling area contributes significantly to total daily intake of mercury. *Journal of Environmental Sciences (China)*, *38*, 83–86. DOI: 10.1016/j.jes.2015.10.003 PMID: 26702970

Peralta, G. L., & Fontanos, P. M. (2006). E-waste issues and measures in the Philippines. *Journal of Material Cycles and Waste Management*, *8*(1), 34–39. https://doi.org/https://doi.org/10.1007/s10163-005-0142-5

Pesaran, M. (2004). *General Diagnostic Tests for Cross Section Dependence in Panels*. Munich: CESifo Working Paper, No. 1229.

Pesaran, M., Ullah, A., & Yamagata, T. (2008). A bias-adjusted LM test of error cross-section independence. *The Econometrics Journal*, *11*(1), 105–127. DOI: 10.1111/j.1368-423X.2007.00227.x

Pesaran, M., & Yamagata, T. (2008). Testing slope homogeneity in large panels. *Journal of Econometrics*, *142*(1), 50–93. DOI: 10.1016/j.jeconom.2007.05.010

Petridis, N. E., Stiakakis, E., Petridis, K., & Dey, P. (2016). Estimation of computer waste quantities using forecasting techniques. *Journal of Cleaner Production*, *112*, 3072–3085. https://doi.org/https://doi.org/10.1016/j.jclepro.2015.09.119

Piscitelli, A., & D'Uggento, A. M. (2022). Do young people really engage in sustainable behaviors in their lifestyles? *Social Indicators Research*, *163*(3), 1467–1485. DOI: 10.1007/s11205-022-02955-0

Pokharel, S., & Mutha, A. (2009). Perspectives in reverse logistics: A review. *Resources, Conservation and Recycling*, *53*(4), 175–182. DOI: 10.1016/j.resconrec.2008.11.006

Polák, M., & Drápalová, L. (2012). Estimation of end of life mobile phones generation: The case study of the Czech Republic. *Waste Management, 32*(8), 1583-1591. https://doi.org/https://doi.org/10.1016/j.wasman.2012.03.028

Pradhan, R., Mallik, G., & Bagchi, T. (2018). Information communication technology (ICT) infrastructure and economic growth: A causality evinced by cross-country panel data. *IIMB Management Review*, *30*(1), 91–103. DOI: 10.1016/j.iimb.2018.01.001

Prasad, P. (2012). *Information communication technology (ICT) - its waste and consequences.* Int. J. Environmental Technology and Management, 15(3/4/5/6), 363-376.

Preethi, P., Deepak, T., Anusha, S. V., & Bhagya, D. (2024). E-Waste Collection and Recycling Best Practices and Innovations. In Sustainable Solutions for E-Waste and Development (pp. 33-43). IGI Global.

Puneeth Kumar, B. S., & Srinivasulu Raju, S. (2021). Smart Villages: IoT Technology Based Transformation. *Journal of Physics: Conference Series*, *2070*(1), 012128. DOI: 10.1088/1742-6596/2070/1/012128

Purchase, D., Abbasi, G., Bisschop, L., Chatterjee, D., Ekberg, C., Ermolin, M., & Wong, M. H. (2020). Global occurrence, chemical properties, and ecological impacts of E-wastes (IUPAC Technical Report). *Pure and Applied Chemistry*, *92*(11), 1733–1767. DOI: 10.1515/pac-2019-0502

Qu, Y., Wang, W., Liu, Y., & Zhu, Q. (2019). Understanding residents' preferences for e-waste collection in China - A case study of waste mobile phones. *Journal of Cleaner Production*, *228*, 52–62. DOI: 10.1016/j.jclepro.2019.04.216

Raheem, I., Tiwari, A., & Balsalobre-Lorente, D. (2020). The role of ICT and financial development in CO2 emissions and economic growth. *Environmental Science and Pollution Research International*, *27*(2), 1912–1922. DOI: 10.1007/s11356-019-06590-0 PMID: 31760620

Rahman, S., & Subramanian, N. (2012). Factors for implementing end-of-life computer recycling operations in reverse supply chains. *International Journal of Production Economics*, *140*(1), 239–248. DOI: 10.1016/j.ijpe.2011.07.019

Rahul, V., & Rajan, U. (2023). Sickle effect: A hybrid model of product life cycle and supply chain management for achieving sustainable development. *Materials Today: Proceedings*. Advance online publication. DOI: 10.1016/j.matpr.2023.06.075

Rajpoot. (2020). A Review on ways to Manage Biomedical Waste at Different Locations in Faizabad. International Journal for Modern Trends in Science and Technology, 6(1), 33-36.

Rajpoot, S. (2021). Impact of Nanotechnology on Environment – A Review. *International Journal for Modern Trends in Science and Technology*, *7*(10), 159–164. DOI: 10.46501//IJMTST0710026

Rajpoot, S. (2024). *Eco-friendly solutions for sustainable waste management: An approach towards a cleaner and greener future*. Waste Management and Life Cycle Assessment for Sustainable Business Practice. DOI: 10.4018/979-8-3693-2595-7.ch003

Rajput, J. E. (2013). E-Waste-Concept, Problems and Solution Policies in India. *International Journal of Management and Business*, *3*(1), 31–36.

Rakhmawati, T., Damayanti, S., Jati, R. K., & Astrini, N. J. (2023). An extended TPB model of waste-sorting intention: A case study of Indonesia. *Management of Environmental Quality*, *34*(5), 1248–1268. DOI: 10.1108/MEQ-11-2022-0309

Ramakrishnan, T., Mohan Gift, M. D., Chitradevi, S., Jegan, R., & Subha Hency Jose, P. (2022). Study of Numerous Resins Used in Polymer Matrix Composite Materials. Advances in Materials Science and Engineering. DOI: 10.1155/2022/1088926

Ramesha, M., & Pai, G. N. (2020). Internet of Things: Internet Revolution, Impact, Technology Road Map, and Features. *Advances in Mathematical Sciences Journal*, *9*(7), 4405–4414.

Ramzan, S., Liu, C., Munir, H., & Xu, Y. (2019). Assessing young consumers' awareness and participation in sustainable e-waste management practices: A survey study in Northwest China. *Environmental Science and Pollution Research International*, *26*(19), 20003–20013. DOI: 10.1007/s11356-019-05310-y PMID: 31102225

Rani, M., Singh, R., & Kumar, A. (2016). Informal electronic waste recycling: A sector review with special focus on India. *Journal of Cleaner Production*, *129*, 431–448.

Rao, S., & Maheshwari, V. (2023). Carbon footprint of electronic medical devices: A comprehensive assessment. *Journal of Environmental Health*, *36*(4), 505–518.

Rashid, A., Asif, F., Krajnik, P., & Nicolescu, C. (2013). Resource conservative manufacturing: An essential change in business and technology paradigm for sustainable manufacturing. *Journal of Cleaner Production*, *57*, 166–177. DOI: 10.1016/j.jclepro.2013.06.012

Rathi, N., & Kameswari, V. L. (2017). Determinants of e-waste disposal behaviour among college students in India: An empirical analysis. *Waste Management & Research*, *35*(6), 650–658.

Rautela, R., Arya, S., Vishwakarma, S., Lee, J., Kim, K.H., & Kumar, S. (2021). Ewaste management and its effects on the environment and human health. *Science of the Total Environment 773*, 145623. .2021.145623DOI: 10.1016/j.scitotenv

Rautela, R., Arya, S., Vishwakarma, S., Lee, J., Kim, K. H., & Kumar, S. (2021). E-waste management and its effects on the environment and human health. *The Science of the Total Environment*, *773*, 145623. DOI: 10.1016/j.scitotenv.2021.145623 PMID: 33592459

Ravi Varma, V., Mathi Krishnan, M., Parayitam, S., Duraisami, S. P. A., & Saravanaselvan, N. R. (2023). Exploring e-waste recycling behaviour intention among households: Evidence from India. *Cleaner Materials*, *7*, 100174. DOI: 10.1016/j.clema.2023.100174

Ravindra, K., & Mor, S. (2019). E-waste generation and management practices in Chandigarh,India and economic evaluation for sustainable recycling. *Journal of Cleaner Production*, *221*, 286–294. DOI: 10.1016/j.jclepro.2019.02.158

Ravisankar, A., Shanthi, A. L., Lavanya, S., Ramaratnam, M. S., Krishnamoorthy, V., & Boopathi, S. (2024). Harnessing 6G for Consumer-Centric Business Strategies Across Electronic Industries. In AI Impacts in Digital Consumer Behavior (pp. 241-270). IGI Global.

Ravi, V., Shankar, R., & Tiwari, M. (2005). Analyzing alternatives in reverse logistics for end-of-life computers: ANP and balanced scorecard approach. *Computers & Industrial Engineering*, *48*(2), 327–356. DOI: 10.1016/j.cie.2005.01.017

Reddy, S. P., & Patwal, P. S. (2022). Data Analytics and Cloud-Based Platform for Internet of Things Applications in Smart Cities. Proceedings of the International Conference on Industry 4.0 Technology (I4Tech), 1-6.

Richard, E. B. (2009). Children's Exposure to Elemental Mercury: A National Review of Exposure Events. The Agency for Toxic Substances and Disease Registry and Centres for Disease Control and Prevention Mercury Workgroup.

Richey, R., Genchev, S., & Daugherty, P. (2005). The role of resource commitment and innovation in reverse logistics performance. *International Journal of Physical Distribution & Logistics Management, 35*(4), 233–257. DOI: 10.1108/09600030510599913

Ritchie, H., & Rosado, P. (2022). *Fossil fuels*. Our World in Data. Retrieved 26/09/2024 from https://ourworldindata.org/fossil-fuels

Ritchie, H., Rosado, P., & Roser, M. (2023). *CO_2 and Greenhouse Gas Emissions*. Our World in Data. Retrieved 26/09/2024 from https://ourworldindata.org/co2-and-greenhouse-gas-emissions

Rizzolatti, G., & Craighero, L. (2004). The mirror-neuron system. *Annual Review of Neuroscience, 27*(1), 169–192. DOI: 10.1146/annurev.neuro.27.070203.144230 PMID: 15217330

Robinson, B. H. (2009). E-Waste: An Assessment of Global Production and Environmental Impacts. *The Science of the Total Environment, 408*(2), 183–191. DOI: 10.1016/j.scitotenv.2009.09.044 PMID: 19846207

Robinson, B. H. (2009). E-waste: An assessment of global production and environmental impacts. *The Science of the Total Environment, 408*(2), 183–191. https://doi.org/https://doi.org/10.1016/j.scitotenv.2009.09.044 PMID: 19846207

Robinson, J. G. (2011). Ethical pluralism, pragmatism, and sustainability in conservation practice. *Biological Conservation, 144*(3), 958–965. DOI: 10.1016/j.biocon.2010.04.017

Romel, M., Kabir, G., & Ng, K. T. W. (2024). Prediction of photovoltaic waste generation in Canada using regression-based model. *Environmental Science and Pollution Research International, 31*(6), 8650–8665. https://doi.org/https://doi.org/10.1007/s11356-023-31628-9 PMID: 38182949

Romer, P. (1990). *Endogenous Technological Change*. NBER.

Roscoe, J. T. (1975). *Fundamental Research Statistics for the Behavioral Sciences* (2nd ed.). Holt, Rinehart, and Winston.

S, K., & Arun, M. R. (2022). Priority Queueing Model-Based IoT Middleware for Load Balancing. Proceedings of the 6th International Conference on Intelligent Computing and Control Systems (ICICCS), 425-430.

S. Chatterjee and A. Kumar, (2023). "Enabling Technologies for Effective E-waste Management,". *IEEE Internet of Things Journal, 10*(2), 2104–2118.

Saad, M., Ahmad, M. B., Asif, M., Khan, M. K., Mahmood, T., & Mahmood, M. T. (2023). Blockchain-enabled VANET for smart solid waste management. *IEEE Access : Practical Innovations, Open Solutions, 11*, 5679–5700. DOI: 10.1109/ACCESS.2023.3235017

Sabbaghi, M. (2025). Perceptions of value in retired smartphones: The role of monetary incentives in influencing end-of-hibernation decisions. *Waste Management (New York, N.Y.), 193*, 199–206. DOI: 10.1016/j.wasman.2024.12.015 PMID: 39671746

Sabbir, M. M., Khan, T. T., Das, A., Akter, S., & Hossain, M. A. (2023). Understanding the determinants of consumers' reverse exchange intention as an approach to e-waste recycling: A developing country perspective. *Asia-Pacific Journal of Business Administration, 15*(3), 411–439. DOI: 10.1108/APJBA-11-2021-0565

Safiullin, M. R., Elshin, L. A., Abdukaeva, A. A., & Burganov, R. T. (2022). Assessment of environmental research in block chain technology. Procedia Environmental Science. *Engineering and Management, 8*(4), 979–986.

Saha, L., Kumar, V., Tiwari, J., Rawat, S., Singh, J., & Bauddh, K. (2021). Electronic waste and their leachates impact on human health and environment: Global ecological threat and management. *Environmental Technology & Innovation, 24*, 102049. DOI: 10.1016/j.eti.2021.102049

Sahin, M., Akcil, A., Erust, C., Altynbek, S., Gahan, C. S., & Tuncuk, A. (2015). A potential alternative for precious metal recovery from e-waste: Iodine leaching. *Separation Science and Technology, 50*, 2587–2595. DOI: 10.1080/01496395.2015.1061005

Sahoo, S., Mukherjee, A., & Halder, R. (2021). A unified blockchain-based platform for global e-waste management. *International Journal of Web Information Systems, 17*(5), 449–479. DOI: 10.1108/IJWIS-03-2021-0024

Sakthivel, U., Swaminathan, G., & Anis, J. (2022). Strategies for Quantifying Metal Recovery from Waste Electrical and Electronic Equipment (WEEE/E-waste) Using Mathematical Approach. *Process Integration and Optimization for Sustainability, 6*, 1-10. https://doi.org/https://doi.org/10.1007/s41660-022-00250-6

Salanie, B. (2003). *The Economics of Taxation*. The MIT Press.

Salhofer, S., Steuer, B., Ramusch, R., & Beigl, P. (2016). WEEE management in Europe and China - A comparison. *Waste Management (New York, N.Y.), 57*, 27–35. DOI: 10.1016/j.wasman.2015.11.014 PMID: 26626812

Sander, W. E., & Zack, S. A. (2023). The veterinarian's role in and attitude to the disposal of unwanted pharmaceuticals. *Environmental Challenges, 11*, 100718. DOI: 10.1016/j.envc.2023.100718

Sangeetha, B., Manjunatha, K., Thirusenthil Kumaran, P., Sheela, A., Yamuna, K. S., & Sivakumar, S. (2024). Performance Optimization in Photovoltaic Systems: A Review. *Archives of Computational Methods in Engineering*, *31*(3), 1507–1518. https://doi.org/https://doi.org/10.1007/s11831-023-10023-0

Sarkis, J., Helms, M., & Hervani, A. (2010). Reverse logistics and social sustainability. *Corporate Social Responsibility and Environmental Management*, *17*(6), 337–354. DOI: 10.1002/csr.220

Savaskan, R., Bhattacharya, S., & Wassenhove, L. (2004). Closed-Loop Supply Chain Models with Product Remanufacturing. *Management Science*, *50*(2), 239–252. DOI: 10.1287/mnsc.1030.0186

Savvilotidou, V., Antoniou, A., & Gidarakos, E. (2017). Toxicity assessment and feasible recycling process for amorphous silicon and CIS waste photovoltaic panels. *Waste Management, 59*, 394-402. https://doi.org/https://doi.org/10.1016/j.wasman.2016.10.003

Schluep, M., Hageluekenb, C., Kuehrc, R., Magalinic, F., Maurerc, C., Meskersb, C., & Wang, F. (2009). *Recycling: From E-Waste to Resources*. United Nations Environment Programme & United Nations University.

Schluep, M., Hagelüken, C., Kuehr, R., Magalini, F., Maurer, C., Meskers, C., & Wang, F. (2019). Sustainable e-waste management: A comprehensive overview. *Environmental Science & Technology*, *53*(15), 8871–8881. DOI: 10.1021/acs.est.8b06916

Schunk, D. H. (2003). Self-efficacy for reading and writing: Influence of modelling, goal setting, and self-evaluation. *Reading & Writing Quarterly*, *19*(2), 159–172. DOI: 10.1080/10573560308219

Scruggs, C., Nimpuno, N., & Moore, R. (2016). Improving information flow onchemicals in electronic products and E-waste to minimize negativeconsequences for health and the environment. *Resources, Conservation and Recycling*, *113*, 149–164. DOI: 10.1016/j.resconrec.2016.06.009

Seitz, M., & Wells, P. (2006). Challenging the implementation of corporate sustainability: The case of automotive engine remanufacturing. *Business Process Management Journal*, *12*(6), 822–836. DOI: 10.1108/14637150610710954

Selvakumar, S., Adithe, S., Isaac, J. S., Pradhan, R., Venkatesh, V., & Sampath, B. (2023). A study of the printed circuit board (PCB) e-waste recycling process. In Sustainable Approaches and Strategies for E-Waste Management and Utilization (pp. 159-184). IGI Global.

Selvakumar, P. (2024). Climate Change. In *Global Challenges for the Environment and Climate Change* (pp. 100–110). Global. DOI: 10.4018/979-8-3693-2845-3.ch006

Selvakumar, P. (2024). Emerging Engineering Technologies: Design, Development, and Industrial Applications of IoT-Based Aquaculture Monitoring System. In El Kacimi, Y., & Alaoui, K. (Eds.), *Emerging Engineering Technologies and Industrial Applications* (pp. 332–341). IGI Global. DOI: 10.4018/979-8-3693-1335-0.ch016

Selvakumar, P. (2024). Reasons for Biodiversity Loss. In *Impact of Societal Development and Infrastructure on Biodiversity Decline* (pp. 37–49). Global. DOI: 10.4018/979-8-3693-6950-0.ch004

Selvakumar, P. (2024). The Role of Corrosion Inhibitors. In Thakur, A., & Kumar, A. (Eds.), *Innovations in Nanomaterials-Based Corrosion Inhibitors* (pp. 130–141). IGI Global. DOI: 10.4018/979-8-3693-3088-3.ch006

Selvakumar, P., & Loganathan, V. (2012). Phytochemical screening and antioxidant activity of red flowered Mirabilis jalapa leaf in different solvents. *International Journal of Pharma and Bio Sciences*, *3*(4), 440–446.

Selvakumar, P., Mariappan, P., Shunmugam, S., Paul, R., & Jabez, S. (2024). Removal of chemicals from effluent using photobioreactor technology to improve environmental and health impacts. *Zeitschrift für Physikalische Chemie*, *2024*(7), 1295–1302. Advance online publication. DOI: 10.1515/zpch-2023-0450

Selvakumar, P., Muthusamy, S., Satishkumar, D., Vigneshkumar, P., Selvamurugan, C., & Satheesh Kumar, P. (2024). AI-Powered Tools. In Satishkumar, D., & Sivaraja, M. (Eds.), *Using Real-Time Data and AI for Thrust Manufacturing* (pp. 20–42). IGI Global. DOI: 10.4018/979-8-3693-2615-2.ch002

Selvakumar, P., Palanisamy, S. K., Cinthaikinian, S., Palanisamy, V., Mariappan, R., & Selvakumar, P. K. (2024). Biosensors and its diverse applications in healthcare systems. *Zeitschrift für Physikalische Chemie*, *2024*. Advance online publication. DOI: 10.1515/zpch-2023-0406

Sengupta, D., Ilankoon, I. M. S. K., Kang, K. D., & Chong, M. N. (2023). Circular economy and household e-waste management in India. Part II: A case study on informal e-waste collectors (Kabadiwalas) in India. *Minerals Engineering*, *200*, 108154. DOI: 10.1016/j.mineng.2023.108154

Sepehrdoust, H., & Ghorbanseresht, M. (2019). Impact of information and communication technology and financial development on economic growth of OPEC developing economies. *Kasetsart Journal of Social Sciences*, *40*, 546–551.

Severo, E. A., de Guimarães, J. C. F., Brito, L. M. P., & Dellarmelin, M. L. (2017). Environmental sustainability and sustainable consumption: The perception of baby boomers, generation x and y in Brazil. *Revista de Gestão Social e Ambiental, 11*(3), 92–110. DOI: 10.24857/rgsa.v11i3.1266

Shahabuddin, M., Uddin, M. N., Chowdhury, J. I., Ahmed, S. F., Uddin, M. N., Mofijur, M., & Uddin, M. A. (2022). A review of the recent development, challenges, and opportunities of electronic waste (e-waste). *International Journal of Environmental Science and Technology*. DOI: 10.1007/s13762-022-04274-w

Shaharudin, M. R., Said, R., Hotrawaisaya, C., Nik Abdul Rashid, N. R., & Azman Perwira, N. F. S. (2023). Linking determinants of the youth's intentions to dispose of portable e-waste with the proper disposal behavior in Malaysia. *The Social Science Journal, 60*(4), 680–694. DOI: 10.1080/03623319.2020.1753157

Shaikh, S., Thomas, K., & Zuhair, S. (2020a). An exploratory study of e-waste creation and disposal: Upstream considerations. *Resources, Conservation and Recycling, 155*, 104662. DOI: 10.1016/j.resconrec.2019.104662

Shaikh, S., Thomas, K., Zuhair, S., & Magalini, F. (2020b). A cost–benefit analysis of the downstream impacts of e-waste recycling in Pakistan. *Waste Management (New York, N.Y.), 118*, 302–312. DOI: 10.1016/j.wasman.2020.08.039 PMID: 32919349

Shamim, A., Mursheda, A. K., & Rafiq, I. (2015). E-waste trading impact on public health and ecosystem services in developing countries. *International Journal of Waste Resources, 5*(4), 1–8. DOI: 10.4172/2252-5211.1000188

Sharma, A., & Gupta, N. (2020). The Circular Economy in Electronic Waste Management: A Global Perspective. *Sustainability, 12*(14), 5678.

Sharma, A., Pandey, S., & Kolhe, M. (2019). Global review of policies & guidelines for recycling of solar PV modules. *International Journal of Smart Grid and Clean Energy, 8*(5), 597–610. https://doi.org/https://doi.org/10.12720/sgce.8.5.597-610

Sharma, H., & Kumar, H. (2022). Sustainable Collection and Classification of E-Waste: A Proposed Computer Vision Technology-Based Framework. Available at *SSRN* 4049693.

Sharma, M., Joshi, S., & Kumar, A. (2020). Assessing enablers of e-waste management in circular economy using DEMATEL method: An Indian perspective. *Environmental Science and Pollution Research International, 27*(12), 13325–13338. DOI: 10.1007/s11356-020-07765-w PMID: 32020449

Sharma, M., Luthra, S., Joshi, S., & Kumar, A. (2022). Developing a framework for enhancing survivability of sustainable supply chains during and post-COVID-19 pandemic. *International Journal of Logistics*, *25*(4–5), 433–453. DOI: 10.1080/13675567.2020.1810213

Sharma, S., Wakode, S., Sharma, A., Nair, N., Dhobi, M., Wani, M. A., & Pottoo, F. H. (2020). Effect of environmental toxicants on neuronal functions. *Environmental Science and Pollution Research International*, *5*(36), 1–16. DOI: 10.1007/s11356-020-10950-6 PMID: 32996088

Sheng, P. P., & Etsell, T. H. (2007). Recovery of gold from computer circuit board scrap using aqua regia. *Waste Management & Research*, *25*(4), 380–383. DOI: 10.1177/0734242X07076946 PMID: 17874665

Shen, L., Si, H., Yu, L., & Si, H. (2019). Factors influencing young people's intention toward municipal solid waste sorting. *International Journal of Environmental Research and Public Health*, *16*(10), 1708. DOI: 10.3390/ijerph16101708 PMID: 31096698

Shittu, O. S., Williams, I. D., & Shaw, P. J. (2020). Global E-waste management: Can WEEE make a difference? A review of e-waste trends, legislation, contemporary issues and future challenges. *Waste Management (New York, N.Y.)*, *7*, 549–563. PMID: 33308953

Shrivastava, A., & Kumar, V. (2023). Strategies for Waste Recycling, Reuse, and Reduction. In Sustainable Approaches and Strategies for E-Waste Management and Utilization (pp. 140-158). IGI Global.

Shruthi, M., & Naveen, K. B. (2020). Internet of Things: A Survey on Devices, Ecosystem, Components, and Communication Protocols. *Proceedings of the 4th International Conference on Electronics, Communication and Aerospace Technology (ICECA)*, 611-616.

Silva, J. A. P., Lima, G. G., Camilo-Cotrim, C. F., Bailão, E. F. L. C., Caramori, S. S., Nabout, J. C., & Almeida, L. M. (2023). Impact of e-waste toxicity on health and nature: Trends, biases, and future directions. *Water, Air, and Soil Pollution*, *234*(5), 320. DOI: 10.1007/s11270-023-06328-2

Simandl, G. J., Paradis, S., & Simandl, L. (2023). Future of photovoltaic materials with emphasis on resource availability, economic geology, criticality, and market size/growth. *CIM Journal*, *14*(3), 133–157. https://doi.org/https://doi.org/10.1080/19236026.2023.2168419

Singh, A., & Verma, S. (2023). Carbon footprint reduction in electronic device recycling. *Journal of Cleaner Production*, *385*, 134688.

Singh, G. K. (2013). Solar power generation by PV (photovoltaic) technology: A review. *Energy*, *53*, 1–13. https://doi.org/https://doi.org/10.1016/j.energy.2013.02.057

Singh, N., Duan, H., & Tang, Y. (2020). E-waste recycling in India: Bridging the gap between formal and informal sectors. *Resources, Conservation and Recycling*, *161*, 104915. DOI: 10.1016/j.resconrec.2020.104915

Singh, N., Li, J., & Zeng, X. (2016). Global responses for recycling waste CRTs in e-waste. *Waste Management (New York, N.Y.)*, *57*, 187–197. DOI: 10.1016/j.wasman.2016.03.013 PMID: 27072617

Singh, S., Chhabra, R., & Arora, J. (2024). A systematic review of waste management solutions using machine learning, Internet of Things and blockchain technologies: State-of-art, methodologies, and challenges. *Archives of Computational Methods in Engineering*, *31*(3), 1255–1276. DOI: 10.1007/s11831-023-10008-z

Singla, A., & Sharma, N. (2022). IoT Group Key Management Using Incremental Gaussian Mixture Model. *Proceedings of the 3rd International Conference on Electronics and Sustainable Communication Systems (ICESC)*, 469-474.

Sinha-Khetriwal, D., Kraeuchi, P., & Widmer, R. (2005). Export of used electronic goods from Switzerland to developing countries: Environmental impacts and their implications. *Environmental Impact Assessment Review*, *25*(5), 525–539.

Sivagami, K., Bose, S., Vinayak, A. K., Sreenivas, M., Ghosh, A., Narasimhan, M., & Gurumoorthy, A. V. P. (2024). Solar Panel Recycling from Circular Economy Viewpoint: A Review. *Applied Solar Energy*, *60*(2), 328–345. https://doi.org/https://doi.org/10.3103/S0003701X23601862

Slade, G. (2007). *Made to Break Technology and Obsolescence in America*. Harvard University Press. DOI: 10.4159/9780674043756

Smith, A., Johnson, M., & Davis, R. (2023). Robotic Kitchen Systems for Food Waste Reduction in Restaurants. *International Journal of Robotics and Automation*, *38*(2), 87–103.

Soetrisno, F. N., & Delgado-Saborit, J. M. (2020). Chronic exposure to heavy metals from informal e-waste recycling plants and children's attention, executive function and academic performance. *The Science of the Total Environment*, *717*, 137099. DOI: 10.1016/j.scitotenv.2020.137099 PMID: 32092800

Souza, V., Rodrigues Figueiredo, A. M., & Santos Bortolocci Espejo, M. M. d. (2024). Challenges and strategies for managing end-of-life photovoltaic equipment in Brazil: Learning from international experience. *Energy Policy*, *188*, 114091. https://doi.org/https://doi.org/10.1016/j.enpol.2024.114091

Stacey, F. G., James, E. L., Chapman, K., Courneya, K. S., & Lubans, D. R. (2015). A systematic review and meta-analysis of social cognitive theory-based physical activity and nutrition behaviour change interventions for cancer survivors. *Journal of Cancer Survivorship: Research and Practice*, *9*(2), 305–338. DOI: 10.1007/s11764-014-0413-z PMID: 25432633

Stanley, T., Doucouliagos, H., & Steel, P. (2018). Does ICT generate economic growth? A meta-regression analysis. *Journal of Economic Surveys*, *32*(3), 705–726. DOI: 10.1111/joes.12211

Sthiannopkao, S., & Wong, M. (2013). Handling e-waste in developed and developing countries: Initiatives, practices, and consequences. *The Science of the Total Environment*, *463-464*, 1147–1153. DOI: 10.1016/j.scitotenv.2012.06.088 PMID: 22858354

Strachala, D., Hylský, J., Jandova, K., Vaněk, J., & Cingel, Š. (2017). Mechanical Recycling of Photovoltaic Modules. *ECS Transactions*, *81*(1), 199. https://doi.org/https://dx.doi.org/10.1149/08101.0199ecst

Streicher-Porte, M., Widmer, R., Jain, A., Bader, H.-P., Scheidegger, R., & Kytzia, S. (2005). Key drivers of the e-waste recycling system: Assessing and modelling e-waste processing in the informal sector in Delhi. *Environmental Impact Assessment Review*, *25*(5), 472–491. https://doi.org/https://doi.org/10.1016/j.eiar.2005.04.004

Sugumaran, K. M. D., Pandiyarajan, V., Y., B S, Selvakumar, P., & M., S. (2024). Analysis of Electromyography Signals for Control of Mechanical Prosthesis using Machine Learning Techniques. 2024 2nd International Conference on Intelligent Data Communication Technologies and Internet of Things (IDCIoT), 1059-1067. DOI: 10.1109/IDCIoT59759.2024.10467707

Suja, F., AbdulRahman, R., Yusof, A., Masdar, & M.S. (2014). E-waste management scenarios in Malaysia. *Journal of Waste Management*, *20*, 14.

Sun, L., & Zhao, X. (2023). Assessing the carbon footprint of internet of things (IoT) devices. *Journal of Environmental Impact*, *46*(3), 124–139.

Swamy, P. (1970). Efficient Inference in a Random Coefficient Regression Model. *Econometrica*, *38*(2), 311–323. DOI: 10.2307/1913012

Swanson, H. L., & Ferrari, J. R. (2023). Predictors of e-waste: Considerations for community psychology prevention and intervention. *Journal of Community Psychology*, *51*(5), 2001–2009. DOI: 10.1002/jcop.22986 PMID: 36586134

Szabo, S., & Webster, J. (2021). Perceived greenwashing: The effects of green marketing on environmental and product perceptions. *Journal of Business Ethics*, *171*(4), 719–739. DOI: 10.1007/s10551-020-04461-0

Szymkowiak, A., Melović, B., Dabić, M., Jeganathan, K., & Kundi, G. S. (2021). Information technology and Gen Z: The role of teachers, the internet, and technology in the education of young people. *Technology in Society*, *65*, 101565. DOI: 10.1016/j.techsoc.2021.101565

T.S. Gaur, V. Yadav, S. Mittal, and Sharma, M.K. (2020) "A Systematic Review on Sustainable E-waste Management: Challenges, Circular Economy Practices, and a Conceptual Framework,". *IEEE Transactions on Sustainable Computing*, *9*(1), 45–58.

Taha, M. M., Mahdy-Abdallah, H., Shahy, E. M., Ibrahim, K. S., & Elserougy, S. (2018). Impact of occupational cadmium exposure on bone in sewage workers. *International Journal of Occupational and Environmental Health*, *24*(3–4), 101–108. DOI: 10.1080/10773525.2018.1518745 PMID: 30222069

Tamar, M., Wirawan, H., Arfah, T., & Putri, R. P. S. (2021). Predicting pro-environmental behaviors: The role of environmental values, attitudes, and knowledge. *Management of Environmental Quality*, *32*(2), 328–343. DOI: 10.1108/MEQ-12-2019-0264

Tang, J., & Li, Q. (2023). Carbon footprint of electronic waste management. *Journal of Environmental Sciences (China)*, *139*, 201–215.

Tan, J., Jia, S., & Ramakrishna, S. (2022). End-of-Life Photovoltaic Modules. *Energies*, *15*(14), 5113. https://doi.org/https://doi.org/10.3390/en15145113

Tan, Q., Duan, H., Liu, L., Yang, J., & Li, J. (2018). Rethinking residential consumers' behavior in discarding obsolete mobile phones in China. *Journal of Cleaner Production*, *195*, 1228–1236. DOI: 10.1016/j.jclepro.2018.05.244

Tansel, B. (2017). From electronic consumer products to e-wastes: Global outlook, waste quantities, recycling challenges. *Environment International*, *98*, 35–45. DOI: 10.1016/j.envint.2016.10.002 PMID: 27726897

Tan, V., Dias, P. R., Chang, N., & Deng, R. (2022). Estimating the Lifetime of Solar Photovoltaic Modules in Australia. *Sustainability*, *14*(9), 5336. https://doi.org/https://doi.org/10.3390/su14095336

Tatariants, M., Yousef, S., Sakalauskaitė, S., Daugelavičius, R., Denafas, G., & Bendikiene, R. (2018). Antimicrobial copper nanoparticles synthesized from waste printed circuit boards using advanced chemical technology. *Waste Management (New York, N.Y.)*, *78*, 521–531. DOI: 10.1016/j.wasman.2018.06.016 PMID: 32559941

Taylor, B., Chapron, G., Kopnina, H., Orlikowska, E., Gray, J., & Piccolo, J. J. (2020). The need for ecocentrism in biodiversity conservation. *Conservation Biology*, *34*(5), 1089–1096. DOI: 10.1111/cobi.13541 PMID: 32424955

Taylor, P., Steenmans, K., & Steenmans, I. (2020). Blockchain technology for sustainable waste management. *Frontiers in Political Science*, *2*, 590923. DOI: 10.3389/fpos.2020.590923

Tembhare, S. P., Bhanvase, B. A., Barai, D. P., & Dhoble, S. J. (2021). E-waste recycling practices: A review on environmental concerns, remediation, and technological developments with a focus on printed circuit boards. *Environment, Development and Sustainability*, 1–83. DOI: 10.1007/s10668-021-01819-w

Terazono, A., Murakami, S., Abe, N., Inanc, B., Moriguchi, Y., Sakai, S., Kojima, M., Yoshida, A., Li, J., Yang, J., Wong, M. H., Jain, A., Kim, I.-S., Peralta, G. L., Lin, C.-C., Mungcharoen, T., & Williams, E. (2006). Current status and research on E-waste issues in Asia. *Journal of Material Cycles and Waste Management*, *8*(1), 1–12. DOI: 10.1007/s10163-005-0147-0

Tetteh, D., & Lengel, L. (2017). The urgent need for health impact assessment: Proposing a transdisciplinary approach to the e-waste crisis in sub-Saharan Africa. *Global Health Promotion*, *24*(2), 35–42. DOI: 10.1177/1757975916686926 PMID: 28353403

Thakur, P., & Kumar, S. (2020). Metallurgical processes unveil the unexplored "sleeping mines" e-waste: A review. *Environmental Science and Pollution Research International*, *27*(26), 32359–32370. DOI: 10.1007/s11356-020-09405-9 PMID: 32533494

Thakur, P., & Kumar, S. (2022). Evaluation of e-waste status, management strategies, and legislations. *International Journal of Environmental Science and Technology*, *19*(7), 6957–6966. DOI: 10.1007/s13762-021-03383-2

Theerthaana, P., & Raj, S. V. (2025). An investigation on adoption of AI-enabled e-waste recycling kiosks in the post-pandemic era. *Electronic Government. International Journal (Toronto, Ont.)*, *21*(1), 68–80. DOI: 10.1504/EG.2025.143120

Thierry, M., Salomon, M., Nunen, J., & Wassenhove, L. (1995). Strategic Issues in Product Recovery Management. *California Management Review*, *37*(2), 114–134. DOI: 10.2307/41165792

Tian, H., Zhang, J., & Li, J. (2020). The relationship between pro-environmental attitude and employee green behavior: The role of motivational states and green work climate perceptions. *Environmental Science and Pollution Research International*, *27*(7), 7341–7352. DOI: 10.1007/s11356-019-07393-z PMID: 31884553

Tian, J., & Chen, M. (2016). Assessing the economics of processing end-of-life vehicles through manual dismantling. *Waste Management (New York, N.Y.)*, *56*, 384–395. DOI: 10.1016/j.wasman.2016.07.046 PMID: 27499291

Tibben-Lembke, R., & Rogers, D. (2002). Differences between forward and reverse logistics in a retail environment. *Supply Chain Management*, *7*(5), 271–282. DOI: 10.1108/13598540210447719

Toader, E., Firtescu, B., Roman, A., & Anton, S. (2018). Impact of Information and Communication Technology Infrastructure on Economic Growth: An Empirical Assessment for the EU Countries. *Sustainability (Basel)*, *10*(10), 1–22. DOI: 10.3390/su10103750

Tokareva, T. (2015). Food Wasting on Consumption Level: Analysis of the Situation in Latvia Compared to Global Tendencies. Economics and Business, 26, 113-118. DOI: 10.7250/eb.2014.028

Torres, R., Segura-Bailón, B., & Lapidus, G. T. (2018). Effect of temperature on copper, iron and lead leaching from e-waste using citrate solutions. *Waste Management (New York, N.Y.)*, *71*, 420–425. DOI: 10.1016/j.wasman.2017.10.029 PMID: 29102357

Trivedi, H. K., Meshram, A., & Gupta, R. (2023). Recycling of photovoltaic modules for recovery and repurposing of materials. *Journal of Environmental Chemical Engineering*, *11*(2), 109501. https://doi.org/https://doi.org/10.1016/j.jece.2023.109501

Tsachtsiris, G., Magoutas, A., & Papadogonas, T. (2022). ICT and Economic Growth in EU: A macro Level Comparison of Estimated ICT Output Elasticities. *Journal of Global Information Technology Management*, *25*(3), 202–216. DOI: 10.1080/1097198X.2022.2094182

Twenge, J. M. (2023). Generations: The Real Differences Between Gen Z, Millennials, *Gen X,Boomers, and Silents—and What They Mean for America's Future*. Simon and Schuster.

Uddin, S., & Karim, M. (2023). Carbon emissions from semiconductor manufacturing: A life cycle perspective. *Journal of Cleaner Production*, *386*, 134705.

Umair, S., Björklund, A., & Petersen, E. (2015). Social impact assessment of informal recycling of electronic ICT waste in Pakistan using UNEP SETAC guidelines. *Resources, Conservation and Recycling*, *95*, 46–57. DOI: 10.1016/j.resconrec.2014.11.008

Uma, K. E., Nwaka, I. D., Nwogu, M. U., & Obidike, P. C. (2020). What are the triggers of household decision-making on waste disposal choices? A gender-differentiated analysis. *Heliyon*, *6*(12), e05588. DOI: 10.1016/j.heliyon.2020.e05588 PMID: 33319089

Umar, U. A., Shafiq, N., & Ahmad, F. A. (2021). A case study on the effective implementation of the reuse and recycling of construction & demolition waste management practices in Malaysia. *Ain Shams Engineering Journal*, *12*(1), 283–291. DOI: 10.1016/j.asej.2020.07.005

UN. (2023). *Global Sustainable Development Report 2023: Times of crisis, times of change: Science for accelerating transformations to sustainable development.* United Nations. https://sdgs.un.org/gsdr

UNEP. (2008). Bali declaration on waste management for human health and livelihood. Ninth meeting of the conference of the parties to the basel convention on the control of transboundary movements of hazardous wastes and their disposal, 23–27 June 2008, Bali, Indonesia. United Nations Environment Programme. Available http://www.basel.int/Portals/4/Basel%20Convention/docs/meetings/cop/cop9/bali-declaration/BaliDeclaration

UNFCCC. (2015). *The Paris Agreement.* United Nations Framework Convention on Climate Change. https://unfccc.int/process-and-meetings/the-paris-agreement

Usman, A., Ozturk, I., Hassan, A., Zafar, S., & Ullah, S. (2021). The effect of ICT on energy consumption and economic growth in South Asian economies: An empirical analysis. *Telematics and Informatics*, *58*, 1–9. DOI: 10.1016/j.tele.2020.101537

Vachon, S., & Klassen, R. (2006). Extending green practices across the supply chain: The impact of upstream and downstream integration. *International Journal of Operations & Production Management*, *26*(7), 795–821. DOI: 10.1108/01443570610672248

Valero Navazo, J. M., Villalba Méndez, G., & Talens Peiró, L. (2014). Material flow analysis and energy requirements of mobile phone material recovery processes. *The International Journal of Life Cycle Assessment*, *19*(3), 567–579. DOI: 10.1007/s11367-013-0653-6

Vann, K. N., Musson, S. E., & Townsend, T. G. (2006). Factors affecting TCLP lead leachability from computer CPUs. *Waste Management (New York, N.Y.)*, *26*(3), 293–298. DOI: 10.1016/j.wasman.2005.06.016 PMID: 16169210

Vasquez, M., & Romero, P. (2023). The environmental impact of telecommunication devices: A carbon footprint analysis. *Journal of Industrial Ecology*, *27*(3), 309–322.

Venturi, S., Zulauf, K., Cuel, R., & Wagner, R. (2025). Trash to treasure: Gamification and informed recycling behavior. *Resources, Conservation and Recycling*, *215*, 108108. DOI: 10.1016/j.resconrec.2024.108108

Vermeulen, W. (2006). The social dimension of industrial ecology: On the implications of the inherent nature of social phenomena. *Progress in Industrial Ecology*, *3*(6), 574–598. DOI: 10.1504/PIE.2006.012754

Vig, S., & Agarwal, R. N. (2021). E-Waste Management in India: Need for a Circular Approach. In Examining the Intersection of Circular Economy, Forestry, and International Trade (pp. 165-173). IGI Global.

Vijayakumar, G., Rajkumar, M., Rajiv Chandar, N., Selvakumar, P., & Duraisamy, R. (2024). Muniyandi Rajkumar, Rajiv Chandar N, Selvakumar P and Ramesh Duraisamy, (2024). Environmentally friendly TDS removal from waste water by electrochemical ion exchange batch-type recirculation (EIR) technique. *Environmental Science. Water Research & Technology*, *10*(4), 826–835. DOI: 10.1039/D3EW00793F

Vijayalakshmi, V., Radha, R., & Sharanya, S. (2024). Modern Technological Innovation in Digital Wase Management. In Computational Intelligence for Green Cloud Computing and Digital Waste Management (pp. 152-171). IGI Global.

Vijayvargiya, A., & Sahu, K. (2016). A Review Paper on “E-Waste” Technology. *International Journal of Engineering Research & Technology (Ahmedabad)*, *4*(32).

Villalba, G., Segarra, M., Fernández, A., Chimenos, J., & Espiell, F. (2002). A proposal for quantifying the recyclability of materials. *Resources, Conservation and Recycling*, *37*(1), 39–53. DOI: 10.1016/S0921-3449(02)00056-3

Vinayagamoorthi, R., Bhargav, P. B., Ahmed, N., Balaji, C., Aravinth, K., Krishnan, A., Govindaraj, R., & Ramasamy, P. (2024). Recycling of end of life photovoltaic solar panels and recovery of valuable components: A comprehensive review and experimental validation. *Journal of Environmental Chemical Engineering*, *12*(1), 111715. https://doi.org/https://doi.org/10.1016/j.jece.2023.111715

Vuong, Q. H. (2021). The semiconducting principle of monetary and environmental values exchange. *Economics and Business Letters*, *10*(3), 284–290. DOI: 10.17811/ebl.10.3.2021.284-290

Wahman, M., Surowiak, A., Ebin, B., & Berent, K. (2024). PV back sheet recovery from c-Si modules using hot knife technique. *Solar Energy Materials and Solar Cells*, *276*, 113067. https://doi.org/https://doi.org/10.1016/j.solmat.2024.113067

Walker, H., Sisto, L., & McBain, D. (2008). Drivers and barriers to environmental supply chain management practices: Lessons from the public and private sectors. *Journal of Purchasing and Supply Management*, *14*(1), 69–85. DOI: 10.1016/j.pursup.2008.01.007

Walton, S., Handfield, R., & Melnyk, S. (1998). The Green supply chain: Integrating suppliers into environmental management processes. *International Journal of Purchasing and Materials Management*, *34*(1), 1–11. DOI: 10.1111/j.1745-493X.1998.tb00042.x

Wang, F., Huisman, J., Stevels, A., & Baldé, C. P. (2013). Enhancing e-waste estimates: Improving data quality by multivariate Input–Output Analysis. *Waste Management*, *33*(11), 2397-2407. https://doi.org/https://doi.org/10.1016/j.wasman.2013.07.005

Wang, F., Yu, L., & Wu, A. (2021). Forecasting the electronic waste quantity with a decomposition-ensemble approach. *Waste Management*, *120*, 828-838. https://doi.org/https://doi.org/10.1016/j.wasman.2020.11.006

Wang, H., & Chen, X. (2023). Carbon footprint assessment of wireless communication systems. *Journal of Cleaner Production*, *387*, 134720.

Wang, H., Zhang, Y., Liu, Q., Wang, F., Nie, J., & Qian, Y. (2010). Examining the relationship between brominated flame retardants (BFR) exposure and changes of thyroid hormone levels around e-waste dismantling sites. *International Journal of Hygiene and Environmental Health*, *213*(5), 369–380. DOI: 10.1016/j.ijheh.2010.06.004 PMID: 20598942

Wang, J., Chen, S., Tian, M., Zheng, X., Gonzales, L., Ohura, T., Mai, B., & Simonich, S. L. M. (2012). Inhalation cancer risk associated with exposure to complex polycyclic aromatic hydrocarbon mixtures in an electronic waste and urban area in South China. *Environmental Science & Technology*, *46*(17), 9745–9752. DOI: 10.1021/es302272a PMID: 22913732

Wang, J., Feng, Y., & He, Y. (2024). The research progress on recycling and resource utilization of waste crystalline silicon photovoltaic modules. *Solar Energy Materials and Solar Cells, 270,* 112804. https://doi.org/https://doi.org/10.1016/j.solmat.2024.112804

Wang, J., & Sun, X. (2015). Olivine LiFePO4: The remaining challenges for future energy storage. *Energy & Environmental Science, 8*(4), 1110–1138. DOI: 10.1039/C4EE04016C

Wang, J., & Zhou, H. (2019). Economic Benefits of Advanced E-Waste Recycling Technologies. *Journal of Economic Sustainability, 15*(5), 502–515.

Wang, M., You, X., Li, X., & Liu, G. (2018). Watch more, waste more? A stock-driven dynamic material flow analysis of metals and plastics in TV sets in China. *Journal of Cleaner Production, 187,* 730–739. https://doi.org/https://doi.org/10.1016/j.jclepro.2018.03.243

Wang, R., Song, E., Zhang, C., Zhuang, X., Ma, E., Bai, J., Yuan, W., & Wang, J. (2019). Pyrolysis-based separation mechanism for waste crystalline silicon photovoltaic modules by a two-stage heating treatment. *RSC Advances, 9*(32), 18115–18123. https://doi.org/https://doi.org/10.1039/C9RA03582F PMID: 35515232

Wang, S., Dong, D., Li, P., Hua, X., Zheng, N., Sun, S., & Li, X. (2020a). Mercury concentration and fatty acid composition in muscle tissue of marine fish species harvested from Liaodong Gulf: An intelligence quotient and coronary heart disease risk assessment. *The Science of the Total Environment, 726,* 138586. DOI: 10.1016/j.scitotenv.2020.138586 PMID: 32481211

Ward, M., & Zheng, S. (2016). Mobile telecommunications service and economic growth: Evidence from China. *Telecommunications Policy, 40*(2-3), 89–101. DOI: 10.1016/j.telpol.2015.06.005

Weckend, S., Wade, A., & Heath, G. (2016). *End of life management: Solar Photovoltaic Panels.* International Renewable Energy Agency (IRENA), International Energy Agency Photovoltaic Power Systems (IEA-PVPS). https://www.irena.org/

Wena, L., Lin, C., & Lee, S. (2009). Review of recycling performance indicators: A study on collection rate in Taiwan. *Waste Management (New York, N.Y.), 29,* 2248–2256. PMID: 19398319

White, C. D., Masanet, E., Rosen, C. M., & Beckman, S. L. (2003). Product recovery with some byte: An overview of management challenges and environmental consequences in reverse manufacturing for the computer industry. *Journal of Cleaner Production, 11,* 445–458.

WHO. (2009). Busan pledge for action on children's health and environment. The 3rd WHO international conference of children's health and the environment, 7–10 June 2009, Busan, Republic of Korea. World Health Organization. Available https://www.who.int/ifcs/meetings/3ce

WHO. (2013). E-waste and child health initiative. Geneva [Online]. Available https://www.who.int/ceh/risks/ewaste/en/

WHO. (2024). *Electronic waste (e-waste).* World Health Organization. Retrieved 10/10/2024 from https://www.who.int/news-room/fact-sheets/detail/electronic-waste-(e-waste)

Widmer, R., Oswald-Krapf, H., Sinha-Khetriwal, D., Schnellmann, M., & Böni, H. (2005). Global perspectives on e-waste. *Environmental Impact Assessment Review, 25*(5), 436–458. https://doi.org/https://doi.org/10.1016/j.eiar.2005.04.001

Williams, J., & Nguyen, M. (2024). Carbon emissions from electronic devices: Trends and future directions. *Journal of Environmental Impact, 47*(1), 145–160.

Wilson, D. C., Velis, C., & Cheeseman, C. (2015). Role of informal sector recycling in waste management in developing countries. *Habitat International, 30*(4), 797–808. DOI: 10.1016/j.habitatint.2005.09.005

Wirani, Y., Eitiveni, I., & Sucahyo, Y. G. (2024). Framework of Smart and Integrated Household Waste Management System: A Systematic Literature Review Using PRISMA. *Sustainability (Basel), 16*(12), 4898–4901. DOI: 10.3390/su16124898

Wood, R., & Bandura, A. (1989). Social cognitive theory of organizational management. *Academy of Management Review, 14*(3), 361–384. DOI: 10.2307/258173

Xavier, L., Ottoni, M., & Lepawsky, J. (2021). Circular economy and e-waste management in the Americas: Brazilian and Canadian frameworks. *Journal of Cleaner Production, 297*, 1–13. DOI: 10.1016/j.jclepro.2021.126570

Xiao, J., Li, J., & Xu, Z. (2017). Recycling metals from lithium ion battery by mechanical separation and vacuum metallurgy. *Journal of Hazardous Materials, 338*, 124–131. DOI: 10.1016/j.jhazmat.2017.05.024 PMID: 28544937

Xu, Y., Li, J., Tan, Q., Peters, A. L., & Yang, C. (2018). Global status of recycling waste solar panels: A review. *Waste Management, 75*, 450-458. https://doi.org/https://doi.org/10.1016/j.wasman.2018.01.036

Xu, L., & Huang, J. (2023). Carbon footprint of virtual meeting technologies. *Journal of Environmental Sciences (China), 140*, 233–248.

Yaacob, N. S., Azman, N. A. N. N., Salleh, S. M., & Alrazi, B. (2025). Assessing the relationship of knowledge and attitudes on e-waste recycling practices among students of higher learning institutions. *Journal of Nusantara Studies*, *10*(1), 353–372. DOI: 10.24200/jonus.vol10iss1pp353-372

Yadav, D., Kumar Saraf, A., & Singh Rathee, N. (2023). Assessment of PV waste generation in India. *Materials Today: Proceedings*. https://doi.org/https://doi.org/10.1016/j.matpr.2023.03.734

Yadav, R., Panda, D. K., & Kumar, S. (2022). Understanding the individuals' motivators and barriers of e-waste recycling: A mixed-method approach. *Journal of Environmental Management*, *324*, 116303. DOI: 10.1016/j.jenvman.2022.116303 PMID: 36191502

Yang, C., Tan, Q., Liu, L., Dong, Q., & Li, J. (2017). Recycling tin from electronic waste: A problem that needs more attention. *ACS Sustainable Chemistry & Engineering*, *5*(11), 9586–9598. DOI: 10.1021/acssuschemeng.7b02903

Yang, J. G., Wu, Y. T., & Li, J. (2012). Recovery of ultrafine copper particles from metal components of waste printed circuit boards. *Hydrometallurgy*, *121–124*, 1–6. DOI: 10.1016/j.hydromet.2012.04.015

Yang, S., & Park, H. (2023). Carbon footprint reduction in the production of electronic displays. *Journal of Cleaner Production*, *388*, 134737.

Yan, H., Li, Q., Yuan, Z., Jin, S., & Jing, M. (2019). Research progress of mercury bioaccumulation in the aquatic food chain, China: A review. *Bulletin of Environmental Contamination and Toxicology*, *102*(5), 612–620. DOI: 10.1007/s00128-019-02629-7 PMID: 31101929

Yeruva, J. R., Guduru, A. K., & Saripalli, B. P. (2024). Technological Advancements in E-Waste Management and Resource Recovery. In Sustainable Solutions for E-Waste and Development (pp. 275-288). IGI Global.

Yilanci, V., & Ozgur, O. (2019). Testing the environmental Kuznets curve for G7 countries: Evidence from a bootstrap panel causality test in rolling windows. *Environmental Science and Pollution Research International*, *26*(24), 24795–24805. DOI: 10.1007/s11356-019-05745-3 PMID: 31240642

Yilmaz, R., & Koyuncu, C. (2023). The Impact of Globalization on the Rate of E-waste Recycling: Evidence From European Countries. *Amfiteatru Economic*, *25*(62), 180–195. DOI: 10.24818/EA/2023/62/180

Yi, Z. (1986). Changes in family structure in China: A simulation study. *Population and Development Review*, *12*(4), 675–703. DOI: 10.2307/1973431

Yoon, H., Jang, Y. C., & Dodson, J. (2016). Extended producer responsibility and product stewardship for electronic waste management: Policy evolution in South Korea. *Waste Management (New York, N.Y.), 56*, 11–19. DOI: 10.1016/j.wasman.2016.06.014

Yoon, J., & Kim, S. (2023). Carbon footprint analysis of photovoltaic panels: A life cycle assessment. *Journal of Environmental Management, 329*, 116519.

Yuan, J., Chen, L., Chen, D., Guo, H., Bi, X., Ju, Y., Jiang, P., Shi, J., Yu, Z., Yang, J., Li, L., Jiang, Q., Sheng, G., Fu, J., Wu, T., & Chen, X. (2008). Elevated serum polybrominated diphenyl ethers and thyroid-stimulating hormone associated with lymphocytic micronuclei in Chinese workers from an E-waste dismantling site. *Environmental Science & Technology, 42*(6), 2195–2200. DOI: 10.1021/es702295f PMID: 18411489

Yuan, W., Li, J., Zhang, Q., & Saito, F. (2012). Innovated application of mechanical activation to separate lead from scrap cathode ray tube funnel glass. *Environmental Science & Technology, 46*(7), 4109–4114. DOI: 10.1021/es204387a PMID: 22385285

Yuan, W., Li, J., Zhang, Q., Saito, F., & Yang, B. (2013). Lead recovery from cathode ray tube funnel glass with mechanical activation. *Journal of the Air & Waste Management Association, 63*(1), 2–10. DOI: 10.1080/10962247.2012.711796 PMID: 23447859

Yuan, W., Meng, W., Li, J., Zhang, C., Song, Q., Bai, J., & Li, Y. (2015). Lead recovery from scrap cathode ray tube funnel glass by hydrothermal sulphidisation. *Waste Management & Research, 33*(10), 930–936. DOI: 10.1177/0734242X15597777 PMID: 26264931

Yu, E. A., Akormedi, M., Asampong, E., Meyer, C. G., & Fobil, J. N. (2017). Informal processing of electronic waste at Agbogbloshie, Ghana: Workers' knowledge about associated health hazards and alternative livelihoods. *Global Health Promotion, 24*(4), 90–98. DOI: 10.1177/1757975916631523 PMID: 27271535

Yu, P. L., Ab Ghafar, N., Adam, M., & Goh, H. C. (2022). Understanding the human dimensions of recycling and source separation practices at the household level: An evidence in Perak, Malaysia. *Sustainability (Basel), 14*(13), 8023. DOI: 10.3390/su14138023

Zellner, A. (1962). An Efficient Method of Estimating Seemingly Unrelated Regressions and Tests for Aggregation Bias. *Journal of the American Statistical Association, 57*(298), 348–368. DOI: 10.1080/01621459.1962.10480664

Zeng, X., Yang, C., Chiang, J. F., & Li, J. (2018). Innovative e-waste recycling technologies: A review. *Sustainable Materials and Technologies*, *15*, 1–12. DOI: 10.1016/j.susmat.2018.02.001

Zhang, B., Du, Z., Wang, B., & Wang, Z. (2019). Motivation and challenges for e-commerce in e-waste recycling under "Big data" context: A perspective from household willingness in China. *Technological Forecasting and Social Change*, *144*, 436–444. DOI: 10.1016/j.techfore.2018.03.001

Zhang, B., He, Y., Zhu, H., Huang, X., Bai, X., Kannan, K., & Zhang, T. (2020). Concentrations of bisphenol A and its alternatives in paired maternal-fetal urine, serum and amniotic fluid from an e-waste dismantling area in China. *Environment International*, *136*, 105407. DOI: 10.1016/j.envint.2019.105407 PMID: 31955035

Zhang, D., Ding, L., & Zhang, L. (2022). The challenges and strategies for e-waste management in the era of the Industrial Internet of Things. *Journal of Cleaner Production*, *275*, 123548.

Zhang, H., & Li, D. (2021). Advances in Resource Recovery Technologies for E-Waste. *Resources, Conservation and Recycling*, *168*(1), 105254.

Zhang, S., Ding, Y., Liu, B., Pan, D., Chang, C., & Volinsky, A. A. (2015). Challenges in legislation, recycling system and technical system of waste electrical and electronic equipment in China. [Y. XU ET AL.]. *Waste Management (New York, N.Y.)*, *45*, 361–373. DOI: 10.1016/j.wasman.2015.05.015 PMID: 26059074

Zhang, S., Gu, Y., Tang, A., Li, B., Li, B., Pan, D., & Wu, Y. (2021). Forecast of future yield for printed circuit board resin waste generated from major household electrical and electronic equipment in China. *Journal of Cleaner Production*, *283*, 124575. https://doi.org/https://doi.org/10.1016/j.jclepro.2020.124575

Zhang, Y., Luo, X. J., Mo, L., Wu, J. P., Mai, B. X., & Peng, Y. H. (2015). Bioaccumulation and translocation of polyhalogenated compounds in rice (Oryza sativa L.) planted in paddy soil collected from an electronic waste recycling site, South China. *Chemosphere*, *137*, 25–32. DOI: 10.1016/j.chemosphere.2015.04.029 PMID: 25974192

Zhao, J., & Lin, M. (2023). Environmental impact of electronic health records: A carbon footprint analysis. *Journal of Cleaner Production*, *389*, 134754.

Zheng, J., He, C.-T., Chen, S.-J., Yan, X., Guo, M.-N., Wang, M.-H., Yu, Y.-J., Yang, Z.-Y., & Mai, B.-X. (2017). Disruption of thyroid hormone (TH) levels and TH-regulated gene expression by polybrominated diphenyl ethers (PBDEs), polychlorinated biphenyls (PCBs), and hydroxylated PCBs in e-waste recycling workers. *Environment International*, *102*, 138–144. DOI: 10.1016/j.envint.2017.02.009 PMID: 28245931

Zheng, X., Xu, F., Chen, K., Zeng, Y., Luo, X., Chen, S., Mai, B., & Covaci, A. (2015). Flame retardants and organochlorines in indoor dust from several e-waste recycling sites in South China: Composition variations and implications for human exposure. *Environment International*, *78*, 1–7. DOI: 10.1016/j.envint.2015.02.006 PMID: 25677852

Zheng, Y., & Li, Q. (2023). The carbon footprint of electronic payment devices: An environmental assessment. *Journal of Environmental Impact*, *48*(2), 177–192.

Zhong, H., Zhou, S., Zhao, Z., Zhang, H., Nie, J., & Simayi, P. (2022). An empirical study on the types of consumers and their preferences for E-waste recycling with a points system. *Cleaner and Responsible Consumption*, *7*, 100087. DOI: 10.1016/j.clrc.2022.100087

Zhou, J., & Fan, T. (2019). Understanding the factors influencing patient e-Health literacy in online health communities (OHCs): A social cognitive theory perspective. *International Journal of Environmental Research and Public Health*, *16*(14), 2455. DOI: 10.3390/ijerph16142455 PMID: 31295924

Zimmerman, B. J. (2000). Self-efficacy: An essential motive to learn. *Contemporary Educational Psychology*, *25*(1), 82–91. DOI: 10.1006/ceps.1999.1016 PMID: 10620383

About the Contributors

Aşkıner Güngör graduated from Gazi University, Faculty of Engineering and Architecture, Department of Industrial Engineering in 1992. In October 1993, he joined Northeastern University in Boston, Massachusetts, USA, where he received his master's degree in 1996 and his PhD degree in 1999 with his thesis on "Algorithmic models for de-assembly systems and processes". At the end of 1999, Dr. Güngör joined the Department of Industrial Engineering at Pamukkale University, where he is currently an Professor and head of the Department of Industrial Engineering. His current research interests include green supply chain, disassembly, disassembly line balancing, product recovery and environmentally conscious manufacturing. He has also been an R&D consultant to companies in various sectors.

Olcay Polat received his Industrial Engineering PhD degree from the Technical University of Berlin (Germany) with the scholarship of German Academic Exchange Service (DAAD) in 2013. He has been working as Professor in the Department of Industrial Engineering at Pamukkale University (Turkey). He has a special interest in supply chain management, logistics and transportation optimization, production management and risk management. He has been published more that 100 papers in respected journals, conference proceedings and books.

***

Şahin Akkaya, having his BA, MA and PhD degrees in the field of Public Finance, has been working at Istanbul University,Faculty of Economics since 1985. He teaches environmental economics, the theory of public finance, tax theory and policy, globalization and fiscal system, taxation and welfare, and international public finance in undergraduate and graduate programs of public finance. Dr. Akkaya, who studies especially in the fields of international public finance, global public goods, tax policy, environmental economics, climate change and carbon tax, has many scientific publications on these subjects. He has published three academic books in the fields of environmental taxes and tax policies in the globalization age.

Maheswari B. is working as an Assistant Professor In RMK Engineering College, Puduvoyal, Tamilnadu, India, in the department of Computer Science and Engineering. She received the B.E degree in Computer Science and Engineering and M.E degree in Software Engineering from Anna University, India. She is currently pursuing PhD in computer science and engineering from Anna University, Chennai. Her area of interest is Internet of things and Big Data.

Sahana B. is a 14-year teaching experience in Assistant Professor, who is presently working at RV College of Engineering since August 2012. She has done B.E. in Electronics and Communication Engineering, M.Tech. in Computer Network Engineering, and Ph.D. in Computer Networks. Her research interests are in the areas of communication systems and networking. She has mentored eight UG projects and has made contributions to academia through publications in one international journal, two international conferences, and two national conferences.

Yamini B is an author and researcher in the Department of Networking and Communications at SRM Institute of Science and Technology, India. She specializes in networking technologies and communication systems, contributing to research and academic excellence. Based in Kattankulathur, Tamil Nadu, she is dedicated to innovation and technological advancements in her field.

Harshitha B. J. is a student at RV College of engineering pursuing her Bachelors in Electronics and Communication. She has significantly been involved in writing many research papers and books since past 2 years and also has been involved in research projects related to AI and robotics.

M. Balaji, Associate Professor in the Department of Electrical and Electronics Engineering, has 19 years of teaching and research experience. He received his B.E (EEE) degree first class with distinction from Annamalai University, M.E. Power Electronics and Drives (Anna University 5th Rank) from Sri Venkateswara College of Engineering, Anna University and Ph.D in the area of Special Electrical Machines from Anna University. He has published 40 papers in international journals and around 50 papers in national and international conferences. He has coordinated international conferences, organized national level workshops and has delivered guest lectures at various conferences and workshops. He is a recognized supervisor for Anna University. He is currently guiding 5 Ph.D scholars and 7 scholars have completed Ph.D under his guidance. His area of interest includes Design and Control aspects of Electrical Machines, Electric Vehicles and Power Converters. He received

SSN Best Teacher Award for the academic year 2011-2012 and 2019-2020. He is a member of IEEE and Life member of ISTE.

Vijay C. R. has 11 years of Experience in Teaching. Specialized in Human Resource Management & Systems. Interested in Learning and Teaching.

Ozan Çapraz received his M.Sc. and Ph.D. degrees from the Department of Industrial Engineering, Pamukkale University. From 2013 to 2022, he worked as a research assistant at the Department of Industrial Engineering, Tekirdağ Namık Kemal University. Since 2022, he has been working as an assistant professor at the Department of Industrial Engineering, Pamukkale University. His research interests include lean manufacturing, multi-criteria decision-making, optimization, artificial intelligence, and their applications in various fields, including energy and product recovery.

Angana Chakraborty is currently working as an Assistant Professor in the Department of Computer Science and Engineering, Haldia Institute of Technology, Haldia-721657, India. Presently she is pursuing her Ph.D. (CSE) from Maulana Abul Kalam Azad University of Technology, West Bengal, India. She received her M.Tech (IT) degree from Indian Institute of Engineering Science and Technology, Shibpur, India in 2014. Her research interest includes Natural Language Processing, Deep Learning, Machine Learning etc. She has published more than 18 technical papers in the referred journals/conferences. She has qualified UGC NET with 97.56 percentile in January, 2019.

Ritu Dahiya is an Associate Professor in the Department of Chemistry at Chhotu Ram Arya College, Sonepat, Haryana, India. With a strong academic background in chemistry, she has contributed significantly to the field through teaching and research. Dr. Dahiya is dedicated to fostering knowledge and understanding of chemistry among her students while engaging in research activities aimed at advancing the discipline. Her expertise and passion for education have made her a respected member of the academic community in Sonepat and beyond.

Namrata Dhanda is presently Professor in the Department of Computer Science and Engineering, Amity School of Engineering & Technology, Amity University Uttar Pradesh, Lucknow campus. She received her Ph.D from University of Petroleum and Energy Studies, Dehradun, India in 2015. She has 23 years of teaching experience in different reputed engineering & technical institutes in India. Her current research interest includes Data Mining, Big Data Analytics, Big Data, Data Wire housing Internet of Things, etc.

Y. Fathima, BBA., MFC., MBA., M.Phil., DGT., Ph.D., is an Associate Professor in the Department of Commerce, School of Commerce at Jain (Deemed-to-be University) Bangalore. An accomplished educator with demonstrated ability to teach, motivate, and direct students while maintaining high interest and achievement. Articulate communicator able to effectively interact with diverse population of students at a variety of academic levels. She has 18 years of Teaching Experience and 10 years of Research experience. She has published papers in reputed journal and authored 4 books.

Iniebiyo Felagha holds a Ph.D. in Biochemistry (Nutrition and Environmental Toxicology) from University of Port Harcourt, Rivers State, Nigeria (2021). My research focuses on nutrition, nutrition toxicology, environmental toxicology and other areas of biochemistry. I have published several articles in learned journals and book chapters. I have been a reviewer with Science Domain Publishing group since 2018. I also have several conference presentations and abstracts.

Likith Gowda B. S. is a student of RV College Of engineering pursuing her Bachelors in Electronics and Communication. He has significantly been involved in writing many research papers and books since past 2 years and also has been involved in research projects related to robotics and communication engineering.

Dankan Gowda V. is currently working as an Assistant Professor in the Department of Electronics and Communication Engineering at BMS Institute of Technology and Management in Bangalore. Previously, he worked as a Research Fellow at ADA DRDO and as a Software Engineer at Robert Bosch. With a total experience of 14 years, including teaching and industry, Dr. Gowda has made significant contributions to both academia and research. He has published over 170 research papers in renowned international journals and conferences. In recognition of his innovative work, Dr. Gowda has been granted six patents, including four from Indian authorities and two international patents. His research interests primarily lie in the fields of IoT and Signal Processing, where he has conducted workshops and handled industry projects. Dr. Gowda is passionate about teaching and strives to create an engaging and effective learning environment for his students. He consistently seeks opportunities to enhance his knowledge and stay updated with the latest advancements in his field.

Nwanne Dike Ijere's base degree is Bachelor of Science (B.Sc) in Plant Science and Biotechnology from Abia State University, Uturu. Her master's and Ph.D degrees in Environmental Health Biology was obtained from Federal University of Technology Owerri, Nigeria. Presently, Nwanne is a lecturer in the College of Arts, Science and Professional Studies, Aba, Abia State, Nigeria, where she teaches general biology, including environmental science and other health related courses. Dr. Nwanne Dike Ijere is very passionate about carrying out qualitative research, and have authored and co-authored several research publications in reputable, peer reviewed, scientific journals. Nwanne is also a peer reviewer in some national and international journals. She has also published a timely international article with The Conversation Africa as well as media publicity {telephonic} which arose from one of her published articles with the station Channel Africa in the show Rise and Shine. Over the years, she has acquired varied and valid knowledge in some relevant areas in her field through workshops and hands on trainings. Nwanne is also a member of some professional societies; Nigerian Environmental Society, Grassroots Environmental Healthcare Initiative, Foundation for African Development through International Biotechnology, Organization for Women in Science for the Developing World, Institute of Safety Professionals of Nigeria, Environmental Health Scientists Association, Nigeria, Environmental Health Council of Nigeria and Society for Environmental Health of Nigeria.

Praveenkumar J. is a dedicated academic and researcher affiliated with Rajalakshmi Institute of Technology, Chennai. With expertise in his field, he actively contributes to research and education, shaping the next generation of professionals.

Subhankar Joardar is presently Professor & Head in the Department of Computer Science and Engineering, Haldia Institute of Technology, Haldia-721657, India. He received his Ph.D degree from Birla Institute of Technology, Mesra, Ranchi, India in 2016. He did his masters (M. Tech and MCA) both from BIT, Mesra, Ranchi in 2009 and 2002 respectively. He has published more than 16 technical papers in the referred journals/conferences. He has served as Organizing Chair of international conference (ICITAM 2017).

Shaji Joseph did his graduation and postgraduate degree in Sociology from Pune University (SPPU) and PhD in Management from Symbiosis International University in 2020. Currently, he is an associate professor at Symbiosis International Deemed University Pune. His area of expertise is in sustainability standards and reporting, Human resources management, and corporate governance and ethics. His research interests include sustainable development, ethical leadership, leadership assessment, and human resources management.

Benu Kesar did her Master's in Business Administration (MBA) from NMIMS Mumbai and is pursuing her PhD from Symbiosis International Deemed University Pune. Dr. Kesar's research interests lie at the intersection of technology, communication, and societal impact, focusing on how digital platforms are reshaping human interaction and influencing organisational strategies.

Sumanta Kuila is currently working as an Associate Professor in the Deptartment of Computer Science & Engineering, Haldia Institute of Technology, Haldia-721657, India. He has 21 years of experience in the field of Computer Science & Engineering. He has completed his Ph.D in Computer Science & Engineering in June 2023 from Amity University Uttar Pradesh, Lucknow campus. He did his masters (M.E) from west Bengal University of Technology in 2005. He has published more than 25 technical papers in the referred journals/conferences.

N. M. G. Kumar, PhD Electrical Power Systems, Sri Venkateswara University, Andhra Pradesh, India, Academic and Research Interests: Area of interest are power system planning, security, and state estimations with data science algorithms, application of FACTS devices in Transmission and Distribution systems for enhancement of network performance; and power system reliability studies for real-time applications

V. Lava Kumar is an accomplished academician with a diverse educational background and over 13 years of enriching experience in the field of education. Holding an impressive array of qualifications including an MBA, M. Com, PhD in Management, PGDTTM, along with certifications from APSET in Management and APSET in Commerce. His area of specialization is Finance and Accounting. Currently serving as an Assistant Professor at GITAM School of Business, GITAM Deemed to be University, Bengaluru, Dr. Lava Kumar is dedicated to nurturing the minds of future business leaders. His teaching methodology is grounded in both theoretical knowledge and practical insights, ensuring that students receive a holistic understanding of their subjects. Dr. Lava Kumar's contributions extend beyond the classroom, evident through his prolific research endeavours. He has authored numerous research papers, with 3 publications in prestigious Scopus-indexed and ABDC journals, and an impressive tally of 10 research articles in international and peer-reviewed journals. He co-authored a Book which is published in Surana publications. His research interests primarily revolve around contemporary issues in management and finance, where he strives to contribute to the academic discourse with valuable insights and innovative perspectives.

Neeta B. Malvi is an Assistant Professor with 12 years of experience, currently teaching at RV College of Engineering since August 2014. She has a B.E. degree in Telecommunication and an M.Tech. degree in Digital Communication and Networking. Her fields of interest are signal processing, image processing, and cryptography. She has mentored 16 UG and one PG project and has various research papers published in international journals, including FOTA update mechanisms, active safety systems, electronic nose design, and software testing automation. She won the 2nd position in her M.Tech. course at GIT, Belgaum.

T. C. Manjunath is a Professor in the Department of Computer Science and Engineering at Rajarajeswari College of Engineering, Bangalore. He specializes in IoT Cybersecurity and Blockchain Technology and currently serves as the Dean of Research and Development (R&D) at the institution. Dr. Manjunath's expertise in these cutting-edge fields has contributed to his reputation as a leading academic and researcher in the areas of cybersecurity and emerging technologies. He is based at the college's campus on Mysore Road, Kumbalgodu Post, Bangalore, Karnataka, India.

Nurul Hidayana Mohd Noor, Ph.D., is Senior Lecturer in Faculty of Administrative Science and Policy Studies, Universiti Teknologi MARA (UiTM), Seremban, Negeri Sembilan, Malaysia (ORCID: https://orcid.org/0000-0003-2262-2524). Her main research activity is entrepreneurship, environmental, nonprofit, organizational management, and youth studies. She has published widely on these subjects in publications such as the Journal of Entrepreneurship in Emerging Economies, Journal of Entrepreneurship and Public Policy, and Human Service Organizations: Management, Leadership & Governance.

Mohit is a research scholar at the Institute of Management Studies and Research, Maharshi Dayanand University, Rohtak, Haryana, India. His research focuses on the dynamic relationship between brand image and purchase intentions, particularly in the context of brand extensions within the FMCG sector. Mohit actively engages in bibliometric research, analyzing publication trends and identifying research gaps in his area of interest. He has co-authored several scholarly works and continues to contribute to the academic community. Mohit can be reached at mohitattri.rs.imsar@mdurohtak.ac.in

Santosh Kumar Nathsharma is a distinguished academic in the field of chemistry, currently serving in the Department of Chemistry at Stewart Science College, Cuttack, Odisha. With a strong dedication to research and teaching, he has contributed significantly to the advancement of chemical sciences. Based in Cuttack (Postal Code: 753001), Dr. Nathsharma is known for his expertise and commitment to nurturing future scientists and scholars.

Odangowei Inetiminebi Ogidi is a Lecturer and Acting Head, Department of Biochemistry, Bayelsa Medical University, Yenagoa, Bayelsa State, Nigeria. He holds a Bachelor of Science degree in Biochemistry, Master of Science and Doctor of Philosophy degrees in Biotechnology. He has also served as a Lecturer, Head of Departments of Science Laboratory Technology and Biochemistry, and Deputy

Director, Centre for Continuing Education in Federal Polytechnic Ekowe, Bayelsa State. He has authored over 80 peer-reviewed academic articles in reputable local and international journals, published over 40 book chapters and presented over 35 conference papers. He is a member of various Journal Editorial Boards. He is indexed in World Scientist and University Ranking (AD Scientific Index ID 4306173) amongst other international bodies. He have supervised over 30 ND and HND students and currently supervising over 10 undergraduate students. His research interest has been Environmental Biotechnology, Phytomedicine, Phytochemistry, Human and Environmental health risk assessments and ecotoxicology amongst others. He is a member of West Africa Society of Toxicology, Nigerian Society of Biochemistry and Molecular Biology, Foundation for African Development through International Biotechnology and Nigerian Environmental Society.

Zeynep Ozsut Bogar received the B.Sc. degree in industrial engineering, in 2012, the M.Sc. degree in industrial engineering from the Anadolu University, Eskişehir, Turkey, in 2015, and, Ph.D. degree in industrial engineering at Pamukkale University, Denizli, Turkey, in 2023. She is currently working on e-waste quantification methodologies. She has been a Research Assistant since 2014. Her research interests include modeling approaches, machine learning, and optimization techniques.

Deepika P. is an Assistant Professor of 8 years and 2 years industry experience and is presently working at RV College of Engineering since August 2014. She has a B.E. in Electronics and Communication and an M.Tech. in VLSI Design and Embedded Systems. Her research interests are digital VLSI design, low-power VLSI, and Verilog. She has supervised 10 UG and 4 PG projects and has published heavily in international conferences and journals. She has also been funded for phase change memory applications research and was awarded the best paper for VLSI design for the research on low dynamic power consumption.

Selvakumar P. is currently an Associate Professor in the Department of Science and Humanities at Nehru Institute of Technology, Coimbatore, Tamil Nadu, India. Since 2023, he has served as the Department IQAC Coordinator at the institute. Prior to joining Nehru Institute of Technology, he held positions as an Assistant Professor in the Department of Science and Humanities various engineering colleges in Coimbatore, Tamil Nadu, India, spanning from 2011 to 2023. Dr.P. Selvakumar's academic and research interests lie primarily in the field of phytochemistry. He has made significant contributions to his field, with an impressive record that includes 30 journal publications, 12 patents, and 20 conference papers. Additionally, he has organized 5 seminars and conferences, further demonstrating his commitment to academic and scientific exchange. Throughout his career, Dr. P.Selvakumar has been recognized for his dedication and excellence in teaching, having been honored with several awards, including the Best Teacher Awards. He holds a prominent position in the academic community, contributing actively to research.

Mikail Pehlivan has his BA degree in the field of Public Administration and Public Finance (double major) at Afyon Kocatepe University in 2015. Subsequently, he received his MA and PhD degrees in the field of Public Finance at Istanbul University. The author currently continues his career at Trakya University, Faculty of Economics and Administrative Sciences, Department of Public Finance, since 2017. Dr. Pehlivan is interested in Fiscal Theory, Fiscal Policy, Environmental Economics and Regulation Economics.

Usha Prabhu is a senior academician having 23+ years of teaching experience – at UG & PG levels, aware of new changes in the education sector like the implementation of the National Education Policy, a member of the Syllabus Revision Committee -PG Level MBA at Bangalore University. She holds three Master degrees - MA, M. Phil, MBA, UGC- NET qualified and holds Ph.D from Bangalore University, presently working at Jain(Deemed-to be University.· She has authored 2 books - Organization Behaviour for MBA andEntrepreneurship Skill for III semester B. Com published by Sapna Book House. She has published 12 research papers in International Journals, presented 9 research papers at International Conferences and 3 research Book Chapters(Scopus indexed).

Siva Subramanian R. is an Associate Professor in the Department of Computer Science and Engineering, RMK College of Engg and Tech, Chennai, India. He received her B.E. in Computer Science and Engineering from Anna University, Chennai in 2009 and M.Tech. in Computer Science and Engineering from Bharath University, Chennai, India in 2013. He has 10 years teaching experience. His research interests include Data Mining, Machine Learning, Big Data Analytics and Networking. He has published many articles in reputed journals.

Varanasi Rahul is an Assistant Professor at JAIN University (Bangalore), a semi-qualified Cost Accountant, holding three postgraduate degrees: M.Com (Accounting & Taxation), M.Com (Management), and an MA (History). He has cleared the UGC NET in Commerce twice and has also cleared the AP SET in Commerce, History, and Management, as well as the TS SET in Commerce. With 3 years of experience as an Assistant Professor and 6 months as a Corporate Tax Associate for Australian Taxation at PwC (Big 4), he brings both academic and professional expertise to his field. Additionally, he is a freelance archaeologist, practicing epigraphist in Brahmi script, and a numismatist. His scholarly and creative contributions include over 30 newspaper articles published in Deccan Chronicle, Times of India, and other local newspapers. He has presented and participated in 25 national and international seminars and has published 1 research article in ABDC Journal, 1 in SCOPUS-indexed journals and 4 in peer-reviewed journals. As a story writer, 5 of his stories have been published in top magazines of Andhra Pradesh. Furthermore, he has authored 3 books. He has also contributed 9 book chapters to various academic publications. In addition to his extensive achievements, he has applied for 6 patents, of which 1 has been granted and 2 have been published.

Renu is currently working as Assistant Professor with Teerthanker Mahaveer University, Moradabad, India. She holds a Master and Doctorate degree in Management from Kurukshetra University, Kurukshetra and Central University in Punjab, India respectively. She has 9 years of experience in Teaching, Research and Mentoring undergraduate and post graduate students of Management department. Her area of research includes Social Media Analytics, Customer Engagement, Customer Satisfaction and Branding. Currently, she is supervising 4 PhD research scholars in Management. She has been associated with Insurance industry for 3 years as consultant. She has published many research papers in reputed international journals/conferences and hold two Indian patents. She has delivered many guest lectures, special talks, communication skills training and career counselling sessions for universities in India. She has been actively involved in organizing international seminars, conferences, and workshops as a key resource person.

Vidhya S. is working as Assistant Professor in the Department of Computing Technologies, School of Computing, SRM Institute of Science and Technology, Kattankulathur, Chennai. She has completed her doctoral degree in the area of machine learning. Her research interests include machine learning, data mining and soft computing techniques.

Safwah Malyanah binti Sahariddun, a bachelor degree student in Faculty of Administrative Science and Policy Studies, Universiti Teknologi MARA (UiTM), Seremban, Negeri Sembilan, Malaysia, has intern experience at Majlis Daerah Tanjong Malim (MDTM), Perak. Her main research activity is environmental.

Jeganathan Gomathi Sankar is a dedicated educator, specializes in marketing and currently teaches at the BSSS Institute of Advanced Studies. With a rich academic background, he served as a research scholar at Pondicherry University, focusing on information diffusion, ICT, Marketing technology, AI, and Data Science. Dr. Sankar's impactful research is published in reputable journals indexed in Scopus and ABDC. Formerly a faculty member at Saveetha School of Management, he contributed significantly to marketing education. Dr. Sankar's pedagogy integrates theoretical insights with practical applications, offering students a holistic understanding of contemporary marketing. Beyond teaching, he is proficient in analytical tools such as SPSS, AMOS, STATA, PowerBI, and Python, enhancing the sophistication of his research. Known for his engaging teaching style, Dr. Sankar fosters an environment encouraging critical thinking. As a member of BSSS Institute of Advanced Studies, he continues shaping future marketing professionals. His commitment to academic excellence, research, and proficiency in modern tools exemplifies his dedication to advancing knowledge in marketing and technology.

Natasha Saqib is currently working as Sr. Assistant Professor at the Department of Management Studies, University of Kashmir, India. She received her B.Sc. (Hon.) Industrial Chemistry from University of Kashmir, MBA from Islamic university of Science and Technology and Ph.D. from University of Kashmir, India. Dr. Natasha has more than 15 years of experience in academia. She has several research publications in internationally acclaimed journals.

Amrita Sarkar is currently working as an Assistant Professor in the Department of Computer Science and Engineering, Haldia Institute of Technology, Haldia-721657, India. She received her M.Tech degree from Maulana Abul Kalam Azad University of Technology, West Bengal, India, India in 2022. Her research interest is AI, Machine Learning, Deep Learning etc.

N. Sathyanarayana has over 15 years of teaching experience at the UG and PG levels. He holds a Master's in Business Administration and Commerce and is pursuing a Ph.D. He currently holds the position of Assistant Professor in the School of Commerce at Jain (Deemed-to-be University), Bangalore. He has published 22 research papers in international journals, 7 research papers in national journals, and presented 14 research papers at international conferences and 11 research papers at national conferences. He has co-authored 2 books at national level and 7 book chapters at international level.

Anish T. P. received his M.E. Degree from the Department of Computer Science and Engineering at Saveetha Engineering College (Affiliated to Anna University), Chennai, India, in 2011. He is currently an Assistant Professor at R.M.K. College of Engineering and Technology, Puduvoyal, India. He is Completed his Ph.D. degree in Information and Communication Engineering from Anna University. His current research interest includes Machine Learning, Deep Learning and Big Data analytics.

Tengku Nurshazwina binti Tengku Sahrum, a bachelor degree student in Faculty of Administrative Science and Policy Studies, Universiti Teknologi MARA

(UiTM), Seremban, Negeri Sembilan, Malaysia, has an internship experience at FGV Kilang Sawit Pasoh, Negeri Sembilan. Her main research activity is environmental.

Kamaraj V. is working as Professor in the Department of Electrical and Electronics Engineering, Sri Sivasubramaniya Nadar College of Engineering. He has 35 years of teaching experience. His area of interest includes Power Electronics and Drives, Electric Vehicles and Power Converters.

Index

B

Basel Convention 96, 108, 115, 140, 145, 262, 264, 267, 279, 284, 287, 293, 301, 416
Biological Processes for E-Waste 151, 159, 170, 171
Blockchain 54, 69, 70, 71, 72, 80, 81, 82, 87, 88, 89, 90, 137, 177, 178, 179, 180, 181, 182, 183, 184, 186, 187, 188, 189, 190, 191, 192, 193, 194, 195, 196, 197, 198, 199, 200, 201, 202, 203, 204, 205, 206, 208, 209, 210, 211, 212, 213, 214, 230, 239, 264, 283, 286, 296, 302, 307, 415, 431
Blockchain Technology 87, 88, 89, 177, 178, 179, 180, 181, 190, 191, 192, 193, 194, 195, 198, 200, 201, 202, 209, 212, 213, 214, 230, 239, 264, 415

C

Cancer 111, 142, 250, 251, 253, 254, 255, 277, 284, 289, 290, 300, 306, 307, 319, 321, 331, 439
Carbon Footprint Analysis 215, 216, 217, 236, 237, 238, 239, 240, 241
Chemical Processes 51, 153, 154, 170, 171, 218, 222
Circular Economy 2, 9, 10, 13, 27, 29, 34, 35, 40, 41, 42, 44, 46, 48, 51, 54, 62, 69, 70, 71, 73, 75, 76, 78, 80, 81, 83, 85, 86, 89, 90, 91, 92, 93, 97, 99, 100, 101, 102, 103, 108, 115, 116, 125, 126, 128, 134, 135, 139, 141, 146, 148, 149, 152, 153, 155, 164, 169, 170, 173, 188, 189, 191, 212, 219, 221, 224, 226, 227, 228, 231, 233, 234, 235, 236, 262, 264, 265, 266, 267, 270, 272, 287, 294, 296, 299, 302, 307, 323, 325, 326, 330, 334, 355, 357, 359, 366, 368, 369, 372, 390, 394, 395, 405, 409, 410, 414, 415, 433, 439
Community Engagement 63, 65, 200, 203

D

Data Analytics 73, 178, 202, 210, 213, 235
Decentralization 177, 178, 181, 192, 197

E

Ecosystem 60, 73, 89, 101, 141, 161, 168, 169, 178, 186, 188, 196, 226, 235, 267, 269, 278, 292, 300, 343, 345
Electronic Products 18, 54, 62, 63, 76, 82, 83, 84, 87, 90, 91, 94, 95, 112, 135, 215, 217, 218, 231, 234, 235, 236, 237, 239, 265, 266, 267, 272, 287, 294, 299, 307, 311, 323, 324, 330, 349, 399, 407, 419, 423
Electronics 31, 53, 59, 60, 61, 63, 64, 72, 73, 76, 77, 78, 79, 80, 86, 90, 91, 92, 93, 94, 95, 97, 98, 99, 100, 108, 109, 111, 112, 116, 119, 125, 134, 135, 137, 139, 141, 143, 144, 145, 155, 177, 218, 219, 220, 225, 226, 230, 231, 232, 233, 234, 235, 236, 237, 238, 239, 244, 245, 247, 248, 249, 251, 252, 257, 262, 265, 267, 270, 274, 283, 284, 285, 286, 287, 294, 295, 296, 297, 299, 301, 302, 307, 309, 310, 311, 312, 313, 314, 315, 319, 321, 331, 332, 333, 338, 339, 342, 392, 410, 415, 418, 419, 420, 433
Electronics Supply Chain 145, 218, 219, 233, 235
Electronic Waste 6, 39, 40, 41, 45, 46, 49, 50, 55, 62, 63, 69, 73, 76, 79, 81, 83, 85, 89, 92, 95, 97, 98, 100, 101, 102, 105, 107, 109, 111, 113, 114, 115, 116, 117, 119, 120, 121, 125, 126, 127, 128, 129, 134, 146, 148, 152, 153, 154, 155, 158, 159, 164, 166, 169, 170, 171, 172, 173, 183, 192, 233, 234, 236, 238, 240, 243, 244, 245, 246, 247, 248, 249, 250, 251, 252, 253, 256, 257, 259, 261, 262, 263, 264, 265, 266, 267, 268, 269,

270, 271, 272, 274, 275, 276, 277, 280, 281, 284, 287, 291, 292, 293, 294, 295, 296, 297, 301, 302, 303, 304, 305, 306, 307, 310, 311, 312, 313, 314, 316, 317, 318, 319, 321, 325, 326, 327, 328, 329, 330, 331, 332, 333, 339, 347, 359, 402, 409, 414, 416, 417, 418, 419, 420, 427, 429, 430, 431, 432, 433, 434, 435, 437, 438, 439

Environment 33, 42, 45, 47, 54, 55, 56, 57, 58, 59, 60, 61, 63, 64, 65, 66, 71, 77, 78, 80, 84, 86, 90, 91, 92, 93, 98, 99, 101, 102, 107, 108, 109, 111, 112, 113, 115, 116, 117, 119, 122, 125, 126, 127, 130, 133, 134, 135, 136, 139, 140, 141, 142, 143, 144, 146, 162, 164, 167, 173, 174, 178, 181, 193, 217, 222, 224, 243, 244, 245, 247, 249, 257, 258, 260, 261, 262, 263, 264, 266, 268, 269, 271, 272, 274, 275, 276, 277, 278, 279, 281, 283, 284, 288, 290, 291, 292, 294, 301, 303, 305, 306, 307, 309, 310, 311, 318, 319, 324, 327, 329, 330, 331, 332, 333, 335, 338, 339, 342, 343, 344, 345, 346, 347, 348, 354, 355, 357, 358, 359, 360, 361, 362, 363, 367, 368, 369, 371, 375, 390, 391, 394, 395, 397, 398, 399, 401, 403, 405, 406, 407, 408, 414, 417, 427, 436

Environmental Impact 1, 10, 12, 13, 28, 29, 30, 37, 39, 49, 50, 51, 58, 73, 78, 89, 99, 118, 122, 124, 125, 129, 130, 146, 154, 157, 158, 160, 164, 169, 178, 188, 208, 211, 216, 219, 220, 221, 223, 224, 225, 226, 227, 228, 229, 230, 231, 232, 233, 234, 238, 239, 240, 241, 256, 271, 280, 286, 287, 297, 300, 306, 313, 324, 325, 327, 331, 366, 439

Environmental Management 44, 72, 119, 146, 238, 239, 241, 304, 338, 358, 359, 363, 372, 404, 407, 408, 438

Environmental Sustainability 9, 27, 31, 70, 81, 122, 148, 154, 156, 158, 168, 169, 178, 182, 220, 224, 235, 237, 267, 268, 276, 294, 301, 324, 343, 354, 361, 371, 419

E-Waste 6, 9, 11, 12, 17, 18, 28, 32, 33, 34, 35, 37, 38, 39, 40, 41, 42, 43, 44, 46, 47, 48, 49, 50, 51, 53, 54, 55, 56, 57, 58, 59, 60, 61, 62, 63, 64, 65, 66, 67, 68, 69, 70, 71, 72, 73, 74, 75, 76, 77, 78, 79, 80, 81, 82, 83, 84, 85, 86, 87, 88, 89, 90, 91, 92, 93, 94, 95, 96, 97, 98, 99, 100, 101, 102, 103, 105, 106, 107, 108, 109, 110, 111, 112, 113, 114, 115, 116, 117, 118, 119, 120, 121, 122, 123, 125, 126, 127, 128, 129, 130, 133, 134, 135, 136, 137, 138, 139, 140, 141, 142, 143, 144, 145, 146, 147, 148, 149, 151, 152, 153, 154, 155, 156, 157, 158, 159, 160, 161, 162, 163, 164, 165, 166, 167, 168, 169, 170, 171, 173, 174, 190, 192, 212, 213, 214, 219, 238, 243, 244, 245, 246, 247, 248, 249, 250, 251, 252, 253, 254, 255, 256, 257, 258, 259, 261, 262, 263, 264, 265, 266, 267, 268, 269, 270, 271, 272, 273, 274, 275, 276, 277, 278, 279, 280, 281, 283, 284, 285, 286, 287, 288, 289, 290, 291, 292, 293, 294, 295, 296, 297, 299, 300, 301, 302, 303, 304, 305, 306, 307, 308, 309, 310, 311, 312, 313, 314, 315, 316, 317, 318, 319, 320, 321, 322, 323, 324, 325, 326, 327, 328, 329, 330, 331, 332, 333, 334, 337, 338, 339, 340, 341, 342, 343, 344, 345, 346, 347, 348, 349, 351, 352, 353, 354, 355, 356, 357, 358, 359, 360, 361, 362, 363, 365, 366, 367, 368, 369, 371, 372, 374, 375, 376, 377, 378, 379, 380, 390, 391, 392, 394, 395, 396, 397, 398, 400, 402, 403, 404, 405, 406, 407, 409, 410, 411, 413, 414, 415, 416, 417, 418, 419, 420, 421, 422, 423, 426, 427, 428, 429, 430, 431, 432, 433, 434, 435, 436, 437, 438, 439, 440

E-Waste generation 6, 17, 57, 61, 79, 91, 94, 121, 123, 125, 135, 136, 247, 249,

276, 277, 287, 297, 307, 311, 314, 317, 323, 359, 375, 376, 377, 395, 416, 419, 420
E-Waste Management 33, 34, 35, 46, 53, 54, 55, 56, 57, 59, 60, 61, 62, 63, 64, 65, 66, 67, 68, 69, 70, 71, 72, 73, 75, 78, 80, 81, 82, 83, 87, 88, 89, 90, 96, 97, 98, 99, 100, 101, 102, 103, 106, 107, 108, 109, 112, 114, 115, 116, 118, 119, 120, 121, 122, 126, 129, 130, 133, 134, 135, 137, 139, 140, 141, 143, 144, 145, 146, 148, 149, 152, 164, 168, 169, 171, 174, 214, 219, 244, 256, 261, 262, 263, 264, 265, 266, 267, 269, 272, 273, 275, 277, 278, 283, 284, 287, 290, 292, 293, 294, 295, 297, 299, 301, 302, 304, 318, 323, 324, 325, 327, 328, 329, 330, 333, 334, 338, 339, 340, 346, 349, 354, 355, 359, 361, 366, 368, 376, 394, 395, 404, 409, 411, 415, 416, 418, 419, 420, 421, 422, 428, 429, 430, 431, 432, 433, 434, 435, 436, 437, 438, 439
E-Waste Recycling 32, 37, 49, 60, 63, 64, 65, 72, 73, 77, 80, 82, 86, 87, 88, 89, 91, 94, 96, 97, 98, 99, 106, 109, 116, 119, 127, 129, 130, 134, 139, 142, 143, 144, 145, 148, 151, 152, 153, 154, 155, 156, 157, 158, 159, 161, 162, 163, 164, 165, 166, 167, 168, 170, 171, 190, 219, 238, 247, 249, 250, 251, 261, 263, 264, 265, 266, 267, 273, 274, 275, 278, 280, 284, 285, 286, 288, 293, 294, 297, 299, 301, 302, 304, 306, 307, 310, 311, 315, 319, 320, 321, 323, 327, 329, 330, 332, 337, 338, 339, 340, 341, 342, 343, 344, 345, 346, 347, 348, 349, 351, 352, 353, 354, 355, 356, 357, 358, 359, 360, 361, 362, 363, 366, 374, 377, 398, 409, 415, 421, 427, 429, 430, 431, 432, 433, 436, 437, 438, 439, 440
Externality 410

G

Gen Z 337, 339, 340, 343, 345, 346, 347, 348, 351, 352, 353, 354, 355, 356, 362, 363

H

Hazardous Substances 9, 10, 33, 51, 54, 60, 65, 111, 136, 142, 160, 170, 171, 217, 218, 245, 252, 268, 284, 285, 286, 288, 290, 291, 292, 294, 295, 299, 307, 320, 321, 327, 329, 330
Heavy Metal 141, 148, 250, 259, 277, 304, 322
Human health 33, 55, 58, 68, 71, 80, 102, 108, 111, 112, 113, 116, 125, 134, 148, 171, 174, 217, 243, 245, 247, 249, 250, 253, 264, 267, 272, 276, 277, 279, 288, 299, 300, 301, 304, 309, 310, 311, 318, 319, 330, 332, 339, 342, 354, 361, 363, 367, 369, 375

I

Industrial Symbiosis 374, 399, 403, 404, 405, 410
Informal Recycling 33, 54, 62, 63, 68, 77, 80, 94, 96, 107, 117, 143, 144, 152, 251, 261, 263, 266, 268, 284, 285, 293, 298, 319, 408, 415, 419, 427, 433
IoT 72, 73, 80, 81, 82, 87, 177, 178, 182, 183, 184, 186, 187, 188, 189, 190, 192, 193, 194, 195, 196, 197, 198, 199, 202, 204, 205, 208, 209, 210, 212, 213, 214, 233, 235, 240, 283, 295, 296, 302, 359, 416

L

Life Cycle Assessment 44, 175, 217, 238, 239, 241, 246, 271, 281, 324, 325, 335, 359
Linear Economy 83, 100, 365, 366, 368, 369, 395, 410

M

Material Recovery 29, 30, 34, 43, 51, 86, 87, 100, 121, 125, 175, 226, 231, 233, 262, 265, 286, 295, 296, 297, 327, 329, 366, 368, 402, 434

O

Observational Learning 414, 417, 418, 422, 423, 425, 426, 429, 430, 431

P

Policy Development 62
Pollution 2, 9, 37, 43, 45, 46, 48, 54, 65, 68, 78, 80, 101, 102, 111, 116, 119, 122, 127, 129, 130, 134, 138, 141, 142, 168, 169, 170, 172, 173, 189, 198, 224, 245, 256, 262, 263, 265, 269, 270, 272, 274, 278, 279, 284, 286, 287, 291, 292, 293, 294, 297, 299, 300, 311, 317, 318, 320, 321, 324, 325, 330, 332, 333, 345, 355, 361, 362, 390, 397, 405, 406, 409, 410, 411, 414, 438
Polybrominated Biphenyl 243
Polychlorinated Biphenyls 243, 250, 282, 290, 306, 319, 342
PV Panels 1, 2, 6, 8, 9, 10, 11, 12, 13, 18, 21, 22, 23, 24, 25, 28, 30, 31, 32, 33, 34, 35, 41, 44, 45, 51
PV Waste Management 11, 18, 22, 30, 32, 33, 34, 35, 36, 51

R

Recycling 1, 2, 9, 10, 11, 12, 13, 23, 25, 27, 28, 29, 30, 31, 32, 33, 34, 35, 36, 37, 38, 39, 40, 42, 43, 44, 45, 46, 47, 48, 49, 50, 51, 54, 55, 56, 57, 58, 59, 60, 61, 62, 63, 64, 65, 66, 67, 68, 71, 72, 73, 74, 76, 77, 78, 80, 81, 82, 83, 84, 85, 86, 87, 88, 89, 90, 91, 93, 94, 96, 97, 98, 99, 100, 101, 105, 106, 107, 108, 109, 110, 112, 113, 114, 115, 116, 117, 118, 119, 120, 121, 122, 123, 125, 126, 127, 128, 129, 130, 131, 134, 136, 137, 138, 139, 140, 141, 142, 143, 144, 145, 146, 148, 149, 151, 152, 153, 154, 155, 156, 157, 158, 159, 160, 161, 162, 163, 164, 165, 166, 167, 168, 170, 171, 172, 173, 174, 175, 178, 179, 181, 182, 183, 184, 185, 188, 189, 190, 191, 192, 193, 194, 196, 197, 198, 200, 201, 202, 203, 207, 208, 209, 210, 214, 219, 220, 221, 224, 225, 226, 231, 233, 236, 238, 240, 244, 245, 247, 249, 250, 251, 252, 256, 258, 259, 261, 262, 263, 264, 265, 266, 267, 268, 269, 270, 271, 272, 273, 274, 275, 276, 277, 278, 280, 281, 283, 284, 285, 286, 287, 288, 290, 291, 292, 293, 294, 295, 296, 297, 298, 299, 301, 302, 303, 304, 306, 307, 310, 311, 312, 313, 314, 315, 316, 318, 319, 320, 321, 322, 323, 324, 325, 326, 327, 328, 329, 330, 331, 332, 333, 337, 338, 339, 340, 341, 342, 343, 344, 345, 346, 347, 348, 349, 351, 352, 353, 354, 355, 356, 357, 358, 359, 360, 361, 362, 363, 366, 367, 368, 370, 371, 374, 377, 395, 398, 399, 400, 401, 402, 403, 406, 407, 408, 409, 414, 415, 416, 417, 418, 419, 420, 421, 422, 423, 425, 427, 428, 429, 430, 431, 432, 433, 434, 435, 436, 437, 438, 439, 440
Recycling Intention 337, 338, 339, 340, 343, 344, 345, 346, 347, 349, 351, 352, 353, 355, 356, 357, 437
Recycling Technologies 34, 44, 51, 56, 58, 72, 73, 94, 98, 115, 126, 140, 143, 152, 153, 158, 165, 170, 172, 233, 236, 274, 280, 299, 304, 323, 414, 415, 431
Resource Conservation 32, 155, 158, 178, 227
Resource Recovery 27, 54, 56, 59, 65, 67, 74, 80, 115, 116, 117, 119, 120, 121, 122, 124, 126, 149, 154, 155, 157, 158, 164, 165, 167, 169, 170, 171, 262, 274, 284, 295, 298, 329, 394, 395, 427

Reverse Logistics 41, 44, 370, 372, 373, 397, 398, 399, 400, 402, 403, 404, 405, 406, 407, 408, 410
Right-to-Repair 100, 283, 286, 297, 302, 307

S

Self-Efficacy 416, 419, 420, 421, 422, 425, 429, 434, 437, 438, 439, 440
Self-Regulation 414, 416, 417, 418, 423, 425, 426, 427, 429, 430, 431, 434, 436
Smart Contracts 178, 181, 182, 186, 188, 191, 192, 194, 195, 196, 197, 199, 200, 202, 203, 209, 210
Social Cognitive Theory 413, 416, 417, 419, 420, 421, 425, 428, 429, 434, 436, 437, 438, 439, 440
Solar PV Waste 11, 28, 33, 51
Sustainability 6, 9, 10, 12, 27, 31, 34, 35, 37, 38, 39, 42, 46, 48, 49, 51, 56, 61, 70, 72, 73, 81, 83, 85, 86, 87, 89, 90, 91, 92, 93, 99, 100, 106, 109, 121, 122, 123, 124, 125, 126, 128, 141, 143, 145, 148, 152, 154, 156, 157, 158, 159, 167, 168, 169, 172, 177, 178, 182, 187, 188, 189, 190, 191, 192, 193, 194, 196, 198, 200, 204, 212, 214, 215, 216, 218, 220, 221, 223, 224, 226, 227, 228, 229, 230, 232, 233, 234, 235, 237, 261, 267, 268, 272, 273, 276, 283, 294, 295, 301, 324, 332, 334, 339, 343, 345, 353, 354, 357, 358, 359, 361, 362, 363, 368, 371, 398, 404, 405, 407, 408, 414, 416, 419
Sustainable Design 54, 56, 59, 61, 67, 69, 71, 72, 116, 231, 266, 399
Sustainable e-waste management 57, 78, 80, 101, 106, 107, 108, 112, 120, 126, 130, 277, 284, 290, 302, 328, 416, 435
Sustainable Management 10, 108, 129, 134, 139, 142, 170, 294, 427
Sustainable Production and Consumption 42, 106, 128, 129, 365, 395, 398, 410

T

Theory of Planned Behavior 337, 341, 357, 359, 360, 364

U

Urban Mining 81, 82, 87, 101, 262, 267, 283, 286, 295, 302, 308, 415, 433

W

Waste Management 1, 10, 11, 12, 13, 18, 22, 30, 32, 33, 34, 35, 36, 37, 38, 39, 40, 41, 42, 43, 45, 46, 47, 48, 49, 50, 51, 53, 54, 55, 56, 57, 59, 60, 61, 62, 63, 64, 65, 66, 67, 68, 69, 70, 71, 72, 73, 75, 78, 79, 80, 81, 82, 83, 87, 88, 89, 90, 94, 96, 97, 98, 99, 100, 101, 102, 103, 106, 107, 108, 109, 111, 112, 114, 115, 116, 118, 119, 120, 121, 122, 126, 128, 129, 130, 131, 133, 134, 135, 137, 139, 140, 141, 143, 144, 145, 146, 148, 149, 152, 153, 162, 164, 168, 169, 170, 171, 172, 173, 174, 175, 177, 178, 179, 182, 183, 184, 185, 186, 187, 188, 189, 190, 191, 192, 193, 194, 196, 197, 198, 199, 200, 201, 202, 203, 205, 206, 207, 209, 210, 212, 213, 214, 219, 222, 224, 226, 238, 240, 244, 245, 247, 256, 261, 262, 263, 264, 265, 266, 267, 268, 269, 272, 273, 274, 275, 276, 277, 278, 279, 280, 283, 284, 287, 290, 292, 293, 294, 295, 297, 298, 299, 301, 302, 303, 304, 318, 323, 324, 325, 326, 327, 328, 329, 330, 331, 332, 333, 334, 335, 338, 339, 340, 346, 349, 354, 355, 358, 359, 360, 361, 362, 366, 368, 376, 394, 395, 398, 401, 402, 403, 404, 409, 411, 414, 415, 416, 418, 419, 420, 421, 422, 428, 429, 430, 431, 432, 433, 434, 435, 436, 437, 438, 439

www.ingramcontent.com/pod-product-compliance
Lightning Source LLC
LaVergne TN
LVHW080351011026
849161LV00009B/2271